Forage for Sustainable Livestock Production

Forage for Sustainable Livestock Production

Editors:
N. DAS, A.K. MISRA, S.B. MAITY
K.K. SINGH & M.M. DAS

SATISH SERIAL PUBLISHING HOUSE
403, Express Tower, Commercial Complex
Azadpur, Delhi-110033 (India)
Phone : 011-47073040, Fax : 91-11-27672046
e-mail : info@satishserial.com, hkjain1975@yahoo.com
Website : www.satishserial.com

Published by :

SATISH SERIAL PUBLISHING HOUSE

403, Express Tower, Commercial Complex, Azadpur, Delhi-110033 (INDIA)
Phone : 011-47073040, Fax : 91-11-27672046
E-mail : info@satishserial.com, hkjain1975@yahoo.com

Print Again, 2023

ISBN : 978-81-89304-63-8

Composed, Designed & Printed in India

Director
Indian Grassland and Fodder Research Institute
Jhansi-284003, (UP) India

Dr K A Singh
Director

FOREWORD

Agriculture is the backbone of Indian economy and livestock sector is an integral part of agriculture. Livestock sector accounts for 24.72 % of the agricultural GDP and about 4.36 % of the total GDP. This is derived from a livestock population of about 185 million cattle, 98 million buffaloes, 61 million sheep, 124 million goats, 13 million pigs and 489 million poultry including 11 million other livestock. Using these resources, India ranks first with 102 million tonnes milk, ranks third in egg production and standing eighth in rank in the world's meat production with 4.8 million tonnes.

Inspite of having huge livestock population, the productivity is low compared to the world average and much below than the developed countries. One of the major limitations to efficient livestock production in the country is the lack of adequate level of quality forages. In the recent years, considerable reorientation of the livestock as well as fodder production systems are intriguing in the country with a perceptible shift from the extensive system of production to semi-intensive or commercial production system.

Ensuring availability of adequate forages both in terms of quantity and quality would be the primary concern for increasing the productivity of livestock economically. Enhancing the availability of green forages either by increasing the area under cultivation or by better utilizing the already available forage resources during monsoon requires adoption of available scientific technologies in the livestock production systems. Forage evaluation is the foremost criteria for scientific feeding of livestock as the balance feeding requires an understanding of broad range of forage nutrients. Since most of the feeding strategies are determined by the nutritional value of forages, therefore, knowledge of recent analytical techniques is of immense importance in the livestock production as well as judicious utilization of feeds and forages.

This compilation is primarily based on the lectures delivered during the winter school on "Recent advances in forage evaluation and utilization for sustainable and economic livestock production" held at Indian Grassland and Fodder Research Institute, Jhansi, India. It covers the important aspects of forages and their role in animal production system, demand and supply scenario, application of recent technologies for nutritional evaluation of various type of feeds and forages, technology option for forage conservation and processing, development of forage based feeding system for livestock, top feed resources, non conventional forage sources including CPR, and voluntary feed intake by the animals.

I admire the editors for bringing this compilation entitled **"Forage For Sustainable Livestock Production"** and I am confident that this publication is a timely addition to the repository of information to be used by the livestock owners, farm managers, scientists, extension workers and those associated with the implementation of animal husbandry programmes in the country.

K.A. Singh

PREFACE

Notwithstanding the contribution of agriculture sector in the Indian economy are steadily declining (from 36.4 per cent in 1982-83 to 18.5 per cent in 2006-07), the agriculture and livestock sectors contribute significantly in the overall economy of the people in rural India. This sector continues to support more than half a billion people providing employment to 52 per cent of the workforce. Within the agriculture sector, the contribution of the livestock sector is however growing steadily over the period (over 4 per cent to the total GDP and about a quarter of the GDP from agriculture and allied activities). In the last three decades, the growth rate of milk and egg production was always higher (> 4 % per annum) as compared to the growth rate of crop production. With the increase in human population, growing income and awareness of the importance of animal food as an integral component of balanced diet, the demand for milk, meat, egg and fish will grow continuously in the coming years. This sector is also a main source of family income in the arid and semi-arid regions. 'Access to asset' is the pre-condition for any self-employment actvity. In comparison to land, livestock are more than proportionately controlled by the poor. In India, marginal farmer (having land less than 1 ha) and small farmer (having land less than 2 ha) are the owner of about 45% and 20% of the total livestock. Animal Husbandry provides job opportunities to about 20 million persons (principal status 11 million and subsidiary status 9 million). In 2006-07, this sector contributed 101 million tonnes of milk, 51 billion eggs, 45 million kg of wool and 2.3 million tonnes of meat. The 17th Livestock Census (2003) has placed the total livestock population at 485 millions and that of poultry birds at 489 millions.

During the last few decades we have seen over emphasis for production of cereals and other crops to meet the growing demand of food for human consumption and this has generated an unequal competition between man, agriculture operations and livestock. Ever increasing demand of the landmass for the sustainability of human population has altered the ratio of human and livestock density per hectare of landmass available. There is appreciable shrinkage of land availability for livestock for their mere sustenance, let alone there optimum productive potential, even though the present production system of livestock is no way demanding on the agriculture production system as animals are not fed with high quality cereals, concentrates and green fodder, but survive on low input nutrients of agricultural by products and naturally growing grass. The effect of Green Revolution on livestock nutrition has been marginal as agricultural by-products and residues, mainstay of Indian livestock, are finding more and more new application for development of other products. Besides high yielding varieties produce less agricultural residue. The cumulative effect of diversification of land for industrialization, use of agricultural products for other non-livestock activities, ever depleting pasture lands and increase in density of livestock population has caused qualitative degradation of the permanent pastureland. It is undeniable that the production of fodder and feed for animals will always be relegated to second rung due to competing demands of agriculture, human population and industry on land. The entire livestock production system would be under tremendous strain for retaining its present position of strength and its large contribution to GDP in next decade unless corrective measures are initiated with holistic approach. If the ecological balance and sustainability of all sectoral growth were to be retained, the problem of animal nutrition would require focused attention.

Current need for the animal fodder and other nutritional requirement falls short by huge margin of which green fodder accounts for about 60% and dry fodder accounts for about 20%. Increasing growth of Livestock particularly that of genetically upgraded animals, further spurs the demand for feed and fodder. The judicial uses of landmass with a right mix of traditional knowledge, technology application and building up of social consciousness of preserving and sharing of the natural resources have to be evolved to meet the nutritional requirement of man and animal. Given the present scenario, it is a gigantic task. The problem needs to be viewed from a holistic angle. Strategies should be prepared which would emphasize enrichment of straw/stover, preparation of hay/silage to overcome fodder scarcities during lean season, conversion of fodder into feed block to facilitate transport of fodder from surplus areas, establishment of fodder banks and promotion of chaff cutter. Due to improper management of common property resources and lack of coordination between different agencies involved, the productivity as well as carrying capacity of public and forestland are decreasing. For sustainable and economic livestock production this problem need to be addressed through scientific utilization of traditional pastures and integration with Watershed Development Programme especially for silvi-pastoral development. For enhancement of grass production, measures should be taken to bring larger areas under joint forest management and treatment of wastelands and areas under problem soils. As scope of increasing areas under cultivated fodder production is limited, efforts should be made to increase productivity through promotion of intensive fodder production technologies and quality fodder seed production by specialized agencies. To implement all these programmes, appropriate training and knowledge updating of the resource persons who are directly working with the farmers becomes imperative.

With this background, the Indian Grassland and Fodder Research Institute (IGFRI), Jhansi (UP), India has organized a Winter School on 'Recent Advances in Forage Evaluation and Utilization for Sustainable and Economic Livestock Production' from 31 October to 20 November 2008. The compilation of all the lecture notes culminates in the publication of the present compendium, which would be distributed to all the concerned organizations involved in the implementation of fodder development programmes in India. We feel that the authors of various chapters have made a good effort in brining out issues of forage production, evaluation and its utilization. We appreciated the commitment of our authors in timely submitting the chapters. The encouragement and guidance of Dr. K.A. Singh, Director, IGFRI, Jhansi are gratefully acknowledged. We also acknowledge the Human Resource Division (HRD) of Indian Council of Agricultural Research (ICAR) for providing the necessary grant for conducting the Winter School and publishing this compilation.

Editors

LIST OF CONTRIBUTORS

1. Dr. K. A. Singh
   Director
   Indian Grassland and Fodder
   Research Institute
   Jhansi- 284003 (UP)
2. Dr. J. P. Singh
   Senior Scientist
   Indian Grassland and Fodder
   Research Institute
   Jhansi- 284003 (UP)
3. Dr. P. N. Dwivedi
   Senior Scientist
   Indian Grassland and Fodder
   Research Institute
   Jhansi- 284003 (UP)
4. Dr. N.S. Ekka
   Scientist (Senior Scale)
   Indian Grassland and Fodder
   Research Institute
   Jhansi- 284003 (UP)
5. Dr. P.K. Pathak
   Senior Scientist
   Indian Grassland and Fodder
   Research Institute
   Jhansi- 284003 (UP)
6. Dr. A.B. Mojumdar
   Principal Scientist
   Indian Grassland and Fodder
   Research Institute
   Jhansi- 284003 (UP)
7. Dr. M. M. Roy
   Principal Scientist
   Indian Grassland and Fodder
   Research Institute
   Jhansi- 284003 (UP)
8. Dr. R. K Bhatt
   Principal Scientist
   Indian Grassland and Fodder
   Research Institute
   Jhansi- 284003 (UP)
9. Dr. G.P.Shukla
   Principal Scientist
   Indian Grassland and Fodder
   Research Institute
   Jhansi- 284003 (UP)
10. Dr. K.K. Singh
    Senior Scientist
    Indian Grassland and Fodder
    Research Institute
    Jhansi- 284003 (UP)

11. Dr. S.K. Mahanta
Senior Scientist
Indian Grassland and Fodder Research Institute
Jhansi- 284003 (UP)

12. Dr. M.M. Das
Senior Scientist
Indian Grassland and Fodder Research Institute
Jhansi- 284003 (UP)

13. Dr. Anil Kumar
Senior Scientist
Indian Grassland and Fodder Research Institute
Jhansi- 284003 (UP)

14. Dr. A.K. Misra
Principal Scientist
Indian Grassland and Fodder Research Institute
Jhansi- 284003 (UP)

15. Dr. A.K. Handa
Senior Scientist
National Research Center on Agro-forestry, Jhansi- 284003 (UP)

16. Dr. B.K. Trivedi
Principal Scientist
Indian Grassland and Fodder Research Institute
Jhansi- 284003 (UP)

17. Dr. S.N. Tripathi
Principal Scientist
Indian Grassland and Fodder Research Institute
Jhansi- 284003 (UP)

18. Dr. N. Das
Principal Scientist and Head
PAR Division, Indian Grassland and Fodder Research Institute
Jhansi- 284003 (UP)

19. Dr. S. B. Tripathi
Principal Scientist and Head
Indian Grassland and Fodder Research Institute
Jhansi- 284003 (UP)

20. Dr. Sultan Singh
Senior Scientist
Indian Grassland and Fodder Research Institute
Jhansi- 284003 (UP)

21. Dr. S.K. Das
Senior Scientist
Indian Grassland and Fodder Research Institute
Jhansi- 284003 (UP)

22. Dr. J. B. Singh
Senior Scientist
Indian Grassland and Fodder Research Institute
Jhansi- 284003 (UP)

23. Sh. Pradeep Behari
Scientist
(Selection Grade)
Indian Grassland and Fodder Research Institute
Jhansi- 284003 (UP)

24. Dr. K.K. Singhal
Principal Scientist and Head
National Dairy Research Institute
Karnal -132001 (Haryana)

25. Dr. Sunil Kumar
Senior Scientist
(Horticulture)
Indian Grassland and Fodder Research Institute
Jhansi- 284003 (UP)

26. Dr. S A Faruqui
Principal Scientist and Project Coordinator AICRP (Forage Crops)
Indian Grassland and Fodder Research Institute
Jhansi- 284003 (UP)

27. Dr. Sunil Kumar
Senior Scientist (Agronomy)
Indian Grassland and Fodder Research Institute
Jhansi- 284003 (UP)

28. Dr. Subir Kumar Nag
Senior Scientist
Indian Grassland and Fodder Research Institute
Jhansi- 284003 (UP)

29. Dr. D. R. Malaviya
Principal Scientist
Indian Grassland and Fodder Research Institute
Jhansi- 284003 (UP)

30. Dr. A. K. Roy
Principal Scientist and Head
Indian Grassland and Fodder Research Institute
Jhansi- 284003 (UP)

31. Dr. P. Kaushal
Senior Scientist
Indian Grassland and Fodder Research Institute
Jhansi- 284003 (UP)

32. Dr. B.K.Bhadoria
Principal Scientist
Indian Grassland and Fodder Research Institute
Jhansi- 284003 (UP)

33. Dr. S.B. Maity
Senior Scientist
Indian Grassland and Fodder Research Institute
Jhansi- 284003 (UP)

34. Dr. Sharmila Roy
Senior Scientist
Indian Grassland and Fodder Research Institute
Jhansi- 284003 (UP)

35. Dr. Pradeep Saxena
Senior Scientist
Indian Grassland and Fodder Research Institute
Jhansi- 284003 (UP)

36. Dr. S.K.Tomar
Principal Scientist
National Dairy Research Institute
Karnal-123001 (Haryana)

37. Dr. S. N. Ram
Senior Scientist
Indian Grassland and Fodder Research Institute
Jhansi- 284003 (UP)

38. A.K. Shukla
Senior Scientist
Indian Grassland and Fodder Research Institute
Jhansi- 284003 (UP)

39. B P Kushwaha
Senior Scientist
Network Project on Bhadawari Buffaloes, Indian Grassland and Fodder Research Institute
Jhansi- 284003 (UP)

40. M.K. Tripathi
Senior Scientist
Central Sheep and Wool Research Institute, Avikanagar (Rajsthan)

41. A.S. Mishra
Senior Scientist
Project Dirctorate on Cattle
Meerut (UP)

42. Deepika Chaturvedi
Senior Research Fellow
Central Sheep and Wool Research
Institute, Avikanagar (Rajsthan)

43. Dr. Sudesh Radotra
Senior Scientist, RRS, Palampur
Indian Grassland and Fodder
Research Institute
Jhansi- 284003 (UP)

44. Dr. Inder Dev
Senior Scientist, RRS
Palampur
Indian Grassland and Fodder
Research Institute
Jhansi- 284003 (UP)

45. Dr. R.K. Agrawal
Senior Scientist
Indian Grassland and Fodder
Research Institute
Jhansi- 284003 (UP)

❑❑❑

CONTENTS

Chapter 1

Feed and Fodder Development Issues and Options

K.A. SINGH
Indian Grassland and Fodder Research Institute, Jhansi-284003 (U.P.), India

Introduction

The livelihood of a large segment of people in India is intricately woven with the livestock sector. With increasing income and changing dietary habits the demand for products from animal origin is increasing. To keep pace with the demand, the livestock sector has to gear itself by increasing the productivity because increasing the number of livestock any further has little scope. We already have a huge livestock wealth in the country (Table 1). The major constraint in the productivity is non-availability of adequate quantity and quality of feed and fodder. The livestock sector offers a secure livelihood to the rural people irrespective of the fluctuations in monsoon rainfall. In difficult regions the world over, livestock have been a source of income and employment. Even in not so difficult regions where livestock have been taken up as a vocation, the prosperity in terms of nutritional well being of people besides income and employment is quite evident as in case of Punjab and Haryana.

What prevents the livestock sector to realize its potential in spite of the overwhelming evidence that it brings prosperity? The feed and fodder, which account for 60 – 70% of the total cost of production of milk and meat has been a neglected lot. Since the growth in livestock sector is propelled by its own momentum by large, there has been a general apathy towards understanding its need and taking concrete steps to support it. Livestock production based on green fodder is not only economical but also good for the general health and well being of the animals, which in turn adds to the income of the livestock owners on a sustainable basis. Moreover, livestock production based on green forage is also environmental friendly.

However, there has been a rapid change in the way agriculture operations are carried out like; agriculture production i.e. cropping system, water resources, diversification of crops, intensification of agriculture, increasing use of mechanical power v/s bullock power, transformation from sustenance farming to market oriented farming, changing food habits etc. They all have their impact on animal husbandry practices including breed character requirements of Indian farmer and thus their feeding.

Table 1. Projected Livestock Population (Million)

Year	Cattle	Buffalo	Sheep	Goat	Equine	Camel	Total
1995	180.5	82.8	4.0	9.2	0.5	0.9	278.0
2000	187.1	87.7	4.1	9.9	0.4	1.0	290.0
2005	192.2	92.6	4.2	10.5	0.3	1.0	30 1.0
2010	197.3	97.5	4.3	11.2	0.3	1.0	312.0
2015	202.3	102.4	4.4	11.8	0.1	1.1	322.0
2020	207.4	107.3	4.5	12.5	0.1	1.1	333.0
2025	212.5	112.2	4.6	13.2	0.1	1.1	344.0

India has over 15% world cattle population and 16% of human population on a land area of only 2%. Due to ever increasing population pressure of human beings, arable land is mainly used for food and cash crops, thus there is little chance of having good quality arable land available for fodder production, unless milk production becomes remunerative to the farmer as compared to other crops.

Scenario of Feed and Fodder Requirement & Availability

There is terrific pressure of livestock on available feed and fodder, as land available for fodder production has been decreasing. Scenario of feed and fodder availability till 2025 is as below:

Table 2. Grazing Resources in India (2000-01)

Resources	Area (million ha)	Percentage
Forest	69.41	22.70
Permanent pasture, grazing land	10.90	3.60
Cultivable wasteland	13.66	4.50
Fallow land	24.99	8.10
Fallow land other than current fallows	10.19	3.30
Barren uncultivable waste lands	19.26	6.30
Total common property resource other than forest	54.01	17.70

Sizeable amount of fodder demand is fulfilled through vast grassland and rangelands (Table 3). Any positive or negative change in its position will have impact on several environmental issues.

Table 3. Fodder Demand and Supply Scenario in the Country (Million Tonnes)

Years	Supply	Demand	Deficit as % of demand		
	Green	Dry	Green	Dry Green	Dry
1995	379.3	421	947	52659.95 (568)	19.95 (105)
2000	384.5	428	988	54961.10 (604)	21.93 (121)
2005	389.9	443	1025	56961.96 (635)	22.08 (126)
2010	395.2	451	1061	58962.76 (666)	23.46 (138)
2015	400.6	466	1097	60963.50 (696)	23.56 (143)
2020	405.9	473	1134	63064.21 (728)	24.81 (157)
2025	411.3	488	1170	65064.87 (759)	24.92 (162)

Source: Draft report of the working group on animal husbandry and dairying for five-year plan (2002-2007, Govt. of India, Planning Commission, August – 2001). Figure in parenthesis indicates actual deficit.

At present the country is facing a net deficit of 61 percent green fodder and 22 percent dry fodder and 64 percent concentrate feeds. This corresponds the deficit of 28 percent CP and 24 percent TDN (Table 4). The demand will reach to 1170 million tons of green fodder and 650 million tons of dry forage and 152 million tons of concentrate feed in 2025. At the current level of growth in forage resource there will be 65 percent deficit in green fodder and 25 percent deficit in dry fodder. Green forage supply situation has to grow at 3.2

percent to meet the projected demand. Diversion of crop residues for packaging and other industrial requirement as well as incorporation of crop residues into soil will worsen the demand and supply situation in near future.

Table 4. Project Availability, Requirements and Deficit of CP and TDN (Million Tonnes)

Year	Requirement		Availability		Deficit (%)	
	CP	TDN	CP	TDN	CP	TDN
2000	44.49	321.29	30.81	242.42	30.75	24.55
2005	46.12	333.11	32.62	253.63	29.27	23.86
2010	47.76	344.93	34.18	262.02	28.44	24.04
2015	49.39	356.73	35.98	273.24	27.15	23.41
2020	51.04	368.81	37.50	281.23	26.52	23.7
2025	52.68	380.49	39.31	292.45	25.38	23.14

Source: IGFRI Perspective plan, vision 2025

Initiatives in Feed Fodder Resource Development by Government of India (Department of Animal Husbandry and Dairying)

Basically, the animal husbandry is a state subject and the State Governments are primarily responsible for the growth of the sector. The Department of Animal Husbandry and Dairying has, however, been operating 30 Central Livestock Organizations and allied Institutions for production and distribution of superior germplasms to the State Governments for cross breeding and genetic up gradation of the stocks. Besides, the Department has been implementing 11 Central Sector and Centrally Sponsored Schemes for the development of requisite infrastructure and supplementing the efforts of the State Governments for achieving the accelerated growth of animal husbandry sector. The Department also advises State Governments/ Union Territories in the formulation of policies and programmes in the field of Animal Husbandry, Dairy Development and Fisheries including feed and fodder resource development.

Besides these activities, department also addresses the feed and fodder developmental issues at state and national level. For this purpose the department has broadly two vehicles: i) Central Fodder Development Organization and ii) Centrally Sponsored Scheme: i.e. Assistance to States for Feed and Fodder Development (A New scheme on Development of Fodder is under preparation).

Central Fodder Development Organization

For forage production & demonstration purpose, seven regional stations located in different agro-climatic zones of the country and one central odder seed production farm, in Hessarghatta, Bangalore are operational. Besides these, one central minikit-testing programme on fodder crops is being funded under this scheme. The details are as under;

A. Regional Stations for Forage Production and Demonstration

For production and propagation of certified seeds of high yielding varieties of fodder crops and pasture grasses/legumes, the Government has established 7 Regional Stations at Mamidipally, Hyderabad (Andhra Pradesh), Gandhi Nagar (Gujarat), Hissar (Haryana), Suratgarh (Rajasthan), Sahema (Jammu & Kashmir), Alamadhi (Tamil Nadu) and Kalyani (West Bengal). These stations are catering to requirement of fodder seeds of the farmers of different agro-climatic regions. These stations also involved in extension activities through field demonstrations and farmers'fairs/ field days.

B. Central Fodder Seed Production Farm, Hesserghatta

In addition to the regional stations, the Central Fodder Seed Production Farm, Hesserghatta, Banglore, is also engaged in production of seeds of fodder crops and pasture grasses/legumes. The farm is fully equipped with modern machinery and equipment as well as seed-storage and processing capacity. During Ninth Plan, the farm produced 783 tonnes of fodder seeds. Besides, the farm also organized demonstrations, training programmes and farmers' fairs/ field days.

C. Central Minikit Testing Programme on Fodder Crops

This programme was started to make farmers aware through field demonstrations about latest technology, high yielding varieties of fodder crops and improved agronomic package of practices to increase production of green fodder. Certified seeds of high yielding fodder crops/grasses/legumes produced at Regional Stations and Central Farm, Hesserghatta, are used under this scheme.

D. Assistance to States for Feed and Fodder Development

Under this scheme central assistance is provided to States Governments to strengthen their activities in the field of feed and fodder development. This scheme has been discontinued during 10th Plan. However, two components of this scheme; i) establishment of fodder banks and ii) enrichment of straw/ cellulosic waste have been revived up to March'2004 and still operational.

E. Establishment of Fodder Banks

Under this scheme, assistance is provided to the State Governments for establishment of fodder banks to conserve fodder for use during drought and scarcity period. Large areas in many States are used for growing grass, which could be preserved in the fodder banks. During Ninth Plan, assistance of Rs.289.49 lakhs was provided for establishment of fodder banks in the different States of the country. During 2002-03, assistance of Rs.123.65 lakhs was given for establishment of 5 fodder banks (2 in J&K, 1 in Karnataka and 2 in Uttaranchal). During 2003-04, Rs. 76.985 lakhs have been provided for two fodder banks (one each in Tripura and Mizoram). Funding pattern of the scheme is in the ratio of 75:25 (Central:State).

F. Enrichment of Straw and Cellulosic Wastes

Under this scheme assistance is provided to the farmers for straw quality improvement by treatment with urea and molasses. This help in increasing productivity and reducing the cost of milk production. During 9th Plan, assistance of Rs.524.30 lakhs was provided to the different States.

G. New Centrally Sponsored Scheme – "Fodder Development"

A new Centrally Sponsored Scheme for Fodder Development during the 10th Plan was initiated. Under this scheme, 7 Regional Stations for Forage Production & demonstration located in different agro-climatic zones of the country and 1 Central Fodder Seed Production Farm, Hessarghatta, Bangalore were started. Besides these, one Central Minikit Testing Programme on Fodder Crops was also funded under this scheme.

Research Institutions

In India, sporadic research work on grasses and fodder crops was initiated as early as the close of the nineteenth Century. Preliminary studies on the subject started in the then Bombay State, Madras, Central Provinces, Bengal, Punjab and at the Imperial Agricultural Research Institute. The Imperial Council of Agricultural Research, established on the recommendations of the Royal Commission on Agriculture in India in 1928, stimulated wider interest in the subject by financing research schemes on fodder and grasses to the Indian Agricultural Research Institute (New Delhi), Rajendranagar (Andhra Pradesh), Dumraon (Bihar), Coimbatore (Madras), Poona (Maharashtra), Mandya and Dharwar (Mysore) and Haringhata (West Bengal), Palampur (Punjab), Anand (Gujarat), Bangalore (Mysore). Animal nutrition studies were also conducted at Indian Veterinary Research Institute (Izatnagar), Veterinary College (Mathura) and National Dairy Research Institute (Karnal), as a departmental activity.

Recognition of the importance of grasses and fodder crops in the agricultural economy of the country, the widening gap between supply and demand of forage necessary for animal production, the limitations of the previous studies, the diversity and the complexity of the problems, and the inadequacy of any national organization for undertaking such studies as a subsidiary activity, led to the realization of the need for establishment of an Institution fully devoted to these crops. The Planning Commission approved the scheme for the establishment of the Indian Grassland and Fodder Research Institute, with its network of regional stations and sub-centres, in July, 1961 mandated to carry out research both of fundamental and applied nature on all aspects grasses, grasslands and fodder crops production and coordinate studies and initiate training programme on the subject.

All India Coordinated Project on Forage Crops

In addition to the specialized institution (IGFRI), ICAR is also strengthening forage development activities through out the country through all India coordinated project on Forage crops. This is a forage research network in India under Indian Council of Agricultural Research (ICAR) started in 1970 as a plan scheme. The research programmes being taken up are multi-disciplinary, multi-locational oriented to crop improvement in forages and generation of forage production technologies at the national level considering the regional socio-economic conditions in different agro-climatic regions in the country. Currently, the AICRP is organizing its forage resource development activities through 18 Coordinated Centers sponsored by the ICAR (Table 5).

There are Many Issues, Which Need to be Addressed to Give Momentum to the Growth in Livestock Sector

A. Organizational Structure

A vibrant organization led by experts will facilitate the accelerated development of feed and fodder sector. The need for such a department was felt long ago, when the 'Cattle Conference' held in 1937 in Shimla adopted a resolution to 'set up a fodder and grazing committee in every province on the lines of agriculture department'. While the agriculture flourished over the years, the livestock sector continues to be a subsidiary activity, in spite of the fact it has the potential to lead the world not by its sheer number but by the productivity as it has got favorable climatic conditions to grow round the year fodder production, which will give it an edge over grain based livestock production.

Table 5. List of AICRP Centers

Center	State	Research focus – Forage Crops
A. Coordinated Center		
Palampur, CSK HPKV	Himachal Pradesh	Setaria, Maize, grasses and legumes for sub-temperate and temperate regions, and grassland management in high altitude areas
Ludhiana, PAU	Punjab	Guinea grass, Berseem, Lucerne, Oats, Maize, Pearl millet, Cowpea and BN hybrids
Hisar, CCS HAU	Haryana	Sorghum, Berseem and Guar
Pantnagar, GBPUAT	Uttaranchal	Cowpea and Berseem
Faizabad, NDUAT	Uttar Pradesh	Grasses and legumes for problem soils and Bajra
Ranchi, BAU	Jharkhand	Range grasses and legumes for problem soils, Maize and minor forage crops
Jabalpur, JNKVV	Madhya Pradesh	Soybean, Sorghum, Berseem and Oats
Bikaner, RAU	Rajasthan	Range grasses and legumes
Anand GAU	Gujarat	Luerne, Bajra, Sorghum and Rabi Maize
Urulikanchan, BAIF	Maharashtra	Maize and range grasses and legumes
Rahuri, MPKV	Maharashtra	Bajra, Sorghum, Maize, Lucerne, range grasses and legumes
Hyderabad, APAU	Andhra Pradesh	Maize, Sorghum and Bajra
Tiptur, USA	Karnataka	Maize, Lucerne and range grasses
Coimbatore, TNAU	Tamil Nadu	BN hybrids, Lucerne, Cowpea, Guinea grass, range grasses and legumes
Vellayani, KAU	Kerala	Guinea grass, BN hybrids, Cowpea and Stylosanthes
Kalyani, BCKV	West Bengal	Rice bean, Maize, grasses and legumes problem (Acidic) soils
Bhubaneswar, OUAT	Orissa	Rice bean, Maize, perennial grasses for problem (Acidic & Lateritic) soils
Jorhat, AAU	Assam	Maize, Ricebean, perennial grasses for problem (Acidic hill) soils
Itanagar, ICAR Research Complex for NEH Region	Arunachal Pradesh	Grasses, legumes and fodder tress for wet sub-temperate and temperate regions of North- East, and grassland management
B. Cooperating Centre		
Almora, VPKAS (ICAR Institute)	Uttaranchal	Testing location for Oats, range grasses and legumes and fodder trees for North-Western Himalayan region and grassland management in sub-temperate and temperate areas
Pithoragarh, DARL	Uttaranchal	Testing location for cultivated forages and range grasses

Contd...

Table 5. Contd...

Srinagar, SKUAT	J & K	Testing location for cultivated forages and range grasses
Rajouri, Jammu, SKUAT	J & K	Testing location for cultivated forages and range grasses
Kanpur, CSAUAT	Uttar Pradesh	Testing location for Berseem, Oats, Sorghum, Cowpea and Bazra
Pratapgarh, RDS KVK	Uttar Pradesh	Testing location for forages on problem (Saline - sodic) soils
Mathura, CIRG	Uttar Pradesh	Testing location for cultivated forages
Jhansi, IGFRI (ICAR Institute)	Uttar Pradesh	Testing location at Jhansi and IGFRI Regional Centres at Srianagar, Palampur, Dharwad and Avikanagar for forage crops, range grasses and legumes
Pusa, RAU	Bihar	Testing location cultivated forage crops
Jodhpur, CAZRI	Rajasthan	Testing location for grasses and legumes for arid and semi-arid regions at Jodhpur, Jalour and Chandan-Jaisalmer
Udaipur, MPUAT	Rajasthan	Testing location for cultivated forage crops
Dhari, GAU	Gujarat	Testing location for range grasses and legumes
Dantiwada, GAU	Gujarat	Testing location for range grasses and legumes
Akola, PKV	Maharashtra	Testing location for Sorghum, Cowpea and Oats
Dapoli, KKV	Maharashtra	Testing location for perennial grasses
Raichur, ARS (UAS-D)	Karnataka	Testing location for cultivated forages
Ooty, NJ&SF, TNCMPFL	Timal Nadu	Testing location for cultivated forages
Koch Bihar, UBKV	West Bengal	Development of forage-based cropping systems for *tarai* region in North Bengal
Gangtok, Sikkim Centre, ICAR Research complex for NEH Region	Sikkim	Research on temperate forage
Barapani (Umiam), ICAR Research Complex for NEH Region	Meghalaya	Testing location for Maize, Cowpea and range grasses, and research on forage-based cropping systems for shifting cultivation areas in North-Eastern India
Basar, Arunchal Centre, ICAR Research Complex for NEH Region	Aruanchal Pradesh	Testing location for range grasses
CARI, Port Blair	Andmans & Nikobar Island	Testing location for cultivated forage crops and grasses

In most of the states, the fodder development programme has been relegated to lowest priority. There are hardly any specialist in feed and fodder department. Even the sanctioned post in this department remain unmanned in most of the States. Very few states have fodder development officers with adequate training in this field. The budget allocation is just sufficient to meet the salary expenditure of the staff and hardly any new programme being taken up. Many of the centrally sponsored schemes are not taken up due to lack of staff.

On the other hand there are few successful models like Punjab where the development of feed and fodder is looked after by a Joint Director, 7 Deputy Directors,121 Assistant Fodder Development Officers. A well managed department for feed and fodder development ensures that the technical know how related to fodder cultivation and utilization reaches the livestock owners, their needs for quality seed are met and the farmers remain updated to the latest development in the field by the Fodder Development Officers posted at Block level. A small country like Bhutan has also a very well organized department for feed and fodder development.

B. Lack of Statistics Related to Feed and Fodder

There is a serious lack of statistics related to area under fodder production and productivity of the different fodder crops. Proper statistics on these aspects will help assess the adoption of technologies and identify the gaps that hamper their adoption and help in taking policy decisions. Therefore, we recommend that the National Sample Survey (NSS) should also take up these aspects in their survey as is being done in case of agriculture.

C. Utilization of Forage Resources from Forest Margin

There is little interaction among the forestry and the livestock departments. The forestry department always looks towards suspicion the livestock in its periphery as destroyer of vegetation. There is need to understand that controlled removal of vegetation/ grass in fact spurs its further growth. Hence, a synergistic approach needs to be adopted for the regulated use of grasses in the forest margin which supply very large quantity of forage.

Further, the requirements of herbivore wild life also need proper attention because degraded forests do not meet their requirement. The forest department have also genuine need for seeds of various grasses which need a fair assessment.

D. The Aforestation

The aforestation drive undertaken by various agencies should invariably take up plantation of trees which have fodder value. This will help bridge the gap in fodder demand –supply in normal years besides coming to the aid during periods of drought and flood and create live fodder banks.

E. Issues Related to Crop Residues

Crop residues provides over 50% of the feed requirement of the livestock and it will continue to be the mainstay of livestock feeding in coming future. The straws of crops in many parts of the country are burnt after the harvest of grain. This not only causes environmental pollution but also severely impacts the fodder availability for the livestock. Policy guidelines to prevent the burning of straw need to be put in place. Besides, suitable machinery should be developed which would harvest these straw quickly, efficiently and economically to let the field open for next crop.

Similarly, diversion of edible crop residues towards packing industry should be checked. A futuristic scenario is also being envisaged, wherein diversion of crop residues for bio-fuel production is being considered. What will happen if diversion of bio-fuel from crop residues becomes more economical? We need to keep a guard towards development in this field because diversion of corn for fuel production has already started whose impact has been felt globally. Prices of crop residues and grain are coming closer which needs proper understanding and policy solutions.

F. Storage and Transport of Feed and Fodder

In periods of natural calamities, the livestock are left to fend off on their own. A fair assessment of the number of livestock in drought and flood prone areas and their feed and fodder requirement has not yet been attempted. There is an urgent need to develop Fodder Banks on the lines of Food Corporation of India to provide succour to the livestock in distress. To facilitate transport and storage, static and mobile type bailing machine need to be encouraged in developing fodder banks.

G. Processing of Forage – Densification and Nutritional Enrichment

Processing of forage – densification and nutritional enrichment including complete feed block: To give an impetus to the livestock production programme, there is need to cater the needs of different categories of animals with varying productivity levels. Peri-urban livestock keeping has emerged as an important vocation. The livestock in peri-urban areas often face difficulty in procuring quality forage. In these areas, complete feed blocks cater the need of animals with different production levels.

H. Addressing the Requirement of Quality Forage Seed

The production and distribution of quality fodder seeds need to be taken up on priority by NSC and State seed corporations. A dedicated department should be established which would take up this work sincerely and improved varieties of fodder crops should be made available to farmers at block level as is being done for various cereal crops.

I. Institutional Mechanisms to Utilize the CPR, Roadside, Railway Lines

There is need to strengthen the village institutions towards development of common property resources. There are many a success story, where vesting the ownership rights to the people has lead to the rejuvenation of the degraded pasture and evolving benefit sharing mechanisms. The vast tract of unused land along roadside, railway line and other wasteland could be utilized productively, if they are leased out to the people residing in the vicinity. The Bangladesh Grameen Fisheries and Livestock (*Motsho O Pashusampad*) Foundation has successfully put to use several stretches along the highway which has lead to increase in livestock and fish production improving the livelihood of people.

J. Pathway: Should the Co-operatives be Made a Major Player?

The success of co-operative sector in the dairy sector in different states is very well known. They have been successful in many states where the government departments have failed to deliver as in Bihar. The members of the co operatives have increasingly asking the co-operatives to supply them quality seeds of fodder crops. The Co-operatives in turn seek help from research Institutes for their requirement.

Will it be more effective to promote fodder production programmes through co-operatives which have a direct stake in their development?

K. Policy Related to Breeding for Crop Varieties

In the pursuit to maximize the yield of grains of cereal crops, the straw component has been neglected. Any breeding programme for cereal crops must realize that crop residues are equally important in Indian context, where they are the mainstay of livestock feeding. Therefore, a minimum straw to grain ratio must be ensured before any new crop variety is released.

Accelerated development of livestock sector requires a coordinated effort by the central and state governments to improve the livestock production and productivity. There is need for vibrant organizational set up at state levels to improve the fodder production activities and to implement the centrally sponsored schemes for fodder development. Improvement in feed and fodder availability will result in improved animal production and hence the livelihood of the farmers will be strengthened.

❑❑❑

Chapter **2**

Forage Research Network in India and Its Role in Enhancing Forage Production

S. A. FARUQUI AND SUNIL KUMAR
Indian Grassland and Fodder Research Institute, Jhansi-284 003 (U.P.), India

Introduction

The forage research network in India is represented as All India Coordinated Research Project for on Forage Crops (AICRP-FC) under Indian Council of Agricultural Research (ICAR). The Project was started in 1970 as a plan scheme, envisaged as multi-disciplinary, multi-location programme to coordinate the crop improvement in forages and generation of forage production technologies at the national level considering the regional socio-economic conditions in different agro-climatic regions in the country. The AICRP on Forage Crops was postulated with following mandate and objectives.

Mandate

- To coordinate multilocational testing programme on forages at the national level with a view to identify appropriate varieties and production technologies for different agro-ecological conditions.
- To coordinate and monitor research related to problems of national and regional importance

- To conduct strategic and applied research for boosting production and productivity of forages (arable fodder crops, range grasses and legumes).
- To function as a major service center for exchange of scientific information and research material related to forages.

Objectives

- Development and conservation of germplasm of forage crops and range species
- Development of superior varieties/hybrids of forage crops and range species for increased animal production
- Development of appropriate forage production technology and forage utilization through multidisciplinary approach for different cropping systems, rangelands and problem soils
- Development of appropriate technology for seed production of forages
- Transfer of technology.

Forage Research Network

To meet the mandate of the Project the following locations have been approved to work as Coordinated centres of the Project (Table 1). Currently the Project has 18 Coordinated centres in SAUs and BAIF(NGO) funded by ICAR. Three new centres have been approved in XI five year plan viz.; Shrinagar, SKUAT; Raipur, IGKV and Imphal, CAU. Besides this, 24 cooperating centres also participate in testing of material and generation of technology for their region.

Table 1 : Coordinated Centres of AICRP on Forage Crops

Location of Coordinated Centres	Agro-ecological Region**
1. Hisar CCS HAU, Haryana	Western Plain & Kutch Peninsula
2. Bikaner, RAU, Rajasthan	Northern Plain & Central Highlands
3. Ludhiana, PAU, Punjab	Central Highlands & Kathiawar Peninsula
4. Anand, AAU, Gujarat	
5. Rahuri, MPKV, Maharashtra	Deccan Plateau
6. Urulikanchan* BAIF, Maharashtra	
7. Hyderabad, ANGRAU, Andhra Pradesh	Deccan Plateau & Eastern Ghats

Contd...

Table 1. Contd...

8. Mandya, UAS (B), Karnataka	Eastern Plateau & Deccan Plateau
9. Coimbatore, TNAU, Tamil Nadu	
10. Pantnagar, GBPUAT, Uttaranchal	Northern Plain, Hot sub-humid
11. Faizabad, NDUAT, Uttar Pradesh	
12. Jabalpur, JNKVV, Madhya Pradesh	Central Highlands (Malwa & Bundelkhand)
13. Ranchi (Kanke), BAU, Jharkhand	Eastern Plateau (Chhotanagpur Plateau) &
14. Bhubneswar, OUAT, Orissa	Eastern Ghats
15. Palampur, CSK HPKV, Himachal Pradesh	Western Himalaya, Warm sub-humid (including humid)
16. Jorhat, AAU, Assam	Assam & Bengal Plains
17. Kalyani, BCKV, West Bengal	Western Ghats & Western Coastal Plains
18. Vellayani, KAU, Kerala	

* Cent-per-cent share of ICAR.

** Based on National Bureau of Soil Survey & Land Use Planning, Nagpur, Pub. No. 24, 1990.

Research Focus : Mandate Forage Crops

The Coordinated and Cooperating Centres of the Project are responsibe for generation of superior forage germplasm and varieties in selected crops. The coordinated centres work with a multidisciplinary team for improvement of forage resources and generation of production technology for their state and region (Table 2). The voluntary centres also contribute in their specific crops and constitute as important partner for regional testing and technology development.

Role of AICRP on Forage Crops in Enhancing Forage Production

The AICRPFC network covering all major agro-climatic situations of the country is engaged on varied aspects of research on forage production and technology development. The project with its multidiscipline scientific manpower and necessary field and laboratory support at coordinated centres, has been given mandate to develop superior forage varieties, region specific production and protection technology. Breeder seed production of forage crops by the centres is also undertaken for the indents received by Department of Agriculture, Govt. of India. The research programme of the project basically attempts applied and strategic research in coordinated, network, location specific mode with need based and problem oriented approach. The project has made a great awareness in the promotion of forages and bridging the gap in demand and supply of feed and fodder in different agro- climatic situations of the country. The transferable production technologies and release of more than 180 improved cultivars on forages for different regions have made stride among the farmers. Major contribution of AICRPFC in field of forage production and utilization in the country are highlighted as under:

Table 2 : Mandate Forage Crops for the AICRP on Forage Crops Centres

Centre	State	Research focus – Forage Crops
A. Coordinated Centres		
Palampur, CSK HPKV	Himachal Pradesh	*Setaria*, Tall fescue grass, White clover, Red clover and grassland management in high altitude areas and disease resistance
Ludhiana, PAU	Punjab	Guinea grass, Berseem, Oats, Cowpea and NB hybrids & disease resistance
Hisar, CCS HAU	Haryana	Berseem, Cowpea, Oat, screening of disease & insect pest
Pantnagar, GBPUAT	Uttarakhand	Cowpea Berseem and Oats
Faizabad, NDUAT	Uttar Pradesh	Bajra, Cowpea and Oat
Ranchi, BAU	Jharkhand	Maize, Cowpea and Dinanath grass
Jabalpur, JNKVV	Madhya Pradesh	Soybean, Sorghum, Berseem, Oats and Ricebean
Bikaner, RAU	Rajasthan	Range grasses and legumes
Anand GAU	Gujarat	Lucerne, Bajra, Sorghum Maize, Cowpea & pest disease
Urulikanchan, BAIF	Maharashtra	Bajra, NB Hybrid, *Stylosanthes* and Lucerne
Rahuri, MPKV	Maharashtra	Sorghum, Maize, Lucerne and Oats
Hyderabad, ANGRAU	Andhra Pradesh	NB Hybrid, Bajra, Lucerne and screening disease & insect pest.
Mandya, UAS	Karnataka	Maize, Lucerne and disease resistance
Coimbatore, TNAU	Tamil Nadu	NB hybrids, Lucerne, Cowpea, Guinea grass, Sorghum, Bajra, Ricebean
Vellayani, KAU	Kerala	Guinea grass, NB hybrids, Cowpea and Rice bean
Kalyani, BCKV	West Bengal	Rice bean, Maize, Coix
Bhubaneswar, OUAT	Orissa	Rice bean, Maize, Lathyrus, range grasses for problem soils
Jorhat, AAU	Assam	Maize, Rice bean, Lathyrus
B. Cooperating Centres		
Almora, VPKAS	Uttarakhand	Testing location for cultivated forages and range grasses
Srinagar, SKUAT	J & K	Testing location for cultivated forages and range grasses
Rajouri, Jammu, SKUAT	J & K	Testing location for cultivated forages and range grasses & Forage Agronomy programme

Contd...

Table 2. Contd...

Kanpur, CSAUAT	Uttar Pradesh	Testing location for Berseem, Oats, Sorghum, Cowpea and Bajra
Meerut, SVBPUAT	Uttar Pradesh	Testing location cultivated forage crops & Forage Agronomy programme
Jhansi, IGFRI	Uttar Pradesh	Testing location at Jhansi and IGFRI Regional Centres at Palampur, Dharwad and Avikanagar for forage crops, range grasses and legumes
Pusa, RAU	Bihar	Testing location cultivated forage crops & Forage Agronomy programme
Raipur, IGKV	Chhatisgarh	Testing location cultivated forage crops & Forage Agronomy programme
Ambikapur, IGKV	Chhatisgarh	Testing location cultivated forage crops & Forage Agronomy programme
Jodhpur, CAZRI	Rajasthan	Testing location for grasses and legumes for arid and semi-arid regions at Jodhpur, and Chandan - Jaisalmer
Udaipur, MPUAT	Rajasthan	Testing location for cultivated forage crops
Jalore, RAU	Rajasthan	Testing location for cultivated forage crops, range grasses
Durgapura, RAU	Rajasthan	Testing location for cultivated forage crops, range grasses
Dhari, GAU	Gujarat	Testing location for range grasses and legumes
Jamnagar, GAU	Gujarat	Testing location for range grasses and legumes
Dantiwada, GAU	Gujarat	Testing location for range grasses and legumes
Akola, PKV	Maharashtra	Testing location for Sorghum, Cowpea and Oats
Dapoli, KKV	Maharashtra	Testing location for perennial grasses
Palghar, KKV	Maharashtra	Testing location for perennial grasses
Raichur, ARS (UAS-D)	Karnataka	Testing location for cultivated forages
Ooty, NJ&SF, TNCMPFL	Tamil Nadu	Testing location for cultivated forages
Koch Bihar, UBKV	West Bengal	Development of forage-based cropping systems for *Tarai* region in North Bengal
Imphal, CAU	Manipur	Testing location cultivated forage crops & Forage Agronomy programme
Gangtok, Sikkim Centre, ICAR Research complex for NEH Region	Sikkim	Testing location for temperate forages and research on forage-based cropping systems
Barapani (Umiam), ICAR Research Complex for NEH Region	Meghalaya	Testing location for Maize, Cowpea, Ricebean and range grasses, and research on forage-based cropping systems for shifting cultivation areas in North-Eastern India

A. Germplasm Collection and Enrichment

Forage genetic resources are an important asset of the country. The centres are engaged in collection of forage germplasm through exploration and their maintenance for utilization in varietal improvement programme. The germplasm are documented for their characters and registered in NBPGR. Suitable lines of forage species are also identified for specific purpose like intercropping, shade tolerance, dual purpose, problem soil etc. Range grasses and legumes evaluated under rangeland conditions in cool sub-tropical and Himalayan Hill region; dry and drought prone areas for performance evaluation. This wealth of material provides strong base for generation of superior material for release. In Table 3, total germplasm collection (crop-wise) by the centres has been given.

Table 3: PGR Status of Forge Crops at AICRP (FC) Centres at the end of X Plan

S.No.	Forage crops	Germplasm holding (Nos.)
1.	Oats *(Avena sativa)*	1163
2.	Sorghum *(Sorghum bicolor)*	1104
3.	Berseem *(Trifolium alexandrinum)*	653
4.	Cowpea *(Vigna unguiculata)*	626
5.	Rice bean *(Vigna umbellate)*	578
6.	Maize *(Zea mays)*	485
7.	Pearl millet *(Pennisetum glaucum)*	426
8.	Lucerne *(Medicogo sativa)*	363
9.	Sewan grass *(Lasiurus sindicus)*	350
10.	Guinea grass *(Panicum maximum)*	315
11.	Hybrid Napier *(P. purpureum x P. glaucum)*	135
12.	Anjan grass *(Cenchrus ciliaris)*	118
13.	Soybean (*Glycine* spp.)	90
14.	Stylosanthes *(Stylosanthes spp.)*	84
15.	Red Clover *(Trifolium pratense)*	82
16.	Marvel ghas *(Dichanthium annulatum)*	71
17.	Tall fescue *(Festuca arundinacea)*	61
18.	Setaria *(Setaria anceps)*	47
19.	Rye grass *(Lolium perenne)*	30
20.	White clover *(Trifolium repens)*	28
21.	Lathyrus *(Lathyrus sativa)*	26
22.	Dinanath grass *(P. pedicilatum)*	25
23.	Dharaf grass *(Chrysopogon fulvus)*	25
24.	Job's tear *(Coix aquatica)*	21
25.	Sen grass *(Sehima nervosum)*	20

Contd...

Table 3. Contd...

26.	Lampa grass *(Heteropogon controtus)*	12
27.	Teosinte(*Zea mexicana*)	11
28.	Lotus *(Lotus corniculatusni)*	8
29.	Clitoria *(Clitoria ternatia)*	5
30.	Fulkara grass *(Bothriochloa intermediate)*	4
31.	Para grass *(Brachiaria mutica)*	4
32.	Rhodes grass *(Chloris gayana)*	2
33.	Dhaman grass *(Cenchrus setigerus)*	2
34.	Nandi ghas *(Setaria sphacelata)*	1
35.	Aparajit / Butterfly pea *(Clitoria ternatia)*	1
36.	Hedge Lucerne *(Desmenthus virgatus)*	1
37.	Karnal grass *((Leptochloa fusca)*	1
	Total	**6973**

Source : QRT document, 2003-07, AICRP on Forage Crops

B. Technological Innovations

I. Varietal Development

A number of superior forage crops varieties have been developed through network of the project at central as well as state level for cultivation in all India / specific regions. Recently developed ones are listed in Table 4.

Table 4 : Superior Forage Crops Varieties Suitable for Different Zones of India

S. No.	*Crop*	*Variety*	Year and level of Release	Adaptability regions / states	Green Fodder Yield (q/ha)
Kharif Forage Crops					
1.	Bajra (*Pennisetum americanum* (L.) Leek)	Giant Bajra	1980/1984 (C&S)	Haryana, Punjab. Gujarat, Rajasthan & UP	290-350
		Raj Bajra Chari-2	1987(S)	Rajasthan, Haryana, Punjab. Gujarat	238-275
		APFB-2	1997(S)	Andhra Pradesh	250-300
		GFB-1	2005 (S)	Gujarat, Haryana, Punjab. Rajasthan	300-325
		UPC-5286	1981(C)	All India	275-300
		UPC-5287	1986(C)	All India	300-400

Contd...

Table 4. Contd...

		UPC-4200	1991(C)	Rajasthan, Haryana, Punjab.	250-320
		Bundel Lobia – 1	1992(C)	Central & North west zone of the country	220-350
		Bundel Lobia-2	1993(C)	North west zone of the country	350-400
		UPC-8705	1997 (C)	Entire Country	300-420
		UPC-9202	1999 (C)	All India	350-420
		Haryana Lobia-88	1997 (C)	North Western regions of the Country	280-350
3.	Guar (*Cyamopsis tetragonoloba*) (L.)	HFG-119	1981 (C)	All India	325-400
		HG-75	1981 (C)	North Western regions of the Country	375-400
		HG-258	1987 (C)	North Western regions of the Country	320-370
		Guara –80	1990 (C&S)	North western zone	350-390
		Bundel Guar-1	1993 (C)	All India	310-325
		Bundel Guar-2	1994 (C)	All India	325-475
		HFG-365	1998 (S)	Haryana, Punjab. Gujarat	300-350
		Bundel Guar-3	1999 (C)	Entire guar growing tract	260-300
4.	Lablab bean (*Lablab purpureus* L.)	Bundel Sem-1	1993 (C)	All India	220-350
5.	Guinea grass (*Panicum maximum* Jacq.)	PGG-9	1986 (C)	Temperate and north west zone	900-1000
		PGG-14	1988 (C)	Central zone	1000-1100
		PGG-19	1989 (C)	Punjab under irrigated conditions	1150-1250
		PGG-101	1995 (C)	All India under irrigated conditions	1000-1100
		PGG-518	1997 (C)	Punjab under irrigated conditions	1120-1300
		JHGG 96-5	2003 (C)	Rainfed areas of Gujarat, Rajasthan Punjab	300-400

Contd...

Table 4. Contd...

6.	Dinanath grass (*Pennisetum pedicellatum* L.)	Bundel –1	1987 (C)	All India	550-850
		Bundel –2	1990 (C)	All India	520-850
7.	Anjan grass (*Cenchrus ciliaris* L.)	Bundel Anjan-1	1989 (C)	All India	225-300
8.	Dharaf grass (*Chrysopogon fulvus)*	GAUD-1	1979 (S)	Central and western India	350-400
9.	Marvel grass (*Dicanthium annulatum)*	GMG-1 (Gujarat Marvel Grass–1)	1980 (C)	Rajasthan, Punjab & Gujarat	325-375
Rabi Crops					
10.	Oats (*Avena sativa* L.)	OS-6	1981 (C)	Temperate & sub tropical regions	450-530
		OS-7	1981 (S)	Tropical/ sub tropical regions	475-500
		UPO-94	1981 (C)	North west and central India	575-650
		UPO-212	1990 (C)	North and central India	550-600
		OL-125	1994 (C)	Northwest zone	580
		Bundel Jai-851	1997 (C)	Northern and North-West regions	400-620
		HJ-8	1997 (S)	All India	550-600
		JHO 99-2	2003 (C)	Gujrat, Rajasthan, Punjab, Haryana	500-650
		JO-1	2004 (S)	North west and Central India	600-650
		RO-19	2005 (C)	All Oats growing states	550 –575
		JHO-99-1	2005 (C)	North west & Central zones	500-650
11.	Berseem (*Trifolium alexandrium* L.)	Mescavi	1975 (C & S)	Central and North West India	700-850
		Wardan	1981 (C)	All Berseem growing regions	726-900
		BL-22	1987 (C)	Temperate and North West Zones	620-700

Contd...

Table 4. Contd...

		JB-5	2004 (S)	Central & North West Zones	1000-1050
		BL-2	1989 (C)	Temperate zone	420-550
		Bundel Berseem-2	1997 (C)	Central and North West zones	580-850
12.	Lucerne (*Medicago sativa* L.)	Anand –2	1975 (S)	Gujarat, Rajasthan and M.P.	900-1000
		RL-88	1995 (C)	All Maharashtra	1450-1600
		LLC-3	1985 (C)	All Lucerne growing states	550-600
		Anand –3	1991 (S)	Cold dry zone	560-750

C = Released by Central Variety Release Committee, S = Released by State Variety Release Committee
Source : Faruqui *et al.* (2005)

II. Forage Production Technologies

i. Intensive Forage Production Systems

The intensive forage crops sequences have been worked out for irrigated conditions of different agro-climatic regions of the country under the aegis of the All India Coordinated Project for Research on Forage Crops (AICRP-FC). Except for high water table areas and sub-tropical and sub-temperate hill regions, perennial crops-based rotations are suitable, whereas in other cases, annual crops in rotation are appropriate (Table 4).

ii. Food-Fodder Production Systems

Integration of food and forage crops in the existing system has been found economical and sustainable in many agro-climatic situations particularly in rainfed/ resource crunch areas. In dryland situations, short duration fodder crops like cowpea, guar, sorghum etc. harvested after 50-60 days of sowing gave 25-30 t/ha of green fodder. After this a normal rabi crop of mustard, gram, linseed and safflower can be grown. In moisture deficit and salt affected areas, recent barley varieties like RD-2035, RD-2552 and RD- 2670 provides fodder and grain both. The crops are harvested at 55-60 days stage which provides 125-225 q/ha green fodder. The regenerated crop is left for grain producing 15-25 q/ha in varied agro-climatic situations.

Table 4 : Intensive Forage Crop Rotations for Different Agro-climate Zones of India

Zones/ Location*	Crop rotation	Green fodder yield (t/ha/year)
North		
Palampur (Himanchal Pradesh) : Sub-temperate, Moist, Red soil		
1. Maize + Cowpea – Lucerne + Oats - Mustard		85
2. NB Hybrid + Velvet bean – Berseem + Mustard		123
Pantnagar (Uttaranchal) : Tarai, Red & yellow soil		
1. Maize + Cowpea – Toria - Oats		177
2. NB Hybrid + Berseem – Cowpea		121
Hisar (Haryana) : Semi-arid, Sandy loam soil		
1. NB hybrid + Berseem		212
2. NB hybrid + Lucerne		176
Central and Western		
Jhansi (UP): Semi-arid, Red soil		
1. NB hybrid + Cowpea - Berseem + Mustard		255
2. Sorghum + Cowpea – Berseem + Mustard – Maize+ Cowpea		176
Jabalpur (MP) : Sub-humid, Black soil		
1. NB hybrid + Cowpea - Berseem		176
2. Sorghum+Cowpea–Berseem+ Mustard–Sorghum+Cowpea		169
Anand (Gujrat) : Semi-arid, Black soil		
1. NB hybrid + Cowpea - Lucerne		253
2. Maize + Cowpea – Oats – Maize + Cowpea – Maize + Cowpea		168
East		
Ranchi (Kanke): Sub-humid, Red acidic soil		
1. Pearlmillet + Cowpea – Maize + Cowpea - Oats		103
2. Maize + Cowpea – Sorghum + Cowpea – Berseem + Mustard		96
Kalyani (West Bengal) : Sub- humid, Alluvial soil		
1. Maize + Cowpea – Dinanath grass - Oats		131
2. Maize + Ricebean – Berseem + Mustard		112
Jorhat (Assam): Humid, Acidic soil		
1. NB hybrid		106
2. Maize + Cowpea – Maize + Cowpea – Maize + Cowpea		85
South		
Coimbatore (Tamil Nadu): Sub humid, Black soil		
1. NB hybrid + Lucerne		225
2. Sorghum + Cowpea – Maize + Cowpea – Maize + Cowpea		111
Vellayani (Kerala): Humid, Red soil		
1. Guinnea grass in coconut plantation		135
2. Congosignal grass in coconut plantation		75

Source : Hazra (1998)

iii. Diversification of Existing Cropping Systems through Forages

Crop diversification as a strategy to bring desirable change towards more potential crops and cropping systems is being pursued to meet ever increasing demand of agricultural products in sustainable manner. In order to support dairy, forages are to be propagated with new dimensions. The research findings of the project carried out in different zones at selected locations have indicated that substitution of less remunerative crops and cropping systems by forages are remunerative. In all the tested situations, forage based sequences or inclusion of a forage crop in most adopted sequence was superior (15 - 500 %) on net return basis over existing system. Perennial NB hybrid system with seasonal legumes like cowpea/ berseem/ Lucerne in NE and South region produce higher biomass and net return (Sunil Kumar and Faruqui, 2007).

iv. Forage from Rice Fallows

In southern plateau, short duration legumes like Pillipesara, Sesbania and Sannhemp have been found promising in the residual moisture of rice fields. Horse gram, moth bean and lathyrus provide a good choice for forage supply during lean period in the rice fallows in eastern and NEH region. The recent results on augmentation of forage legume yield through zero tillage after rice are encouraging. In east region after harvest of late maturing rice, berseem sown with zero tillage (sowing in standing crop before15 days of harvest of rice) and 40 kg/ha seed gave one additional harvest of berseem and 15-20 % increase in green fodder yield (Anonymous, 2005).

v. Forage Production from Problem Soil

Problem soils are invariably poor in productivity. Utilization of marginal or problem soils for diversification of rice based mono-cropping through forages will lead to ameliorative effects on soil acidity, sodicity and sustainability of agricultural production system. Diversity of forage resources in problem soils largely depend upon soil type, rainfall and other climatic-edaphic factors. Growing of forage crops tolerant to acidity and salinity-sodicity can ensure reasonably good returns in the initial phases of reclamation. Under AICRP-FC, AAU, Jorhat, BCKV, Kalyani, OUAT Bhunbaneswar and BAU Ranchi for acid soil and NDUAT, Faizabad for sodic soils, have evaluated the performance of number of crops and their improved varieties. The summary of suitable forage species and their varieties under different category of problem soils is presented Table-5 and shrubs and trees are given in Table 6 (Faruqui and Sunil Kumar, 2007).

Growing of acid tolerant forage species and varieties is recommended for acidic soils. For higher yield and better soil health, liming is standard management practice to be adopted for increasing productivity of such soils. In Acidic soil (pH < 5.5), higher forage yield (30%) may be obtained with the application of lime along with recommended dose of Phosphorous and Potassium (PKR) and VAM to Rice bean-Oat sequence with 21% increase in net return over recommended dose of fertilizer. For highly sodic soils (pH > 10.0), suitable and sodicity tolerant grasses are Para > Karnal > Rhodes with higher biomass production, increase in the Ca content and infiltration rate and marked reduction in soil pH, EC and ESP after 4 years of establishment. Under moderate pH (9.0 - 9.5), forage sorghum is superior over bajra and productivity increases by 2 – 3 times by application of pyrite @ 5 t/ha.

Table 5 : Potential Forage Crops and their Improved Varieties for Problem Soils

Problem soil	Crop	Variety	Green fodder yield (t/ha)
1. Acidic		**(i) Cultivated**	
	Cowpea (*Vigna unguiculata*)	EC-4216, UPC-5286, UPC-4200	25 – 30
	Rice bean(*Vigna umbellata*)	K - 1, K - 16, Bidhan - 1, Bidhan - 2	20 – 25
	Oats (*Avena sativa*)	Kent, JHO-99-1, JHO-99-2, JHO-851	35 – 40
	Maize (*Zea mays*)	African Tall	30 – 35
	Deenanath grass (*Pennisetum pedicellatum*)	Bundel Dinanath-1 Bundel Dinanath-2 PS-3, Pusa-19	50 – 60
	Thin napier (Pennisetum polystachyon)	Selection	40 – 60
	Guinea grass (*Panicum maximum*)	Hamil, PGG-9, PGG-14	80 – 90
	Napier Bajra Hybrid	NB-21, IGFRI-3, IGFRI-6, CO-2, CO-3	90 – 120
		(ii) Perennial range grasses	
	Setaria (*Setaria anceps*)	Nandi, Narok, PSS-1, Kazungula	70 – 90
2. Sodic		**(i) Cultivated**	
	Sorghum (*Sorghum bicolor*)	MP Chari, PC-6, PC-9, UP Chari-2, UP Chari-3	35 - 40

Contd...

Table 5. Contd...

	Bajra (*Pennisetum glaucum*)	Giant Bajra, Raj Bajra Chari-2	35 – 45
	Napier Bajra Hybrid	IGFRI - 6	80 – 100
	Oats (*Avena sativa*)	Kent, JHO-822	35 – 45
	Berseem (*Trifolium alexandrinum)*	Mescavi, Bundel Berseem-3 Wardan	60 – 80
	Shaftal (*Trifolium resupinatum)*	SH – 48	50 – 70
	(ii) Perennial range grasses		
	Karnal grass (*Leptochloa fusca*)		25 – 30
	Rhodes grass (*Chloris gayana*)		20 – 25
	Para grass (*Bracharia mutica*)		40 – 45
	Gutton Panic grass (*Panicum maximum*)		45 – 60
	Desmenthus (*Desmenthus virgatus*)		35 – 55
3. Saline	**(i) Cultivated**		
	Sorghum (*Sorghum bicolor*)	HC-136, HC-171, SSG-59-3	35 – 50
	Teosinte (*Zea mexicana*)	Improved Teosinte, TL -1	30 – 40
	Napier Bajra Hybrid	IGFRI-6, IGFRI-10	90 – 110
	Lucerne (*Medicago sativa*)	T-9, RL-88	60- 80
	(ii) Perennial - As in sodic		
4. Coastal saline	**(i) Cultivated**		
	Oats (*Avena sativa*)	Kent, JHO-851, JHO-822	35 – 40
	Cowpea (*Vigna unguiculata*)	EC-4216, UPC-5286	25 – 35
	Lathyrus (*Lathyrus affaca*)	Local	20 – 25
	Coix (*Coix aquatica, Coix lachroyna* Zobi)	Local	25 – 30
	Napier Bajra Hybrid	IGFRI-3, IGFRI-6, CO-1, CO-3	70 – 90
	(ii) Perennial grass		
	Congosignal grass (*Bracharia ruziziensis*)		30 -40
	Para grass (*Bracharia mutica*)		25 - 35
	Sabi grass (*Urochloa mosambicensis*)		25 – 30

Contd...

Table 5. Contd...

5. Marshy/ Wet land	Teosinte, Coix (*Coix aquatica*),Rhodes / Para grass, *Congosignal, Paspalam* sp., *Iseilema laxum*		25 – 45
6. Ravenous / Degraded lands	Dinanath grass, Guinea grass, *Stylosanthes hamata, S. guianensis*, *Chrysopogon fulous, Sehima nervosum*		20 – 40
7. Desert/ Arid soils	Anjan grass (*Cenchrus ciliaris*)	CO-1, CAZRI-75	15 – 30
	Yellow Anjan (*C. setigerus*)	CAZRI-358, CAZRI-76	25 – 40
	Sewan grass (*Lasiurus sindicus*)	Selection	18 – 25

Table 6 : Trees and Shrubs Suitable for Problem Soils

Problem soil	Trees/ Shrubs
Acidic	*Leucaena leucocephala, Bauhinea variegata, Glyriciridea, Moringa, Sesbania grandiflora, Morus serrata, Albizia chinensis, Alnus nepalensis*
Saline -sodic	*Acacia tortilis, A. nilotica ,Prosopis juliflora, A. amara, Salvadora spp., Sesbania spp. Zizyplus sp., Desmenthus virgatus, L. leucocephala*
Coastal saline	*L. leucocephala, Bahunia, Gliriciridea sepium, Desmenthus virgatus, Sesbania gradiflora, Casurina equisetifolia, A. nilotica, A. tortilis*
Wetland	*Delbergia sisso, A. catechu, A. nilotica, Salix spp., Zizyphus* spp.
Ravenous/ forest land	*A. tortillis, A. nilotica, A. amara, D. sisso, Dendrocalamus strictus, Eugina jambolana, Azadirachta indica, Madhuca latifolia, Hardwickia binata*

vi. Forage Production Under Plantation Crops

Considerable area in coastal areas, southern plateau and in NEH region has potential to grow forages in partial shade of plantaion crops. Forage crops like guinea grass, congosignal grass, signal grass and *Desmenthus virgatus* are partial shade tolerant forages are successfully grown in humid and hot-pre-humid eco-regions. The inference of multilocations researches indicate almost equal green forage yield of guinea grass (70-130 t/ha) was obtained when supplemented with 50-100 % higher recommended doses of NPK in coconut shade than open. In NE regions, the forage yield was 90 % of open when grown under Silkworm tree (*Machilus bombycini*). In general, the productivity of these plantation crops was enhanced by 10-15 % due to input management in understorey forage crops. In Kerala, congosignal grass under coconut plantation crops gave 40-50t/ha green fodder when supplemented with150 kg N/ha (in three splits) and harvesting at 45 days interval. Application of *Azospirillum/Azotobacter* to these grasses under shade saves 25-30 % of nitrogen requirement. Benefit : cost ratio of Coconut / banana intercropped with guinea grass & *Desmenthus virgatus* (3:1) ranged 2.2-2.8 while sole coconut / banana stands obtained B: C ratio 1.6-1.9 (Anonymous, 2005).

III. Forage Seed Production

The forage cultivation is highly influenced by the availability of seed of cultivated species and other range grasses and legumes. The forage seeds are short in supply not due to physical factors but certain genetical, biochemical and physiological reasons associated with these crops. These crops are developed for higher forage yield than seed yield. The seed productivity is also very low as they are shy seeders. To resolve this situation a strategy for seed production from breeder seed to certified level has been found helpful. The efforts are made in this direction under AICRP on Forage Crops which suggests that seed productivity may be raised if they are grown in specific agro-climatic situations. The list of crops with suitable agro climatic region for seed production has been given in Table 7. Recently variety wise potential region for seed production of cowpea has been identified. The semi spreading type forage cowpea cv. Bundel Lobia-1 with narrow spacing of 30 cm produces 17 per cent higher seed yield as compared to other popular cultivars of cowpea viz., IGFRI-95 (erect type), UPC-9202 & UPC-4200 (spreading type). Location wise, forage cowpea varieties like Bundel Lobia -1 in central and North West, UPC-9202 in North West and South; UPC-4200 in Central and East were superior for seed yield. Application of some chemicals in grasses in arid region has given superior seed yields. The spray of Thiourea @ 0.05% to *Lasiurus sindicus* increases the seed yield by 13-15 percent.

R&D Set up Available for Forage Production and Utilization

Forage research and development activities in India are governed by different organizations. As apex institution, Indian Grassland and Fodder Research Institute (IGFRI) govern its basic and strategic research on all aspect of forage improvement, production, post harvest and utilization to all most all agro-climatic situations of India. The regional requirements are maintained through its three Regional Research Stations located Palampur, Avikanagar and Dharwad. Besides this, the All India Coordinated Research Project on Forage Crops (AICRP-FC) with 18 coordinating centres and 22 collaborating centres and testing locations is operative for germplasm evaluation and varietal performance testing, generation of production and protection technology for specific regions. The programme of IGFRI and AICRP-FC are adequately been supported with other agencies. Inter-project linkages are being maintained for better output for specified target approach. Certain NGO's organizations like NDDB and State Dairy Federations are playing active role in promoting farmer friendly forage production technology and seed production and marketing. Seven Regional stations for forage production and demonstration (RSFPD) located at different regions provide strong base for demonstration of released varieties and technology in their respective area of jurisdiction. The centers are also engaged in multiplication of seed for distribution among the farmers.

Table 7: Regional Suitability of Forage Crops for Seed Production

S. No.	Forage crops	Suitable region
A. Cultivated forages		
1.	Sorghum	South west, West
2.	Pearl millet	North west
3.	Maize	East
4.	Cowpea	North west, Central, South west
5.	Rice bean	East
6.	Oats	North and North central
7.	Berseem	Central, North west, West
8	Lucerne	West
9.	Guinea grass	South, North west
10.	Dinanath grass	East
B. Range grasses and legumes		
11.	*Cenchrus, Dicanthium, Sehima, Chrysopogon, Lasiurus*	West, Central, South central
12.	Congosignal, Signal grass	Southern plateau
13.	*Stylosanthes, Siratro, Dolichos*	South, West
14.	Hedge Lucerne, *Leucaena, Sesbania*	South, Southern plateau

Linkages

The Project has developed linkages and strengthened the existing linkages with various Government and non-government organizations for effective research, development and extension of the forage programme at national and global level. At national level the linkages exist with Ministry of Rural Areas and Employment (Department of Wasteland Development); Ministry of Environment and Forests (National Afforestation and Eco-development Board); Department of Science and Technology; Department of Agriculture, Animal Husbandry and Dairying; U.P. Council of Agricultural Research; State Animal Husbandry Departments; Forest Departments; Tribal Development; North Eastern Council; Dairy Farms and State Dairy Boards /NDDB. Presently, 17 SAUs are contributing as partners in the AICRP on Forage Crops. Efforts will be made to motivate other SAUs and selected traditional Universities for collaborative research on forage crops, range grasses and legumes, and rangelands through voluntary contribution and ad-hoc research. Among NGO's, BAIF and its sister organizations; KRIBHCO, NGO's working for rural enlistment through animal husbandry and agriculture. In ICAR system, main partners are; IGFRI and its regional centres; CIRG. Makhdoom; CAZRI, Jodhpur; CSWRI, Avikanagar; CSWCRTI, Dehradun; VPKAS, Almora; CRIDA, Hyderabad; CARI, Port Blair; NRCAF, Jhansi; NRC Sorghum, Hyderabad; NBPGR, New Delhi; DMR, New Delhi; AICRP on Pearl Millet, Arid legumes (Guar and Cowpea), Small millets (Finger millet), DWR, Karnal, Network project Barley, IIPR, Kanpur, IIVR, Varanasi (Pea, Cowpea and Lathyrus) and Net work project on under-utilized plants (Amaranthus, Rice bean and other minor forages). At global level the germplasm exchange programme for specific traits and attributes through NBPGR is being strengthened. Among those important organization/ institutes are; FAO, Rome; CSIRO, Australia; IGER, U.K.; IPGRI (South Asia Centre), New Delhi; USAID, USA; CIMMYT, Mexico; ICRISAT, Hyderabad; Cornell University, Ithaca, USA; Colorado State University, Fortcollins, USA; ICRAF. Kenya; ICRA, The Netherlands, ILRI.

Future Thrust Areas

- Exploration, collection and conservation of the indigenous germplasm of forage crops, range grasses, legumes and fodder shrubs and trees.
- Acquisition of global germplasm for forage improvement research
- Breeding for biotic and abiotic stress tolerant cultivars suitable for non-arable agriculture.
- System research approach for ameliorating the conditions of small and marginal land holders.

- Introduction of short duration forage crops in existing farming systems to maximize forage production.
- Exhaustive migratory grazing system to improve the forage requirements.
- Improvement in the production potential of fodder trees, bushes, range grasses and legumes.
- Post harvest technologies and value addition to surplus forage in main season.
- Strengthening forage seed network at regional level.

References

Anonymous. 2005.Proceeding of the AICRP on Forage Crops, Annual Group Meet, (Rabi).

Faruqui, S.A., Singh, D.N. and Sunil Kumar. 2006. Technological innovation for augmenting forage resources for dry regions. In : Pathak, P.S. and Kundu, S.S. (eds.) *Livestock feeding strategies for dry regions*. International Book Distributing Co. Lucknow : pp 549-573

Faruqui , S.A. and Sunil Kumar. 2007. Forage production from problem soils, Technical bulletin no. 04 / 2007, AICRP on Forage Crops, IGFRI, Jhnasi.

Hazra, C.R. 1998. Advances in fodder production system In : Strategy for maximization of forage production ED(s) A.K. Mukherjee, S. Maith & M.K. Nanda, Directorate of Research, BCKV, Kalyani, West Bengal pp. 40-56.

Sunil Kumar and Faruqui, S. A. 2007.Crop diversification through forages: A new paradigm towards sustainable agriculture. In: *Abstracts: National Symposium on "A new vistas to forage crop research"*, September 10-11, BCKV, Kalyani : pp.79.

❑❑❑

Chapter **3**

Trees as a Source of Fodder for Livestock in India

N. DAS
Indian Grassland and Fodder Research Institute, Jhansi-284003 (U.P.), India

Introduction

Tree leaves and pods form a natural part of the diet of many ruminant species and have been conventionally used as sources of forages for domesticated livestock in Asia, Africa and the Pacific (Skerman, 1977). The part of tree fodder commonly used as feed is referred to as browse and it includes tender shoots, twigs and leaves of shrubs and woody plants and also fruits and pods. As leguminous species provide more protein and minerals, particularly calcium, they are preferred over other species. More than 200 leguminous tree species are reported to be used as forage, with most being tropical or sub-tropical in origin. Most of the leguminous fodder trees come from the genera *Accacia, Albizia, Calliandra, Desmanthus, Desmadium, Gliricidia, Leucaena, Prosopis* and *Sesbania* (Mandal, 1977). Non-leguminous tree species are also used for livestock feeding. Tree-fodder is important in providing nutrient to grazing ruminant in arid and semi-arid region; they supply major part of the protein supply during the driest months. Tree leaves serve as a concentrate to goat and sheep reared by poor farmers at a very low cost (only the labour cost of harvesting fodder). Trees and shrubs are an effective insurance against seasonal feed shortage or against the risk of drought and floods. Fodder trees and shrubs become potential source of livestock feed during the month of December-June when grasses are either grazed away or become dry with poor palatability and digestibility. However, it needs to mention that excessive and indiscriminate lopping of trees is causing depletion of valuable fodder resources, destruction of forestlands and consequent soil erosion. Many forest areas are suffering due to over presence of the livestock and destruction of the young seedlings by them. Overgrazing accompanied with trampling is often a cause of degradation of the ranges.

Importance of Fodder Trees in Arid Regions

Everist (1969) points out that "Palatable trees on a property can make the difference between survival and disaster during drought". The importance of trees as a source of fodder increases with the severity of drought and the progress of dry seasons. Generally livestock owner often caught in the vicious circle of having either too few livestock to make a sufficient emergency stock reserve (to be used as economic reserve for hard times) and to produce enough manure to support soil fertility or having stock at the expense of crop land resulting in an immediate short fall of the minimum requirement of crop production in spite of increased manure availability. The introduction of fodder trees that, on one hand do not, or marginally, compete with crop production and, on the other help to bridge the dry-season fodder gap, can be one of the possible ways to break the circle. Although the importance of woody perennials in the arid and semi-arid rangelands is well recognized but they are managed poorly.

Importance of Fodder Trees in Hilly Areas

Trees are equality or even more important in mountainous region and flood prone areas. The role of woody perennials (commonly known as multipurpose trees and shrubs, MPTS) in fodder production is well recognized particularly for their higher nutrient content as compared to dried grass during the cold dry season in hilly areas. Fodder tree leaves are the most important forage materials available during winter dry period of 5-6 months in the most of the hills. With a good protein and calcium level, tree leaves are suitable for milk producing animals. The shoots, leaves and young twigs from trees are fed to animals during winter dry months (November to March) in most of the high land areas. Nevaro, dubhilo, bur, pakhri, barbar, kutmero, lute khanyum and rai Khanyum trees are commonly used for fodder purpose in hilly areas. The fodder yield from these trees varied from about 300-700 kg/tree. In Sikkim, fodder trees are also grown in cardamom plantation. The farmers regulate shade by partial copping but do not allow lopped twigs to fall directly on cardamom plants. Fodder trees in the proximity of farmhouses serve as readily available green fodder store. They just cut few twigs and put before cattle and goat. Fodder is collected by lopping the fodder trees - leaves and smaller soft twigs are chopped off the branches. After the flush of new leaves, trees are not touched for six months. Sometimes young trees of fodder trees such as Artocarpus lakoocha are deleafed by hand without any cutting of branches at all. Most of these trees are grown between 600 to 2000 meter altitudes. Hailstorm during new flush and excessive lopping are the important factors affecting fodder yield. In Kashmir Himalaya, leaves and shrubs are commonly used as fodder. During summer months, a thorny tree, Robinia pseudoacacia is lopped and its fodder is highly palatable for sheep and goat. Other important fodder trees are celtis, cotoneaster, populus, Ulmus etc. At altitude of above 2000 meter salix trees are grown. Small branches are lopped and stacked on the tree itself for use of fodder during winter. The dried leaves remain palatable even after their exposure to sun, rain, and snow.

Nutritive Values of Tree Fodder

Green leaves, fallen dry leaves, seeds, seed-pods as well as seed cakes and fruit peelings are used as animal feed. Some of the products like sal seed cake, mahua seed cake, karanji seed cake etc. have already been introduced commercially as component of the concentrate feed mixture. Nutritive qualities of top feeds obtained from different tree species vary widely (Table 1 and 2). The crude protein in browse may be twice as high as grass. The crude protein (CP) content ranged from 8 to 25%, while cell wall content ranged between 20 to 40%. One problem of feeding tree leaves to livestock is that most of these contain antinutritional factors like tannins which hamper the availability of the protein. The soluble phenolic varies from 5 to 50%. In the early stages of growth, tree leaves contain fairly high amounts of crude protein and a comparatively low percentage of crude fibre. As maturity progresses, there is a gradual decline in protein content with increase in crude fibre. As leguminous tree leaves contain high amount of CP, they can be useful as protein supplement to low grade roughages. For example when sesbania (Agastha) and Subabul leave are added at 25% level to paddy straw, wheat bhusa, oat bhusa, the dry matter digestibility is increased considerably. Leaves of some trees (like pipal and Burgad) have high amount of calcium, magnesium and nitrogen. Generally tree forages are rich in calcium but poor in phosphorus content. Presence of oxalate in higher amount in some plants affects calcium utilization. Leaves also contain higher amount of trace minerals (Fe, Cu, Mn). Rapid weight gain and improved conception rate in females of small ruminants staying in arid and semi-arid region is a proof of the fodder value of woody species.

Table 1: Nutritional Quality of some of the Important Fodder tree in Tropical India

Tree species	Period of availability		Palatability		Crude Protein		Crude Fibre		Nitrogen free extract	
	Leaf	Pod	Leaf	Pod	Leaf	Pod	Leaf	Pod	Leaf	Pod
Acacia Nilotica (Babul)	May-Feb	May-June	Good	Good	13.9	13.6	9.2	14.9	69.8	61.0
Ailanthus excelsa (Arya, Maha nim, Maharukh)	May-March	-	Good		19.8	-	14.3	-	41.9	-
Albizia amara (Tugli, Chikreni)	June-March	-								
Albizia lebbek (Sirish)	April-Oct	-	Moderate		29.2	-	25.3		43.8	-
Albizia Procera (Safed Sirish)	July-March									
Bauhinia Varieqata (Rakta Kanchan)	April-Nov		Fair		13.3	-	29.1		46.3	-
Dalbergia Sisso (Sishu, Sisham)	Feb.-Dec.		Fair		16.6	-	22.3		49.4	-
Dichrostachys cinerea (Kunlai, Sigamkati)	July-April	Aug-Dec.	Good	Good	15.2	11.4	28.1	25.6	50.4	56.4
Ficus glomerata (Gular, Umar)	Dec.-Aug	March-July								
Hardwickia binanta (Anjan)	April-Jan	-								
Leucaena Leucocephala (Subabul)	July-Dec	-	Good	-	21.4	-	14.2	-	49.5	-
Parkinsonia aculeate (Vilayati Kikar)	Throughout year	Dec.-March	Moderate	-	13.1	-	24.9	-	40.9	
Pitheceolobium dulce	Throughout year	April-June								
Prosopis cinerea (Shami)	Throughout year	June-Aug	Good		13.9		20.3		59.2	
Prosopis Juliflora	-	March-May	Fair	Good	21.4	30.8	20.8	35.0	50.0	-
Sesbania grandiflora (Basna)	July-Nov	-	Good		26.2		8.8		51.4	
Ziziphus nummularia (Kokan ber, Bhor, Paragi)	Throughout year	Dec-Jan	Good		11.7		16.0		65.0	

Table 2: Digestibility Coefficient of Nutrients of some Important Top Feeds

Tree	Animal	DM	OM	CP	CF	EE	NFE
Ailanthus exelsa (Arya, Mahanim, Maharukh)	Sheep	72.1	75.1	81.3	55.1	31.7	81.7
Zizyphus nummularia (Kokan ber, Bhor, Paragi)	Goats	52.4	-	41.9	43.5	37.5	66.4
Moringa oleifera (Sainjna, Sujina))	Sheep	64.3	65.3	77.3	34.9	65.8	65.1
Moringa oleifera	Goats	56.6	61.4	65.8	29.6	61.5	65.0
Symingtomia populnea	Goats	64.7	66.5	27.9	44.3	61.3	76.5
Giricidia spp.	CB calves	56.7	-	62.2	50.1	52.8	65.3
Pithecellobium dulce	Sheep	-	62.2	76.9	57.7	57.2	57.6
Vicus cunia	Goats	40.2	51.1	52.0	24.4	36.8	61.1
Morus alba (Kala Shehtoot)	Goats	75.5	76.8	74.1	72.3	55.5	82.9
Morus alba (Mulberry)	Sheep	72.9	77.3	83.2	82.3	52.3	79.0
Thespesia populnea (Parash, papal)	Sheep	71.8	76.5	73.4	48.3	43.8	85.5
Thespesia populnea	Goats	73.0	76.7	73.3	51.5	40.5	84.7
Acacia leucopholoea (Rinj, Arinj)	Calves	46.7	49.6	49.7	43.9	54.0	51.1
Boswellia serrata (Tad, Salai, Saler)	Sheep	70.2	75.9	58.4	63.9	62.0	73.1
Boswellia serrata	Goats	71.1	75.7	77.5	63.1	63.9	77.0
Leucaena leucocephala (Subabul)	Sheep	57.7	74.5	56.3	58.9	64.7	59.4

Advantage of Tree Fodder over Grazing Grasses and Shrubs

Browse plants (small size trees) have some advantage over grazing grasses and shrubs.

I. Due to deep root system, trees can use larger nutrient resources as well as limited water resources available in the deep soil during drought.

II. New leaf generally arrives before rain (i.e. during summer) and thus provides, at the end of the dry season, the only source of fodder available.

III. Besides leaves, trees have some other edible parts like pods which are generally riche in protein. The pods can be stored for use during scarcity.

IV. As the animal can easily reach foliage of shrubs, the chance of overgrazing is high. Whereas the foliage of trees may be inaccessible to animals and should be lopped. This requires manpower but fits well with cut-and-curry strategy. However if equalities between the demand of fodder and the regeneration capacity of the tree cannot be maintained, the ecological system quickly deteriorates.

V. Trees and not shrubs provide shelter for animal.

VI. Trees have a low level of competition with grass and often have a positive impact on the growth of grasses, (specially the N- fixing trees).

Important Fodder Trees in India

The commonly available fodder trees used in different agro-climatic zones of India is given in the Table 3. The yield of leaf fodder from different trees is mentioned in the Table 4. Promising combination of trees, grasses and legumes suitable for different parts of the country are presented in the Table 5. Although large number of tress species are used as source of fodder in our country, but little information is available on their productivity and usefulness.

Khejdi

It is most commonly found in Rajasthan and adjoining areas in the North- West. Khejdi trees grows on their own in Rajasthan and farmers maintain them, lopping their branches from time to time for fodder and fuel. On deep alluvial soils, the number of Khejri per hectare varies from 20 to 40 giving a 4-10% cover. In districts Nagaur and Ganganagar, 40-120 trees/ hectare can be grown with rainfall from 150-350 mm. In Sikar district (rainfall 400 mm), 150 trees per hectare are grown. However, in the natural rangeland of Rajasthan (rainfall 30-200 mm; area like Jaisalmer), not more than 10 trees/ hectare are found. An adult Khejri tree may give upto 5kg air dried pod per year. However, with the advancement of age, pod production decreased; from 10 to 30 years of age, the tree will give 2-3 kg pod per year.

Prosopis Specis

Prosopis species are drought resistant, N- fixing and palatable to ruminants. In one hectare land, prosopis can produce up to 4 ton of pods per year, which contain about 9% protein, 47% non-nitrogenous extract, 6% fat and 25% cellulose. Their fodder value is close to barley. The root system is strong, so the production of leaves and pods may be used from the third or fourth year and in some cases after two and a half year.

Table 3: Important Top Feeds Growing in Different Agro-climatic Zones of India

Tree species	Region	Fodder availability period		Preference to livestock
		Leaf fodder	Pod fodder	
Acacia albida	SAR	May-Feb	Mar-June	Goat, Sheep, Camel
Acacia nilotica (Babul, Kikar)	AR, SAR, IGP, CZ, COZ	May-Feb	Apr-June	Cattle, Goat, Sheep, Camel
Acacia Senegal (Khor, Kumta)	AR, SAR	Apr-Dec	May-June	Goat, Sheep, Camel
Acacia tortilis	AR, SAR	May-Feb	May- June	Goat, Sheep, Camel
Ailanthus excelsa (Arua, Maharukh)	AR, SAR	May-Mar	-	Goat, Sheep
Albizia lebbek (Indian walnut, Siris)	AR, SAR, IGP, CZ, NER	Apr-Nov	Aug-Nov	Goat, Sheep, Cattle
Albizia procera (Safed siris)	SAR, CZ, NER	Jul-Mar	Jan-Mar	Cattle, Goat, Sheep
Anogeissus latifolia (Dhaura)	SAR, AR	Apr-Oct	Jan-Mar	Cattle, Goat, Sheep and camel
Artocarpus lakoocha (Dhew, Lakuch)	IGP,NER, STZ	Apr-Nov	-	Goat, Sheep, Cattle
Azadirachta indica (Neem)	AR, SAR, IGP, CZ, NER	Through out the year	July-Aug	Goat, Sheep, Cattle
Bauhinia purpurea (Krairwal, Gairal)	STZ, CZ, NER, IGP	Apr-Nov	-	Goat, Sheep, Cattle
Bauhinia variegata (Kachnaar, Rakta Kanchan)	IGP, NER, CZ, STZ	Apr-Nov	-	Goat, Sheep, Cattle
Dalbergia Sissoo (Shisham, Sissu)	SAR, IGP	Feb-Dec	Apr-Sep	Goat, Sheep, Cattle
Dichyostachys cinerea (Kunlai, Sigamkati)	AR, SAR	July-April	Apr-Aug	Goat, Sheep, Cattle, Camel
Ficus reliogosa (Ashwatha, Pipal)	SAR, CZ, IGP	July-Nov	-	Cattle, Sheep
Gliricidia sepium	SAR	Mar-Nov	-	Goat and Sheep

Contd...

Table 3. Contd...

Gmelina arborea (Gumhar, Khamara)	SAR, CZ, IGP	July-Nov	-	Cattle, Sheep, Goat
Grewia optiva (Bihul)	STZ	May-Nov	-	Cattle
Hardwickia binata (Anjan)	SAR, CZ	Apr-June	Feb-Mar	Cattle, Sheep, Goat
Leucaena leucocephla (Subabul)	SAR, IGP, CZ	Through out year	Dec-May	Cattle, Goat, Sheep
Melia azedarach (Bakain, Drek)	IGP,SAR, STZ	July-Nov	-	Goat, Sheep, Cattle
Morus alba (White mulberry, Shehtoot)	IGP, CZ,STZ	Through out year	-	Cattle, Goat, Sheep
Pithecolobium dulce	SAR	Though out year	-	Cattle, Goat, Sheep
Prosopis cineraria (Shami)	AS, SAR	Through out year	May-June	Goat, Sheep, Camel, Cattle
Quercus incana (Oak tree, Baanj)	NER	Mar-Nov	-	Cattle, Sheep, Goat
Robinia pseudocassia (Robinia)	STZ	Mar-Nov	-	Cattle, Sheep, Goat
Sesbania grandiflora (Basna)	IGP, CZ, NER	July-Nov	-	Cattle, Sheep, Goat
Sesbania sesban	SAR, IGP, CZ, NER	Through out the year	Dec-Mar	Cattle, Goat, Sheep
Tecomella undulata (Lahura, Rohera)	AR,SAR	Mar-Nov	Apr-June	Cattle, Sheep, Goat, Camel
Terminalia bellerica (Bahera, Bhaira)	COZ, STZ	July-Nov	-	Cattle, Sheep
Terminalia Tomentosa (Sain, Saj)	COZ, STZ	July-Nov	-	Cattle, Sheep
Ziziphus nummulasia (Kokan ber, Bhor, Malla)	AR, SAR	Through out year	-	Cattle, Goat, Sheep

(AR=Arid region, SAR= Semi arid region, IGP= Indogangetic plains, CZ= Central zone, COZ= Coastal zone, NER= North eastern region, STZ= Sub-temperate zone)

Babul

It is a very useful tree for an Indian villager with multipurpose uses like firewood and fodder. It is a medium size tree. Both leaves and pods are used for fodder in the dry and arid zones of India. It thrives best in areas with annual rainfall of about 500 to 1200 mm. It grows on a wide variety of soils provided sufficient moisture is available for its growth. Acacia nilotica helps in efficient nutrient cycling as annual production of leaf litter/ ha is reported to be 5.27 tonnes. An average tree yields about 18 kg of pods in a year.

Anjan

Leaves of this tree are commonly used in the semi- arid region of Karnataka. The foliage is conspicuous in hot weather when most other trees are leafless. In general annual yield of the tree fodder varies from 50-200 kg per tree depending upon the age and site condition.

Acacia Feruginea

It is a common tree of Karnataka, Andhra Pradesh, Tamil Nadu and Maharashtra. It is extensively found in black cotton soil. The leaves and its long pods are a reach nutritive feed to sheep and goats.

Sirish

It grows wild in the sub-Himalayan region and its leaves are also reported to be good fodder. This is also suitable for planting in dry areas.

Gular

Leaves are extensively lopped for fodder and are generally rated as good. It grows will in the sub-Himalayan tract and outer Himalayan valleys. It is a characteristic tree of tropical hill valley swamp forest in Dehradun.

Neem

The neem tree is famous for various biological activities such as insecticidal, nematicidal, spermicial, antiviral, antifungal, diuretic, antipyretic etc. Fresh and mature neem fruits have 60% dry matter which in turn have 48% pulp, 25% husk and 23% Kernel. The Kernel gives about 45% oil and 55% neem cake. The neem cake is comparable to groundnut cake and can be used for feeding of poultry, buffaloes and sheep upto 25% of the concentrate mixture.

Basna

Sesbania with its profuse root nodulation and nitrogen rich foliage make it a valuable green manure and source of extra fodder during lean period.

Mulberry, Shehtoot

It is found in the northern India and the leaf fodder is reported to be of good quality.

Aswatha

Leaves of this tree are liked by goats and are commonly used by the landless people in the villages.

Ber

It is identified to be most promising trees for use as fodder in the semi-arid region.

Plantation of Fodder Trees

Two groups of agro-forestry systems are linked with fodder production.

- Silvopastoral system, which combines woody perennials with animal production. Planting of multipurpose trees (for fuel, fodder and timber) with grasses and legumes in an integrated manner and their utilization through 'cut and carry of forage' in early years followed by in-situ grazing is known as silvipastoral system.
- Agro-silvopastoral system, which combines woody perennials with both crops and animals.

The basic aim of both the systems is optimizing productivity of land (mostly degraded land), conserving plants, soil and its nutrients and producing forages, timber and firewood on a sustainable basis.

Table 4: Pasture Production (in t/ha/yr) from 5- year-old Silvipastoral Combinations Semi-Arid Areas (Bundelkhand Region).

Tree	Grass	Dry forage yield (t/ha/yr)
Acacia tortilis	Natural pasture	3.2
	Cenchrus ciliaris	3.4
Albizia lebbeck	*Cenchrus ciliaris*	7.4
(Sirish)	*Chrsopogon fulvus*	7.8
	Sehima nervosum	7.8
Albizia procera	*Cenchrus ciliaris*	
(Safed Sirish)	*Dolichos lablab*	7.9
Dalbergia sissoo	Natural pasture	2.9
(Sisham)		

Contd...

Table 4. Contd...

Dichrostachys cinerea	*Cenchrus ciliaris*	4.3
(Kunlai, Sigamkati)	*Chrysopogon fulvus*	5.2
Hardwickia binata	*Cenchrus ciliaris*	4.2
(Anjan)	*Chrysopogon fulvus*	
	Sehima nervosum	3.7
Leucaena leucocephala	*Cenchrus ciliaris*	4.1
(Subabul)	*Cenchrus setigerus*	3.2
Albizia amara	*Cenchrus ciliaris*	4.1
	Chrysopogon fulvus	
	Sehima nervosum	3.3

Source: Deb Roy 1988: IDRC 1990: Singh & Pathak 1992

The factors that need to be considered while selection of trees are: -

i) Quick regeneration capacity after lopping

ii) High growth rate

iii) Palatable leaves, twigs and pods having high nutritive value but low toxic elements

iv) Short rotation

v) High fuel value of timber

vi) Ability to be easily established (by seed, cutting etc.) and fast growing

vii) Tolerate drought and cold

viii) Tolerate lopping shock

The techniques for establishing trees in silvi-pastoral system are similar to those followed in the tree plantation program - seed collection, raising of nursery stock, soil working, transplanting (at a distance of 4-6 m from row to row and 4-5 m from plant to plant), and disease / pest control etc. Immediately after completion of planting of tree seedlings, grass seeds may be sown/ transplanted in lines; 90-100 cm from row to row and 45-60 cm apart within the row. In between two lines of grasses 1-2 lines of legumes may be sown. It is possible to get 3.5 to 4.0 ton/hectare of dry forage from silvipastoral system without any adverse effect on the growth of tree species. In well-managed silvipastoral system, production up to 8-10 ton/ hectare is possible.

Table 5: Suitable Grasses, Legumes and Trees for Different Situations in India

S.No.	Region	Grasses	Legumes	Trees
1.	High hills of J&K, HP and U.P.	Lolium perenne Dactylis Glomerata Brumus spp. Festuca arundinacea	Trifolium pratense Trifolium repens Lupinus hirsutus	Robinia pseudoacacia Grewia oppositifolia Fagus sylvatica Quercus Incana Morus alba
2.	Eastern Highlands	Lolium perenne Lolium multiflorum Dactylis glomerata Festuca Arundinacae	Trifolium pratense Lotus corniculatus Trifolium repens Lupinus Hirsutus	Pterocarpus indicus Albizia falcataria Acacia mangium Albizia Stipulata Alnus nepalensis Parkia Roxburghii
3.	Low Hills of J. & K., H.P. & U.P.	Dichanthium annulatum Pennisetum pedicellatum Chrysopogon fulvus Cenchrus ciliaris Bothriochola intermedia Bothriochola Pertusa	Pueraria phaseoloides Stylozobium deeringianum Macroptilium artopurpureum Stylosanthes hamata Stylosanthes scabra	Albizia stipulata Grewia optiva Leucaena leucocep- hala Melia azedarach Robinia pseudoacacia Ficus sp. Bauhinia sp. Mimosa
4.	Eastern Low Hills	Dichanthium annulatum Panicum maximum Pennisetum Polystachyon Pennisetum sp. Borthichloa Intermedia	Stylosanthes guinensis Lablab purpureus Alysicarpous regosus Glycine javanica Desmodium intortum	Ficus Sp. Albizia Sp. Artocarpus lakcocha Maringa oleifera

Contd...

Table 5. Contd...

5.	Southern Plateau	Dichanthium annulatum Panicum antidotale Borthichola pertusa Sehima nervosum Chrysopogon fulvus Cenchrus ciliaris Cenchrus se tigerus	Lablab purpureus Dolich os axilliaris Stylosanthes hamata Macroptilium artopurpureum Stylosanthes hamata	Hardwickia binata Leucaena leucocep - hala Albizia amara Acacia albida Acaia tortilis
6.	Eastern Plains decumbense	Bracharia mutica Setaria sphacelata Dichanthium annulatum Bracharia purpureus Pennisetum polystachyon Pennisetum pedicallatum Cynodon dactylon	Phaseolus calcaratus Glycine javanica Lablab morus alba	Albizia sp. Leucaena leucocep - hala Bauhinia sp. Melia azedarach Moringa oleifera Dalbergia sissoo
I.	Dry Regions of Rajasthan, Gujrat, Haryana and H.P.	Dichanthium annulat um Panicum antidotale Cenchrus ciliaris Lasiurus sindicus	Macroptillium artopurpureum Stylosanthes hamata Clitoria ternatea Lablab purpureus	Prosopis cinerarea Acacia tortilis Azadirach - tha indica Ailanthus excelsa Albizia lebbeck Albizia amara Zizyphus Mauritiana
8.	Coastal Region	Setaria sphacelata Panicum maximum Bracharia mutica Bracharia decumbense Paspalum notatum Cenchrus ciliaris Cenchrus setigerus Panicum antidotale	Lablab purpureus Glycine Javanica Medicago sativa Centrosema pubescens Glycine wightii Stylosanthes Guianensis	Albizia falcataria Albizia lebbeck Ailanthus malabarica Ficus retusa Erythrina indica Anacardiun occidentale

Contd...

Table 5. Contd...

9.	Water Logged Area	Bracharia mutica Iseilema laxum	Sesbania aculeate	Eucalyptus globules Terminalia arjuna Prosopis juliflora Sesbania Sesban
10.	Usar Soils	Sporobolos sp. Chloris gayana Ciplachena fusca Cynodon dactylon Bracharia mutica Bothichola Intermedia Bothriochola pertusa Setaria sphacelata	Slysicarpous rugosus Rhynchosia minima Desmenthus virgatus Indigofe ra sp. Desmodium sp. Cassia tora	Prosopis juliflora Acacia tortilis Acacia nilotica Albizia Lebbeck Sesbania sesban Balanites Aagyptiaca

Harvesting of Tree Leaves and Pods

Based on the regenerating ability of trees through sprouting mainly five ways of harvesting are used for fodder collection.

i. *Thinning:* The poor and diseased trees are removed which facilitate the healthy trees grow faster. The leaves and small branches of the removed trees are used for fodder.

ii. *Coppicing :* Individual trees are cut at their base (15-75 cm above ground level). This is commonly used in Subabul trees (*Leucaena leucocephala*).

iii. *Pollarding:* All the branches including top of the tree are removed at a height of 2-3 meter above ground level. After that, repeated harvesting of fodder is made till the tree loses its sprouting vigor. Example: Subabul, poplar. Sprouting of stumps is depended on the stump diameter. The percentage of stumps that bore sprouts and number of sprouts decrease with the decrease of stump diameter, while the higher stump height favors sprouting. The sprouts arising from stump with medium diameter (15-45 cm) grow better than those emerged from the stumps of smaller or larger diameter.

iv. *Pruning*: It involves removal of small branches and mainly practiced in alley cropping systems for fodder and firewood.

v. *Lopping*: Here major branches of the trees are removed for supply of green fodder and firewood during the fodder scarcity period. This is a very common practice in India; hence it is important that a scientific lopping management is followed which would ensure optimum utilization of tree tops.

The important factors, which influence the post-lopping regeneration through spouting and total lifetime fodder yield of a tree, are: -

i. *Stage of growth*: This may be correlated with stem girth or canopy spread or both depending on the utility and situation.

ii. *Lopping period*: The yield and quality of harvested leaves vary with species and season. Lopping cycle needs to be planned keeping in view the time of arriving of new leaves and rainfall pattern. Lopping of trees just before rains is not desirable.

iii. *Lopping intensity*: It has been observed that two-third lopping is better than complete lopping. It is advisable to leave a few top shoots, which may help the tree in early recovery from injury. Pruning of branches from all sides of the tree will help quick regeneration.

iv. *Lopping interval*: It depends on the recovery capacity of tree species and climate. In favorable climate conditions (plain areas having adequate rainfall) annual lopping can be practiced without much damage; whereas in adverse climate (arid region) lopping after interval of 2-4 years is recommended.

v. *Fodder yield and quality*: It varies considerably (from 5kg to 50 kg/ tree/ year). The quality of fodder is primarily dependent on the species of tree, season of harvesting, maturity level of leaves/ pod and the location of trees.

vi. *Regeneration capacity*: The cut ends of lopped branches develop new shoots (sprouting). From a cut end normally 2-6 new shoots develop depending on the initial diameter of the cut branch.

The equation, which is used to estimate biomass of browse species grown at 260-600 mm of rainfall, is given below.

Log biomass= 2 log diameter + 1

For trees and shrubs stem, having a diameter of 5, 10 and 15 cm, the biomass would be 0.25, 2.2 and 6.2 kg respectively. An average tree produces

1 kg of leaves, 0.250 kg of fruit and 4-5 kg of branches providing a yield of 1 ton/ha. Yield capacity varies depending upon the browse regimen and local conditions.

Table 6. Leaf Fodder Yield to Pruning and Removal of Top Feed Trees Species Under Agrisilvicultural System

Species	Green leaf yield (kg/plant)+				Foliage yield (t/ha)#	
	2 years	3 years	4 years	5 years	Fresh	Dry
Acacia nilotica (Babul)	2.79	7.21	1.50	4.26	3.00	1.75
Acacia nilotica var cupressiformis	2.90	6.16	1.66	3.99	2.19	1.24
Casuarina equisetifolia	3.01	2.32	2.54	3.74	3.00	1.38
Madhuca latifolia	0.36	0.51	0.91	1.86	3.64	0.23
Melia azedarach (Bakain)	2.90	2.58	1.59	7.56	4.57	1.87
Leucaena leucocephala (Subabul)	7.96	6.10	6.31	14.56	2.03	0.95
Dalbergia sissoo (Shisham)	2.87	4.27	3.47	5.66	4.39	1.67
Albizia lebbek (Sirish)	2.23	4.81	4.85	10.39	3.89	1.70
Syzygium cumini	1.42	2.74	2.96	5.63	2.90	1.19
Eucalyptus tereticornis	5.19	7.13	5.31	3.72	11.52	5.74
Emblica officinalis (Amla, Aonla)	3.21	3.04	2.51	7.24	2.08	0.82
Hardwickia binata (Anjan)	2.05	1.45	1.58	4.22	1.94	0.74

+ Green leaf yield (kg/plant) due prunings up to 50% height
# Foliage yield (t/ha) due to 50% removal of top feed trees at the age of 4 years

Tree leaves are generally utilized through browsing by sheep and goat or as a freshly lopped fodder offered to both large and small ruminants. However, not all animals use wood species in the same way. The pattern of use (%) of different forages by various classes of animals (Table 7) is given below (Lusigi, 1981).

Table 7. Preference of Forages by Different Class of Livestock

Forages	Camel	Cattle	Goats	Sheep
Tree, Shrubs & Bushes	30	0	30	9
Sub-Shrubs	48	2	23	24
Non-grassy plant	11	11	22	30
Grasses	11	87	25	37
Total	100	100	100	100

In another study, Bell (1978) Indicated that sheep consumes 60% grasses, 30% weeds and 10% browse materials whereas goats consume 60% browse materials, 20% grasses and 20% weeds.

Anti-Quality Factors

When trees leaves are fed as major fodder (75% and above), one must consider the effect of different anti-quality factors that are present in most of the tree leaves. The important anti-nutritional compounds found in leaves of trees and shrubs are given in Table 8. Tannins, oxalic acid and HCN are present at varying levels in tree leaves and pods. The problem of mimosine toxicity in Subabul is well known. Conservation of Subabul leaves as silage has been found to reduce the mimosine concentration by 50%. Some of the tree leaves also contain reasonably higher amount of estrogen, which may affect the reproductive performance of livestock. Many tree species contain variable amount of secondary compounds such as phenolics, tannins (proanthocyanidins) and other compounds with anti-nutritional qualities. Tannin is an important constituent in vascular plants. The presence of phenolic hydroxyl groups in tannin lead to the formation of complexes primarily with proteins and a lesser extent with metal ions, amino acids and polysaccharides. The effect of tannin can be good or bad depending upon their concentration, molecular structure, animal species, physiological state of the animal and composition of the diet.

Table 8. Anti-Nutritional Factors in Leaves of Trees and Shrubs used in Livestock Feeding

Anti-nutritional substances	Tree Species
Non-protein Amino acids	
Mimosine	*Leucaena leucocephala*
Indospecine	*Indigofera spicta*
Glycosides	
Cyanogens	*Acacia giraffae, A. cunninghamii, A. Sieberiana Bambusa bambos, Barteria fistulosa, Manihot esculenta*
Saponins	*Albizia stipulata, Bassia latifolia, Sesbania- sesban*
Phytohemagglutinins	
Ricin	Ricinus communis
Robin	Robinia pseudoacacia
Polyphenolic compounds	
Tannins	All vascular plants
Lignins	All vascular plants
Alkaloids	
N-Methyl-B-phenethylamine	Acacia berlandieri
Sesbanine	*Sesbania vesicaria, S. drummondii, S. punicea*
Triterpenes	
Azadirachtin	*Azadirachta indica*
Limonin	*Azadirachta indica*
Oxalate	*Acacia aneura*

References

Bell, H.M. 1978. Rangeland Management for Livestock Production, 2nd ed. University of Oklahome, Press, Norman, USA

Brandis, D. 1971. Book: Indian Trees. Publisher: Bishen Singh Mahendra Pal singh. Dehradun.

Carlowits, P.G.V. 1988. Fodder production oriented agroforestry technologies with particular reference to woody perennials.

Dev Roy, R. 1980. Silvipastoral system of management for arid zones of India (Vol. III, CAZRI, Jodhpur, 101-106.

Dev Roy, R. 1988. Strategy for increased forage production through silvipastoral system. In: Forage Production In India (ed. Panjab Singh), RMSI, Jhansi (India): 164-169.

Deb Roy, R. 1990. Agroforestry in twenty first century in India. National Seminar on Agro-technology and Bio- energy for sustained crop and Agro-industrial productivity, Vasantdada Sugar Institute, Manjari, Pune, May 10-11.

Everist, S.L. 1969. Use of fodder trees and shrubs. Queensland Department of Primary Industries, Division of Plant Industry Advisory Leaflet No.1024.

IDRC. 1990. Report on Silvipasture Research Project. Indian Grassland and Fodder Research Instutute, Jhansi.

Lusigi, W. 1981. Combating desertification: rehabilitating degraded production systems in Northern Kenya, Paris. UNFSCO, IPAL.

Mandal, L. 1997. Nutritive values of tree leaves of some tropical species for goats. *Small Ruminant Research* 24, 95-105.

Pathak, P.S. and Roy, M.M. (1994). Silvipastoral system of production. Research Bulletin. Indian Grassland & Fodder Research Institute.

Pathak, P.S., Gupta, S.K., Singh, Punjab (1995). Rehabilitation of degraded land. Technology Bulletin. Indian Grassland & Fodder Research Institute, Jhansi

Proceeding: 3rd International Rangeland Congress, New Delhi 7-11, November, 1988.

Singh, K.A. and Rai, R.N. (1990). Important fodder and fuel tree species of Sikkim Hills. Proceeding of the National Workshop, Feb. 1-3, 1990, IGFRI, Jhansi. Trees used as fodder in hilly areas of NE Region.

Singh Panjab (1990). Multipurpose trees for agroforestry systems. Proceedings of a National Workshop held at Jhansi during Feb. 1-3, 1990.

Singh Panjab and Pathak, P.S. 1992. Revegetation of degraded lands. In: Status of Indian Forestry - Problems and Perspective. (ed. P.K. Khosla). ISTS, Solan: 60-81.

Skerman, P.J. 1977. Tropical Forage Legumes. FAO Plant Production and Protection, Series No. 2. FAO Rome. p. 609.

❑❑❑

Chapter 4

Use of Remote Sensing in Assessing Forage Resources Availability

J. P. SINGH, P. N. DWIVEDI, N.S. EKKA, R.K. AGRAWAL AND S. RADOTRA
Indian Grassland and Fodder Research Institute, Jhansi (U.P.), India

Introduction

Remotely sensed satellite imagery and Geographic Information System (GIS) are being widely used for the assessment of rangelands / grazinglands and estimation of forage availability as well as monitoring of the range resources (Singh *et at,* 2004). Grazinglands remain an important land use in India and supply about 40% of the total forage need. Since, the grazing resources are limited and natural grazinglands are becoming scarcer, there is a greater need to effectively manage grasslands for optimum biomass production and eco-development. The inventory of grazing/ forage resources, using remote sensing and GIS techniques and its management requires detailed information and thematic maps related to landforms, physiography, soil plant community association/combination and density and pattern of natural vegetation. Geomorphological features exert an indirect yet powerful influence on forage resources. Their effect on soil - water conditions, slope and gradient are major factors influencing the habitat of plants. The application of GIS and remote sensing techniques in the analysis of satellite data and generation of thematic maps can provide reliable, accurate and timely information related to the extent and conditions, and periodic assessment of present and potential forage production from rangelands.

In the past, several attempts have been made for the study of grazinglands. Tuller (1988), Li Bo (1988) and Isavwa (1988) have used remotely sensed data for preparation of an inventory, monitoring and management of rangelands. Tyagi and Patil (1982) have studied wastelands of Bundelkhand region for forage resource development. Naugraiya and Pathak (1987) have emphasized the role of topo sequence in planning for higher biomass production. Singhal and Rawat (1987) have described the relationship between soil and vegetation in the Himalayan region. At present about 40 per cent of the forage supply comes from 120 million ha or 40 per cent of total land in India which is available for grazing livestock population estimated to be over 400 million. This projection is based on secondary data mostly from revenue records, agricultural statistics and livestock census. Precise estimates based on primary data from satellite imagery and ground survey are lacking.

Previous work shows that a detailed inventory of grazinglands and grazing resources of different agro-ecological zone of India has not yet been prepared. In India some institutes and organizations, like ISRO, NRSA, NATMO, IIRSC, NBSS & LUP and CAZRI etc., are engaged in the studies of land degradation, soil resource management and land use/cover planning etc., but till date accurate knowledge about grazinglands and forage resources in the form of data based information and thematic maps is not available. Although, from the ancient times cattle breeding and milk production has been the second most important profession in India, next only to agriculture (Pandeya, 2000). Precise information on the extent, conditions and forage availability from grazinglands is essential for developmental planning of sustainable livestock production. There are evidences of excessive degradation of tropical grazinglands. Excessive grazing and deforestation further convert them into heavily degraded wastelands. The current data shows that about 158 million ha area in India has turned into wastelands. This is a worldwide phenomenon as over 88 per cent grazinglands of the world are at least moderately degraded and the rate of degradation is 17.7 m ha annually. (Van Dyne *et al*,1987).

The Study Area

Bundelkhand region is spread over 71602.50 km^2 area located between $23^{0}8'$-$26^{0}31'$latitude and $78^{0}11'$-$81^{0}31'$E longitude. It is bounded by the Yamuna in the north, ranges of Vindhyan Plateau in the south, the Sind in the northwest and Bhander ranges in the south-west (Pathak and Singh, 2003). The topography of the region consists of northern plain and the southern plateau area dotted with hillocks and a number of Vindhyan ranges particularly in the north and southeast. The Vindhyan constitute the chief mountain range of the region, providing mineral wealth in form of precious stones, bauxite, silica and lime along with building stones (Singh, 1961). Administratively, it covers 7 districts of Uttar Pradesh (UP) viz., Jhansi, Jalaun, Lalitpur, Hamirpur, Mahoba, Banda and Chitrakut and 6 districts of Madhya Pradesh (MP) viz., Datia, Tikamgarh, Chhatarpur, Damoh, Sagar and Panna and Lahar & Bhander tehsils of Bhind and Gwaliar districts respectively (Fig.1). The drainage

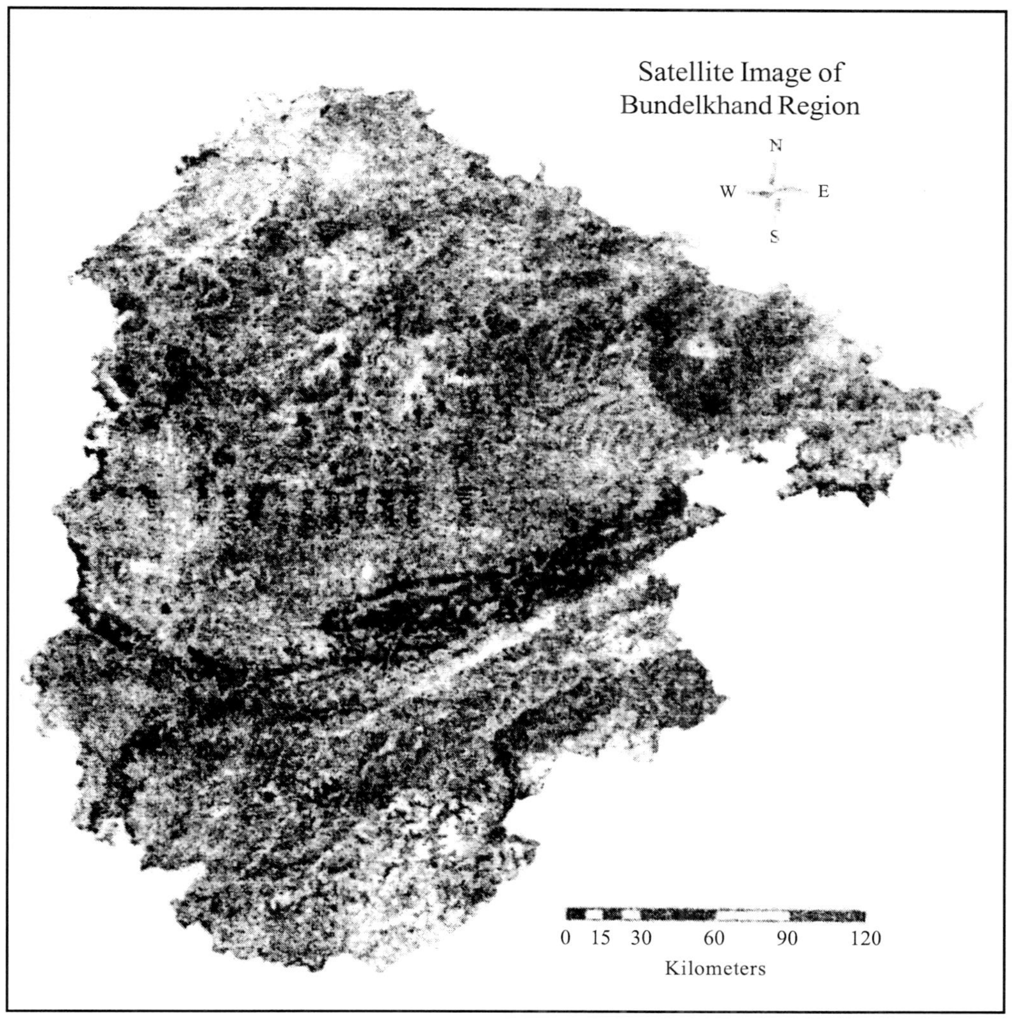

Fig. 1.

system of this region is mainly distributed in to channels of rivers like Betwa, Ken, Dhasan, Pahunj, Jamani, Baghain and Paisuni (most of them are tributaries of Yamuna) which flow from south-west to north-east.

Materials and Methods

Bundelkhand region was selected for the present study. Survey of India (SOI) toposheets number 54K, L, J, P, O; 63 A, B, C, D, H, G; 55I and 65M were used for the generation of base map. Indian Remote Sensing Satellite data (1A, 1B and P6) dated 1989, 1995 and 2005 was used for the identification of grazinglands and estimation of forage yield using ERDAS Imagine (ver. 8.7 & 9.1) software. ArcGIS- ArcInfo workstation (ver.8.1 & 9.2) were used for the geo-spatial analysis. Total 64 sites were selected from the different part of the study region for intensive ground truthing and sample survey.

Livestock Population

In year 1982, there were 5141456, 1490511, 1699874, 361833 cattle, buffalo, goat and sheep respectively in the region. Buffalo and goats were more in UP part while cattle and sheep were more in MP part. In 1997 the population of cattle, buffalo, goat and sheep changed to 4327750, 1764373, 1837052 and 408269 respectively. Compared to 1982, there has been a decrease of 813706 in cattle population, where as buffaloes, goat and sheep have increased by 273862, 137178 and 46436 respectively. Except in Banda and Chhatarpur buffalos population was increase from 1982 in all the districts of the region. In Bundelkhand only about 0.25 per cent cattle are crossbred. Both sheep and goat have increased in number (Pathak *et al,* 2005). The comparison of livestock population from 1982 to 2003 (Table 1) reveal that there is decline in cattle population by 15.87 per cent, which was more in UP part. At the same time there is almost matching increase in buffalo population in Bundelkhand. The goat was more preferred animal in UP part as compared to MP part. There has been an increase in goat and sheep population in UP part while in MP part it has decreased. The total livestock population of UP part has decreased by 0.26 per cent while in MP part it has been decreased by 7.30 per cent resulting in overall decrease of livestock population in Bundelkhand by 4.11 per cent. The livestock (ACU) density in the region is presented in table 2.

Table 1: Per cent Change in Livestock Population (1982-2003)

Districts	Cattle	Buffalo	Goat	Sheep	Total
Jhansi	-7.87	64.33	21.32	13.62	11.62
Lalitpur	1.13	54.70	40.62	26.56	16.68
Jalaun	-2.45	55.02	42.93	-13.41	21.99
Hamirpur & Mahoba	-15.80	6.71	-1.08	-17.82	-8.29
Banda & Chitrakut	-47.06	-4.24	14.23	-17.06	-25.54
Sub. Total	-19.69	25.03	19.41	35.55	-0.26
Datia	16.07	115.79	36.52	19.96	39.25
Tikamgarh	-31.36	4.96	-27.95	-25.01	-25.32
Chhatarpur	-19.64	-14.30	-13.82	5.03	-16.25
Panna	2.67	17.14	-0.57	11.36	4.76
Damoh	7.73	27.69	37.46	-24.16	13.55
Sagar	-22.50	8.08	18.63	37.66	-13.74
Sub.Total	-13.12	11.58	-3.39	-5.66	-7.30
Grand Total	-15.87	18.44	8.10	12.88	-4.11

Table 2: Livestock Density in Bundelkhand Region

Districts	Potential area) 2004 -05 ha	No of animals / ha of potential land				
		Cattle	Buffalo	Sheep	Goat	Total
Jhansi	427091	0.79	0.39	0.17	0.50	1.85
Lalitpur	458566	0.87	0.30	0.07	0.32	1.56
Jalaun	407907	0.59	0.53	0.09	0.48	1.69
Hamirpur	643126	0.61	0.26	0.07	0.35	1.28
Banda	628639	0.57	0.42	0.06	0.39	1.44
Sub. Total	2565329	0.68	0.37	0.09	0.40	1.53
Datia	241604	0.70	0.47	0.12	0.39	1.68
Tikamgarh	377857	0.91	0.33	0.17	0.41	1.80
Chhatarpur	747438	0.77	0.27	0.08	0.33	1.46
Panna	595265	0.82	0.23	0.03	0.18	1.25
Damoh	639935	0.71	0.15	0.01	0.16	1.03
Sagar	931178	0.60	0.16	0.01	0.12	0.89
Sub.Total	3533277	0.73	0.23	0.05	0.23	1.25
Grand Total	6098606	0.71	0.29	0.07	0.30	1.37

Grazing Resources

Grasslands /grazinglands/rangelands are essentially the non-arable terrestrial ecosystems and play vital role in the agricultural economy and environmental conservation. Grazinglands or grasslands are naturally occurring areas of grasses, forbs and open stands of trees with in situ grazing as the predominant mode of land use. Management of grazinglands for both, human livelihood through pasture based livestock industry and environmental conservation, requires an understanding of grazing resources, their production potential and natural limitations to the sustained use. In India grazinglands generally occupy class IV and V lands and this is why they are euphemistically passed on as wastelands (Gupta and Shankar, 1992). The present study reveals that forage production from the rangelands of the Bundelkhand region was approximately 2.57 m t DM/ year. Based on the intensive field survey, study of IRS imagery and geo-spatial analysis geo-referenced information on range forages have been generated at district level.

Grazinglands

The grazinglands include the area under natural grasslands including permanent fallows, ravines affected and shrubs infested lands. There is an evidence of reduction of area (-14.60%) under grazinglands during past 25 years. It was 21.08% in 1980 where as it was only 6.48% in 2005. Major portion of these lands has been brought under cultivation. The land available for grazing animals was found highest in the northwestern part of the region especially in Lahar (31.32%), Bhander (18.50%), Jalaun (12.84%) and Jhansi (12.51%). Moderately distributed grazinglands were recorded in Lalitpur (7.84%), Chhatarpur (7.06%), Datia (6.73%) and Sagar 6.35%) where as grazinglands was found in very limited patches (0.83% to 4.93%) in Damoh, Banda, Mahoba and Hamirpur districts of the region (Fig. 2&3). Based on satellite imagery district wise area under grasslands, grasslands affected by ravines and grasslands infested with shrubs was grnerated and presented in table 3. The important grass/ legume/ shrub species were *Eragrostis sp., Chrysopogon sp.& Heteropogon sp./ Indigofera sp., Atylosia sp. / Carissa sp., Zizyphus sp.* and *Cynodon sp./ Tephrosia sp.* and *Acacia sp.*

Grazinglands Affected by Ravines

About 170943 ha or 36.89% land under grazing lands was affected by the ravines (Table 3). It was about 2.39% of the total area of the Bundelkhand region. The severity of ravines was found more along the middle and lower reaches of rivers and streams in the northern part of the region (Fig. 2 & 4). The result shows that the situation was alarming in Lahar and Bhander tehsils

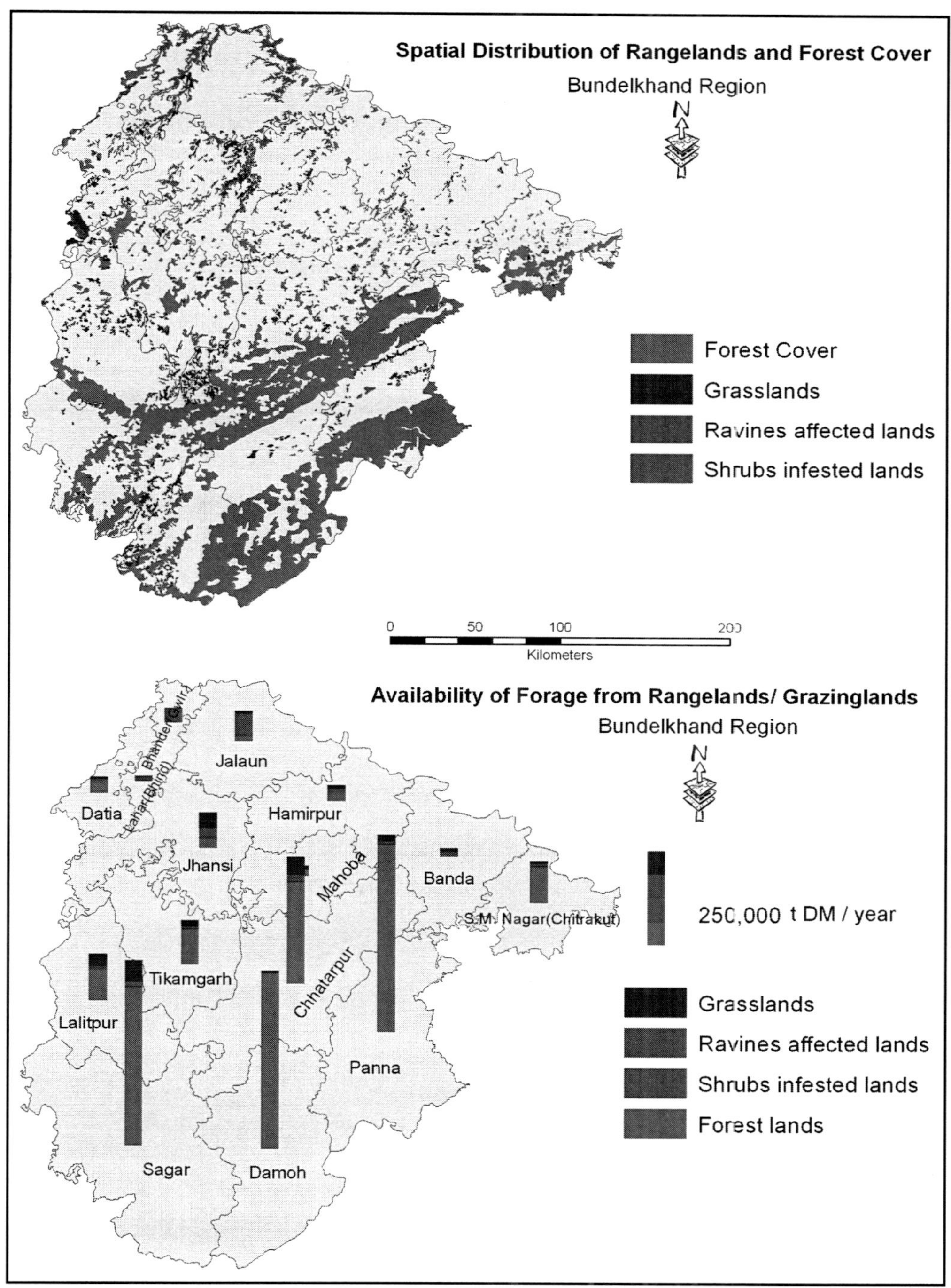
Spatial Distribution of Rangelands and Forest Cover
Bundelkhand Region
N
Forest Cover
Grasslands
Ravines affected lands
Shrubs infested lands
0
50
100
200
Kilometers
Availability of Forage from Rangelands/ Grazinglands
Bundelkhand Region
N
Jalaun
Datia
Hamirpur
Jhansi
Mahoba
Banda
S.M. Nagar(Chitrakut)
Tikamgarh
Chhatarpur
Lalitpur
Panna
Sagar
Damoh
250,000 t DM / year
Grasslands
Ravines affected lands
Shrubs infested lands
Forest lands

Fig. 2.

where as Jhansi, Datia Hamirpur and Banda were badly affected by ravines. There was an urgent need to restore these lands through proper land treatments and reseeding of suitable grass and tree species. It has also been observed that fertile cropped lands were being affected by rill erosion in Jalaun and Hamirpur districts.

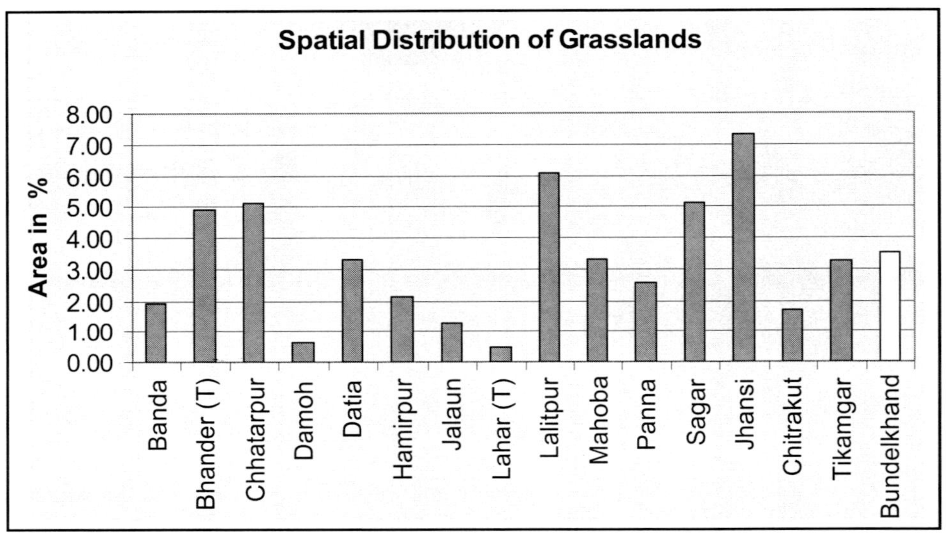

Fig. 3. Spatial Distribution of Grazinglands

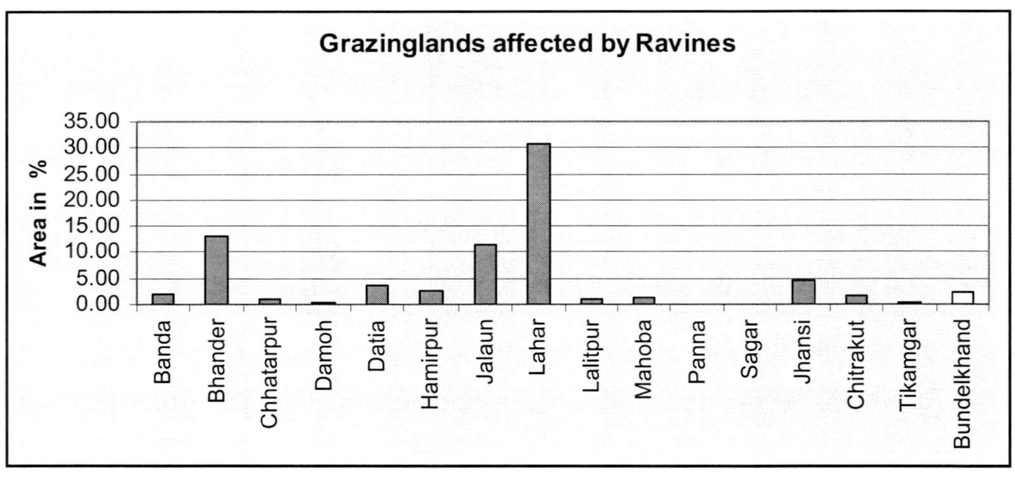

Fig. 4. Grazinglands Affected by Ravines

Table 3: Distribution of Grasslands (Km^2)

District/ Tehsils	Geographical Area	Grazinglands		
		Grassland	Ravines	Shrubs
Banda	4542.27	88.31	95.68	3.30
Bhander	658.26	32.21	86.6	2.95
Chhatarpur	8679.86	443.46	98.05	71.27
Damoh	7260.63	46.8	13.78	0
Datia	2016.83	66.18	69.54	0
Hamirpur	4171.79	89.42	113.13	3.09
Jalaun	4556.46	58.08	525.83	1.30
Lahar	1075.87	5.45	331.5	0
Lalitpur	5016.64	304.35	42.35	28.74
Mahoba	3015.23	100.17	37.05	0
Panna	7065.71	179.77	5.52	60.24
Sagar	10354.75	530.38	0.23	127.10
Jhansi	5039.87	369.36	222.6	38.49
Chitrakut	3107.81	52.36	55.37	21.06
Tikamgarh	5040.52	164.95	12.2	35.88
Bundelkhand	71602.50	2531.25	1709.43	393.42

Grazinglands Affected by Shrubs

Infestation of Grazinglands with shrubs is considered as a threat to the existing grazinglands in the region. The study reveals that about 0.55% (393.42 km^2) land (mostly grazinglands) in the region are affected by shrubs. The situation is more alarming in Panna (0.85%), Chhatarpur (0.82%), Sagar (0.76%) and Tikamgarh (0.71%) districts. It is evident from the satellite image (Fig. 5) that these lands are closely associated with middle and lower reaches of rivers.

Landform–Vegetation Cover Relationship

The effect of landform on soil conditions, water relationships, slope and gradient are major factors influencing the habitat of plants. GIS and remote sensing techniques are being used in the analysis of satellite data and generation of thematic maps to generate information related to the extent and conditions of grazinglands and landform - vegetation cover relationship. Based on survey of India toposheets and satellite imagery total 9 major landform units (LFU) as presented in table 4 were identified and grassland data from analyzed imagery was transferred to landform map.

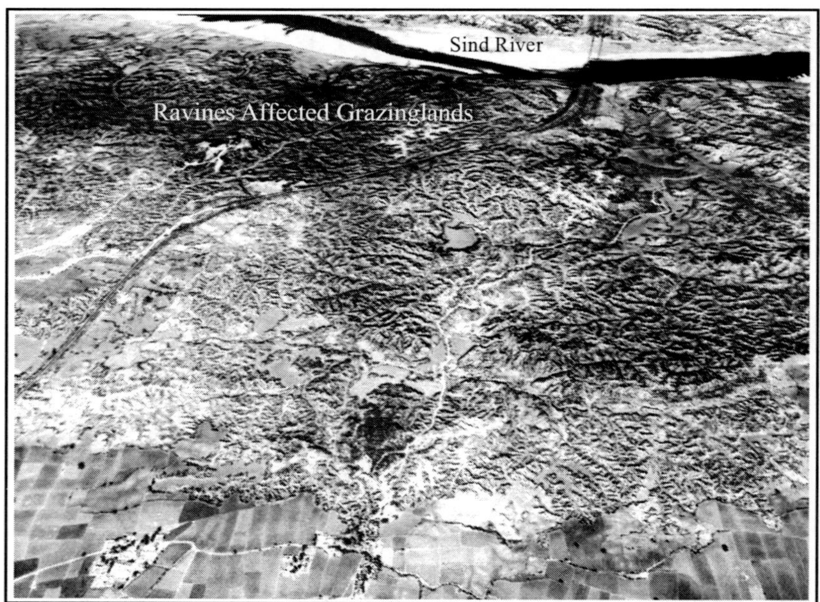

Fig. 5. High Resolution Satellite Imagery of Ravines Affected Lands

Table 4: Landform Units

Land Form Units (LFU)	Area (ha)	% of total area	Slope category	Average slope (%).
Hills(LFU1)	224425	9.55	Xe,E,D,C	>40
Upland Dissected Plateau(LFU2)	243225	10.35	B, C	20-30
Monadnoks (isolated residual hillocks / out rock crops) (LFU3)	66975	2.85	E, C	30-40
Pediments(LFU4)	131130	5.58	C, B	10-20
Undulating plain(LFU5)	834250	35.50	A, R	10-15
Weathered and Buried Rocky Surface(LFU6)	144525	6.15	A, B,C	00-10
Gully and Eroded land(LFU7)	212675	9.05	Xe,C,D,Fr	10-35
Gentle slopping plain and Recent Alluvium(LFU8)	478225	20.35	A,F	00-05
River Valley(LFU9)	14570	0.62	A,F	00-05
Total	2350000	100.00		

The grasslands of the study area were in various stages of degradation (Table 5). It is evident from the table 5 that maximum area under heavily degraded grasslands was under LFU 6 and 7, while moderately degraded grasslands was recorded in LFU 1 and partially degraded grasslands in LFU 9.

Table 5: Level of Degradation (2005)

LFU No.	Grasslands / Grazinglands Area (in % of total area)		
	Partially degraded	Moderately degraded	Heavily degraded
1	26.42	47.17	26.41
2	45.45	33.77	20.78
3	16.13	29.03	54.84
4	36.14	31.63	32.23
5	23.24	33.24	43.52
6	10.00	16.67	73.33
7	11.11	20.37	68.52
8	48.75	32.50	18.75
9	60.00	40.00	0.00

Estimation of Present and Potential Forage Production

Two methods. viz., (I) transfer of analyzed satellite data of 1989, 1995 and 2005 on landform map and, (ii) ground truthing using quadrant and line transect techniques were applied for the estimation of forage production in MgDm (Mega gram Dry matter). The forest types and grazinglands exhibit a marginal degradation trend in terms of production (2.88 per cent and 4.44 per cent respectively) within a span of 6 years (1989-1995). The estimation of potential production is based on the existing vegetation types and their ambient biophysical features. The study reveals that the existing forest production level of 0.60 MgDm/ha can be raised up to 1.91 MgDm/ha and in grazinglands the existing production level of 1.07 MgDm/ha can be raised up to 3.84 MgDm/ha without hampering the ecological resilience. The existing and potential production level in different degraded stages of forest and grazinglands has been depicted in detail in table 6.

Table 6: Forage Production in Bundelkhand Region

Land use/ Land cover	Area Total (ha)	% of total area	Existing Production MgDm/ha	Potential Production MgDm/Ha
1.Forest				
Dense	5528	30.10	0.4-0.6	0.5-0.8
Mod. Degd.	9331	50.80	0.5-1.0	0.8-2.0
Heav. Degd.	3503	19.00	<0.5	0.5-2.5
Total	18362	100.00		
2.Grazinglands				
Partialy Degd.	47650	50.60	1.3-1.5	4.0-6.5
Mod. Degd.	100977	16.20	1.0-1.3	3.0-5.0
Heav. Degd.	32313	09.90	0.3-1.0	1.5-3.5
S.S.R.	18512	100.00	<0.3	1.0-2.5
Total	199452			

Forage Resources Availability

Forage Availability from Grasslands

Hardly 0.25m ha land (3.54%) was covered and dominated with native grass species. The result shows that the spatial distribution of grassland in the region was not even (Fig. 2). The total forage production was estimated as 0.278 m t DM/year. It was found maximum in Sagar district (58342 t DM/y) followed by Chhatarpur (48781 t DM/y) and Jhansi (40630 t DM/y).The important edible perennial grasses were *Bothriochloa pertusa, Chrysopogon fulvus, Dichanthium annulatum, Heteropogon contortus, Iseilema laxum* and *Sehima nervosum* where as other annul grasses were *Andropogon pumilus, Apluda varia, Aristida reducta, Brachiaria ramose, Digitaria spp., Setaria glauca* and, *Themeda quadrivalvis.* Important legumes and forbs found in the region were *Alysicarpus vaginalis, Atylosia scarabaeoide, Cassia pumila, C. tora, Indigofera hirsute, Tephrosia villosa* and *Borreria stricta, Eclipta alba, Evolvulus alsinoides, Lepidagethes trivervis, Merremia , marginata, Rungia sp.* respectively. District wise information on forage availability from grasslands is presented in table 7.

Table 7: Availability of Forages from Grazinglands / Rangelands (including forest lands) in Bundelkhand Region

District/Teh.		Area in ha		Forage Availability (t DM/year)				Total Forage		
	Total Area	Grassland	Ravines	Shrubs	Forest	Grassland	Ravines	Shrubs	Forest	(t DM/ year)
Banda	454227	8831	9568	330	1930	9714	10716	304	2606	23339
Bhander	65826	3221	8660	295	0	3543	9699	271	0	13514
Chhatarpur	867986	44346	9805	7127	205259	48781	10982	6557	277100	343419
Damoh	726063	4680	1378	0	349918	5148	1543	0	472389	479081
Datia	201683	6618	6954	0	21157	7280	7788	0	28562	43630
Hamirpur	417179	8942	11313	309	15228	9836	12671	284	20558	43349
Jalaun	455646	5808	52583	130	11191	6389	58893	120	15108	80509
Lahar	107587	545	33150	0	28	600	37128	0	38	37765
Lalitpur	501664	30435	4235	2874	62037	33479	4743	2644	83750	124616
Mahoba	301523	10017	3705	0	9479	11019	4150	0	12797	27965
Panna	706571	17977	552	6024	375519	19775	618	5542	506951	532886
Sagar	1035475	53038	23	12710	316368	58342	26	11693	427097	497158
Jhansi	503987	36936	22260	3849	19708	40630	24931	3541	26606	95708
SM Nagar(Chitrakut)	310781	5236	5537	2106	73295	5760	6201	1938	98948	112847
Tikamgarh	504052	16495	1220	3588	71872	18145	1366	3301	97027	119839
Bundelkhand	7160250	253125	170943	39342	1532989	278438	191456	36195	2069535	2575623

Forage Availability from Grazinglands Affected by Ravines

About 170943 ha (36.89%) land of the total area under grazinglands (463410 ha) was affected by ravines. It was about 2.39 percent of the total area of the Bundelkhand region. The severity of ravines was found more along the middle and lower reaches of rivers and streams in the northern part of the region. The situation was alarming in Lahar and Bhander tehsils where as Jhansi, Datia, Hamirpur and Banda were badly affected by ravines (Table 3). The total forage production from this land was found as 191456 t DM/y. It was recorded maximum in Jalaun (58893 t DM/y) followed by Lahar (37128 t DM/y) and Jhansi (24931 t DM/y). There is an urgent need to restore these lands through proper land treatments and reseeding of suitable grass and tree species.

Forage Availability from Grazinglands Affected by Shrubs

Infestation of grazinglands with shrubs has considered as a threat to the existing grazinglands in the region. The study reveals that about 39342 ha (0.55%) land (mostly grazinglands) in the region was infested with shrubs. The situation was more alarming in Sagar (1.23%), Panna (0.85%), Chhatarpur(0.82%), and Tikamgarh(0.71%) districts. These lands are closely associated with fringe area of forest and along the channel in middle and upper reaches. Though forage production from shrubs infested land was recorded as only 36195 t DM/y but it provides valuable forage to small ruminants during lean period.

Forage Availability from Forest Lands

The total forest cover (including degraded forest land) in the Bundelkhand region was about 1532989 ha or 21.41 percent out of which good forest cover (>40% canopy cover) shows hardly 11 percent. The important shrub and tree species found in the region were *Acacia leucophloea, Butea monosperma, Dalbergia sissoo, Ficus religiosa, Madhuca indica, Diaspyros melanogylon, Zizyphus nummularia* and *Prosopis julifera.* Two district of the region Damoh and Panna (forest area 53.15 and 48.19% respectively) have more than the targeted national forest cover. The spatial distribution of forests in the region was directly related to the amount of rainfall, geomorphic controls and to some extent human and livestock pressure. The study reveals that the share of forage production from the forest land to the total range forages was about 80 percent (2.07 m t DM/y) in Bundelkhand region.

Conclusions

Today there are evidences of grazinglands degradation throughout the country. Excessive grazing has done massive harm to grazinglands, converting them into highly degraded areas. About 63.7 million ha area under forest cover in India provides grazing resources but more than half of it is in various stages of degradation providing no useful biomass for foraging. The trend of degradation is upward. The need of the geo-referenced information on the extent of area under rangelands or grazinglands their present and potential carrying capacity and forage supply-demand balance sheets for the different region of India has been felt long back. The knowledge of habitat conditions and vegetation is also limited. With the help of Satellite Remote Sensing and Geographic Information System it will be possible to identify, analyze and characterize the area under grazinglands and also to categories the grazinglands types and assess the present and potential carrying capacity. This geo-referenced digital information can be projected through thematic maps that will be useful in developmental planning of forage resources in various agro-ecological regions of India.

References

Isavwa, Luka A. 1988. Inventory and Monitoring of Rangeland Resources in Kenya Using Remote Sensing Techniques. Proceedings of Third International Rangeland Congress. Vol.1, *Rangeland management Society of India*, IGFRI, Jhansi, 36.

Li Bo, and Chain kai 1988. Remote Sensing for Range Resource Inventory in Inner Mongolia, China. Proceedings of Third International Rangeland Congress, Vol.1, *Rangeland Management Society of India*, IGFRI, Jhansi, 30.

Naugraiya, M.N. and. Pathak P.S 1987. Effect of Gibbrellic acid on growth of Atylosia scarabioides Benth. *Indian Journal of Range management.* Vol. 8, Number 1, Range Management Society of India, IGFRI, Jhansi. 55-57.

Pandeya, S.C. 2000. The Grazingland Resources and Fodder Cultivation in India *Foundation Day Guest Lecture.* Indian Grassland and Fodder Research Institute, Jhansi.1

Pathak,P.S. and Singh J.P. 2003. Bundelkhand: Perspective and Agroforestry Prospects. Souvenir. National Symposium on Agroforestry and Sustainable Production & Silver Jubilee of Agroforestry in India. Nov.7-9 . NRCAF, Jhansi. 28-40.

Pathak, P.S., Singh J.P., Sharma P., Singh K.K., Dwivedi P.N. and Singh J.B. 2005. Bundelkhand Region: Agricultural Perspective - Status, Constraints and Prospects.Indian Grassland and Fodder Research Institute, Jhansi.

Singh J.P., Trivedi B.K., Ekka N.S., Khan T.A. and Ashok P. 2004. GIS and RS as a decision making tools for the rangelands development in ravines affected area. *Indian Cartographer*, Hyderabad. Vol. 24, 414-421.

Singh, R.L. (eds) 1961. India A Regional Geography. *National Geographical society of India*, Varanasi. 599-601.

Singhal, R.M., Rawat B.R.S. and Kumar P. 1987. Soil-Vegetation Relationship in some Forest Ranges of Chakaranta Himalaya, India. *Indian Journal of Range Management*, Vol.8, Number 2, Range Management Society of India, IGFRI, Jhansi, 91-101.

Tueller, Paul T. 1988. Remote Sensing Procedures for Analysis of Shrub dominated Rangelands. Proceedings of Third International Rangeland Congress, Vol.1, *Rangeland Management Society of India*, IGFRI, Jhansi, 27.

Tyagi, R.K. and Patil B.D. 1982. Watelands of Bundelkhand: Its Potential for Forage Resource Development, *Indian Journal of Range Management*, Vol.3, Number 1&2, Range Management Society of India, IGFRI, Jhansi, 53.

Van Dyne, G.M., Smith, F.M., Czplleski, R.L. and Woodmansee, R.G. 1978. Analysis and Synthesis of Grassland Ecosystem Dynamics" in J Singh and B Gopal (eds)*Glimses of Ecology* . International Science Publishers, Jaipur. India 1-80.

Vinod Shankar and Gupta, J.N. 1992. Restoration of Degraded Rangelands. in J.S. Singh edited *Restoration of Degraded Land: Concepts and Strategies*, Rastogi Publication, Meeruth. 115-155.

❑❑❑

Chapter 5

Seed Production for Forage Crops and Grasses and National Network

R. K. BHATT
Indian Grassland and Fodder Research Institute, Jhansi-284003 (U.P.), India

Introduction

Availability of forage seeds is the base of fodder and pasture development programme in India. Current demand for seeds of cultivated fodder is estimated at 3,5,5000 tonnes / annum based on the area under cultivation (8.3 m ha). There exist a huge deficit in fodder availability to the extent of 80 million tonnes dry and 660 million tonnes green fodder as per different reports. The country has about 8.3 m ha of land under cultivated fodder, which accounts for the 5.9 percent of the total cultivated area. Propagation of improved fodder production technologies particularly the improved varieties suitable for their areas have not reached to the farmers mainly because of the non-availability of good quality seed. The availability of quality seed is only 15-20% for fodder. With the development of a number of improved and high yielding varieties in forages, it has become important that quality seed should be readily available and supplied to the farmers, so that benefit of breeding and research efforts should go to the users.

Commercialisation of forage seed is possible through the development of strong linkages between producers and users. There is need to reorganised the routine activity of research towards the prioritization of need based, target oriented and time bond achievements. The seed production technology must go in the field. As the new varieties released, its seeds must be made available along with seed production, harvesting, processing and storage technology to the users.

Major Constraints

By and large the forage crops are low seed yielder and poor in seed setting. There are a numbers of constraints in the way of fodder promotion and seed production. The major constraints are:

1. Forage crops are given low priority
2. The allocation of area under quality fodder production is very less
3. Non availability of quality seeds
4. Standard seed production techniques of forage crops are lacking
5. Indeterminate growth habit affects the seed yield in forages
6. Forage is basically bred for herbage yields and thus is low in seed yield.
7. Harvesting of crops at vegetative stage which do not allow the crop to reach to seed stage leading to seed production
8. High cost involved in seed collection and processing.
9. Uncertainty of seed demand
10. Poor extension work for the popularization of improved varieties
11. Transfer of technology of improved verities, package of practices, suitable species for various types of lands and niches is not reaching to the farmers and other needy clients.
12. No organized net work of forage seed production, distribution and marketing
13. Lack of Govt. Department dealing specially in fodder crops
14. Policy support lacking

Status of Breeder Seed Production During Xth Five Year Plan

Forage Crop	2001- 02	2002- 03	2003-04	2004-05	2005-06
Sorghum	**52.98**	70.30	37.58	63.98	15.45
Bajra	1.82	3.20	3.20	2.60	4.20
Maize	90.75	51.00	102.00	68.00	72.00
Teosinte	10.00	3.75	8.00	-	5.00
Cowpea	17.00	16.30	29.98	29.90	19.85
Guar	48.15	-	130.50	6.26	271.80
Ricebean	4.50	3.70	3.76	1.25	2.25
Sunhemp					12.00
Sub-Total	**225.20**	**148.25**	**315.02**	**171.99**	**402.55**
Oats	152.66	141.40	174.00	344.00	274.50
Berseem	51.95	40.38	71.60	51.70	36.80
Lucerne	28.44	21.90	12.66	8.75	10.70
Sub-Total	**233.05**	**203.68**	**258.61**	**404.45**	**322.0**
Grand Total	**458.25**	**351.93**	**573.63**	**576.44**	**724.55**

Source- PCFC, IGFRI, Jhansi Report

Seed Yield of Forage Crops

Berseem and sorghum occupy maximum area under cultivation, whereas most of the range grasses and legumes occupy little area (Table 1). Among annual fodder crops, lucerne and berseem produce maximum green fodder, whereas in perennial cultivated forages *Panicum maximum*, *Brachiaria mutica* and napier grass are high- yielding forages. Cultivated fodder cereals are high seed yielders as compared to legumes.

Major problems of seed production in range grasses and legumes

1. Indeterminate growth
2. Apical dominance
3. Prolonged head emergence

4. Uneven maturity
5. Low seed setting
6. Seed shedding
7. Blank seed
8. Seed dormancy
9. Lodging of seed crop
10. Low density of panicle producing tiller
11. Site effect

Table 1. Seed and Forage Yield of Major Forage Crops

Crops	**Area** (1000 ha)	**Seed yield** (Kg/ha)	**Green forage** Yield (t/ha)
Legumes			
Trifolium alexandrianum	1900	400-700	60-100
Medicago sativa	1000	150-260	60-110
Vigna unguiculata	300	700-100	25-45
Cy nopsis tetragonoloba	200	500-900	15-30
Cultivated cereals			
Avena sativa	100	1200-2200	35-50
Sorghum bicolor	2600	600-1700	55-70
Pennisetum typhoides	900	700-1000	20-35
Zea mays	900	1800-2200	30-55
Pasture grasses			
Pennisetum perureum	10	120-350	70-100
Panicum maximum	100	30-550	70-140
Setaria sphacelata	40	60-150	50-95
Panisetum pedice llatum	20	500-800	50-90
Bracharia mutica	100	35-60	100-190
Bracharia bryzantha	10	25-60	50-80
Cenchrus ciliaris	200	60-150	15-40
Cenchrus setigerus	30	50-90	10-30
Dichanthium annulatum	-	60-110	20-35
Pasture legumes			
Lablab perpureus	10	300-400	15-25
Macroptelium atroperpureum	-	150-250	15-25
Stylosathes hamata	-	500-1200	30-55
Clitoria ternatea	-	60-100	20-30

Technology of Seed Production

The technology of seed production programme involves number of steps for the production of genetically pure and good quality seeds. This requires a high level technical skills and timely operations.

The International Crop Improvement Association and member agencies recognised four classes of seeds-

Nuclear seed	:	Seed supplied by plant breeder of a improved variety.
Breeder seed	:	Seed multiplication of nuclear seeds by breeder and specialized seed technologist.
Foundation seeds	:	Seed multiplication by seed farms/ seed sectors under the control of seed technologists/breeder.
Certified seed	:	Multiplication of seeds by registered seed grower, GOs, NGOs, AHDs, NDDB for distribution to the farmers.

Seed production is major component of seed technology and mainly governed by genetic principles and agronomical practices.

A. Genetic Principles

Maintenance of genetic purity during seed production is an important aspect which bases on seed source. Haan (1953) and Hartman and Kester (1968) suggested various ways to maintain genetic purity of seeds. Hartman and Kester (1968) suggested the following steps for maintaining genetic purity.

- Providing adequate isolation to prevent contamination by natural crossing or mechanical mixtures.
- Rouging of seed fields prior to the stage at which they could contaminate the seed crop.
- Periodic testing of varieties for genetic purity.
- Avoiding genetic shifts by growing crops in areas of their adaptation only.
- Certification of seed crops to maintain genetic purity and quality of seed.
- Adopting the generation system.
- Grow out test.

Maintenance of Genetic Purity

The genetic purity of a crop can be maintained by following ways and means:

I. Control of Seed Source

i) Use of certified seed for further seed multiplication
ii) Nucleus seed
iii) Breeder seed
iv) Foundation seed
v) Certified seed

II. Field and Crop Requirement

i) Location /area selection
ii) Field selection
iii) Crop sequence/rotation
iv) Isolation
v) Field Inspection
vi) Rouging
vii) Seed harvesting, processing and storage

III. Testing of Variety for Genetic Purity

i) Seed certification and issuing of tag
ii) Seed testing for true to the type
iii) Grow out test

B. Agronomical Principles

To produce quality seeds the following agronomical package of practices should be followed. In major fodder crops the agronomical package of practices have been standardized and given in table 3. The following steps have to be taken while taking the quality seed production programme of a particular crop:

1. *Selection of suitable Agro-climatic region as per recommendation :* In majority of forage crops, seed production depends on photoperiod, thermo period, humidity, soil texture, structure, reaction and moisture. These crops can give better seed yield if grown on ideal sites. Each forage crop is, therefore, suited to only specific area for forage and/or seed production such as berseem in the northern plains and lucerne in the north-west India. Similarly pasture grass like Congo Signal grass and guinea grass in the highly humid regions of Kerala. The studies carried out by Indian Grassland and Fodder Research Institute and All India co-ordinated Research Project (AICRP), Jhansi on forage crops over the years and on

the basis of multiplication trials some important fodder crops have been categorised according to agro-climatic locations for higher seed production (Table 2).

Table 2. Suitable Areas for Different Fodder Crops for Increased Seed Productivity

Crops	Area
Sorghum	South and South -West
Cowpea	North-West and South-West
Dinanath grass	East and Central
Pearl millet	North-West
Buffel grass	Western, Southern
Marvel, Sehima	Central, South-Central
Siratro, Stylo	Central, South-Central
Rice bean	East
Berseem	Central and West
Lucern	West and South-Central
Oats	North and North-Central
Guinea grass	South and North-West

2. Selection of field. Field for seed production should be selected keeping in view the soil type, soil texture, fertility level, free from soil borne disease and insect-pest, free form weed plants, crop sequence over the previous years and seasonal crop rotation. An ideal seed production area differs from the areas suitable for commercial cultivation of the same crop species for grain or fodder. The land should be fertile, uniformly leveled and assured of irrigation. It should possess the capacity of drain out excess water to avoid water logging. Seed production area should be kept free from weeds. Fields infected with soil born diseases should be avoided. Area should be free from other genotypes of cultivated and cross pollinated wild species
3. Isolation distance of the crop – The seed crop of a particular variety must be isolated from other fields of the same variety and from other contaminated crops as per requirement of seed certification standards.
4. Preparation of field

5. Seed treatment
   i) Chemical treatment
   ii) Bacterial inoculation
   iii) Seed treatment for breaking of seed dormancy
6. Time of sowing
7. Seed rate and row spacing

   It is advisable to grow a seed crop at sufficient row distance to allow inter culture, removal of off type plants, allow better light interception of the flowering shoots and also facilitate seed harvesting without any hindrance.
8. Method of sowing
9. Depth of sowing
10. Use of fertilizers

Use of high level of fertilizer is usually economic for pasture seed crop under high rain fall situation and relatively less so with lower rain fall. For grass or legume pasture seed crop, the fertilizers which are required in the tropics are phosphorus (P), potassium (K), nitrogen(N), sulphur(S) and often small quantities of molybdenum (Mo), copper(Cu) and Zinc (Zn). The response of nitrogen may vary with the crops and with the season of crop growth. The nitrogen response may be better in a pasture grass seed crop in the establishment year than in the latter years of growth when the plants have developed an exhaustive root system. High nitrogen application may not be effective under the situation of excessive moisture shortage and in fact may result in further depletion of the soil moisture due to excessive vegetative growth.

Legumes seed need to be actively modulated at the time of sowing. With desirable inoculums, the nitrogen fixing bacteria in the root nodules readily fixes atmospheric nitrogen in the soil for utilization of the plant and for enriching the soil with nitrogen.

- Use of proper N P K fertilizers and other micronutrients, dose of nutrients and time of application. (Table 3)
- Integrated nutrient management
- Foliar application of nutrients (Table 4).

Table 3. Cultural Practices for Quality Seed Production of Forage Crops.

Crop	Isolation Distance (mt.)	Seed Rate (kg/ha)	Inter-row Spacing (cm)	Fertilizer (kg/ha) N	P	K	Seed Yield (q/h)
Sorghum	200	12-15	40-50	75-80	30-40	25-30	18-20
Pearl millet	400	25-30	40-50	45-50	15-25	20-25	10-15
Sudan grass	100	12-15	40-50	25-60	20-25	25-30	6-8
Oat	20	60-80	25-30	80-100	35-40	30-35	20-30
Teosinte	200	15-20	40-50	80-100	30-40	25-40	10-20
Guar	20	20-25	40-50	25-30	30-40	30-35	8-10
Cowpea	20	20-25	40-50	25-30	30-40	30-35	10-12
Lucerne	400	10-15	30-35	25-30	55-60	40-45	2.5-3.5
Berseem	400	20-25	20-25	25-30	55-60	30-35	3-5
Dichanthium	20	3-5	50-75	40-60	30-40	15-20	.60-.75
Panicum	20	3-5	50-75	75-80	55-60	35-50	.80-1.5
C. ciliaris	20	3-5	50-75	40-60	25-30	25-30	.75-1.2
C. setigerus	20	3-5	50-75	40-60	25-30	25-30	.75-1.7
Rhodes	20	5-6	50-75	40-60	25-30	25-30	.50-.80
Setaria	20	2-3	50-75	40-60	25-30	20-25	.70-1.4
Chrysopogon	20	3-5	50-75	40-60	25-30	20-30	.75-1.5
Stylo	20	6-8	40-50	20-30	30-40	30-40	8-12
Clitoria	20	10-15	40-50	20-30	30-40	30-40	3-5

Table 4. Seed Production Through Foliar Application of Growth Regulators and Nutrients.

Cowpea	Supper phosphate + KNO_3	1+2 Kg/ha	Flower initiation stage
	GA_3	10-50 ppm	Pre anthesis stage
	Phosphon -D	50 ppm	Vegetative and anthesis stage
	CCC	50 ppm	Vegetative and anthesis stage
M.P. Chari	CCC	0.5 Kg/ha	At 45 days after sowing
Berseem	KNO_3 + Supper P	2 Kg+2.5 Kg/ha with 60 Kg P_2O_5 as basal	Flower initiation stage
	GA_3	10-50 ppm	Flower initiation stage
Cluster bean	KNO_3	4 Kg/ha	Flower initiation stage
Clitoria	KNO_3+$CuSO_4$	4 Kg+0.5 Kg/ha	Flower initiation stage
Siratro	NAA	100 ppm	Flower initiation stage
	MH	50 ppm	1st spray at vegetative and
	2nd spray		at flower initiation stage
Dolichos	NAA	100 ppm	Flower initiation stage
	MH	50 ppm	1st spray at vegetative and 2nd
spray			at flower initiation stage
Cenchrus	KNO_3	2 Kg/ha	1st spray at vegetative and 2nd at preanthesis stage
Stylosanthes	NAA	100 ppm	1st spray at vegetative and 2nd
spray	+DAP	2 kg/ha	at flower initiation stage
	+K2O	2 kg/ha	

11. *Irrigation* : To ensure good and uniform germination and healthy crop growth irrigation is needed as per crop requirement and soil condition.

12. *Weed control* : Weed control is essential so that they do not interfere with crop prodution. The amount of control is usually balanced between the cost involved and the possible damage to the crop caused by seeds themselves. There are three basic measures of weed control, A-prevention, B – eradication and C – control;

    Prevention of weed is better than control as it is the cheapest and most effective method and includes the following steps; (1) Ensure that seeds for sowing are free from weed seeds (2) Do not allow weeds to grow to seeding stage and (3) Do not allow the spread of vegetative parts of perennial weeds.

    Weed eradication involves total eradication of all live plants parts or seed of the weedy plants from an area. It is applicable in dealing with a small localised area of a potentially dangerous weed. Both mechanical and chemical (non selective soil sterilent) means could be effectively used. Clean fallow and following crop rotation are also help in weed control.

13. *Disease and insect control* : A wide range of insect pests and diseases are involved which affect the crop at the different stages of crop growth. A few major species, which are crop specific may account for the major losses in productivity. Insect pest and disease attack is accentuated by high moisture and temperature during crop growth and seed storage. The most practical measure for use against field infestation are to reduce the reservoir of infestation by treating or destroying the infected commodities.

    To produce the healthy seed crop it is essential to control disease and insect damage. Various fungicides are available to control the soil borne diseases. Application of appropriate fungicides and insecticides in right quantities and at the right time helps in disease and insect control.

14. *Cutting management* : Tropical pasture crops of perennial nature should not be cut, grazed or slashed lower than 4" to 6" above crown level. Buds for new growth are quite high on the plants. This is important for continuous renewal of the pasture and production of the following tillers.

15. *Rouging* : At vegetative/pre-flowering stage, flowering stage and maturity stage

16. Supplementary pollination – use of honey bees

17. Harvesting of seed crop
    i) Time of harvesting influence the seed yield and quality of seeds. Seed moisture is good indication of optimum time of harvesting the seeds.
    ii) Method of harvesting- To avoid mechanical damage of seeds and mixture of other seed crop harvesting should be done properly.
18. *Processing* : Thrashing and cleaning of seed: The process involves separation of undesirable material from the main seed. Drying involve processes to bring down seed moisture levels in the range of 8 to 12 percent for medium to long term storage. The quality of seeds could be upgraded by proper seed drying , removing of foreign materials and under sized seeds, grading of seeds and application of chemicals to protect the seeds. Seed drying is generally required to lower the seed moisture content before storage. The drying is done by sun drying and forced air drying. The safe temperature for drying is 45^{0}C but it varies with crop to crop. Seed temperatures for oats 43-65 ^{0}C, maize 40-45 ^{0}C, rye grass 40-70 ^{0}C and for clovers 27-40 ^{0}C are safe. The seeds should be dried to keep the moisture below 10% for efficient storage. Seed lots are mainly processed for removal of foreign materials like weeds, inert materials, damaged seeds and grading for size and treatment with protecting chemicals.
19. *Storage of seeds* : Storage of seeds depends on seed moisture and temperature. Seed treatments with fungicides and insecticides are essentially required to control the seed born diseases and insects during storage. Chemicals like Bavistine, Thiram, Vitavax, Captan and Dithan and their suitable combinations in the ratio of 1-2 g per kg of seeds are used for proper storage. Seed should be packed in bags of standard size and shape and suitably labelled. Tags of different colour such as golden yellow tags for the breeder's seed, white tag for foundation seeds and blue tag for certified seeds are in the general use.
20. *Seed Certification* : Seed certification is the measure to control the quality and purity of seeds during multiplication and production. It includes the inspection at various level of seed production programme like pre and post control tests, field and bin inspection, inspection of seed fields, inspection of seed processing, seed sampling, tagging, seed quality and health tests. In a seed quality control programme the seed stocks are verified as per minimum seed certification standards(Table 5 & 6).
21. *Seed Quality* : Seeds of a particular variety can be released after testing for its quality with respect to purity, germination percentage, seed

viability, vigourness test, and seed health on the basis of certain minimum standards. A Central Seed Testing Laboratory at IARI, New Delhi and State Seed Laboratories at state level were established for seed testing.

Steps Involved in Quality Seed Production

Seed Production

Genetical Principal	**Agronomical practices**
-Seed source	
A. Nucleus seed	Specific site
B. Breeder seed	Selection of varieties & seed quality
C. Foundation seed	Presowing treatment
D. Certified seed	Proper planting time
-Quantity of seed	Seed rate
-Isolation distance	Seeding depth
-Roughing of seed plots	Optimum fertilizer application
	Timely irrigation
	Disease and insect control
	Timely harvesting

Seed certification Processing

Seed drying, grading and protection

Storage

RH & Temperature

Seed quality & Seed test for

Purity, Germination

Seed viability, Seed vigour test

Seed health, Moisture content

Cultivar purity

Marketing and Distribution

National Network

The production and multiplication of fodder seed are governed by public and private sectors. The formal seed sectors are having framework of institutions which are involved in the development, multiplication, processing and distribution of seeds of released or certified varieties. Some time the seed production is also carried out by other means which can be called the informal sector.

Table 5. Minimum Field Standards in Forage Crops

Crop	Seed Class	Minimum Isolation Dis. (m)	No. of field inspection	Maximum permissible limit (%) Off type	Objectionable weeds	Plants/heads diseased
Cenchrus	FS	200	3	0.1	-	-
	CS	100	3	1.0	-	-
Chrysopogon	FS	200	3	0.1	-	-
	CS	100	3	1.0	-	-
Penesetum	FS	20	3	0.1	-	-
	CS	10	3	1.0	-	-
Panicum	FS	200	3	0.1	-	-
	CS	10	3	1.0	-	-
Dicanthium	FS	200	3	0.2	-	-
	CS	100	3	1.0	-	-
Setaria	FS	400	3	0.1	-	-
	CS	200	3	1.0	-	-
Napiar grass	FS	10	2	0.1	-	-
	CS	10	2	0.3	-	-
Sorghum	FS	400	3	0.1	-	0.1
	CS	400	3	0.2	-	0.2
Oat	FS	3	2	0.05	0.01	0.1
	CS	3	2	0.20	0.02	0.5
Maize	FS	400	2	1.0	-	-
	CS 200	2	1.0	-	-	
Berseem	FS	400	2	0.2	none	-
	CS 200	2	1.0	0.05	-	
Lucerne	FS	400	2	0.2	none	-
	CS	100	2	0.1	0.05	-

Government of India has been continuously launching and revamping policies to create a favorable environment for the growth of seed industry both in the public and the private sector. National Seeds Corporation (NSC) and State Farms Corporation of India(SFCI) and Tarai Development Corporation (TDC) were established in 1963 and 1969. The Seed Review Team (1968) was enacted through which a system of valuntary seed certification was adopted for almost all the crops and the minimum seed certification standards for most of the crops were accepted. Currently, National Seed Policy is being reviewed in view of changing global scenario.

Indian seed industry having two national sectors (NSC and SFCI), 13 state seed corporations, 19 state seed certification agencies, 63 notified seed testing laboratories and more thane 140 private seed companies. Recently some private multinational seed companies have come up and they have their own Research and Development programme (Tiwari 1996). ICAR is having the responsibility to produce the breeder seed as per the requirement given by

Table 6. Minimum and Maximum Seed Standards for Quality Control and Seed Certifications in Forage Crops

Crop	Seed class	Minimum				Maximum				Moisture	
		Germination (%)	Pure (%)	Innert seed (%)	Other Crop matter seed (No/kg)	Weed (No/kg)	Objectionable seed (No/kg)	Diseased weed (% by No)	Other seed	Ordinary variety seed (No/kg)	Vapour pack proof
Cenchrus ciliaris	FS	30	80	20	20	20	-	-	-	10	8
	CS	30	80	20	40	40	-	-	-	10	8
Chrysopogon fulvs	FS	15	80	20	20	20	-	-	-	10	8
	CS	15	80	20	40	40	-	-	-	10	8
Penesetum pedicelatum	FS	50	95	5	20	20	-	-	-	10	8
	CS	50	95	5	40	40	-	-	-	10	8
Panicum maximum	FS	20	80	20	20	20	-	-	-	10	8
	CS	20	80	20	40	40	-	-	-	10	8
Dicanthium annulatum	FS	40	90	10	10	-	-	-	10	10	8
	CS	40	90	10	20	-	-	-	20	10	8
Seteria sphacelata	FS	50	95	5	20	-	-	-	-	10	8
	CS	50	95	5	40	-	-	-	-	10	8
Sorghum	FS	75	97	3	5	-	-	0.02	10	12	8
	CS	75	97	3	10	-	-	0.04	20	12	8
Oat	FS	85	98	2	10	10	2	-	10	12	8
	CS	85	98	2	20	20	5	-	20	12	8
Maize	FS	90	98	2	5	-	-	-	10	12	8
	CS	90	98	2	10	-	-	-	20	12	8
Guar	FS	70	98	2	10	-	-	-	10	9	8
	CS	70	98	2	20	-	-	-	20	9	8
Stylosanthes hamata	FS	40	90	10	10	10	-	-	10	10	8
	CS	40	90	10	20	20	-	-	20	10	8

seed division in the Ministry of Agriculture. In ICAR system the breeder seed production is being undertaken with the help of ICAR institutions, All India Coordinated research Projects on forages and SAUs. The breeder seed production is also sponsored by ICAR to State Farm Corporation, State Seed Corporations and NGOs. Breeder seed which is produced through breeder seed production programme is supplied to the NSC, SCFIS SSCS and well organised private companies and some time to the SAUs for the production of certified and foundation seeds.

For the range grasses and legumes where specific varieties are not available and not tested through the National Coordinated Programme, the local seed certification may be done by breeder/scientist which could be utilised as Truthfully labeled Seeds. The State Forest Department, should take a lead role in organising production of seeds of range grasses and legumes in forest lands. Specific responsibility should be given to the Forest Deptt. For undertaking production of these plant species along with fodder tree species.

In context of fodder seeds, the public sector is the dominant player in plant breeding and varietal development. Since 1970, research on fodder crops has operated under the umbrella of AICRP (FC). Regarding the supply of fodder seeds, the role of formal sector is quite limited and the informal sector is a dominant player. An effective synthesis of these two sectors is needed to ensure timely supply in adequate quantities of quality fodder seed.

Table 7. National Net Work Organization

1.	**Central level**	**Functions**
1.1	**Enforcement Agency**	
	1.1.1 Central Seed committee	To advice Central and State Govts. on matters arising out of the administration of the Seeds Act 1966 and the seeds Rules 1968 as amended from time to time and to carry out other functions assigned by or under the act.
	1.1.2 Central seed certification board	To advise the Central and State Govt. on all matters relating to certification, and to coordinate the functioning of agencies in the State.
	National Seeds Board (NSB)	The **National Seeds Board** (NSB) will be established in place of existing Central Seed Committee and Central Seed Certification Board
	1.1.3 Central Seed testing laboratory	Testing programme in collaboration with the state seed laboratories for uniformity in test and surveillance on the quality status of seed found in the country.
	1.1.4 Seed Division Ministry (Ministry of Agril. & Co-operation)	Nodal unit at the National level for the enforcement and administration of acts and rules and proposal for amendments from time to time addition, deletion in the minimum standard list. Coordination and monitoring of the activities in this regard.

Contd...

Table 7. Contd...

1.2	**Production**	
	1.2.1 Seed Division (Ministry of Agril. & Co-operation)	1. Collection of indent regarding of breeder seed from different organizations and passing consolidated information to crop Division of ICAR for its production.
		2. Collection of information on production from ICAR, allotment to indenting agencies.
	1.2.2 Crops Division, ICAR	1. Release and with drawal of cultivars through varietal release committee.
		2. Collection of information on production from ICAR, allotment to indenting agencies.
		3. Research on seeds and related aspects.
	1.2.3. Animal Husbandry Division (Ministry of Agril. And Co-operation)	Foundation and certified seed production through its 7 regional stations and one large seed farm.
	1.2.4 N.S.C.	Multiplication of foundation and certified seeds.
		Multiplication of Breeder, foundation and certified seeds
	1.2.5. S.F.C.I.	Multiplication of foundation and certified seeds.
	1.2.6. N.D.D. B.	
		through its growers.
	1.2.7 I.G.F.R.I.	1. Research on seed aspects of cultivated forages, grasses & range legumes.
		2. Production of Nucleus and Breeder seeds of cultivated fodder and pasture crops
	1.2.8 C.A.Z.R.I	Breeder seed of pastures.
	1.2.9.I.A.R.I	Breeder seed of its sorghum cultivars.
2.	**State Level**	
2.1	**Enforcement Agency**	
	2.1.1 State seed certification agency	Certification of seeds of notified cultivars, field inspections, out lining procedures for submitting application, and for growing, harvesting, labeling of seeds ensure that seeds lots finally approved are true to cultivar and meet prescribed standards, maintain list of growers, samples and inspect seed lost, processing of inspected seed, ensure action at all stages e.g. field inspection, seed processing, analysis a sample, issue of certification marks, levels and seals, education and extension programme to promote use of certified seeds.
	2.1.2 State Seeds Corporation	These have been established in many states. It undertakes foundation and certified seed production programme for the states.
	2.1.3 Agril. Universities	A.U. at Hissar, Ludhiana, Pantnagar, Anand & Rahuri are participating in Breeder seed programme. Also in the production of other categories of seed including other S.A.U.'s
	2.1.4. State Departments	U.P. Animal Husbandry Deptt. Is producing foundation and certified seeds. Also intending to participate in Pasture Seed Multiplication. In other states, too, A.H. Deptt. Is involved in seed production at their farms.

Contd...

Table 7. Contd...

	2.1.5. State Corporations	Like "Tarai" Development & Seeds Corporation, Dairy Corporation are producing foundation and certified seeds.
3.	**Private Level**	Private seed companies, registered and some voluntary organisations are involved in foundation and certified seeds production.
		Farmers in Anantpur Distric of Andhra Pradesh have taken Stylosanthes seed production since 1976-77, after trying *S. hamata, S. scabra & S. humilis,* due to high susceptibility of *S. humilis* to Anthracnose, only *S. hamata & S. scabra* have been taken up for seed production.
4.	**Major Clients**	The major clients of fodder seeds have been government agenceies like Forest Departments, Wasteland Development Organisation & Watershed programmes undertaken at Govt. & NGO's level.

Strategies to Sort out the Problem of Forage Seed Production

1. Improvement of fodder crops should not only be based on fodder production but also on adequate seed yielding capacity.
2. Genotypes environment interaction for seed production – Site-specific seed production.
3. Identification of traditional and non-traditional areas for seed production.
4. Pilot programme has to be launched for seed production - Participatory seed production of high priority varieties with the involvement of farmer through NGO.
5. Development of potential areas for forage seed production.
6. Use of pre-sowing seed treatments for invigoration and better crop establishment in range grasses and legumes.
7. Integrated nutrient management for better seed production.
8. Studies on the economics of seed multiplication and interventions to reduce the cost of seed production are important.
9. There should be an apex body in GOI to monitor exclusively the programme of production, distribution and marketing of fodder seeds.
10. Advance indenting and forecasting of the forage seeds as well as creation of storage facility for the left over seeds are required.
11. Seed production through decentralized schemes - An efficient and sustainable net working identification of organizations/agencies including farmers for effective seed production and distribution.

12. Institutional support for forage seed production programme.
13. The seed production, processing, storage and marketing should get the status of industry. The private entrepreneurs may be motivated to involve in the enterprise with more dedication.
14. HRD at different levels needs to be tailored.

References

Hann, H. De. 1953. Maintaining verieties of self -fertilized crop plants. *Euphytica* 2, 37-45.

Hartmann, H.T. and Kester, D. E. 1968. *Plant Propagation Principles and Practices*, Prentice - hall, Inc.

Tiwari, S. P. 1996. Seed Production in India. Indian Farming. 46(8), 16-20.

❑❑❑

Chapter 6

Varietal Improvements in Forage Crops

G. P. SHUKLA
Indian Grassland and Fodder Research Institute, Jhansi-284003 (U.P.), India

Introduction

An initation of varietal improvement programme in forage crops has been from the year 1960's onwards. Prior to this basic researches on cytogenetics and related disciplines were conducted. An evolvement of "Pusa Giant Berseem" in the year 1965, development of some ecotypes in Lucerne for example T9 and Sirsa type and traditional cultivars like MP Chari in Sorghum are certain examples. In self pollinated crops breeding of a variety is a tedious process on one hand and to assure longevity of the developed variety in open pollinated crops is another tedious task. Perhaps, because of uncontrolled pollination and cytogenetical reasons the earlier developed varieties in Berseem and Lucerne degenerated. Precised breeding programmes came later into being looking into the problems of crop specific, situation specific and soil specific reasons. Testing of the developed materials has been another problem which could be mitigated on the onset of All India Coordinated Project for Research on Forage crops in the 1970's. The project has 18 coordinated locations spread over the country and distributed in 5 Agro ecological zones. Consequent upon the incessant breeding researches over 140 forage crop varieties have been released through Central Varietal Release Committee. Apart from this a number of varieties have been evolved in a State Released status exemplified as JB-Series in Berseem from Jabalpur (M.P.), Anand-Series from Gujarat and BL-Series in Punjab possess several varieties of Berseem as State released. Besides, high forage yield being a major criteria for varietal Improvement Programme some specific varieties released also showed their important and prove to be an assets in the forage production and there by feeding need of the cattle. For example a variety Anand-3 of Lucerne developed in Gujarat state has proved its potentials in the Lahul and Spiti Valley of Himanchal Pradesh at the elevation of about 10,000 ft. from mean sea level and at the temperature of –15°c to –20°c. Similarly few varieties of grass say Buffel grass/white Anjan grasses (*Cenchrus ciliaris*) was found to be tolerant to the saline alkali soils in the pH range of from 7.5 to 9. Similarly, alfalfa has been proved its wider adaptability over a range of environmental conditions. Its survive at the temperature as low as - 25°c and as high as 50°c. Alfalfa is highly drought tolerant. It becomes dormant during period of severe drought and resume growth when moisture condition becomes favorable (Hall *et al.*, 1988). Alfalfa is moderately tolerant to acidic soil and in turn all the varieties developed so far in this species show acid tolerant properties (Lanyon and Griffith, 1988). Some of the species which have little human value could be proved to be of substantial importance for forage. Example is the Teosinte crop, which provides 30-40 tonnes of green forage per hectare. TL–1 is a released variety which has gained popularity in Punjab and in a part of Haryana. In the era of intensification of breeding programme in1960's and subsequent later in 1970's maize breeding got several leaps in producing higher grain yielding varieties. A boon in developing forage type varieties started in the 1980's when a variety African tall in maize was developed in Maharashtra state out of the African materials. The programme was intensified at Kolhapur and Rahuri and breeding lines developed at two places were used to synthesize the variety African Tall which is still unconquered for forage production potentials only the draw back in this varieties that this is a long duration variety and takes over 130 days for maturity. Secondly there is intravarietal variation in the cob shape from cylindrical to conical. Now the breeding objectives for forage maize breeding programme has to be in close proximity to these parameters. Breeding objectives in case of sorghum today have to rest around dual purpose sorghum on one hand and multicut forage materials on the other. HC-308, ICSV-15 and several varieties from the PS-series are the examples of dual type varieties. In multicut types selections for slender stem and leafy characteristics are the important criteria. In dual purpose types apart from forage attributes ear characteristics like nondrupy sparse panicle is important. For example we prefer MP- Chari type earhead in dual type sorghum in order to facilitate natural bids scaring at the time of maturity. In forage bajra breeding programme multicut types possessing characteristics of late flower, fast growth, high tillering type and broad and leafy genetic materials are preferred. The multicut types are recommended to be sown in the on set of the spring season, in February. These materials fill up the gap of forage availability in summers. JHPM- Series developed at IGFRI is an example of such materials. JHPM-05-2 recommended has been for release from this Institute for entire country except for South Zone for single cut kharif season variety is having such a multicut property. In case of single cut types apart from old varieties exemplified as CO-1 and UUJ IV-1, the latest developed variety Avika Bajra Chari–19(AVKB-19) is a worth mentioning example. Avika Bajra Chari has released status and recommended for cultivation for entire country. Development of forage hybrids has been in a very slow pace. Only one forage hybrid, FMH-3, could be released as yet through CVRC so far. This is a contribution from Proagro Seed Company Limited and the said hybrid was tested in our All India Coordinated Breeding Trials as a prerequisite for release. In forage production programme the breeding approach is more centered towards availability for green forage during summers. Lucerne as such is a permanent answer to this problem. Only the thing is that in the availability of sufficient moisture in summers the crop continues to yield cut after cuts. In the irrigated areas berseem appears to be better alternative. Only the disadvantage with berseem is that it matures in April in tropical regions. Thus two major months of summers May and June remain, without berseem fodder availability. Major emphasis in berseem breeding programme is, therefore, happens to be to increase longevity of the crop that is to say summer persistence. In adjoining areas of tropical and temperate region in the Northern part of the country summer persistent materials in Berseem have been evolved as a consequence of natural selection. Some of the varieties developed at PAU, Ludhiyana for example BL-2, BL-10 and BL-22 are the examples of such summer persistent varieties. Other objectives in berseem improvement are to breed for *Rhizoctonia* root rot and *Sclerotinia* stalk rot. A few species of berseem have been identified having disease resistance characteristics for example *Trifolium apertum*, *Trifolium lapaceum* and *Trifolium vesiculosum*. Frost susceptibility is another problem in berseem which may be tackled through introgression of favorable genes from *Trifolium pratense*. Another aspects in breeding is to develop varieties for drought prone areas. The grasses eg. Buffel grass and yellow Anjan are such species. The varieties of Buffel grass Bundel Anjan-1 and CAJRI-75 and of yellow Anjan CAJRI-76 are the examples. Before we discuss details of varieties developed so for we would like to mention forage resources and grazing resources in the country.

Forage Resources

Low productivity of livestock is a matter of concern which is mainly due to poor fodder and feed resources. The country accounts for 15 per cent of the total world livestock population with only 2 per cent of the world geographical area. This is an indicative of tremendous animal pressure on this limited land resources. The major reason for the low productivity of the animals is the low availability of forages and poor in quality. To compensate for the fall in productivity, the villagers generally resort to maintain large herd of livestock, but the desirability of maintaining such a big and poor quality livestock herd is of questionable proposition.

Pastures and forests are often regarded as great reserves of land which can be ploughed up and put into cultivation; specially in areas, where population pressure become sufficiently great. This trend in agriculture is not scientifically correct and environmentally sound, as large part of the grasslands is not only unsuitable for crop cultivation but hazardous from the stand point of ecological aspects. A greater proportion of land, 150m ha has been eroded severely as a result of overgrazing which destroyed the vegetative cover and substantially reduced the absorptive capacity of the soils by the loss of top fertile soil.

Adequate and balanced feeding of animals is a major limiting factor in realizing increased livestock production. Some of the major feed resources are the herbages from cultivated fodders, grazing materials from grasslands and crop residues. In this country, animals get only about 25 per cent of their feed from grazing in woodlands and other uncultivated areas; the balance comes from crop residues unsuitable for human consumption. At least 80 per cent of the total world livestock feed consists of roughages, hay, stover and silage which are unsuitable for human food.

The present livestock population in this country as has been projected is around 420 million heads. The growth of livestock population between the periods of 1951 to 1982 was about 42 per cent that is 1.4 per cent annual growth rate. Density of livestock population over the States is presented in table 1 which indicates a relative pressure in different areas. Low animal productivity is mainly ascribed due to poor nourishment. The fodder produced in the country is ill matched to such a huge livestock population. It is difficult to find out fully reliable figures of fodder production in India, however, working estimates are available. According to the estimates given by Indian Council of Agricultural Research the availability of green and dry fodders was about 214.1 and 231.0 million tones, respectively. The present availability is around 550 mt of dry forage as derived from residues of agricultural crops and also through pasture and grazing resources and about 370 mt of green forage derived from cultivated fodders, tree loppings and crop residues on dry weight basis, this means a total available dry forage of about 643 mt. This indicates an overall availability of forages is to the extent of 60 per cent of the desired need.

Table 1 : Density of Livestock Population

State	Number of livestock units		
	Per ha of Cropped Area	**Per ha of Geographical Area**	**Per capita**
Andhra Pradesh	9.83	4.40	0.67
Assam	9.59	4.21	0.49
Bihar	10.17	6.54	0.58
Gujarat	5.77	3.28	0.51
Haryana	5.56	7.12	0.58
Himachal Pradesh	16.00	5.07	1.16
Jammu & Kashmir	16.80	3.50	0.97
Karnatka	7.12	3.99	0.64
Kerla	6.24	4.59	0.22
Madhya Pradesh	7.44	3.60	0.82
Manipur	6.93	0.69	0.36
Meghalaya	12.16	1.21	0.76
Nagaland	4.75	0.69	0.62
Orissa	7.90	4.45	0.82
Punjab	5.88	7.91	0.54
Rajasthan	7.31	3.70	1.44
Sikkim	9.57	1.21	1.00
Tamil Nadu	11.35	5.65	0.51
Tripura	8.92	3.19	0.56
Uttar Pradesh	8.80	7.27	0.51
West Bengal	11.31	9.74	0.54

Source: Hazra, 1995

As a result of continuous degradation of the forests and grassland, dry fodder production in the form of grass is expected to decrease further. The possibility of increasing dry fodder production form agricultural area is also limited. There is also no likelihood of any increase in the agricultural area devoted for cultivated fodder crops for green fodder production from the present 6.9 million hectare. The state wise area and production of cultivated fodder crops (Table2) indicated that some of the States like Rajasthan, Gujarat, Haryana, Punjab, Uttar Pradesh and Madhya Pradesh have the larger acreage under cultivated fodder crops. The total area under cultivated fodders is 8.3 m ha on individual crop basis which indicates a cropping intensity of cultivated fodders is to the extent of about 120 percent. Sorghum amongst the kharif crop (2.6 m/ha) and berseem amongst the rabi crops (1.9 m ha) occupy about 54 percent of the total cultivated fodder cropped area (Table3). Amongst the annual fodder crops, berseem and Lucerne are the highest green forage producer (60-130 t/ha in a season). Amongst perennial cultivated forages, napier- pearl millet hybrid, guinea and para grass are high yielding forages (90-200 t/ha annually). In addition to this, about 5 m t of green forage is made available from tree leaves for the animals. The gap between the demand and supply of fodders in the country is expected to further widen if effective steps are not taken immediately. However, the productivity of forages could be substantially increased by adopting suitable crop production technology.

Table 2. Area Under Fodder Crops and Estimated Fodder Production in Different States.

State/Union Territory	Area under fodder crops (thousand hectares)	Estimated green forage (million tones)
Andhra Pradesh	147	4.41
Assam	3	0.12
Bihar	14	0.56
Gujarat	877	52.62
Haryana	683	47.81
Himachal Pradesh	8	0.32
Jammu & Kashmir	28	0.98
Karnataka	86	3.44
Kerala	2	0.12
Madhy Pradesh	876	26.28
Maharastra	801	32.04
Manipur	-	-
Meghalaya	-	-

Contd...

Table 2. Contd...

Negaland	-	-
Orissa	-	-
Punjab	697	48.79
Rajasthan	3100	62.00
Sikkim	-	-
TamilNadu	124	4.34
Tripura	-	-
Uttar Pradesh	861	34.40
West Benga;	4	0.20
Andaman & Nicobar Islands	-	-
Arunachal Pradesh	-	-
Chandigarh	4	0.40
Dadra & Bagar Haveli	5	0.20
Delhi	6	0.60
Goa, Daman & Diu	-	-
Lakshadweep	-	-
Mizoram	-	-
Pondicherry	-	-
All India	8323	319.63

Source: Hazra, 1995

Table 3. Forage Crops Grown in India with Area and Productivity

Name	**Area (000 ha)**	**Productivity green forage (tones/ ha)**
1	**2**	**3**
Berseem (Egyptian clover) *Trifolium alexandrinum*	1900	60-110
Lucerne (Alfalfa) *Medicago sativa* L.	1000	60-130
Senji (Sweet clover) *Melilotus indica* L.	5	20-30
Shafttal (Persian clover) *Trifolium resuipnatum* L.	5	50-75
Metha (Fenugreek) *Trigonella foenum graecum* L.	5	20-35
Lobia (Cowpea) *Vigna unguiculata* (L.) *Walp.*	300	25-45
Guar (Cluster bean) *Cyamopsis tetragonaloba* (L.) Taub.	200	15-30

Contd...

Table 3. Contd...

Rice bean (Rice bean) *Vigna umbellate* (Thunb.) Ohwi & Ohashi	20	15-30
Jai (Oats) *Avena sativa* L.	100	35-50
Jau (Barley) *Hordeum vulgare*(L.) emend. Bowden	10	25-40
Chari (Sorghum) *Sorghum bicolor* (L.) Moench.	2600	35-70
Bajra (Pearl Millet) *Pennisetum glacum* (L.)	900	20-35
Makka (Maize) *Zea mays* L.	900	30-55
Makchari (Teosinte) *Zea mexicana* Schrad.	10	30-50
Chara sarson (Chinese cabbage) *Brassica pekinensis* (LOUR.) Rupr.	10	15-35
Napier grass (Elephant grass) *Pennisetum purpureum Schum.*	100	70-100
Guinea grass (Green panic) *Panicum maximum Jacq.*	100	70-140
Setaria grass *Setria anceps* Stapf ex Massey.	40	50-95
Deenanath grass *Pennisetum pedicellatum* Trin.	20	50-90
Para grass (Angola grass) *Bracharia mutica* (Fordk.) Stapf.	100	100-190
Ruzi grass (Congo singal grass) *B. riziziensis* Germain & Evrard	15	70-90
Singal grass(Goria)*Chrysopogon fulvus* (Spreng) Chiov.	10	50-80
Dharaf grass (Goria) *Chrysopogon fulvus* (Spreng) Chiov.	-	30-50
Anjan grass (Buffel) *Cenchrus ciliaris* L.	200	15-40
Yellow anjan grass (Birdwood) *Cenchrus setigerus Vahi.*	30	10-30
Sawan grass *Lasiurus hirsutus* (Forsk.) Boiss	100	7-25
Rhodes grass *Chloris gayana* Kunth	-	20-35
Marvel grass (Karad grass) *Dichanthium annulatum* (Forsk) Stapf.	-	20-35

Contd...

Table 3. Contd...

Sabi grass (Little para) *Urochloa mosambicensis* [Hack] Dandy	-	40-60
Sudan grass [Sudan grass] *Sorghum cudanense* [Piper] Stepf.	100	50-80
Shem [Lablab beans] *Lablab purpureus* [L.] Sweet.	10	15-30
Kulthi [Horse gram] *Dolichos biflorus* Lam.	10	15-25
Moth [Moth been] *Vigna aconitifolius* Jacq.	10	15-30
Subabool *Leucaena leucocephala* [Lam.] De Wit	-	60-120
Shevri *Sesbania sesban* [L.] Merr.	-	50-110
Agati *Sesbania grandiflora* [L.] Poir.	-	40-60
Dasrath grass [Hedge Lucerne] *Desmanthas virgatus* [L.]Willd	-	50-90
Aparjit [Butterfly pea] *Clitoria ternatea* L.	-	30-50
Siratro *Macroptilium atropurpureum* [DC.] Urb.	-	15-25
Stylo *Stylosanthes hamata* [L.] Taub	-	30-55
S. guianensis[Aubi.] SW	-	30-55
S. humilis H.B.K.	-	40-50
Asillaris *Macrotyloma axillare* [E. May.] Verde.	-	25-40
Desmodium [green leaf] *Desmodium intortum* [Mill] Urb.	-	25-45
[silver leaf] *D. uncinatum* [Jacq.] DC.	-	25-35
Centro Centrosema pubescens Benth.	-	15-35
Glycine *Glycine wiqhtill* [R. Grah. Ex Wight & Arn] Verse.	-	20-35

Forage Crop Varieties Developed

Since the inception for a breeding programme in 1960's a number of forage crop varieties have been developed. A bird's eye view of these varieties worth to specify has already been concisely presented in the beginning of this text. In this part variety developed so far crop wise, area of adaptation and their green forage yield potentials have been tabularized so as to provide an entire look to readers on the subject.

Table 4: Forage Crops Varieties Developed Along with the Area Cultivation and their Yield Potential

Crop	Variety	Area of adaptation	Green forage yield [q/ha]
1	2	3	4
	I Cultivated fodder - Legumes		
Berseem	Mescavi Wardan	Entire growing area	700-900
	(S- 99-1)	Entire growing area	800-1000
	BL –1, BL -10	Punjab, H.P. and Jammu	800-1000
	BL-2	Hill and sub- temperate	800-1050
	BL-22	Hill, subtemperate North West	900-1150
	JB-1, JB-2 & JB-3	Central India	700-950
	Bundel Berseem-2(JHB-146)	Central and North West Zone of India	800-1100
	Bundel Berseem-3	North East Zone	600-800
	UPB-10	South Zone	700-900
	JHB-ISB-86	South Zone	700-900
Lucerne	Co- 1(Perennial)	Tamilnadu and Karnataka	600-800
	T-9 (Perennial)	Entire growing area	800-950
	LLC-3 (Annual)	Entire growing area Punjab	800-1050
	LLC-5 (Annual)	H.P. and Gujrat	700-1050
	Anand-3 (Annual)	Gujrat, Rajasthan, M.P.& H.P.	600-950
	Anand -2 (Perennial)	Entire growing area	800-950
	RL-88 (Perennial) Chetak	Entire growing area	900-1100
	(Perennial)	Entire growing area	700-800
Indian clover	Safed-76, Fos-1, YSL –106	Punjab, Haryana and Himachal Pradesh	275-350
Fenu greek	T-8	Gujrat	200-350
Persian clover (Shaftal)	S H-48	Punjab	600-800
Cowpea	Russian giant	North India	300-450
	EC- 4216,Bundel lobia –2	North, West and Central India	350-400
	UPC-287, UPC-5286 & Bundel Lobia –1	Whole country	300-450
	UPC-4200		
	GFC-1, GFC-2 & GFC-3	North- East region	350-450
	Shewta (No. 988)	Gujrat state	250-350
	UPC-5287	Maharastra	300-350
	C-88	North India & Punjab	350-400
			250-350
Field bean	JLP-4	Whole India (Annual/ Biannual)	150-350
Cluster bean	Guara-80	Punjab and Haryana	200-300
	FS-277, HFG- 119, HFG-156, IGFRI –212 & IGFRI-2395-2	Entire guar growing tract	175-350
Rice bean	K-1	Bihar, West Bengal, Orissa North- East region and A.P.	200-350

Contd...

Table 4. Contd...

	II. Cultivated fodder - Cereals		
Oats	Kent, OS-6 & OS-7	Whole country (single cut/ double cut)	400-575
	OL-9	North, North- West and South hills (single/double cut)	400-525
	JHO-822	Central India (Single cut)	425-500
	UPO-94 & UPO-212	Whole country (Multi – cut)	450-575
	OL-125	Central & North – West (single cut)	400-550
	JHO-851	Entire country	500-600
	JHO-99-1	Entire country	400-550
Sorghum	PC6, PC9, PC-23 HC-171 & HC-260	Whole country (Early to medium, single cut)	350-500
	SL-44, M.P.Chari, UP. Chari 2, (Sel -278)	North India (single cut)	375-450
	UP. Chari -1 (IS-4776)	U.P., Maharastra, Tamilnadu, A.P. (single cut)	350-450
	HC-136 & Raj, chari-1 Raj- chari-2 (SU-45)	Whole country (Late single cut)	375-500
	CO-27	Tamil Nadu (two cut)	400-550
	Pant chari (UPFS-23)	U.P.(Single cut)	350-450
	Gujrat Forage	Gujrat (Two cut)	450-650
	sorghum (AS-16)	Whole country (Multi cut)	650-850
	Meethi sudan (SSG-59-3) X-988,MFSH-3 & Hara sona (855 F)	Whole country (Multi cut)	750-1050
Pearl millet	Giant bajra	Maharastra & Central India (single cut)	350-450
	Giant bajra (Multi- cut)	Entire country	700-800
	UUJ-IV-M & TNSC-1 Rajko	Entire growing tract	300-475
	JHPM-05-02	Gujrat & Rajasthan	325-450
		Entire country except South Zone	750-850
Maize	African tall	Whole country	550-800
		Whole country	350-475
	J-1006	Punjab	350-550
	A-de cuba	North East	250-440
Teosinte	Improved teosinte	North, North- West & Central	350-450
	TL-1	Punjab	400-450
Barley	DL-157,DL-417, DL-454	North & North West	150-250(Fodder) 15-50 (grain)
	Azad & DL-36	North, North- East & Central	150-200 (Fodder) 15-35(gain)

Contd...

Table 4. Contd...

III. Cultivated fodder – Other than cereals and legumes			
Brassica	Chinese cabbage & Japan rape	Entire country	150-350
IV. Cultivated fodder- Perennial grasses			
Napier bajra hybrid	Pusa giant & NB-21	Whole of India & tropics	1000-1600
	CO-1	TamilNadu, Karnataka & Gujrat	1200-1700
	IGFRI –3 & IGFRI-6	Central India, North-East hills and Northern hills	900-1600
	IGFRI-7	Hilly, Sub- humid & subtemperate region of India	1400-1700
	PBN-83	Punjab	1250-1700
	IGFRI-10	Whole country	1500-1800
	Yeshwant [RBN-9]	Maharastra	1350-1700
Guinea grass	Macuenni	Kerala (rainfed)	600-800
	Hamil	South, North, Northe-West & Central India	900-1300
	PGG-1 & PGG-9	Hill, North, North-West & Central India	850-1300
		Central India & Hills	
	PGG-13& PGG-14	Punjab	950-1400
	PGG-19	Entire country	900-1300
	JHGG-04-01		950-1450
V. Cultivated fodder- Annual grasses			
Deenath grass	PS-3, IGFRI-4-2-1 & IGFRI-43-19	Whole country	600-900
VI . Perennial Range Grasses			
Setaria	Nandi, Kazungula,	Whole of India	700-1100
	Narok PSS-1	Himachal Pradesh	550-700
Anjan grass	Bundel Anjan (IGFRI-3108) Marwar Anjan (CAZRI-75)	Arid & semi arid areas of entire country	150-300
Blue anjan	Co-1 (FS-391)	Tamil Nadu (Semi arid areas)	280-470
Yellow anjan	Marwar Dhaman (CAZRI-76)	Arid & semi arid of entire country	90-200

Contd...

Table 4. Contd...

Marvel anjan	GMG-1	Gujrat, Central & Western India	150-250
Dharaf grass	GAU-D-1	Gujrat, Central and Western India	200-350
Sen grass	Bundelsen Ghass -1	Entire Country	300-400
Dhawalu grass	Bundeldhawalu Ghass-1	Grass growing regions	250-300
Lampa grass	BundelLampa Ghass-1	Grass growing regions	250-300

References

Halden, J. B. 1930. Theoretical genetics of autopolyploids. *J. Genet,* 22:349-372.

Hazra, C. R. 1995. Advances in forage production technology. IGFRI, Jhansi.pp.2-14.

❑❑❑

Chapter 7

Forage Production Technology for Sustaining Animal Productivity

R. K. AGRAWAL, S. N. TRIPATHI AND J. P. SINGH
Indian Grassland and Fodder Research Institute, Jhansi – 284 003 (U.P.), India

Introduction

Livestock rearing is an important supplementary occupation of Indian farmers. In India the crop and livestock enterprises constitute the two functional components of mixed farming system in determining the agricultural economy and the ecological balance. Forage production is the backbone of the livestock industry and therefore green nutritious forages are essential for economic production of animal products. It has been estimated that fodders are 5-14 times cheaper source of important feed constituents than traditional practice of feeding concentrates (Agrawal *et al.*, 2008). Therefore, to maintain the current growth trend in livestock products, strategies should be orientated towards promotion of forage based cropping systems or inclusion of forages as sole, mixed or intercrop in existing cropping situation to narrow down the gap in demand and supply matching with agro – economic and environmental considerations. However due to ever-increasing pressure of human population and rising demand for food and commercial crops only about 5.5% (8.6 m/ha) of the cultivated land is devoted for fodder production in India. With this meager land allocation country meets only about 46% of the fodder requirement. Since there is very little scope for diverting extra arable land for fodder production, Therefore efforts are needed to produce adequate amount of green nutritious fodder from the available cultivated lands along with food, vegetable and commercial crops without jeopardizing the productivity of the items of human importance. This article explains the techniques for producing forage for intensive dairy farming areas and also for mixed cropping system with arable crops (food and cash crops) for proper land utilization and sustaining the animal productivity.

In India the important gifts of nature are abundant sunshine, diversified soil types, wide range of climatic conditions and a large group of forage species – all of which contribute to create congenial environment for intensive forage production programmes. The multicut nature of several forage species and flexibility of manipulating their duration are desirable traits to increase cropping/harvesting frequency.

Salient Features of Forages Crops

The knowledge of salient features of forages would be useful in understanding the techniques of their management and tailoring the cultural and fertilizer schedules for increased herbage production. These are:

- Short growth period.
- Grown in closer spacing with high seed rate.
- Dense stand, smoother weeds and prevent soil erosion.
- Improve soil health through addition of higher amounts of organic residues in the soil.
- Crop duration can be adjusted and risk due to aberrant weather conditions minimized.
- High persistency and regeneration capacity reduce the need for frequent sowing and tillage.
- Crop management differs with the purpose of growing forages and mode of their utilization.
- Wider adaptability with capacity to grow under stress conditions.
- High nutrient and water requirement under intensive cropping.
- Multi cut nature with capacity to provide regular income and employment.

- Economic viability depends on secondary production (livestock products).
- Storage, transport, processing and conservation are cumbersome.
- Shy seed producer, poor harvest index and narrow seed multiplication ratio.
- The cost of cultivation goes down in subsequent cuts in case of multi cut and/or perennial forages as well as in forage-cum-seed crops.

I. Intensive Forage Production Systems

Multiple Cropping

It consists of growing 3 to 4 appropriate annual forage crops as sole crops are in mixed stands in a calendar year to enhance forage productivity per unit area. The degree of its success depends upon agro-climatic conditions, crop and soil management practices followed and availability of inputs. It has been found that multiple crop sequences produced 120 to 190 tones of green forage/ha/year. Selection of appropriate crops/varieties and adoption of scattered sowing and harvesting schedules ensure the regular supply of the quality forage.

Combination of Perennial and Annual Forages for Year Round Forage Production

Considering the needs of dairy farmers for green fodder round the year and also small cultivators requiring maximum forage from a small piece of land, overlapping cropping system was developed at IGFRI, Jhansi which consisted of raising berseem, inter-planted with hybrid Napier in spring and intercropping the inter row spaces of the grass with cowpea during summer after the final harvest of berseem. This system was found superior to multiple crop sequences both in terms of production and economic returns. In this cropping system hybrid napier could be successfully replaced with relatively soft and palatable perennial grasses like Setaria and guinea and berseem with Lucerne wherever required.

Association of Both Perennial Grass and Legume Components

Attempts were made to select ideotypes of perennial grass and forage-legume components in order to reduce the necessity of repeated sowing and tillage and to economize irrigation water in the system. This resulted in the identification of an erect, leafy and compact hybrid napier-IGFRI No. 3 and K8 variety of *Leucaena leucocephala*. These crops when grown together in alternate paired rows (2:2) yielded around 200 tones of nutritious green forage/ha/year. This system is less sensitive to fluctuations in soil moisture and more suited to southern region of the country where both the components grow throughout the year.

These forage production techniques besides giving higher yields maintain the flow of green fodder through out the year and ensure effective utilization of land and other production inputs. The associated legumes improve the herbage quality in terms of protein and minerals and help economize on fertilizer nitrogen and sustain soil fertility over a longer period. Moreover, such production systems are less expensive and offer continuous employment potential. The component crops of the system can be changed depending upon water availability and the yield indices of the crops in a region. Similarly, cultural management practices like crop geometry, spacing, planting pattern etc. could be adjusted to facilitate use of appropriate farm machinery and effective utilization of irrigation water.

The intensive cropping systems when managed properly using modern techniques of soil and crop management are able to yield 180 to 300 tones of green fodder (30 to 55 t dry forage) per ha/year. Some of the intensive cropping systems have been suggested for different regions.

Intensive Forage Sequences Recommended for Different Regions

North Zone

1. Maize + Cowpea - Sorghum + Cowpea (two cuts) - Berseem + Mustard.
2. Sudan grass + Cowpea - Maize + Cowpea - Turnip - Oat (two cuts).
3. Hybrid napier or Setaria inter-planted with cowpea in summer and berseem in wither (9 to 10 cuts/year).
4. Teosinte + Cowpea (two cuts) - Carrot - Oat + Mustard/Senji (two cuts).

Western and Central Zone

1. Bajra + Guar (two cuts) - Annual Lucerne (6 cuts).
2. M.P. Chari + Cowpea (two cuts) - Maize + Cowpea _ Teosinte + Cowpea (two cuts).
3. Hybrid napier or Guinea or Setaria grass inter-planted with Cowpea in summer + Berseem in winter (8-9 cuts/year).
4. Hybrid napier or Guinea grass or Setaria inter planted with Lucerne (8-9 cuts/year).

Southern Zone

1. Sorghum + Cowpea (3 cuts) - Maize + Cowpea - Maize + Cowpea.
2. Hybrid napier or guinea or *setaria* grass inter-planted with lucerne (8-9 cuts) or Hybrid napier + subabul/ *sesbania* (9-11 cuts/year).

3. Sudan grass + cowpea (3 cuts) – M.P. Chari + Cowpea (three cuts).
4. Para grass + *Centrosema pubescens* (9-11 cuts/year).

Eastern Zone

1. Maize + cowpea – teosinte + rice bean (two cuts) – berseem + mustard (3 cuts).
2. M.P. Chari + cowpea – dinanath grass (two cuts) – berseem + mustard (3 cuts).
3. Para grass + *Centrosema pubescens* (8-9 cuts/year).
4. Hybrid napier or setaria inter-planted subabul or *sesbania sesban* (9-10 cuts/year).

II. Integration of Forage Production with Food and Other Crops

This is an important aspect of fodder production, which can be utilized by all classes of farmers both under irrigated and rainfed situations. In this system appropriate forage species are included in the crop production program of the farmers in the following ways.

Inclusion of Forages in Crop Sequences

On medium and heavy soils in rainfed areas, only Rabi crops are taken on conserved moisture through fallowing in rainy season. In this situation short duration forage crops like cowpea, rice bean, and pearl millet can be taken. This will conserve soil and suppress weeds besides providing forage. Under irrigated conditions inclusion of fodder crops in sequences improves the soil productivity due to dense canopy and addition large amounts of stubble biomass. Leguminous forages enrich the soil nitrogen also.

Examples: Paddy – berseem; maize + cowpea (F) – wheat/potato.

As Catch Crops

Short duration forages can be grown in gap periods of main season crops and on residual moisture. *Examples*: Maize + Cowpea/Sorghum/Pearl millet after harvest of Rabi season crops and before planting of rainy season crops. Similarly, turnips, carrot and mustard can be taken as catch crops after the harvest of early rainy season crops. Leguminous forages like horse gram, moth and lathyrus are suitable for cultivation on residual moisture after the harvest of late paddy.

As Intercrops in Widely Spaced Row Crops

Cotton, sugarcane a grain crops of maize and sorghum offer scope for growing short statured forage like cowpea, moth and guar in the inter spaces without affecting their yields. Some of these forage benefit the main crop indirectly for example moth reduces root rot disease while cowpea controls spotted bollworm. Similarly, pangola grass in tobacco has been reported to control root knot nematode.

Through Ratooning

Crop like pearl millet (cv. BJ-104) and barley (cv. DL-36, 157, 457) have good regeneration capacity. These crops can be harvested 40-50 days after sowing to harvest green forage yields (100-150 q/ha) without adversely affecting the grain yields. Like wise maize (grain) can be planted at 30 cm spacing, alternate rows may be harvested for forage 45-50 days after planting to get around 200 q of green forage/ha.

Dual-Purpose Crops

These crops give forage as byproduct. The farmers should give preference to such crops in their cropping schemes so as to get substantial amounts of forage along with main product.

Among vegetables -Turnip, carrot, pea, cauliflower, cowpea, guar etc. are the important examples while in commercial crops maize for cobs and sugarcane are important. These crops besides main product yield 90 to 200 q/ha green fodder for animal feeding.

Mixed Intercropping System of Forage Production

It consists of growing together two or more plant species with different growth habit, canopy structure, rooting pattern and offering little or no mutual competition. Mixed cropping of cereals/grasses and legumes, is commonly practiced to improve herbage quality, increase biomass production and economize on fertilizer usage. Besides, it also ensures efficient land utilization, suppression of weeds and insurance against aberrant weather conditions.

In India with warm climate most of the grasses and cereals do not have enough protein to support animal growth and production. It has been found that by associative cropping of graminaceous and leguminous forages the nutritive value of the mixed herbage can be greatly improved. Thus, besides enriching the soil productivity and ensuring better utilization of plant nutrients from different soil layers, mixed cropping produces nutritious herbage since the legume component contains higher amounts of protein, calcium and phosphorus. Investigations on intercropping systems showed that Sorghum/ Maize/ *Pennisetum pedicellatum*/ Pearl millet with Cowpea/Rice bean and Oat with India clover/pea in paired rows (2:2) recorded well forage yields along with higher forage protein (Table 1).

Table 1 : Forage and Crude Protein Yields Under Pure and Mixed Cropping of Different Fodder Crops.

Crop/Crop combination	Nutrient level (Kg/ha)		Forage yield (q/ha)		Crude protein yield (Kg/ha)
	N	P_2O_5	Green forage	Dry matter	
Oat (Kent) pure	90	30	433	90.4	564
Oat (Kent) + Senji	55	45	412	87.8	1034
Oat (Kent) + Pea	55	45	398	86.8	1067
Sorghum (Pioneer 988) pure	120	50	490	101.8	680
Sorghum (Pioneer 988) + cowpea	70	55	493	101.0	1024
Sorghum (M.P. Chari) pure	120	50	384	82.0	490
Sorghum (M.P. Chari) + Cowpea	70	55	391	81.0	886
Maize (African tall) pure	120	50	448	93.3	722
Maize (African tall) + Cowpea	70	55	454	93.0	1049
Maize (Manjri)) pure	120	50	323	65.0	553
Maize (Manjri) + Cowpea	70	55	362	71.8	872
Bajra (Rajko) pure	120	50	372	82.7	632
Bajra (Rajko) + Cowpea	70	55	430	89.2	971
Pennisetum pedicellatum pure	90	30	463	77.2	658
P.pedicellatum + Cowpea	55	45	435	76.8	942
P.pedicell atum + guar	55	45	431	75.4	836

III. Forages on Bunds

Bunds around fields are a common feature on cultivated lands and occupy 2 to 10% of the cultivated area. Bunds are formed to demarcate field boundaries, conserve water and soil and to use as pathways. If not properly protected, the bunds are eroded by rains and damaged by animals. Growing of appropriate forages/bushes stabilizes the bunds through protective cover and soil binding action of roots, besides providing considerable amount of forage for feeding animals.

Experiments conducted at IGFRI, Jhansi showed that sunhemp yielded maximum forage among annual species while *Sesbania* gave highest forage among perennials (Table 2). Similarly under rainfed conditions planting of hybrid napier + stylo along with fruit trees like *aonla, ber* and karonda recorded maximum forage yields ranging from 44 to 48 kg dry matter/15 meter length of bund [Khan *et. al.,*2001].

Table 2 : Productivity of Forages on Rice Bunds

S.No.	Crops	Yield of forages (kg/100 rm)			Rice grain
		Green	Dry	C.P	(t/ha)
1.	Pure rice	-	-	-	4.5
2.	Rice with *S.sesban*	80.25	18.05	3.98	4.2
3.	Rice with *Leucaena*	58.15	17.57	3.61	4.0
4.	Rice with *Pigeonpea*	67.50	23.08	4.40	4.2
5.	Rice with *Sunhemp*	146.50	38.76	3.10	4.6
6.	Rice with *Patsan*	144.50	41.40	2.90	4.2
7.	Rice with *ricebean*	126.50	18.28	4.40	4.4
8.	Rice with *S.sphacelata*	39.00	6.60	0.30	4.3
	CD at 5% P	26.80	11.70	1.50	N.S.

IV. Principles and Practices of Growing Important Irrigated Fodder Crops

Fodders are the plant species that are cultivated and harvested for feeding the animals in the form of forage (cut green and fed fresh), silage (preserved under anaerobic condition) and hay (dehydrated green fodder). Crop production is concerned with the utilization of plant morphological and physiological responses within an agro-climatic environment to produce maximum yield per unit area and time. The development of crop production technologies is the master key to unlock the yield potential of crops. Fodders as a group of crops differ from food and commercial crops in several aspects; the principles and practices of their cultivation vary accordingly.

Water is the most important input for crop production especially in fodder crops where maximum vegetative growth is desired within a short period of time. Provision of irrigation permits maximum utilization of resources for intensive forage production, which is very important in our country with small land holdings. There are certain fodder crops like berseem, lucerne, turnip etc. which can be grown only with adequate moisture while in case of others the production potential can not be fully realized without irrigation.

V. Critical Inputs in Forage Crop Production

The forage production per unit area is a consequence of the interactions between genotypes and environment. A crop environment may be regarded as having two components, the gross environment, which takes into environmental factors affecting crop growth, and the current environment, which takes into account the general soil and atmospheric conditions outside the crop and also the changes caused by the plant community (Eddowes, 1969). However, the phenomenal increase in productivity is possible mainly through better seeds, fertilizers and better agronomic technology (Menhi Lal *et. al.*, 1994).

Improved Varieties

The forage crop production aims at increasing yield along with improvement in quality traits to enhance ultimately the secondary productivity (animal products). Therefore, the forage crop varieties should possess the following desirable characteristics (Menhi Lal, 1993).

- Growth rhythm with early bulkiness and matching physiology for higher rate of dry matter accumulation and longer leaf area duration.
- Compatibility with associated crops in mixed/intercropping systems.
- High palatability, greater digestibility, better nutritive value and free from anti-quality constituents.
- Quick regeneration potential, flexibility to defoliation management and greater persistency.
- Ability to withstand aberrant weather and adverse soil conditions such as drought, water logging, saline-sodic and acidic soils.
- Multiple resistances to diseases and insect pests.
- High nodulating and non-structural carbohydrates reserve abilities in forage legumes for subsequent regeneration following frequent cuttings.
- Desirable characteristics for making good quality silage and hay.
- Suitability for fitting in the prevalent cropping systems.
- Acceptable seed production ability and/or dual-purpose types.

Cultural Management

The following points need to be given due consideration while adopting appropriate cultural practices (Tripathi, 1990).

1. Efficient forage crops of the area should be identified and grown on properly prepared seedbeds well in time. Timely sowing is very important in forage crops as these have very limited growth period.

2. Since forage crops need higher amounts of seeds and/or planting materials, these should be obtained from reliable sources and germination should be tested before sowing.
3. The seed rate and spacing should be adjusted depending upon the purpose and growing conditions. For instance, higher seed rates are used when the crop is raised for hay or the field is likely to be infested with weeds.
4. During rainy season, sometimes, it becomes difficult to re-sow the field due to incessant rains and wetting of the soil. Under such conditions, multicut forage crops can be sown in the beginning for continuous supply of green forage throughout summer and rainy seasons. Moreover, growing of multi cut forage crops/varieties results in the saving of seeds, fertilizers and tillage operations.
5. Mixed cropping of graminaceous and leguminous fodder crops is better from the view point of herbage quality and maintenance of soil fertility. Both the crop components of the system should be of similar duration to facilitate harvesting at one time.
6. Maximum crop-weed competition occurs upto 4-5 weeks in most of the seasonal forages. Weed menace should be checked from very beginning through proper land preparation, use of clean seed and well decomposed FYM, pre-emergence application of herbicides and growing of smoother intercrops.
7. Nutrient management practices should always be based on soil test values and cropping systems being followed. The fertilizer recommendations are of general nature. Use of bulky organic manures @ 15-20 t/ha takes care of soil health and micronutrient deficiencies to a large extent. Economy in nutrient use can be ensured though efficient weed management, use of bio-fertilizers and inclusion of legumes in crop sequences.
8. Irrigation management practices vary with season, soil type, crop growth stage and fertilizer practices. Water requirement of fodder crops is usually greater because of dense stand and higher vegetative growth in a short period. Similarly, adequate drainage is important for crops like maize, pearl millet and cowpea etc. grown in the rainy season. Forage crops respond well to sewage irrigation.
9. For controlling insect pests and diseases, it is better to avoid the use of chemicals particularly when the harvesting is due. However, if any pesticide is applied on the crop, the forage should not be harvested for 3-4 weeks to avoid the risk of residual toxicity in animals.

10. Cutting management influences not only the yield but also the forage quality. The fodder crops, should therefore, be harvested at appropriate growth stage to obtain adequate fresh fodder with acceptable dry matter nutrients particularly the crude protein.

11. Multi cut forages especially those regenerating from auxiliary buds should be harvested leaving 8-10 cm stubbles to promote quick regeneration and adequate stand of the ratoon. Cultural management practices of important fodder crops are given in table 3.

Table 3 : Package of Practices for Important Forage Crops

crop	Sowing time	Seed rate (kg ha^{-1}) spacing (cm)	Irrigation (nos.)	Harvesting schedule (Days)	Green forage yield (t ha^{-1})
Sorghum Multicut	March - April	40-50 kg 20-25 kg in Sudan type sown in 25 cm apart rows	Summer 5-6 Monsoon-depending on rains	1st : 55-60 Subsequent 35-40	80-90
Single cut	June-July	40-50 kg. sown in 25 cm apart rows	Monsoon-depending on rains	70-75	35-40
Maize Single cut	March – April June – July	50-60 sown in 25 cm apart rows	Summer 7-8 Monsoon depending on rains	60-75	35-45
Bajra Multicut	March – April	8-10,sown in 25 cm apart rows	Summer 4-6 Monsoon depending on rains	50-55 Subsequent 30 – 35	90-100
Single cut	June – July	8-10,sown in 25 cm apart rows	depending on rains	60-70	40-45
Teosinte Multicut	March – April	35-40 sown in 25 cm apart rows	Summer 7-8 Monsoon depending on rains	75-80 Subsequent 35-40	80-85
Single cut	June – July	35-40 sown in 25 cm apart rows	depending on rains	90-95	30-35
Cowpea summer crop	March – April	40-45 sown in 25 cm apart rows	Summer 6-7	70-75	30-35

Contd...

Table 3. Contd...

Rainy Crop	June – July	40-45 sown in 25 cm apart rows	depending on rains	60-65	30-35
Rice bean	March – July	30-35 sown in 25-30 cm apart rows	Summer 5-6	75-90	30-35
Winter Season Forage Crops					
Berseem	October	25 kg berseem + 2 kg mustard sown broadcast on muddy seedbed	12-15	1st : 50 – 55 Subsequent 30 – 35	100 - 120
Lucerne	September – October	15-20 sown in 25 cm apart rows	15 – 20	1st : 60 – 75 Subsequent 30 – 40	75 – 85
Oat Multicut	October – Last Nov.	100 sown in 25 cm apart rows	7-8	1st : 65 days Subsequent 30 – 40	
Single cut	October – Last Nov.	100 sown in 25 cm apart rows	3-4	75 days	40 – 45
Perennial Forage Crops					
Hybrid napier	Mid Feb. – July	34000 rooted slips planted at 100 x 30 cm spacing	During spring & summer 8-9	1st : 75 – 80 Subsequent 30– 40	150 – 160
Guinea grass	Mid Feb. – July	4-5 kg Seed /34000 rooted slips planted at 100 x 30 cm spacing	During spring & Summer 9-10	1st : 75 – 80 Subsequent 30– 40	100 – 120
Para grass	March – July	12 – 15 quintals stem cuttings planted at 50 x 25 cm spacing	During summer 10-12	1st : 70 – 80 Subsequent 30– 40	130 - 140

Fertilizer Management

The requirement of fodder crops for nutrients particularly nitrogen is comparatively higher. This is due to the fact that fodder crops are desired to produce luxuriant vegetative growth with succulent and nutritive herbage in a short period. Thus, the fertilizer management strategies in fodder crops aim at increasing the herbage production per unit area and time along with improvement in quality parameters. Based on the experimental evidences at IGFRI, Jhansi, the fertilizer recommendations for important fodder crops is given in Table 4 (*Menhi Lal and Tripathi*, 1987).

Water Management

Scientific water management will hold the key to intensive agriculture in the years to come. Irrigated agriculture, water management aims to provide suitable moisture environment to the crops to obtain optimum yields commensurate with maximum economy in irrigation water an maintenance of soil productivity. The limited availability of water enables only 30 per cent of country's cropped area under irrigation and that too lion's share goes to food/cash crops. In view of mounting human and livestock population, the forage farming is faced with twin problems of meager allocation of land and water resources.

Table 4 : Fertilizer Recommendation for Important Forage Crops

Crop	Nutrient requirement (kg ha^{-1})			Basal dose (kg ha^{-1})			Top dressing of N (kg ha^{-1})	Time of top dressing (days after sowing)
	N	P_2O_5	K_2O	N	P_2O_5	K_2O		
Sorghum (single cut)	90	30	-	60	30	-	30	40
Sorghum (two cuts)	120	30	-	70	30	-	50	After 1st cut
Sorghum multicut (4 cuts)	210	60	60	60	60	60	50+50+50	After 1st, 2nd & 3 cuts
Maize	120	40	-	80	40	0	40	40
Teosinte	130	30	-	50	30	-	40+40	After 35 & 60 days of sowing
Oat (single cut)	90	30	-	60	30	-	30	40
Oat (two cuts)	120	40	-	80	40	-	40	After 1st cut
Oat multicut (4 cuts)	180	60	40	60	60	40	40+40+40	After 1st, 2nd & 3rd cuts
Berseem	20	80	-	20	80	-	-	-
Lucerne	20	120	40	20	120	40	-	-
Cowpea	20	60	-	20	60	-	-	-
Hybrid napier	250	50	50	50	50	50	50+50+50+50	After each cut-
Guinea grass/Para grass	250	50	50	40	50	50	40+40+40+40	After each cut

Schedules offers an important approach to water saving because timely supply of water to crops in adequate quantity is the crux of the efficient water management for ensuring optimum yield and quality of produce. The suitable soil moisture environment for forage crops where objectives tend to be luxuriant vegetative growth, higher dry matter yield and better forage quality have been worked out in table 5. The optimum soil moisture regime for berseem, oat, maize, *teosinte*, sorghum, hybrid napier and cowpea has been found to be 75% available soil moisture (ASM). Lucerne, cluster bean and barley may be irrigated at 50% ASM, whereas, triticale can be grown at 25% ASM with acceptable yield.

Table 5 : Irrigation Requirement and Water use Efficiency of Important Forage Crops.

Crop	*Optimum soil moisture regime for irrigation (% ASM)	Irrigation interval (days)	No. of irrigati on	Irrigation req. (Delta of water, mm)	Water use efficiency (Kg dm ha^{-1} mm^{-1})
Berseem	75	10-12	16-18	710	22
Lucerne	50	14-18	11-13	600	14
Oat	75	12-14	6-8	340	32
Sorghum (M.P. Chari)	75	9-12 (S)	8	380	30
		12-15 (S)	1-2	40-80	45
Teosinte	75	9-12 (S)	8	380	20
Cowpea	75	9-12 (S)	8	380	18
Clusterbean	50	14-18 (S)	5	270	22
Hybrid napier	75	11-15	10	380	68
Triticale	25	26-30	2-3	110	30
Barley	25	26-30	2-3	110	30
Turnip	75	10-12	3-4	140	34
Carrot	75	10-12	7-8	380	27

*Optimum moisture regime as worked out in loam soil having 11% available water for top 45 cm depth. ASM-Available soil moisture, S-Summer season, R-Rainy season

References

Agrawal, R. K., Sunil Kumar, Tripathi, S. B., Sharma, R. K. and Singh, K. A. 2008. Agro-economic analysis of fodder crop based cropping systems. Indian journal of fertilizers 4:121-131.

Eddowes, M. 1969. *Crop Technology*. Hutchinson Educational Ltd. London, U.K.

Menhi Lal, 1993. Recent advances in agronomy of forage crops. Paper presented during training programme on "*Agroforestry, Forage Production and Animal Nutrition*" organized at IGFRI, Jhansi from 5th May to 4th June, 1993.

Menhi Lal, Shukla, N.P., Tripathi, S.N., Sinsinwar, B.S. Niranjan, K.P., Gupta, S.D. and Arya, R.L. 1994. *Forage Production Technology*, Technical Bulletin IGFRI, Jhansi.

Menhi Lal and Tripathi, S.N. 1987. Role of fertilizer in the production of forage and fodder crops. *Fertilizer News* 32:77-83.

Tripathi, S.N. 1990. Principles and practices of forage crop production. Lecture Notes, Summer Institute on "*Forage Production Technology*" organized at IGFRI, Jhansi from 2-21 May.

Khan, T.A., Sharma, S.K., Suresh, G. and Ranjitha, P., 2001. Performance of some fruit and forage species on bunds. *Range Management & Agroforestry* 22:43-50.

❑❑❑

Chapter **8**

Forage Production Strategies for Problematic Soils

S. B. TRIPATHI
Indian Grassland and Fodder Research Institute Jhansi-284 003 (U.P.), India

Introduction

Agriculture and animal husbandry both are interrelated enterprises. Forages are one of the important components of farming system, which provides nutritionally adequate diet for improving animal productivity. With increased demand of food to the vast human population the more area under arable lands for forage production is not possible. The option is only to cultivate forages/grasses in problematic soils. These lands are economically not suitable for the raising of food crops and can be put under forage production. In India about 173.65 m ha lands (50% of the total area) suffers from salinity/alkalinity, acidity, calcareousness, water logging, ravines and gullies due to erosion and other physical and chemical constraints etc (Table-1).

Table 1: Problem Soils in India

Problem soils	Million ha.
Water and wind erosion	144.13
Reverine and terrain	2.73
Alkali soil	3.88
Saline and Coastal saline	5.50
Waterlogged	8.53
Ravine and gullies	3.97
Shifting cultivation	4.91
Total	173.65

(Department of Agriculture, Govt. of India 1990).

Problem Soils and Their Limitations

Saline soils: Electrical conductivity of soluble salts is more than 4 mmhos/cm and exchangeable sodium less than 15 per cent. The higher content of salt (more than 1%) depresses water availability to crop plants due to higher osmotic pressure of the soil solution. Soil pH is less than 8.5.

Alkali soils: Exchangeable sodium is more than 15 percent, electrical conductivity less than 4 mmhos/cm. The carbonate and bicarbonate of sodium salts are present in excess. Kanker layers are found in the lower soil horizons and create waterlogging condition. The growth of plant suffers due to excess concentration of sodium and impeded drainage. Soil pH ranges from 8.5 to 10.0.

Saline-Sodic Soils: Exchangeable sodium and electrical conductivity of soluble salts is found greater than 15 percent and 4 mmhos/cm, respectively. Soil pH is more than 8.5. The excess of sodium salts like sodium carbonate and sodium bicarbonates have problem in field preparation and soil aeration.

Acidic soils: Soil pH ranges from 4.5-6.5. The toxicity of Al, Fe and Mn and deficiency of Ca, Mg, P and Mo at soil pH below 5.5 adversely affects crop growth and yield. The soil microorganisms do not flourish well. The physical properties of soil are not involved in limiting crop yield.

Waterlogged soils: The continuous availability of excess water in root zone causes poor aeration, which in turn results into poor growth and nutrient uptake. The nitrification process is found to be low.

Calcareous soils: The production of certain crops is restricted because of less availability of plant nutrients viz: P, S, Fe, Mn in presence of high concentration of calcium carbonate of the soil The presence of kankar pan in lower soil layers restrict drainage. Soils are very hard and have aeration problem.

Desert soils: Soils are sandy, covers with sand sheets of variable thickness, dunes and sand hummocks. Evaporation exceeds the mean annual rainfall, which creates problem of low moisture availability in surface soil layer. Wind erosion is a serious problem. Soils are single grained, non-coherent and structure less. Their infiltration rates are higher (30 to 60 mm/hr). The available water capacity is low from (3 to 4.5 percent). Such soils are generally deficient in plant nutrients and have poor water holding capacity. Some soils are in high percentage of soluble salts.

Hill soils: Soils are sloppy and easily drained. Sub-mountain and sub tropical low hill soils are shallow and light textured with pH range from 6.5 to 7.5. The mid hill sub-humid soils are deficient in N, P and have poor water holding capacity. Hill wet zone soils are characterized as forest soils, light in texture with pH 5.5-8.0, rich in organic matter, N, P and K and deficient in micronutrients. Soils of high hills temperate zone are neutral to alkaline, sandy loam, low in fertility and respond to N and P application. Soil erosion through water is a serious problem.

Forage Production Strategies

Soils have different capabilities for different uses. Depending upon textural classes and other limitations like erosion, compaction, salinization, slope, depth, nutrient deficiency, stoniness and out growth of rocks etc the following practices are required for increasing production potential of these lands.

- Selection of suitable crop /grass /tree species according to soil environments.
- Land management through use of suitable agro-techniques like alternate crop planning, soil and water conservation measures, balanced use of plant nutrients application and the use of amendments etc.

Selection of Crop Genotypes and Their Varieties

Selection of crop genotypes and their varieties according to soil environment is important criteria for improving soil productivity .The research work carried out by the various scientists at IGFRI, Jhansi and other places the following crop genotypes are recommended for producing higher forage yield (Hazra, 1994, Singh *et al,* 1998 and Shukla and Tripathi, 1996).

Saline Soils

Fodder crops: Sudan grass (SSG-59-3), sorghum (HC-136,171), oat (Kent,), Teosinte (TL-16), Napier-bajra hybrid (IGFRI-6 & 10), Lucerne (T-9) and Senji (FOS-1).

Grasses: Cenchrus ciliaris, Saccharum spontanium, Penesetum pedicellatum, Dicanthium annulatum, Stylosanthes hamata, Phaseolus atroperpureus.

Tree shrubs: Prosopis juliflora, Acacia farnesiana, Acacia nilotica, A. tortilis, Tamarix articulata, Grevillia robusta, Azadirachta indica, Leucaena leucocephala.

Coastal Saline Soils

Bracharia decumbense, B. brizantha, B. mutica, Coix lacrymajobi, Panicum sps. (Hamil, PGG-4), oat (Kent, JHO-851 & 822) and Barley (DL-36 & Azad).

Sodic Soils

Fodder crops: Berseem (Mescavi), shaftal (SH-69), sorghum (PC-6 & 23), oat (Kent, JHO-822), Senji (FOS-1, Safed-16). Napier-bajra hybrid (IGFRI-6).

Grasses: Leptochloa fosca, Cynodon dactylon, Bracharia mutica, Pannisetum sps., Chloris gayana etc.

Shrubs/Trees: Prosopis juliflora, Acacia nilotica, Tamarix anticuta,Casurina equisetifolia, Terminalia arjuna, Dalbergia sissoo, Sesbania sesbans, Albizia lebbeck, Zizyphus mauritiana etc.

Acidic Soils

Fodder crops: Cowpea (EC4216, UPC 5286 & 4200), Rice bean (K-1 & 6), Oat (Kent, JHO-810), Maize (African Tall), *Pennesetum pedicellatum (Bundel 1 & 2), Napier bajra hybrid (IGFRI-7 & 10).*

Trees/Shrubs: Sesbania grandiflora, Leucaena leucocephala, Albizia lebbeck, Dalbergia sissoo, Grevilla robusta and Casuarina sps. etc.

Grasses: Setaria sphacelata (Set 125, Narok, Nandi), Pennesetum pedicellatum, *Panicum maximum*, Kikuii grass and *Paspalum notatum* etc.

Calcareous Soils

Fodder crops: Sorghum (PC-6, HC-136), Pearl millet (L-72, Rajko), Napier-bajra hybrid (IGFRI-6 & 7), Teosinte (TL-16), Lablab bean (JLP-4).

Grasses: Cenchrus ciliaris, Panicum antidotale, Dicanthium annulatum, Heteoropogon contortus, Penesetum pedicellatum, Stylosanthes hamata, Macroptilium atropurpurium, Clitiria ternatea, Cenchrus setigerus, lasiurus sindicus, Saccharum munja and Sehima nervosum.

Marshy/Wetlands

Fodder crops: Teosinte (TL-6), Sorghum (PC-9).

Grasses: *Brachiaria mutica, B. brizantha, Coix sps., Iseilema laxum, Chloris gayana, Almon grass (Echinochloa polyptachya,) and Vitiveria Zizadoides etc.*

Degraded Forests Ravines

Pennisetum pedicellatum Panicum maximum (PGG-1, 14&Hamil), Stylosanthes hamata, S.guianensis, Chrysopogon fulvus, Dichrostachys nutans, Leucaena leucocephala, Prosopis juliflora, Acacia nilotica etc.

Desert/Arid Soils

Fodder crops: Pearlmillet, guar, moth bean.

Grasses: *Cenchrus ciliaris, Cenchrus setigetus, Lasiurus sindicus, Coligonum polygonoida, Saccharum munja, Clitoria ternatea, Eulaliopsis binata, Stylosanthes hamata.*

Shrubs/Trees: Acacia tortilis, Acacia senegal, Prosopis juliflora, Dichrostachrys cineria, Hordwickia binata, Zizyphus mauritiana, Tocomella undulata, Albizia lebbeck.

Shallow Soils and Rock Terrains

Hilly Terrains

Grasses: *Chrysopogon fulvus, Sehima nervosum, Heteoropogon contortus, Pennesetum pedicellatum, Themada quadrivalis, Stylosanthes hamata, Atylosia, Indigofera astragalina.*

Shrubs/Trees: *Anona squasoma, Anoglissus pendula, Butea monosperma, Holoptelia*

Dry Plain

Grasses: *Cenchrus ciliaris, C. setigerus, Dicanthium annulatum, Panicum maximum, Setaria sphacelata, Stylosanthes hamata, Clitoria ternatea, Macroptilium atropurpurium etc.*

Shrubs/Trees: *Dalbergia sissoo, Albizia amara, Leucaena leucocephala, Tectona grantis etc*

Moist Plain

Grasses: Dicanthium annulatum, Iseilema laxum, Bracharia mutica, B. intermedia Cynodon dactylon, Stylosanthes hamata, S. humilis, S. scarba, Macroptilium sps.

Shrubs/Trees: Leucaena leucocephala, Dichrostachys cinerea, Emblica officinalis, Dalbergia sissoo etc.

Mine Spoil Soils

Grasses : Penesetum purpureum, Saccharum spontaneum, Cenchrus setigerus, Cynodon dactylon, Dicanthium annulatum.

Shrubs/Trees: *Grevillea pteridifolia, Acacia auricutiformis, A. nilotica, A. tortilis, Dalbergia sissoo, Leucaena leucocephala, Prosopis juliflora, Zizyphus nummularia, Tamarix articulata etc.*

Land Management Practices

The soil, water and vegetation are the basic components of land resources for the agriculture and livestock production. Therefore, land management through various agro-techniques like crop planning, soil- water conservation measures, conservation tillage (energy management, integrated nutrient supply systems including green manure crops and pulses and use of soil ameliorators etc are important land management practices for improving land productivity.

Soil Moisture Conservation

Mulching : It is useful against crust formation by absorbing rainwater drops for the longer period and checks the growth of fungal which adversely affects growth and establishment of legume seedlings. The use of dry grass mulch @ 5 ton/ha increased seedling emergence and forage productivity of pearl millet, maize and sorghum in calcareous red soils (Hazra, 1986)

Hydrophilic polymer : Jalshakti absorbs and releases water for the growth of plants and its application @ 2.5 – 5.0 kg/ha is found quite useful for sorghum, oat and barley grown in degraded soils Menhi Lal and Shukla, 1992)

Less tillage and minimum tillage : Coarse textured red soils is superior than normal tillage for higher yield of sorghum oat and barley .The preparation of bunds across the slope, land leveling, summer deep ploughing opening of dead furrow fodder etc. are most important practices of run off collection *in situ* and watershed areas under degraded lands (Punjab Singh, 1989).

Use of non-monetary inputs : Sowing dates, adoption of intercropping of range legumes, sowing across the slope, cutting management according to soil moisture availability and selection of early maturity crop variety etc. are important non monetary inputs.

Alternative Crops Planning

Cultivation of Forage/ Grasses

In limited irrigation water and drought conditions, the most tolerant to drought and less water requiring crops are barley, senji, mustard and turnip during Rabi season and cowpea, sorghum, guar and *pennesetum sps* etc during kharif season (Hazra, 1992). Similarly,in Kharif season suitable agronomic practices have been selected i.e. early sowing at proper soil depth, optimum seed rate and row spacing, balanced dose of nutrients application on soil test basis.

The inclusion of legume fodder crops in cereal-cereal based crop rotation not only promotes soil fertility but improves quality of fodder also. The crop rotation paddy-berseem is superior to paddy-wheat in saline/ sodic soils for higher benefit(Singh,*et.al.*, 1998).

Para grass performed well under standing water than other grasses. For controlling erosion, *Eulaliopsis binata, Pancum antidotale, Dicanthium annulatum* and *Cenchrus ciliaris* are more important grasses. *Colligonum polygonoida* is superior for higher sand binder than other grasses (Shukla and Tripathi, 1996)

Agroforestry/ Silvipostoral Production System

Tree legumes have been intensively identified for their feed value *viz Leucaena leucocephala, Calliandra calothyrsus, Sesbania sesban and Gliricida sepium* etc for high crude protein (20-30%) and thus have been use for optimizing animal production (Deb Roy, 1974). The importance of silvipastural production system for higher biomass forage yield and soil fertility improvementwas also emphasized by Hazra, 1992).

Agroforestry/ Silvipostoral production system and their various facets when compared with traditional land use have been found to yield more than 7 times with year round forage availability in degraded lands. The practice for the adoption of Silvipostoral system along with soil and water conservation measures in sloppy lands of arid and hill region is more ruminative by giving 2-3 folds higher total yields than natural grassland (Hazra and Singh, 1992).

Small addition of fertilization (20-40kgN\ha) helped to a great extent in realizing the higher yield of *Cenchrus ciliaris*+ *Clitoria ternatea* under without tree (8.38t/ha) and with tree (7.76t/ha) plantation over *Cenchrus ciliaris* (4.85t/ha) in pure grassland of semi tropics(Hazra and Tripathi, 1994). According to Deb Roy (1986) the over grazing and unchecked continuous movement of animals in rangelands the carrying capacity has gone down. On well-developed pastures in semi arid *Cenchrus setigerus* could support to have up to 2.1 cow units /ha than *Cenchrus ciliaris (1.4), Dicanthium (1.2) and Chrysopogon fulvus (0.8).*

The tree lopping for top feeds during lean periods gave higher yield of 2.1-3.0t/ha fodders and 4.5 - 6.0t/ha firewood in case of *Albizia amara* than 1.0t/ha forage and 1.74t/ha firewood in case of *Hardwickia binata.* Thus lands could produce sufficient forage and fire wood biomass to improve the rural economy (Singh, 1988)

The cultivation of paddy and berseem under *Acacia* tree plantation in sodic soils and Cenchrus ciliaris + Stylosanthes hamata under Hardwickia tree plantation in degraded land of semi tropics are the most promising production systems (Kaur *et.al.*, 2000).

In hilly region of sloppy soils, the practice for the growing of annual crops at less slope silvipostoral / hortipastoral at medium slope and fuel-fodder trees at high slope is recommended (Hazra, 1994)

Use of Amendments and Fertilizer Application

The degraded lands are generally poor in organic matter, available nutrients and physical properties of soils like compaction, porosity and water holding capacity. The addition of organic amendments like FYM, decomposed crop/ grass waste residues would be useful for improving soil physical, chemical and biological properties. The use of suitable amendments is important in minimizing the toxicities and deficiencies of the plant nutrients (Singh *et.al.*, 1998)

Salts in saline soils are removed by leaching but salts in alkali/ sodic soils, these are removed through use of amendments like gypsum, pyrite and sulphur (chemical amendments) and FYM and press mud (organic amendments). The application of gypsum @ 5-10 t/ha is recommended (Tripathi and Hazra, 1996). Depending upon alkalinity and sodacity the application of FYM @ 10-20 t /ha is recommended for higher biomass yield of forages/ grasses (Singh *et.al.*, 1998). The combined use of gypsum (5 t/ha) and FYM (10t/ha) gives higher grass productivity as well as considerable reduction of the sodicity/

alkalinity in the soil. They also reported that the application of pyrites 2-4 t/ ha increases berseem yield by 16-26 percent. Some important fodder crops and grasses in problematic soils are recommended (Table-2)

Table 2: Some Important Fodder Crops and Grasses in Problematic Soils

Problem Soil	Crop	Varieties
Acid soils	Cowpea	EC 4216, UPC 5286
	Rice	K-1, K-16
	Dinanath	IGFRI-4-2-1, S 56-1
	Maize	African tall, Ganga-5, PLP local
	Setaria	Set.-125, Narok
	Napier bajra hybrid	IGFRI-7
	Guinea grass	Hamil, PGG-9, 14
Saline soils	Sudan grass	SSG-59-3
	Sorghum	HC-136, S-172
	Oat	Kent, OS-6
	Napier bajra hybrid	IGFRI-6, 10
	Lucerne	T-9
	Senji	FOS-1
	Hedge lucerne	
	Shevri	
	Para grass	
Coastal saline soils	Coix lacryma-jobi	
	Para grass	
Calcareous soils	Sorghum	PC-6, HC-136
	Pearl millet	L-72, Rajkc
	Napier-bajra hybrid	IGHRI-6
	Shevri	
Sodic soils	Berseem	Mescavi
	Shaftal	
	Napier-bajra hybrid	IGFRI-6
	Karnal grass	
	Shevri	
	Hedge lucerne	

Similarly the production of fodder crops like berseem maize, teosinte, lucerne in acid soils was higher with application of lime @ 5.0-8.5 t/ha (Tripathi etal, 1988 & 1989). The application of lime is recommended at 5 years interval. According to Tripathi etal (1988)) the combined use of P @ 80kg P_2O_5/ha and 1.0 kg/ha molybdenum in acid soil (pH 5.5) proves better in producing higher yield of berseem (15-20 percent). The fodder quality is also improved with higher content of P, Ca, Mg, and Mo and low content of Al & Mn due to application of lime.

The crust formation in red calcareous soils is a common phenomenon if rains come just after seed sowing. The use of dry grass @ 5 t/ha or 10t/ha FYM is recommended for increasing 3-4 times seed germination and 95-265 percent fodder yield of various crops (Hazra, 1986 a and 1986 b). The use of N half as organic N (50 kg/h) and rest half as fertilizer- N (50kg N) along with 40kg P_2O_5/ha gave higher fodder yield of oat (Tripathi et al. 1984). Similarly, Tripathi etal (2004) reported that the combined use of 40kg N (20kg as urea-N and 20kg as FYM slurry-N) and 80 kg K20/ha along with bacteria culture produces higher yield of *Cenchrus ciliaris* by 15%. The combined application of P@ 80 kg/ha and Mo @ 1.5 kg/ha is found more advantageous in giving 10-15% higher fodder yield of berseem compared to P-application alone (Tripathi et al. 1988 a).

In most of the grasses gave significant increase in the production with the addition of N dose (Tripathi and Hazra, 1995 and Tripathi, 2003). Similarly the application of 20kgP could also give higher yield of *Heteropogon* and *Sahima nervosum* grasslands. The production potential of *Cenchrus ciliaris* +*Stylosanthes hamata* mixed pasture was highest (7.8t/ha) when 80kg/ha K_2O and 40kg/ha N (20kg/ha urea N and 20kg/ha FYM slurry N) was applied together along with bacteria culture (Tripathi et al. 2004).

References

Deb Roy, R and Pathak, P.S. 1974. Silvipastoral management. *Indian Farming*. Nov. 41: 142.

Deb Roy, R (1986). *Silvipastoral research and development in the Country- an over view*. Proc. Nat. Sym. Livestock feed resources, IVRI, Izatnagar, pp 77-78.

Hazra, CR (1992. Watershed - A way to sustainable agriculture. *Intensive Agriculture* 30, 11-39.

Kaur,B; Gupta, SR and Singh, G (2000).Soil carbon, microbial activity and nitrogen availability in agroforestry system on moderately alkaline soils in Northern India. *Applied Soil Ecology* 15,283-284

Hazra, C.R. and Singh D. P. 1992. Soil and water conservation and silvipasture for waste land and barren hill improvement and management. In proceeding second International Symposium on Integrated land use management. ISCO, Deptt. of primary Industries, Brisbane, Australia, key paper No. 28, 1-8.

Hazra, C.R. 1992.Fertilizer use in forage pasture and grassland production. Published non traditional sectors for fertilizer use, FDCO, New-Delhi pp 30-47.

Hazra, C.R. and Tripathi, S.B. 1994. Effect of N fertilization to Anjan grass (Cenchrus ciliaries) with and without pasture legumes in silvipasture. *Forage Research* 20(4), 287-290.

Hazra, C.R.1994. Rehabilitation of problematic and degraded lands through forages. *Journal of Soil and water conservation* 38, 115-130.

Hazra, C.R. 1986 a. Effect of soil management practices on soil crusting, seedling emergence and forage yield of pearl millet. *Forage Research* 12, 37-41.

Hazra, C.R. 1986 b. Seedling emergence and herbage yield of maize and cowpea in relation to soil crusting. *Journal of Indian Society of Soil Science* 34, 398-400.

Mehi lal and Shukla, N.P.1992. Effect of Jalshakti and moisture regimes on productivity of graminaceous forage crops. *Haryana Journal of Agronomy* 6(2), 85-87.

Singh, P. 1988. Forage Research present status and future strategy. In: Pasture and Forage Research. A State knowledge Research edited by Punjab Singh. In third International Rangeland Congress, New Delhi, pp 1-15

Singh, A., Yadava, R.B. and Tripathi, S.B. 1998. Problem soil techniques for higher biomass production alongwith amendments & fertilizers. *Fertilizer News* 43, 45-50.

Shukla, N.P. and Tripathi, S.B. 1996. Judicious utilization of waterlogged, saline–alkdi calcareous and acid soils for forage production. *Journal of Environmental Research* 6, 99-112.

Tripathi, S.B., Hazra, C.R. and Tyagi, R.K. 1988. Soil management practices for higher forage production. *Indian Farming* 37, 160.

Tripathi, S.B. and Hazra, C.R.1989. Forage production potential of acid soils. *Range Management. Society of India.*

Tripathi, S.B., Mannilcor, M.D. and Hazra, C.R.1989. Effect of molybdenum and phosprate on berseem and residue effect on fodder maize in acid soil. *Journal of Indian Society of Soil Science* 37, 832-834

Tripathi, S.B. and Hazra, C.R. and Srivas, N.C.1994. Soil fertility, forage yields and nuyrient uptake of oats as influenced by nitrogen sources and phosphate application. *Indian Journal of Agriculture* 8(2),70-74

Tripathi, S.B. and Hazra, C.R. 1995. Nutrient management and fertilizer use in forages. In: Forage production & utilization (Eds Dr. R.P.Singh) , ICAR, New Delhi. 201-232.

Tripathi, S.B and Hazra C.R. 1996. Forage Production on problem soils. Indian Farming, 6, 9-13.

Tripathi, S.B. 2003. Nutrient management in forage based cropping systems. Fertilizer News 48, 41-45 & 47-48.

Tripathi, S.B., Pahwa, M.R., Singh, A. and Patra. A.K. 2004. Effect of N&K fertilization on herbage field of *Cenchrus ciliaris* + *Stylosanthes hamata* mixed pasture and soil fertility. Range Mangement & Agroforestry 25, 92-97.

❑❑❑

Chapter 9

Soil Health for Forage Production Soil Enzyme Activities as Bio-Indicators

S. K. DAS
Indian Grassland and Fodder Research Institute, Jhansi-284003 (U.P.), India

Introduction

Interest in evaluating the quality and health of our soil resources has been stimulated by increasing awareness that "soil" is a critically important component of the earth's biosphere, functioning not only in the production of food and fibre but also in the maintenance of local, regional and global environmental quality (Glanz, 1995). The thin layer of soil covering the surface of the earth represents the difference between survival and extinction of most land-based life. Like water, soil is a vital natural resource, essential to civilization. But, unlike water, soil is non-renewable on a human time-scale (Jenny, 1980, 1984).

Development of sustainable agricultural management systems has been complicated by the need to consider their utility to humans, their efficiency of resource use, and their ability to maintain a balance with the environment that is favourable both to humans and most other species (Harwood, 1990). We are challenged to develop management systems, which balance the needs and priorities for production of "food and fibre" and those for a safe and clean environment.

A recent call for development of a "soil health index" was stimulated by the perception that human health and welfare is associated with the quality and health of soils (Habern, 1992). However, defining and assessing soil quality or health is complicated by the fact that soils perform multiple functions in maintaining productivity and environmental well-being. Identifying and integrating the physical, chemical and biological soil attributes, which define soil functions is the challenge (Rodale Institute, 1991; Papendick and Parr, 1992).

In the following sections, various aspects of soil quality and soil health will be discussed, which are equally relevant and important for forage production like any other crop.

Concept of Soil Quality

Blum and Santelises (1994) describe a concept of sustainability and soil resilience based on six main soil functions – three ecological functions and three are linked to human activity. Ecological functions include: (i) biomass production (food, fibre and energy); (ii) the soil as a reactor which filters, buffers, and transforms matter to protect the environment, groundwater, and the food chain from pollution; and (iii) soil as biological habitat and genetic reserve for many plants, animals and organisms which should be protected from extinction. Functions linked to human activity include: (i) the soil as a physical medium, serving as a spatial base for technical and industrial structures and socio-economic activities such as housing, industrial development, transportation systems, recreation and refuse disposal; (ii) soil as a source of raw materials supplying water, clay, sand, gravel, minerals etc., and (iii) soil as a part of our cultural heritage, containing palaentological and archaeological treasures important to preserve the history of earth and humankind.

Warkentin and Fletcher (1977) discussed the evolution of soil quality concepts in intensive agriculture. The oldest and most frequently used concept was one of "suitability for chosen uses", with emphasis on capability to support crop growth or engineering structures. This evolved to a "range of possible uses" concept which is ecologically based and recognizes the importance of soil to biosphere function and its multiple roles in enhancing biological productivity, abating pollution, and even serving to enhance human health and aesthetic and recreational uses of landscapes. Another stage in this evolution was development of the "intrinsic value" concept of soil as a unique and irreplaceable resource, of value apart from its importance to crop growth or ecosystem function. Historically, soil has been used as a waste disposal system; it was conceived to be a biological incinerator destroying all the organic wastes deposited on or in it over time. But in 1960s and 1970s, it became increasingly apparent that soils were receiving wastes of a type, and at a rate, that overwhelmed their assimilative capacity. This trend threatened soil function and called for a major responsibility by agriculturists in defining soil quality criteria (Alexander, 1971).

Quality of soil, as distinct from health, is largely defined by the ability of soil to perform various intrinsic and extrinsic functions. "Quality" is represented by a suite of physical, chemical and biological properties that together (i) provide a medium for plant growth and biological activity; (ii) regulate and partition water flow and storage in the environment and (iii) serve as an environmental buffer in the formation and destruction of environmentally hazardous compounds (Larson and Pierce, 1991, 1994).

Much like air or water, the quality of soil has a profound effect on the health and productivity of any given biome and the environments and ecosystems related to it. However, unlike air or water for which we have quality standards, the definition and quantification of soil quality is complicated by the fact that it is not directly ingested or respired by humans and animals, as are air and water. The perceptions of what constitutes a good soil vary depending on individual priorities for soil function.

Concept of Soil Health

Soil health can be defined as the continued capacity of soil to function as a vital living system, within ecosystem and land-use boundaries, to sustain biological productivity and maintain their water quality as well as plant, animal, and human health The concept of soil health includes the ecological attributes of the soil, which have implications beyond its quality or capacity to produce a particular crop. These attributes are chiefly those associated with the soil biota; its diversity, its food web structure, its activity and the range of functions it performs. A fundamental shift is taking place worldwide in agricultural research and food production. In the past, the principal driving force was to increase the yield potential of food crops and to maximise productivity. Today, the drive for productivity is increasingly combined with a desire and even a demand for sustainability. Sustainable agriculture involves the successful management of agricultural resources to satisfy human needs while maintaining or enhancing environmental quality and conserving natural resources for future generations. Soil health is the result of continuous conservation and degradation processes and represents the continued capacity of soil to function as a vital living ecosystem. A unique balance of chemical, physical and biological (including microbial) components contribute in maintaining soil health. Evaluation of soil health therefore requires indicators of all these components. Historically, the term *soil quality* described the status of soil as related to agricultural productivity or fertility (Singer *et al.* 2000). In the 1990s, it was proposed that soil quality was not limited to soil productivity but instead expanded to encompass interactions with the surrounding environment, including the implications for human and animal health. Doran & Safley (Doran *et al.* 1997) proposed the following definition of soil health:

"The continued capacity of soil to function as a vital living system, within ecosystem and land-use boundaries, to sustain biological productivity, promote the quality of air and water environments, and maintain plant, animal and human health."

Soils contain an intricate network of plants and microbes in a heterogeneous solid medium in which chemical and physical conditions vary at the scale of the molecule and the cell. It is therefore, difficult to understand the variations in soils in the absence of knowledge derived from both chemical and biological approaches, because microorganisms affect the environment and vice versa. Soil microorganisms are key-players in the global cycling of organic matter, reworking organic residues or mineralizing them to CO_2, H_2O, nitrogen, phosphorus, sulfur and other nutrients (Bloem *et. al.*, 1997). Intensive agriculture is practiced in many countries with adverse consequences on biodiversity and soil organic matter (SOM) status. Changes in the nutrient status of the soil, such as a decrease in the organic matter content, directly affect its microbial biodiversity. Some of the more resilient soils are found in grassland pastures, where bacteria are associated with root material and are attached to clay particles (Lynch and Poole, 1979). In well-drained soil sustained by a healthy bacterial microbiota, much of the space between soil aggregates is filled with air, which is necessary for soil productivity (Colwell, 1997).

Soil Health and its Biological Indicators

Agricultural research in soils must be challenged by some of the fundamentals that are not measured completely by criteria including no more than yields as bushels or cash returns per acre or man-hours. Qualities that deal with life, not quantities of materials alone must be emphasized. In that research the farmer too must share some of the thinking responsibilities. The experimenter in research may well think with but he dare not think for the farmer. As more folks, farmers and all of us, think about the fundamental processes of creation by which the soil supports all that we call agriculture, we will not be contented with the mechanics of it, or complain of the high costs, and the low prices when only the speculative aspects of agriculture are considered. We will invest ourselves more in understanding the production of food for health, or our major national wealth, according to the fertility of the soil. The cow, then, as the symbol of all that is living in agriculture, along with all humans dependent on it, will be convincing us that not only animal health but human health too is dependent on the fertility of the soil.

Soil health provides an overall picture of soil functionality. Healthy soils maintain a diverse community of soil organisms that help to: (i) control plant diseases as well as insect and weed pests, (ii) form beneficial symbiotic associations with plant roots, (iii) recycle plant nutrients, (iv) improve soil structure with positive repercussions for its water and nutrient-holding capacity, (v) improve crop production. One of the most important objectives in assessing the health of a soil is the establishment of indicators for evaluating its current status. A list of indicators is given in table 1.

Table 1. Biological, Physical and Chemical Indicators used for Determining Soil Health, and Standard Analytical Procedures.

Indicator	Measurement*
Microbial biomass	Direct microscopic counts
	Chloroform fumigation
	SIR
	CO_2 production
	Microbial quotient
	Fungal estimation
	PLFA
Microbial activity	Bacterial DNA synthesis
	Bacterial protein synthesis
	CO_2 production
Carbon cycling	Soil respiration
	Metabolic quotient (qCO_2)
	Decomposition of organic matter
	Soil enzyme activity
Nitrogen cycling	N-mineralization
	Nitrification
	Denitrification
	N-fixation
Biodiversity and microbial resilience	Direct counts
	Selective isolation plating
	Carbon and nitrogen utilization patterns
	Extracellular enzyme patterns
	PLFA
Bioavailability of contaminants	Plasmid-containing bacteria
	Antibiotic-resistant bacteria
Physical and chemical properties	Bulk density
	Soil physical observations and estimations
	pH
	EC
	CEC
	Aggregate stability and soil slaking
	Water holding capacity
	Water infiltration rate
	Macro/micronutrient analysis

*Acronyms: SIR, substrate induced respiration; PLFA, phospholipid fatty acids; EC, electrical conductivity; CEC, cation exchange capacity. [Source: Arias *et. al.*, 2005]

Most researches in this subject have focused on the development of environmental quality indicators and soil management practices that restore and/or improve vegetal and biological production aiming at minimizing negative environmental impacts (Doran *et. al.*, 1994), and contributing for the onset of a more sustainable agriculture (Dick 1994). It was argued that a lower productive capacity in a soil was the result of a neglected microbiological functioning within that soil. Therefore, there is growing evidence that soil microbiological and biological parameters may possess potential as early and sensitive indicators for soil ecological stress or reparation (Dick 1992, Dick 1994, Dilly *et. al.* 1998), as is the case of soil enzyme activities and exopolysaccharides, soil microbial biomass, composion of soil microflora, that were used as potential biochemical/biological indicators of soil quality (Dick, 1994, Fauci *et. al.* 1995, Filip *et. al.* 1998).

Definition of policy-relevant end points is very important as monitoring programmes are developed. End points should be pragmatic in the sense of providing logical categories for regulatory decisions and they should be integrated for indicators that are ecologically related. An end point matrix was suggested (Figure 1) that, when integrated together, provide a comprehensive and effective assessment of soil health (Nielsen and Winding, 2002).

Soil Enzymes as Indicator of Soil Quality/Health

Soil enzymes play an essential role in catalyzing reactions necessary for organic matter decomposition and nutrient cycling in ecosystem. They are involved in energy transfer, environmental quality and crop productivity under the influence of edaphic factors (Deng *et. al.*, 1996). Soil enzymes also take part in important soil processes such as nutrient cycling and energy transformation by catalyzing numerous chemical, physical and biological reactions. Soil enzymes are sensitive to variations induced by natural and anthropogenic disturbances (Gupta *et al.*, 1988, Dick 1997). Vegetation, soil type, land use history and soil management strategy affect soil enzymes (Green and Oleksyszyn 2002, Wyszkowska et al. 2005). Many studies have suggested that soil enzymes can be used as indices of soil contamination, soil fertility and soil health (Masciandaro et al. 1998, Saviozzi et al. 2001). Soil enzyme activity is sensitive in discriminating soil qualities among soil management treatment effects (Saviozzi et al. 2001, Sadowsky et al. 2006). Soil enzyme activity is an effective indicator of soil quality resulting from environmental stress or management practices (Dick 1997).

As known, many currently used agricultural technologies and practices such as monocrops, residue management, mineral fertilization, overuse of pesticides, agricultural heavy machinery, inadequate management practices of the soil and irrigated water can significantly affect soil quality by changing the physical, chemical and biological properties of the soil (Fauci and Dick, 1994). Besides, the low input of organic matter into the soil in conventional production systems has generated disbalances in the edaphic ecosystems, with a decrease in the microbial activities of mycorrhizas, endobacteria, and plant growth promoting rhizobacteria (PGPR). This effect contributes significantly to modify the plant growth by holding plant nutrition and by making plants more susceptible to pests (Agrawal et. al., 1999; Frighetto et. al., 1998; García et. al., 1994; Valarini et. al., 1999). Most researches in this subject have focused on the development of environmental quality indicators and soil management practices that restore and/or improve vegetal and biological production aiming at minimising negative environmental impacts (Doran and Perkin, 1994), and contributing for the onset of a more sustainable agriculture (Dick, 1994). The use of animal manure, crop diversification and rotation, and application of biological control agents are appropriate practices recommended to restore and/or improve the quality of agricultural soils. Historically and until recently, chemical and physical properties have been used as crude measures of soil productivity. However, only a static type of data can be obtained in this way. Besides, organic matter content in soils changes very slowly over time, and therefore, many years may be required to detect changes in it caused by intensive agriculture or other sort of disturbances. As far back as 1916 (Lipman et. al., 1916), it was argued that a lower productive capacity in a soil was the result of a neglected microbiological functioning within that soil. Therefore, there is growing evidence that soil microbiological and biological parameters may possess potential as early and sensitive indicators for soil ecological stress or reparation (Dick, 1992, 1994, Dilly and Blume, 1998), as is the case of soil enzyme activities and exopolysaccharides, soil microbial biomass, composion of soil microflora, that were used as potential biochemical/biological indicators of soil quality (Dick, 1994, Fauci and Dick, 1994, Filip, 1998).

Microorganisms possess the ability to give an integrated measure of soil health, an aspect that cannot be obtained with physical/chemical measures and/or analyses of diversity of higher organisms. Microorganisms respond quickly to changes, hence they rapidly adapt to environmental conditions. The microorganisms those are best adapted will be the ones that flourish. This adaptation potentially allows microbial analyses to be discriminating in soil health assessment, and changes in microbial populations and activities may therefore function as an excellent indicator of change in soil health (Kennedy et al., 1995; Pankhurst et al., 1995).

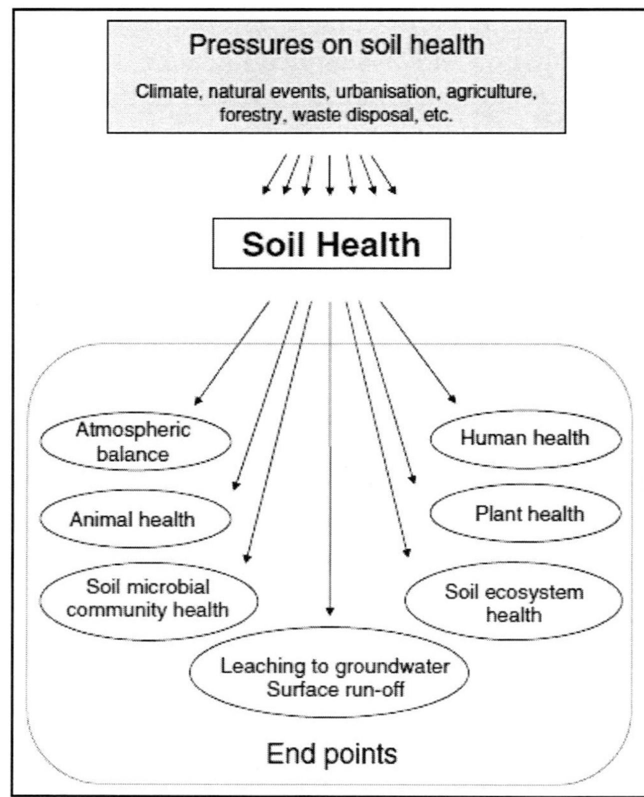

Fig. 1. Policy-relevant end points of soil health monitoring. Several examples of pressures on soil health are presented (grey box) and this may impact several end points of soil health (elliptical boxes). [Source: Nielsen and Winding, 2002]

The bioavailability of chemicals, e.g. heavy metals or pesticides, is also an important issue of soil health because of its connection with microbial activities. The impact of such chemicals on soil health is dependent on microbial activities. For example, the concentration of heavy metals in soil will not change over small time periods, but their bioavailability may. It has thus been shown that the bioavailability of poly-aromatic hydrocarbons was lower in autumn compared to early spring due to a higher microbial activity after the growing season (H. Harms, pers. comm. 2001). Therefore, the total content of chemicals in soil is not a reliable indicator of its bioavailability (Logan, 2000) and thereby soil health. Instead, bioavailability has to be measured in relation to bioassays and specific microbial processes. In context of this, microbial responses also integrate the effect of chemical mixtures, an information not obtained by studying the chemical mixtures themselves.

It is also very important to consider the possible impact of pesticides on soil quality and health, under more realistic conditions and with the formulation applied to the field. Enzyme activities have been considered to be sensitive indicators of soil quality (Doran et. al., 1994; Dick, 1994; Nannipieri, 1994). The importance of soil enzymes resides in their relationship to soil microbiology, ease of measurement and rapid response to change in soil management. It is conceptually wrong to rate a single enzyme activity as criteria of soil quality or soil microbial activity (Nannipieri, 1994). Thus many enzyme activities should be considered. Phosphatase activities are considered especially useful indicators of both positive and negative effects of soil management practices on soil quality (Dick, 1994; Jordan *et. al.*, 1995). Glucosidase activity is often included in studies searching for sensitive biomarkers of soil quality (Bucket and Dick, 1998). Its activity is considered to be a useful index for measuring side effects of pesticides on microbial activity in soil (Schäffer, 1993). Dehydrogenase activity in soil provides an index of the overall microbial activity (Nannipieri *et. al.*, 1990).

Soil degradation through loss of organic matter and structural integrity is a well-known outcome of an intensive cropping regime. From experimental findings, it was observed that urease, protease and phosphatase activities were higher under no tillage than under ploughed treatments (Klein and Koths, 1980). The enzyme activities in all aggregate size fractions were decreased due to long-term cultivation; the Michaelis Constant (K_m) and maximum reaction rate (V_m) for arylsulfatase activity in all cultivated soils (Gupta et. al., 1988; Farrell et. al., 1994). In the conventional tillage, soil xylanase activity reduced greatly, because a large proportion of this enzyme activity was located in the coarsely textured organic fraction (Figure 2) and conventional tillage had decreased soil organic-C content in the coarsest (.200-mm) fraction (Kandeler *et. al.*, 1999). The other enzymes, invertase and alkaline phosphatase, were more closely aligned with the finer soil fractions and were less affected by tillage, although the proportions in the coarses soil fraction also were diminished (Fig 2). We can say that, if a soil enzyme can tell us something about the location and perhaps the quality of soil organic matter in cropped soils, e.g., by measurement of xylanase activity in soil particle-size fractions, then it may be possible to use its activity as an early warning of potential structural degradation. Soil invertase activity was significantly correlated with soil organic matter but not with other soil physical-chemical variables. The indirect effects of soil organic matter on invertase activity through other variables were mostly negative but the effects were marginal, which reduced the direct effect in general. Soil total P, alkali-hydrolysable N and clay had a positive indirect effect on invertase via soil organic matter.

Soil available P, pH, and EC did not show any direct or indirect effect on invertase. Invertase activity increased by soil organic matter but was inhibited by soil total N.

It has been estimated that about 1600 ´ 10^9 m^3 of mine spoils had accumulated on the Earth's surface upto 1980 and had increased by about 40 x 10^9 m^3 per year by 1998 (Kiss et. al., 1998). Rehabilitation of these spoils and degraded landscapes is an integral part of mining operations in many parts of the world. These constructed or "technogenic" soils may have a "topsoil" composed of entirely subsurface materials or the stockpiled original topsoil or some intermediate combination. Stockpiling of topsoil leads to a decline of soil biological activity. In a pot trial, it was observed that the protease, sulfatase and urtease activities declined markedly in fallow soils over 5 months, but those activities remained unchanged or increased where soils were planted with rye grass (Speir et al., 1980).

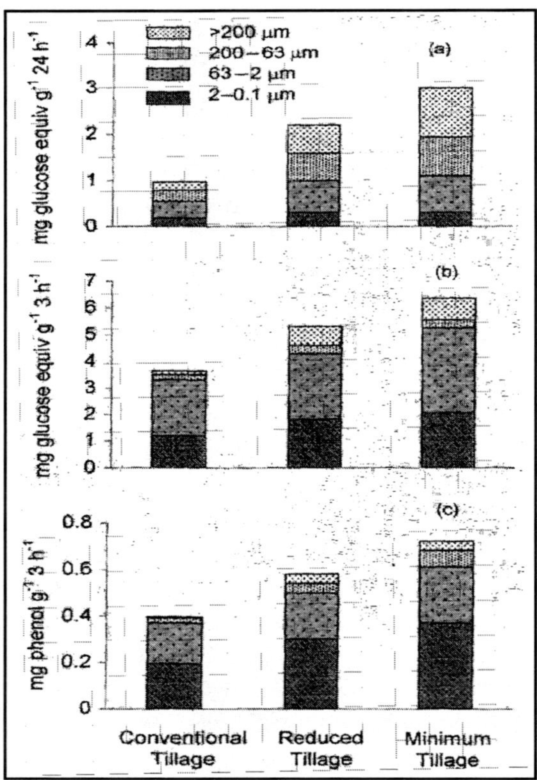

Fig. 2. Xylanase (a), Invertase (b), and Alkaline Phosphatase (c) Activities in Particle-size Fractions of the 0- to 10-cm Layer of a Haplic Chernozem soil [Source: Kandeler *et. al.*, 1999]

The assessment of the health of soils, through the identification of key soil properties, which can serve as indicators of soil health, has become a major issue for land managers and the food and agricultural sector through the world. FAO has recently been identified as the executing agency for conducting, in close cooperation with multiple partners, the GEF/UNEP Land Degradation Assessment in Drylands (LADA). Soil biota and its functions should be a key component of such assessments. In particular, there is a need to determine short- and long-term effects of agricultural management practices on soil biological community populations, biodiversity, functioning and resilience. Relating soil quality/ health to productivity, in terms of crop yield and profitability, and environmental effects from drainage, leaching, runoff and erosion is essential in order to evaluate the sustainability of various land management strategies. There is recent progress in realising that soil health, by its broadest definition, is inseparable from issues of sustainability. The challenge ahead is to develop holistic approaches for assessing soil quality and health that are useful to producers, specialists and policy makers in identifying agricultural and land use management systems that are profitable and will sustain finite soil resources for future generations. Benefits of paying more attention to soil health and its assessment include its potential use in the evaluation of land-use policy and of practices that degrade or improve the soil resource in the identification of critical landscapes or management systems and of gaps in our knowledge base and understanding of sustainable management. The forage crops are also drawing more attention from the government, policy-makers and researchers in recent days and dairy industries started showing interest in quality of forages. Good quality forage can be produced only quality is maintained at every aspects of its growth, including a "healthy" soil. Therefore, the topic is very much contemporary in the background of present global situation and needs further research for answering many questions.

References

Agrawal, A.A.; Tuzun, S.; Bent, E. 1999. *Induced plant defenses against pathogens and herbivores. Biochemistry, Ecology and Agriculture.* APS Press, St. Paul, 390p.

Alexander, M., 1971. Agricultures responsibility in establishing soil quality criteria. In: Environmental Improvement Agricultures Challenge in the Seventies, National Academy of Sciences, Washington DC, pp 2-7.

Arias, M. Enriqueta, González-Pérez, José Francisco, González-Vila, A. J. and Ball, Andrew S., 2005. Soil health—a new challenge for microbiologists and chemists. *International Microbiology* , 8, 13-21

Bloem, J. de Ruiter, P., Bouwman, L.A., 1997. Soil food webs and nutrient cycling in agro-ecosystems. In: van Elsas J.D., Trevors, J.T., Wellington, H.M.E. (eds) Modern soil microbiology, Marcel Dekker, New York, pp 245-278.

Blum, W.E.H. and Santelises, A.A., 1994. A concept of sustainability and resilience based on soil functions. In: Greenland, D.J. and Szabolcs, I. (eds.). Soil Resilience and Sustainable Land Use, CAB International, Wallingford, UK, pp 535-542.

Bucket, J.Z. and Dick, R.P., 1998. Microbial and soil parameters in relation to N mineralization in soils of diverse genesis under differing management systems. *Biology and Fertility of Soils*, 27, 430-438.

Colwell, R.R., 1997. Microbial biodiversity and biotechnology. In: Reaka-Kudla, M.L., Wilson, D.E., Wilson, E.O. (eds). Biodiversity II: Understand and protecting our biological resources. Joseph Henry Press, University of Washington, Washington DC, pp 279-288.

Deng S P, Tabatabai M A. 1996. Effect of Tillage and Residue Management on Enzyme Activities in Soils. I1. Glycosidases. *Biology and Fertility of Soils*, 22, 208-213.

Dick, R.P.,1992.A review: long-term effects of agricultural systems on soil biochemical and microbial parameters. *Agricultural Ecosystem and Environ*ment, 40, 25-60,

Dick, R.P., 1994. Soil enzyme activities as indicators of soil quality. In: Doran, J.W.; Coleman, D.C.; Bezdicek, D.F.; Stewart, B.A. (eds). Defining soil quality for a sustainable environment. Minneapolis, Soil Science Society of America, p.107-124. (SSSA Special Publication, 35).

Dick R.P., 1997. Soil enzyme activities as integrative indicators of soil health. In: Pankhurst C.E., Doube B.M.,Gupta V.V.S.R. (eds.): Biological Indicators of Soil Health. CAB International, Wallingford: 121–156.

Dilly, O., Blume, H.P. 1998. Indicators to assess sustainable land use with reference to soil microbiology. *Advances in Geology and Ecology*, 31, 29-36.

Doran, J.W., Parkin, J.B., 1994, Defining and assessing soil quality. In: Doran, J.W.; Coleman, D.C.; Bezdicek, D.F.; Stewart, B.A. (eds). Defining soil quality for a sustainable environment. Minneapolis, Soil Science Society of America, cap.1, p.3-22. (SSSA Special Publication, 35) Madison, Wisconsin, USA.

Doran, J. W. and Safley, M. (1997). Defining and assessing soil health and sustainable productivity. In: Pankhurst, C. E., Doube, B. M., and Gupta, V. V. S. R. (eds.)., Biological Indicators of Soil Health,.CAB International, pp. 1-28.

Farrell, R.E., Gupta, V.V.S.R. and Germida, J.J. 1994. Effects of cultivation on the activity and kinetics of arylsulfatase in Saskatchewan soils. *Soil Biology and Biochemistry*. 26, 1033-1040.

Fauci, M.F., Dick, R.P. 1994. Microbial biomass as an indicator of soil quality: effects of long-term management and recent soil amendments. In: Doran, J.W.; Coleman,

D.C.; Bezdicek, D.F.; Stewart, B.A. (eds). Defining soil quality for a sustainable environment. Minneapolis, Soil Science Society of America, cap.17, p.229-234. (SSSA Special Publication, 35).

Filip, Z.K. 1998. Soil quality assessment: an ecological attempt using microbiological and biochemical procedures. *Advances in Geology and Ecology*, 31, 21- 27.

Frighetto, R.T.S., Valarini, P.J., Tokeshi, H., Filizola, H.F., Felipe, A., Oliveira, D.A. 1998. *Effects of management practices on soil physicochemical, biochemical and microbial parameters under irrigated agriculture.* Conferência Internacional IFOAM 98, S 9, n.158.

García, C., Hernandez, T., Costa, F., Ceccanti, B. 1994. Biochemical parameters in soils regenerated by the addition of organic wastes. *Waste Manage. Res.*, 12, 457-466,

Glanz, J.T., 1995. *Saving our soil: Solutions for Sustaining Earth's Vital Resource.* Johnson Books, Boulder, Colorado.

Green D.M., Oleksyszyn M., 2002. Enzyme activities and carbon dioxide flux in a sonoran desert urban ecosystem. *Soil Sci. Soc. Am. J.*, 66, 2002–2008.

Gupta, V.V.S.R.and Germida, J.J. 1988. Distribution of microbial biomass and its activity in different soil aggregate size classes as affected by cultivation. *Soil Biology and Biochemistry*. 20, 777-786.

Gupta V.V.S.R., Lawrence J.R., Germida J.J., 1988. Impact of elemental sulfur fertilization on agricultural soils. I. Effects on microbial biomass and enzyme activities. *Can. J. Soil Sci.*, 68, 463–473.

Habern, J., 1992. Viewpoint: a soil health index. *Journal of Soil Water Conservation*, 47, 6

Harwood, R.R., 1990. A history of sustainable agriculture. In: Edwards, C.A., Lal, R., Madden, P., Miller, R.H. and House, G. (eds.) sustainable Agricultural Systems, Soil Water Conservation Society of America, Ankeny, Iowa, pp 3-19.

Jenny, H., 1984. The making and unmaking of fertile soil. In: Jackson, W., Berry, W. and Coleman, B. (eds.) Meeting the Expectations of the Land, North Point Press, San Francisco, California, pp 44-55.

Jenny, H. 1980. The soil resource: origin and behaviour. *Ecological Studies*, 37, Springer-Verlag, New York.

Jordan, D., Kremer, R.J., Bergfield, W.A., Kim, K.Y.and Cacnio, V.N., 1995. Evaluation of microbial methodsas potential indicators of soil quality in historical agricultural fields. *Biology and Fertility of Soils*, 19, 297-302.

Kandeler, E., Palli, S., Stemmer, M., Gerzabek, M.H. 1999. Tillage changes microbial biomass and enzyme activities in particle-size fractions of a Haplic Chernozem. *Soil Biology and Biochemistry.*, 31, 1253-1264.

Kennedy, A. C. and Papendick, R. I.,1995. Microbial characteristics of soil quality. *Journal of soil and water conservation,* May-June, 243-248.

Kiss, S., Pasca, D. and Dragan-Bularda, M. (eds.) 1998. *Developments in Soil Science*, 26: Enzymology of Disturbed Soils. Amsterdam: Elsevier,

Klein, T.M. and Koths, J.S.,1980. Urease, protease and acid phosphatase in soil continuously cropped to corn by conventional or no-tillage methods. *Soil Biology and Biochemistry*, 12, 293-294

Larson, W.E. and Pierce, F.J., 1991. Conservation and enhancement of soil quality. In: Evaluation for Sustainable Land Management in the Developing World, Vol. 2: Technical Papers IBSRAM Proceedings No. 12(2), International Board for Soil Research and Management, Bangkok, Thailand.

Larson, W.E. and Pierce, F.J., 1994. The dynamics of soil quality as a measure of sustainable management. In: Doran, J.W., Coleman, D.C., bezdicek, D.F. and Stewart, B.A. (eds.) Defining Soil Quality for a Sustainable Environment, Soil Science Society of America, Special Publication No. 35, Madison, Wisconsin, pp 37-51.

Lipman, J.G.; McLean, H.; Lint, H.C. 1916. Sulphur oxidation in soils and its effect on the availability of mineral phosphatases. *Soil Sci*ence, 1, 533-539.

Logan, T. J. 2000. Soils and environmental quality. In: Sumner, M. E. (eds.). Handbook of Soil Science, CRC Press, Boca Raton, pp. G155-G169.

Lynch, J.M. and Poole, N.J.,1979. *Microbial Ecology: a conceptual approach*. John Willey, New York.

Masciandaro G., Ceccanti B., Gallardo-Lancho J.F., 1998. Organic matter properties in cultivated versus set-aside arable soils. *Agricultural Ecosystem and Environment*, 67, 267–274.

Nannipieri, P., 1994. The potential use of soil enzymes as indicators of productivity, sustainability and pollution. In: Pankhurst, C. E., Doube, B. M Gupta, V. V. S. R., Grace, P.R. (eds.). CSIRO, East Melbourne, pp 238-244.

Nannipieri, P., Ceccanti, B., Gregos, S., 1990. Ecological Significance of the biological activity in soil. In: Bollag, J.M., Stotzky, G. (eds). *Soil Biochemistry*, Mercel Dekker, New York, Vol. 6, pp 293-355.

Nielsen, Mette Neiendam, and Winding, Anne, 2002. Microorganisms as Indicators of Soil Health, Technical Report No. 388, National Environmental Research Institute, Denmark.

Pankhurst, C. E., Hawke, B. G., McDonald, H. J., Kirkby, C. A., Buckerfield, J. C., Michelsen, P., O'Brien, K. A., Gupta, V. V. S. R., and Doube, B. M.,1995. Evaluation of soil biological properties as potential bioindicators of soil health. *Australian Journal of Experimental Agriculture*, 35, 1015-1028.

Papendick, R.I. and Parr, J.F., 1992. Soil quality – The key to a sustainable agriculture. *American Journal of Alternative Agriculture*, 7, 2-3.

Rodale Institute, 1991. Conference reports and abstracts. International conference on the Assessment and Monitoring of Soil Quality. Rodale Press, Emmaus, Pennsylvania.

Saviozzi A., Levi-Minzi R., Cardelli R., Riffaldi R., 2001. A comparison of soil quality in adjacent cultivated, forest and native grassland soils. *Plant Soil*, 233, 251–259.

Schäffer, A.. 1993. Pesticide effects on enzyme activities in soil ecosystems. In: Bollag, J.M., Stotzky, G. (eds). *Soil Biochemistry*, Mercel Dekker, New York, Vol. 9, pp 273-340.

Shi, Z.J., Lu, Y.,. Xu, Z.G, Fu, S.L., 2008. Enzyme activities of urban soils under different land use in the Shenzhen city, China. *Plant Soil Environment* 54(8), 341–346.

Singer, M. J. and Ewing, S. 2000. Soil quality. In: Sumner, M. E. (eds.). Handbook of Soil Science. CRC Press, Boca Raton, pp. G271-G298.

Speir, T.W., Lee, R., Pansier, E.A. and Cairns, A. 1980. A comparison of sulfatase, urease and protease activities in planted and in fallow soils. *Soil Biology and Biochemistry*., 12, 281-291.

Valarini, P.J., Frighetto, R.T.S., Tokeshi, H., Scramin, S., Silveira, A.P.D., Valarini, M.J., Oliveira, D.A. 1999. Physico-chemical and microbiological properties of soil as affected by irrigated agricultural systems. *In*: Foguelman, D.; Lockeretz, W. (eds). Organic agriculture the credible solutionfor the XXIst Century: proc. of the 12ª Intern. IFOAM Scientific Conference, Mar del Plata, p.165-171.

Warkentin, B.P. and Fletcher, H.F., 1977. Soil quality for intensive agriculture. In: Proceedings of an International Seminar on Soil Environment and Fertility Management in intensive Agriculture, Tokyo, Japan, pp 593-598.

Wyszkowska J., Kucharski J., Lajszner W., 2005. Enzymatic activities in different soils contaminated with copper. *Poland Journal of Environmental Studies*, 14, 659–664.

❑❑❑

Chapter **10**

Potential Forage Production Systems for Different Agro Climatic Situation of India

SUNIL KUMAR
Indian Grassland and Fodder Research Institute, Jhansi-284 003 (U.P.), India

Introduction

Production efficiency of the Indian livestock in most of the regions of the country is very low as compared to some of the states like Gujarat, Punjab and Haryana. Balanced nutrition through feed and fodder to livestock is available in selected milk shed areas of the country where intensive fodder production systems are practiced. Rest of the farming community maintains uneconomic large herds on grazing supplemented with stall feeding. Urbanization has brought a marked shift in the lifestyle of people in feeding habits towards milk products, meat and eggs with resultant increase in demand of livestock products. Peri-urban livestock farming and emerging fodder markets are indicators of fast changing economic scenario in livestock sector. Livestock population of India is around 500 million and is expected to grow at the rate of 1.23% in the coming years. To meet this challenge, concentrated efforts are to be made for bridging the large gap in demand and supply of the fodder in this country.

The success of dairy as a profitable enterprise mainly depends upon the quality of breeds and supply of nutritious fodder. Despite of one-sixth of cattle population in India, per capita availability of milk is 210 g. With the introduction of cross-breeds, there lies ample scope for increased milk production per lactation. The maintenance and production ration of these animals is 30-40% more than the local breeds. The productivity of these animals can be maintained by providing nutritious fodder. Presently, the gap in demand and supply of green and dry fodder can be reduced by increasing the production of green fodder three times and dry fodder by two times (Table 1). Effective breakthrough in country's milk production can be achieved only if fodder production is sustained at an adequate level throughout the year. Hence, in present scenario, intensive fodder production system holds the promise especially for milk- shed areas (Sunil Kumar, 2007).

Table 1. Availability Status of Feed and Fodder Resources in India

Feed Resources	Availability (Million Tonnes)
Greens	
Forest Area	89.37
Fallow lands	23.21
Permanent pastures, grazing area	28.70
Cultivable wastelands, misc. tree tops	17.51
Cultivated fodders	303.26
Total	**462.05**
Crop Residues	
Coarse Straw-from coarse grains, sugarcane tops	154.83
Fine straw (Paddy, wheat straws)	194.11
Leguminous straws (pulses, others)	44.94
Total	393.88
Concentrates	
Oil cakes	15.76
Brans	13.26
Grains for feeding livestock	5.74
Chwins	0.53
Total	**35.29**

Source: XI Plan Proposals: A Report of the Sub-Committee on feed, fodder and pasture development to the Planning Commission.

Forage Production Scenario

At present, the country faces a net deficit of 61.1% green fodder, 21.9% dry crop residues and 64% feeds. To meet the current level of livestock production and its annual growth in population, the deficit in all components of fodder, dry crop residues and feed has to be met from either increasing productivity, utilizing untapped feed resources, increasing land area (not possible due to human pressure for food crops) or through imports. In animal feed supply, coarse cereals have a major role and four major cereals viz. maize, barley, sorghum and pearl millet account for about 44% of the total cereals. At the current level of growth in forage resources, there will be 65% deficit in green fodder and 25% deficit in dry fodder. This situation indicates that green forage supply has to grow at 3.2%, to meet the deficit.

Forage Resources

The forage resources in India are mainly derived from crop residues, cultivated forages and grazing from pastures and grasslands. The crop residues mainly constitute the major feed material in most of the states. The country has about 4.9% of the total cropped area under cultivated forages. In India, cattle of intensive cropped areas obtain only about 25 percent of their feed from grazing in nearby forests and other uncultivated lands; the balance comes from crop residues unsuitable for human consumption. In different, agro-climatic regions, the productivity of some prominent cultivated forage is highly variable. Among the *Kharif* forage crops, sorghum, maize, cowpea, NB hybrid and guinea grass have wide amplitude. However, during *Rabi*, the choice is limited up to oat, lucerne and Berseem (Pathak and Sunil Kumar, 2004).

Forage production systems

A. Intensive forage production systems

Intensive forage production systems aims at efficient utilization of land and other inputs for maximum fodder production per unit area per unit time with equal quantum of increase in animal productivity. The pre- requisites of intensive forage production systems are- fertile land, assured inputs viz. fertilizers and irrigation and improved cultivars. Intensive forage production systems include multiple cropping, intercropping, overlapping cropping and relay cropping. This system is practically useful to the farm or lands which are exclusively devoted to the cultivation of fodder crops. Based on research results of AICRP on Forage crops and on farm trials intensive forage production systems have been identified for different agro-climatic situations of India. Under assured irrigated condition, the fodder based crop sequences provide not only nutritious green fodder but also take care of overall productivity over a long period of time (Hazra, 1988). Intensive fodder crop rotations for different agro-climate zones have been given in Table-2. Except for high water table areas and sub-tropical and sub-temperate hill regions, perennial crops-based rotations are suitable, whereas in other cases, annual crops in rotation are appropriate.

Table 2. Intensive Forage Crop Rotations for Different Agro-climate Zones of India.

Zones/ Location* (Climate & Soil)	Crop rotation	Green fodder yield (t/ha/year)
North		
Palampur (Himanchal Pradesh) : Sub-temperate, Moist, Red soil		
1. Maize + Cowpea – Lucerne + Oats - Mustard		85
2. NB Hybrid + Velvet bean – Berseem + Mustard		123
Pantnagar (Uttarakhand) : Tarai, Red & yellow soil		
1. Maize + Cowpea – Toria - Oats		177
2. NB Hybrid + Berseem – Cowpea		121
Hisar (Haryana) : Semi-arid, Sandy loam soil		
1. NB hybrid + Berseem		212
2. NB hybrid + Lucerne		176
Central and Western		
Jhansi (UP): Semi-arid, Red soil		
1. NB hybrid + Cowpea - Berseem + Mustard		255
2. Sorghum + Cowpea – Berseem + Mustard – Maize+ Cowpea		176
Jabalpur (MP) : Sub-humid, Black soil		
1. NB hybrid + Cowpea - Berseem		176
2. Sorghum+Cowpea–Berseem+ Mustard–Sorghum+Cowpea		169
Anand (Gujrat) : Semi-arid, Black soil		
1. NB hybrid + Cowpea - Lucerne		253
2. Maize + Cowpea – Oats – Maize + Cowpea – Maize + Cowpea		168
East		
Ranchi (Kanke): Sub-humid, Red acidic soil		
1. Pearlmillet + Cowpea - Oats - Maize + Cowpea		103
2. Maize + Cowpea – Sorghum + Cowpea – Berseem + Mustard		96
Kalyani (West Bengal) : Sub- humid, Alluvial soil		
1. Maize + Cowpea – Dinanath grass - Oats		131
2. Maize + Ricebean – Berseem + Mustard		112
Jorhat (Assam): Humid, Acidic soil		
1. NB hybrid (Perennial)		106
2. Maize + Cowpea – Maize + Cowpea – Maize + Cowpea		85

contd....

South	
Coimbatore (Tamil Nadu): Sub humid, Black soil	
1. NB hybrid + Lucerne	225
2. Sorghum + Cowpea – Maize + Cowpea – Maize + Cowpea	111
Vellayani (Kerala): Humid, Red soil	
1. Guinnea grass in coconut plantation	135
2. Congosignal grass in coconut plantation	75

*Centres are the testing locations in each region.

B. Mixed / Intercropping Systems of Forage Production

The combination of graminaeous and leguminous fodder crops improves the herbage quality substantially in terms of protein and mineral balances as the legumes component contains higher amount of protein, calcium and phosphorus. This system of cropping also helps to maintain soil fertility over a long period due to addition of root organic matter and better utilization of plant nutrients from different soil depth. The results of a field study presented in (Table 3a, 3b) Intercropping of Oat + Senji / Pea or *Pennisetum pedicellatum* + Cowpea/Guar supplements 35 kg N/ha with and additional application of 15 kg P_2O_5/ha (Anonymous, 1998) which can replicated in other agro-climatic situations as well.

Table 3a. Forage and Crude Protein Yield Under Different Mixed Cropping Systems.

Crops/Crop Combination	***Nutrient level (kg/ha)***		***Forage yield (t/ha)***			***Land Equivalent Ratio (LER)***
	N	**P_2O_5**	**Green**	**Dry Matter**	**Crude Protein**	
Sorghum (Sole)	120	50	49	10.2	0.69	-
Sorghum + Cowpea (F)	70	55	49	10.1	1.02	1.14
Maize (Sole)	120	50	44	9.3	0.72	-
Maize + Cowpea (F)	70	55	45	9.3	1.05	1.12
Pearl millet (Sole)	120	50	37	8.2	0.63	-
Pearl millet + Cowpea (F)	70	55	43	8.9	0.97	1.14
P. pedicellatum (Sole)	90	30	46	7.7	0.66	-
P. pedicellatum + Cowpea (F)	55	45	44	7.6	0.94	1.80
P. pedicellatum + Guar (F)	55	45	43	7.5	0.84	1.29
Oat (Sole)	90	30	43	9.0	0.56	-
Oat + Senji	55	45	41	8.8	1.03	1.14
Oat + Pea	*55*	*45*	*40*	*8.6*	*1.07*	*1.15*

C. Food-Fodder Components in Sequential Cropping

The rainfed areas often fail to support two successive grain crops in rotation because of their long duration. However, there is possibility of accommodating short duration forage species in sequence with food crops to increase the intensity of cropping in time. This is feasible on black soils consisting of Kabar and Mar, which retain sufficient residual moisture for growing winter crops. In this system, forage crops may be included to precede or succeed the food component depending upon food and fodder needs. It has been proved that forage sorghum – gram/ lentil/ mustard sequences can successfully be grown to meet the twin objectives. Among these (Table-3), forage sorghum – lentil has been found consistently more productive and remunerative food-fodder system (Ali & Rawat, 1986).In semi-arid tract with less irrigation facility, barley as fodder-cum-grain crop could be grown in the northern, north-western and central part of India by harvesting the crop at 55 days and subsequently manage it for grain. The crop requires fertilization of 30 kg N and 30 kg P_2O_5/ha basally and an additional 30 kg N/ha applied after the first cut along with an irrigation of 5.0 cm. About 1.6- 4.0t/ha of grain and 9-15 t/ha of green forage could be produced from barley. RD-2035, RD-2035, RD-2552, RD- 2670, DL-36 DL-157, DL-454 and Azad are suitable dual purpose barley varieties for the entire growing tracts.

Table 3b. Food-Fodder Production in Double Cropping Systems Under Rainfed Conditions.

Cropping system	Grain (t/ha)		Forage/ Stover (t/ha)
	Main Crop	Inter/ sequential Crop	
Sorghum + pigeonpea (2:1)	1.610	0.614	9.78
Forage sorghum - chickpea	-	1.012	17.28
Forage sorghum - mustard	-	0.129	18.23
Forage sorghum - lentil	-	0.593	18.43
Soyabean + pigeonpea (4:1)	0.633	0.360	-
Soyabean	0.939	-	-
Soyabean + pigeonpea (2:1)	0.473	0.742	-
Sunflower - chickpea	1.003	0.838	-
Sunflower – mustard	1.090	0.095	-
Sunflower - lentil	1.068	0.466	-

D. Remunerative Forage based System for Diversification of Existing Cropping Systems

Fodder based crop sequences have potential for realization of higher monetary returns over food based crop rotation in majority of the agro-ecological regions. Inclusion of forages in any of the season was equal or superior than food based systems (Agrawal *et al.*, 2008). Crop diversification as a strategy to bring desirable change towards more potential crops and cropping systems is being pursued to meet ever increasing demand of agricultural products in sustainable manner. The research findings of the project carried out in different zones at selected locations have indicated that substitution of less remunerative crops and cropping systems by forages are remunerative (Sunil Kumar & Faruqui, 2007). In all the tested situations, forage based sequences or inclusion of a forage crop in most adopted sequence was superior (15 - 500 %) on net return basis over existing system (Table 4 a & b).

Table 4a. Forage Production Economics of Remunerative Forage based Sequences at Selected Locations in Different Regions of the Country.

Cropping sequence	Forage equivalent yield (t/ha/yr)	Gross return (Rs./ ha/yr)	Cost of cultivation (Rs./ha/yr)	Net return (Rs./ha/yr)	B:C ratio
Hill region : Palampur					
Maize (G)- Wheat*	39.84	49734	28251	21483	1.76
Maize (F)- Wheat	57.63	59241	29565	39676	2.34
Maize(F) + cowpea(F) – Oat + Sarson(F)	60.72	57795	28690	39105	2.36
North West region : Pantnagar					
Rice – Wheat*	192.74	77928	38704	39224	2.01
Rice – Wheat – Maize(F) + Cowpea(F)	248.16	98401	47949	50452	2.05
Rice – Berseem – Maize(F) + Cowpea(F)	208.73	75763	36894	38869	2.05
Central region : Jhansi					
Groundnut – Wheat- Mung*	115.69	80616	41250	39366	1.95
Groundnut – Berseem- Maize(F)	155.76	92306	42362	49944	2.18
NB hybrid + Berseem	109.06	78831	34540	44291	2.28

contd....

Table 4 (a) contd....

Eastern region : Bhubaneswar					
Groundnut- Mustard- Mung*	34.58	56164	24340	31824	2.31
Pearl millet(F) – Oat - Maize (F)+ Cowpea(F)	45.35	89534	32373	57161	2.77
Groundnut – Oat - Maize(F) + Cowpea(F)	32.36	86273	32638	53635	2.64
South region : Coimbatore					
Cotton -Rabi –Sorghum(F)*	37.54	72159	48647	23512	1.48
Cotton + Green gram- Maize(F)- Sunflower	54.55	89257	50792	38465	1.76
Maize(F)- Cowpea -Tomato	73.59	85346	40993	44353	2.08

* Most adopted food based sequence in the region

Table 4 (b) : Remunerative Forage Based System under Irrigated Condition for Different Agro-climatic Situation

Zones / Location	Remunerative forage based crop sequence	Most adopted crop sequence	Percent increase in net return over most adopted sequence
Hill zone : Himachal Pradesh			
Palampur	Maize (F)- wheat	Maize (G)- wheat	85
North West zone : Uttarakhand, Haryana, Rajasthan			
Pantanagr	Rice- wheat- maize (F) +cowpea (F)	Rice- wheat	29
Hisar	Pearl millet (F)–wheat-mung	Cotton- wheat	40
Bikaner	Guar (F) + Pearl millet (F)-oats (F) cowpea (F)	Guar (G)- wheat	100
Central zone ; Uttar Pradesh, Madhya Pradesh, Gujrat, Maharashtra			
Jhansi	Groundnut– berseem- maize (F) + cowpea (F)	Groundnut– wheat- mung	44
Jabalpur	Sorghum (F)– berseem– maize (F) + cowpea (F)	Rice – wheat- mung	94
Rahuri	Lucerne (Perennial)	Soybean– wheat- mung	84
Urulikanchan	Maize (F)– berseem– sorghum (F)	Soybean– wheat- mung	54
Anand	NB hybrid (P) + cowpea / lucerne	Maize (F)– potato- pearl millet (G)	58
North East zone : Eastern Uttar Pradesh, Jharkhand, Orissa, West Bengal, Assam			
Faizabad	NB hybrid (P) + berseem	Rice – wheat- mung	48

contd....

Table 4 (b) contd....

Ranchi	NB hybrid (P) + berseem	Rice – fallow	1770
Bhubaneswar	Pearl millet (F)– oats (F)– maize (F) + cowpea (F)	Groundnut– mustard- mung	80
Kalyani	Rice – oat - sesame	Rice- mustard- groundnut	63
Jorhat	NB hybrid (P)	Sesame– mustard- mung	120
South zone			
Tiptur	NB hybrid (P)	Maize (F)- + cowpea (F)- maize (F) + cowpea (F)	142
Hyderabad	NB hybrid (P) + berseem / cowpea	Sorghum + pigeonpea (G) – pigeonpea- tomato	10
Coimbatore	NB hybrid (P)	Maize (F)- + cowpea (F)- maize (F) + cowpea (F)- sorghum + cowpea (F)	300
Vellayani	NB hybrid (P)	Rice – cowpea (Veg.)- Okra	180

Perennial NB hybrid system with seasonal legumes like cowpea/ berseem/ Lucerne in NE and South region produce higher biomass and net return. Forage based sequences realized more profit in hill and south zone. Under Tarai region, rice- wheat and rice-berseem – forage maize + cowpea system have been found to be at par.

E. Forage Production Systems for Rice Fallows

In southern plateau, short duration legumes like Pillipesara, Sesbania and Sannhemp have been found promising in the residual moisture of rice fields. Horse gram, moth bean and lathyrus provide a good choice for forage supply during lean period in the rice fallows in eastern and NEH region (Sunil Kumar & Melkania, 2003). The recent results on augmentation of forage legume yield through zero tillage after rice are encouraging. In east region after harvest of late maturing rice, berseem sown with zero tillage (sowing in standing crop before15 days of harvest of rice) and 40 kg/ha seed gave one additional harvest of berseem and 15-20 % increase in green fodder yield.

F. Forage Production under Plantation Crops

Considerable area in coastal areas, southern plateau and in NEH region has potential to grow forages in partial shade of plantaion crops. Forage crops like guinea grass, congosignal grass, signal grass and *Desmenthus virgatus* are partial shade tolerant forages are successfully grown in humid and hot-pre-humid eco-regions. The inference of multilocations researches indicate almost equal green forage yield of guinea grass (70-130 t/ha) was obtained when supplemented with 50-100 % higher recommended doses of NPK in coconut shade than open. In NE regions, the forage yield was 90 % of open when grown under Silkworm tree (*Machilus bombycini*). In general, the productivity of these plantation crops was enhanced by 10-15 % due to input management in understorey forage crops (Table 5a). Benefit : cost ratio of Coconut / banana intercropped with guinea grass & *Desmenthus virgatus* (3:1) ranged 2.2-2.8 while sole coconut / banana stands obtained B: C ratio 1.6-1.9 (Table 5b). In Kerala, congosignal grass under coconut plantation crops gave 40-50t/ha green fodder when supplemented with150 kg N/ha (in three splits) and harvesting at 45 days interval. Application of *Azospirillum/ Azotobacter* to these grasses under shade saves 25-30 % of nitrogen requirement (Anonymous, 2005).

Table 5 a : Performance of Guinea Grass under Different Shade Levels at Multilocations

Treatment	Performance of guinea grass					
	Coimbatore		Vellayani		Jorhat	
	GFY	NMR	GFY	NMR	GFY	NMR
A. Shade levels						
Open	133.0	29770	82.0	28747	67.3	30538
Under shade	113.3	27832	70.7	23382	58.1	26013
B. Fertlity shade						
RDF (NPK)	97.8	20992	64.1	21850	49.6	22709
25 % above RDF	107.8	23314	72.1	24464	55.9	25597
50 % above RDF	125.4	30919	76.3	26031	64.1	29230
75 % above RDF	136.9	32839	81.7	27751	68.6	30775
100 % above RDF	148.0	35944	87.7	30229	75.1	33066

RDF- Recommended dose of fertilizer, GFY- Green fodder yield (t/ha/yr), NMR- Net monitory return (Rs/ha/yr)
Note : Plantation crop – Coconut (Coimbatore & Vellayani), Silkworm tree (Jorhat)

Table 5 b : Forage Production Potential in Plantation Crops at Coimbatore

Crops	Guinea grass equivalent yield (t/ha/yr)	Benefit : cost ratio
Banana (sole)	312.5	1.91
Guinea grass (sole)	94.0	1.12
Banana + Guinea grass + Desmenthus (3:1)	378.0	2.11
Coconut (sole)	141.8	1.66
Guinea grass (sole)	122.3	2.12
Coconut + Guinea grass + Desmenthus (3:1)	222.3	2.80

G. Forage Production from Problem Soil

After proper amelioration, the forage production in problem soils may be adopted successfully and most economically. The productivity of forage production systems under problem soils have been evaluated for acid soil at AICRPFC centres viz. BAU, Ranchi, AAU, Jorhat, OUAT, Bhubaneswar and BCKV, Kalyani and saline sodic soils at NDUAT, Faizabad. In Table 6 suitable cropping systems for different land use and management options have been given (Faruqui and Sunil Kumar, 2007).

Table 6 : Potential Forage Production System for Problem Soils after Amelioration

	Cropping system	Productivity (Green fodder) (t/ha)
(A) Acid Soils	**I. Intensive cropping**	
	Maize/ Dinanath grass/ Teosinte – Oats – Cowpea	120 – 135
	Ricebean/ Cowpea – Oats/ Maize – Cowpea	80 – 95
	Teosinte/ Maize+Cowpea – Oats – Maize+Cowpea	105 – 120
	Dinanath grass – Oats – Maize+Cowpea	110 – 130
	II. Food-forage cropping	
	Upland	
	Ricebean/ Cowpea – Oats/ Maize – Rice (Direct seeded)	100 – 115*
	Medium land	
	Rice – Oats – Fallow	60 – 65
	Rice – Oats – Cowpea/ Ricebean	120 – 135
	Rice – Oats – Rice + Cowpea	120 – 145
	III. Mixed/ Inter cropping	
	Maize+Cowpea – Oats+Lathyrus – Maize+Cowpea	100 – 110
	Sorghum+Cowpea– Oats+Pea – Maize/Teosinte +Cowpea	85 – 100
	Maize/ Teosinte + Cowpea – Rapseed/ Niger + Oats - Rice+Ricebean(F)	90 – 115
	IV. Perennial grass based system	
	NB Hybrid/ Setaria/ Guinea – Oats - Cowpea	115 – 135
(B) Saline - Sodic Soils	**I. pH>10.5**	
	Dhaincha (GM) – Rice – Wheat	100 – 115
	Rice - Berseem	125 – 145
	II. pH<9.0	
	Sorghum(F) – *B. juncea*	80 – 95
	Pearlmillet(F) – Barley (F&S)	75 – 95
	III. ECe 5-10 mmhos/cm	
	Sorghum(F) – B. Jnncea/ Wheat	70 – 85
	Pearlmillet(F) – Wheat/ Barley/ B.Jnncea	80 – 95
	IV. Pere nnial grass based system	
	NB hybrid + cowpea – berseem	110 – 135

*Forage equivalent yield

Assured supply of green forage can be achieved through adoption of intensive cropping system involving forage crops managed through staggered planting and harvesting at proper stage. Generally well managed perennial forage components *viz.* NB hybrid, *Guinea, Setaria* etc. gives higher assured production of green fodder. Problem soils are invariably poor in productivity. Utilization of marginal or problem soils for diversification of rice based mono-cropping through forages will lead to ameliorative effects on soil acidity, sodicity and sustainability of agricultural production system. Diversity of forage resources in problem soils largely depend upon soil type, rainfall and other climatic- edaphic factors. Growing of forage crops tolerant to acidity and salinity-sodicity can ensure reasonably good returns in the initial phases of reclamation.

Conclusions

The current information on forage production systems indicate that the forage cultivation is equally or more remunerative than grain, cash or commercial crops, besides having positive effects on soil health and environment. On regional basis the production economics of forage (based) via- a-vis prevalent crop sequences of the region indicated the advantages of the inclusion of forage crops in the system. Farmers' decision on land use, in future, therefore would be based on comparative advantages, rather than family needs due to changing economy from agrarian to service sector oriented and awareness and market forces. Livestock based mixed farming sustaining on forage based feeding has great potential in future.

References

Agrawal R.K.,Sunil Kumar, Tripathi S.B., Sharma, R.K. and Singh, K.A. 2008. Agro-economic analysis of fodder based cropping systems. *Indian Journal of Fertilizers.* 4, 121-132.

Ali M. and Rawat, C.R. 1986. Production potential of different food-fodder cropping system under dryland conditions of Bundelkhand In : Proceedings of National Seminar on '*Advances in Forage Agronomy and future strategy for increasing Biomass Production*. Indian Grassland and Fodder Research Institute, Jhansi Oct. 6-7, pp 65-68.

Anonymous. 1998. Annual Report, IGFRI, Jhansi

Anonymous. 2005. Proceeding of the AICRP on Forage Crops, Annual Group Meet, (Rabi).

Faruqui, S.A. and Sunil Kumar. 2007. Forage production from problem soils, Technical bulletin no. 04 / 2007, AICRP on Forage Crops, IGFRI, Jhnasi.

Hazra, C.R. 1998. Advances in fodder production system In : Strategy for maximization of forage production ED(s) A.K. Mukherjee, S. Maith & M.K. Nanda, Directorate of Research, BCKV, Kalyani, West Bengal pp. 40-56

Pathak, P.S. and Sunil Kumar. 2004. Forage and grazing resources in different agroclimatic regions in India In : S.S. Kundu, A.K. Mishra and P.S. Pathak (eds.). *Buffalo production under different climatic regions.* International Book Distribution Co. Lucknow : pp 1-41.

Sunil Kumar and Faruqui, S. A. 2007.Crop diversification through forages: A new paradigm towards sustainable agriculture. In: *Abstracts: National Symposium on "A new vistas to forage crop research"*, September 10-11, BCKV, Kalyani : pp.79.

Sunil Kumar and Melkania, N.P. 2003. Improving forage production for sustaining livestock in India. *In:* S.P. Mittal (ed.) Alternative land use systems for sustainable production, Oriental Enterprises. Dehradun pp. 308-330.

Sunil Kumar. 2007. Advances in forage production technology: An Indian perspective In: *Special Summer School Training on Recent Advances in Agricultural Sciences* on 23 May, 2007 CSAUAT, Kanpur.pp: 125-140.

❑❑❑

Chapter **11**

Analytical Techniques for Determination of Mineral in Feed and Forages

K. K. SINGH, S. K. MAHANTA, M. M. DAS, ANIL KUMAR AND A. K. MISRA
Indian Grassland and Fodder Research Institute, Jhansi-284003 (U.P.), India

Introduction

All the biological samples viz., feed, fodder, blood, faeces and urine contain one or the other mineral elements. Determination of mineral from these samples need destruction of organic matter either by dry or wet digestion.

Dry Digestion (Ashing)

Weighed amount (0.5-1.0 g) of biological material is put into a silica basin/ crucible and ash at 550^{0}C for 3 hours in a muffle furnace until a whitish -grey residue appear. When all the organic matter has been destroyed remove the crucible from the furnace. After cooling add 10 ml of 25% hydrochloric acid taking care that losses due to effervescence does not occur. Filter the content through Whatman's No. 42, 100-125 mm filter paper into a distilled water rinsed volumetric flask (25-50 ml) and make the final volume with de-ionised water. Now the sample is ready for the analysis of mineral elements. In this procedure there is loss of some of the volatile element viz., phosphorus, sulphur and fluorine.

Wet Digestion

Weighed amount (0.5-1.0 g) of biological material is put into digestion tubes (100-400 ml capacity) or in a conical flask (150-200 ml capacity) and add di-acid (HNO_3: $HClO_4$) or tri-acid mixture (HNO_3: $HclO_4$: H_2SO_4= 15:2:4) and digest on a digester at 120⁰C for 2 hours. There after, increase the temperature of digester up to 270⁰C for next one and half hours. Cool the test tubes. Filter the content through Whatman's No. 42, 100-125 mm filter paper into a distilled water rinsed volumetric flask (25-50 ml) and make the final volume with de-ionised water. Now the sample is ready for the analysis of mineral elements.

A. Determination of Phosphorus (Colorimetric)

Phosphorus stock standard solution (1 mg/ml): Weigh 4.3913 g of potassium di-hydrogen orthophosphate, which has been dried for 1 hour at 105⁰C. Dissolve the content in de-ionized water and make up the volume to 1 liter.

Phosphorus working standard solution (0-8 μg/ml): Prepare solutions containing 0, 1, 2, 3, 4, 5, 6, 7 and 8 μg/ml of phosphorus.

Ammonium molybdate/ ammonium vanadate reagents: Dissolve separately 20 g ammonium molybdate in 400 ml de-ionized water and 1g ammonium vanadate in boiling water. Cool the ammonium vanadate solution and add 140 ml concentrated nitric acid. Now, add the molybdate solution to the acid vanadate solution and dilute to 1 liter.

Preparation of standard graph: Take 5 ml of the working standard solution into 50 ml volumetric flask, add 10 ml of vanadate reagent and dilute to 50 ml with de-ionized water. Put stopper and shake well and allow to stand for 30 minutes for complete colour development. Now, measure the absorbance at 420 nm.

For unknown samples, repeat the above steps as has been done for the standard graph preparation and read the absorbance at 420 nm.

Calculation of results: The phosphorus concentration in samples are calculated by taking 0.8 μg/ml of phosphorus is equal to the 0.8% of P.

B. Determination of Minerals using Flame Photometer

Flame photometer is a device used in inorganic chemical analysis to determine the concentration of certain metal ions such as sodium, potassium, lithium, calcium and barium. In principle, it is a controlled flame test with the intensity of the flame colour quantified by photo electronic current circuit. The sample is introduced to the flame at a constant rate. The filter selects the colour depending upon the photometric settings and determine the concentration while excluding the influence the others.

Determination of Sodium

Sodium stock standard solution (1 mg/ml): Weigh 2.541 g of sodium chloride, which has been dried for 1 hour at 105°C. Dissolve the solute and make up the volume to 1 liter with 0.5N acetic acid.

Sodium working standard solution (0-5 µg/ml): Prepare solutions containing 0, 1, 2, 3, 4 and 5 µg/ml in 0.5N acetic acid solution.

Preparation of standard graph: Set the flame photometer according to the manufacturers instructions for measure the sodium. Nebulise the sodium working standard solution containing 0-5 µg/ml sodium and adjust the controls until a steady zero and maximum readings are obtained. Then check the flame photometer by nebulising the intermediate sodium working standard solutions and construct the graph relating meter readings to µg/ml sodium in all the standard solutions.

Examination of unknown samples: Pipette out 2.5 ml of digested sample solution into a 50 ml volumetric flask and dilute to 50 ml with de-ionized water. Now Nebulise the diluted sample solutions and note the meter readings. Retain the diluted solutions for further analysis of potassium, calcium etc.

Calculation of results: Read from the standard graph the number of µg/ml sodium equivalent to the meter readings of the sample and blank determinations. The difference gives the µg/ml of sodium solutions. Express the result as percent i.e., 5 µg/ml sodium is equivalent to 1% sodium in sample.

Determination of Potassium

Potassium stock standard solution (1 mg/ml): Weigh 1.9066 g of potassium chloride, which has been dried for 1 hour at 105°C. Dissolve the solute and make up the volume to 1 liter with 0.5N acetic acid.

Sodium working standard solution (0-50 µg/ml): Prepare solutions containing 0, 10, 20, 30, 40 and 50 µg/ml in 0.5N acetic acid solution.

Preparation of standard graph: Set the flame photometer according to the manufacturers instructions for measure the sodium. Nebulise the sodium working standard solution containing 0-50 µg/ml sodium and adjust the controls until a steady zero and maximum readings are obtained. Then check the flame photometer by nebulising the intermediate sodium working standard solutions and construct the graph relating meter readings to µg/ml sodium in all the standard solutions.

Examination of unknown samples and calculation of results: This is same as sodium. However, in case of potassium 50 µg/ml potassium is equivalent to 10% potassium in sample.

Determination of Calcium

Calcium stock standard solution (1 mg/ml): Weigh 2.4971 g of calcium carbonate, which has been dried for 1 hour at 105°C. Dissolve the solute with 1N hydrochloric acid and make up the volume to 1 liter.

Calcium working standard solution (0-100 µg/ml): Prepare solutions containing 0, 10, 20, 40, 60, 80 and 100 µg/ml with de-ionized water.

Examination of unknown samples and calculation of results: Other procedures are same as sodium. However, in case of calcium 100 µg/ml calcium is equivalent to 1% calcium in sample.

C. Determination of Mineral by Atomic Absorption Spectrophotometer

This is a technique for elemental analysis in liquids. Metallic elements can be determined in both organic and inorganic samples. It is a sensitive technique that can determine the concentration of most elements at the part - per - million (ppm) level. If lower detection limits are required then a graphite furnace is used as the excitation source (GFAAS), replacing the standard flame. GFAAS is for ultra-trace analysis. It has detection limits that exceed the conventional flame by several orders of magnitude. Through the use of calibration curves, prepared from suitable standards, a high level of accuracy and precision (±1 to 3 %) is achieved for flame AAS. This allows for compositional as well as trace impurity analysis. Most spectral interferences and matrix effects are overcome by utilizing well-characterized conditions and matrix matching of samples and standards. GFAAS has an average accuracy of ±20% (relative), which is suitable for reporting at the ultra-trace impurity level.

Reagents and apparatus: The purity of the reagents may determine the accuracy of the analysis and reagents should therefore be of the highest purity available. Prior to any analysis all the apparatus should be thoroughly soaked in dilute nitric acid solution (0.5 N HNO_3) and rinsed several times with distilled water.

Sample preparation: The sample to analyze can be brought in to solution by a number of means e.g., by acid attack, alkali fusion and dissolution in solvent. However, normally acid mixture digestion is preferred over other methods of sample preparation as discussed earlier.

Preparation of standard solutions: For the preparation of standard solution of a particular element, recommended chemical or material should be used for high accuracy of analysis (Table 1). The ready-made standard solutions (1000 ppm) are also available.

Table 1 : Recommended Chemical/ Material for Standard Preparation

Element	Material/ chemical	Solvent used	Amount dissolve to get 1000 µg/ml of element (g)
Calcium	$CaCO_3$	Nitric acid	2.4970
Cobalt	Cobalt metal strip	Nitric acid	1.0000
Copper	Copper metal strip	Nitric acid	1.0000
Iron	Iron metal strip	HCl	1.0000
Mercury	$HgCl_2$	Nitric acid	1.3540
Potassium	KCl	Water	1.9070
Magmesium	Magnesium metal strip	Nitric acid	1.0000
Manganese	Manganese metal strip	Nitric acid	1.0000
Molybdenum	Molybdenum metal strip	Nitric acid	1.0000
Selenium	Selenium metal pellet	Nitric acid	1.0000
Zinc	Zinc metal granules	HCl	1.0000

AAS manual (1989)

Recommended standard concentration: One must use the recommended levels of standard concentration (Table 2). Because below this level and above this level the sensitivity of the instrument is reduced significantly.

Table 2 : Recommended Concentration (µg/ml) of Few Standard Solutions

Elements	1st Standard	2nd Standard	3rd Standard	4th Standard
Calcium	1	2	3	4
Cobalt	5	10	15	20
Copper	2	4	6	8
Iron	5	10	15	20
Mercury	200	400	600	800
Potassium	0.5	1.0	1.5	2.0
Magmesium	0.2	0.4	0.6	0.8
Manganese	2	4	6	8
Molybdenum	25	50	75	100
Selenium	50	100	150	200
Zinc	05	1	1.5	2

Recommended working conditions: The atomic absorption spectrophotometer should be set as per the recommendations given in Table 3 for optimum performance for dilute aqueous solutions. The upper limit of the working range concentration will give about 0.8-10. absorbance. The lower limit of the working range is about 10 times the defined detection limit.

The optimum wavelength has been selected on the basis of sensitivity (Table 3). However, an alternative wavelength is also used to give a variety of analytical range.

Table 3 :

Elements	Wavelength (nm)	Slit width (nm)	Optimum working range (µg/ml)
Calcium	422.7	0.5	0.01-3
Cobalt	240.7	0.2	0.05-15
Copper	324.7	0.5	0.03-10
Iron	248.3	0.2	0.06-15
Mercury	253.7	0.5	2-400
Potassium	766.5	1.0	0.03-2
Magmesium	285.2	0.5	0.003-1
Manganese	279.5	0.2	0.02-5
Molybdenum	313.3	0.5	0.2-100
Selenium	196.0	1.0	5-250
Zinc	213.9	1.0	0.01-2

After setting the instrument (atomic absorption spectrophotometer), first nebulise the standard solutions followed by extracted aliquot of samples and record the absorbance.

Calculation

$$\text{Element concentration (ppm)} = \frac{\text{OD of sample x Concentration of standard}}{\text{OD of standard x aliquot (ml) x Wt. of sample}} \text{ x dilution}$$

References

AAS Manual. 2003. *Analytical Methods*. Flame Atomic Absorption Spectrometry (Varian). Pp. 1-173.

AOAC. 1980. *Official Methods of Analysis*. 15th ed. Association of Official Analytical Chemists, USDA, Washington, DC.

□□□

Chapter **12**

Forage Evaluation and Animal Production under Grazing

S. K. MAHANTA, M. M. DAS, K. K. SINGH AND N. DAS
Indian Grassland and Fodder Research Institute, Jhansi-284003 (U.P.), India

Introduction

Grazing animals constitute a significant proportion of livestock population in India. These animals not only provide milk, meat and fiber, but also they can be incorporated into a crop rotation to take the advantage of nutrient cycling. They can be utilized to control weeds or to harvest crop residues. Grazing animals can also be an added source of income, diversifying farm enterprises and thereby rendering a farm more sustainable from an economic point of view.

Much of what we understand about animal nutrition has been developed from studies and experience with confinement feeding operations, where concentrated nutrients in the form of grain, oilseed cakes, and harvested forages are given to animals in a dry-lot/ stall. These types of practices leave out many of the biological and climatological variables that accompany grazing situations: plant species, herbage stage of maturity, soil fertility and water holding capacity, annual and seasonal precipitation and mean temperature, etc. As Livestock producer/farmers plan for the nutritional needs of their grazing animals, they need to take each of these variables into consideration.

In fact, determination of the amount eaten and characteristics of the diet of grazing animals remains one of the most difficult tasks in research. While several lifetimes have been devoted to developing techniques to 'measure' intake and diet quality, they are laborious, expensive, and often lack both precision and accuracy.

How the Animal Meets Her Nutrient Demands

Herbivore animals eat to satisfy a need and desire for nutrients, the most prominent being energy and protein. Demands for maintenance, lactation, growth, and conceptus sum to the total requirements (Fig. 1). Energy requirements may be modified by requirements for locomotion or thermal stress. Feed bulk density, ease of consumption, ease of communition, palatability, and ease of digestion and passage once in the rumen all interact with the needs and desires of the animal. Maximum or optimum nutrient intake occurs under ideal conditions and when the nutritional needs are met, anatomical sensors send a signal to the brain that satiety has occurred. Under many grazing situations, satiety does not occur because the animal is unable to satisfy its nutrient requirements with the forage on offer. Diet selection by large herbivores such as cattle requires the solution of two opposing problems: obtaining adequate quantity of forage with maximal quality. Reduced dry matter (DM) intake is the major impact of inadequate nutrition of grazing animals, whether constrained by deficient herbage quality or quantity.

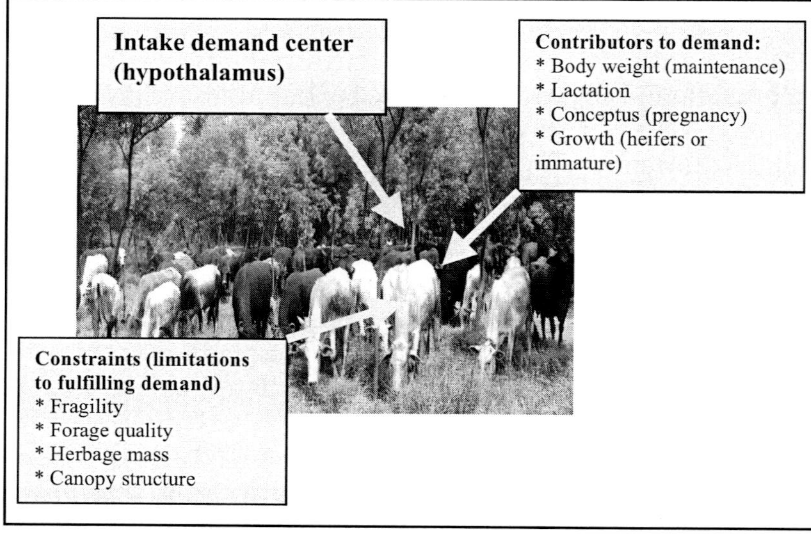

Fig. 1. The Animals Demand for Energy based on Components which Drive the Desire to Consume Herbage and Constraints to *ad Libitum* Intake by the Pasture Characteristics

Daily Requirements Make up

The energy requirements are met when the animal consumes sufficient energy to meet the summative requirements. The energy consumed is a function of the amount of feed consumed and the concentration of utilizable energy in the feed. Utilization of energy (Fig 2) is slightly different for different functions (e.g., maintenance, growth or milk). Because the largest and most variable loss of consumed energy is in the feces, digestibility may be used as a proxy for the entire process of utilization (digestion, absorption, assimilation). For most purposes, the digestibility of the energy is similar to that for the total dry matter, so dry (or preferably organic) matter digestibility (OMD) is a good proxy from which net energy can be calculated using equations from NRC (1996). There has been good agreement between the intake of digestible organic matter (DOMI) and gain.

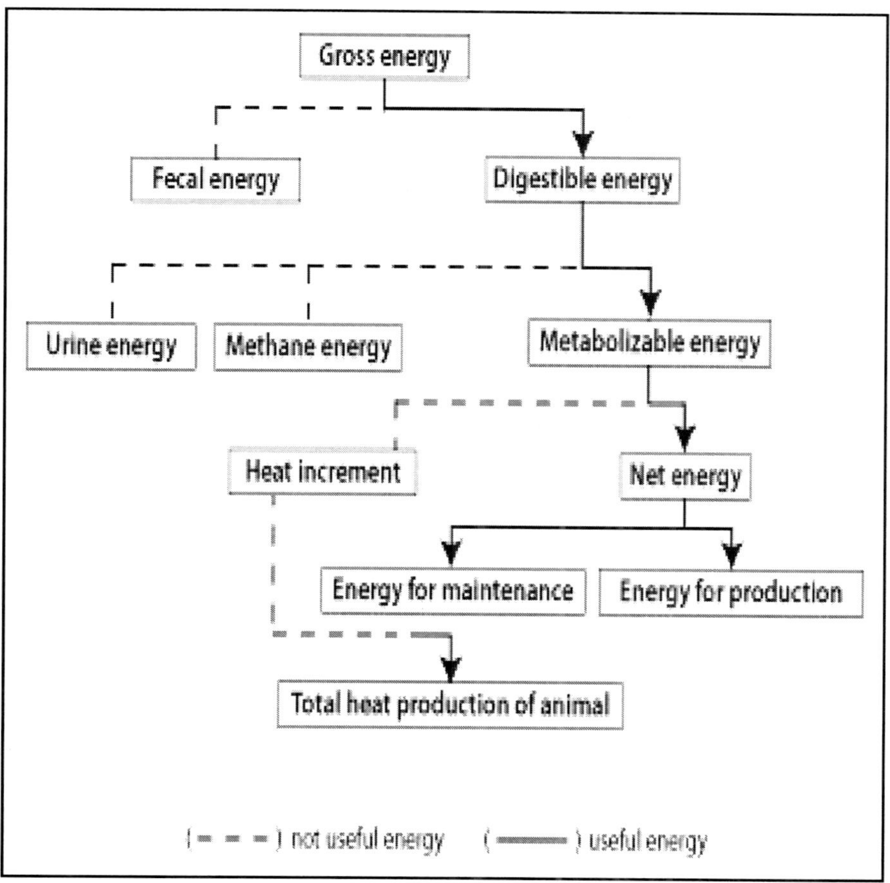

Fig. 2. Energy Partitioning in Grazing Animals

Forage Evaluation from a Vegetation

Forage Production

Terms such as biomass, herbage, browse, forage, cover and density are used to describe the ecological condition of grazing land and its ability to contribute nutrients to a grazing animal. Biomass is one of the most important characteristics of grazing land vegetation and it is defined as the total amount of living plants and animals above and below ground in an area at a given time. Herbage is "herbs taken collectively or total above ground biomass of herbaceous plants regardless of grazing preference or availability"; browse is "that part of leaf and twig growth of shrubs, woody vines and trees available for animal consumption"; and forage is "browse and herbage which are available and may provide food for grazing animals or be harvested for feeding". Availability and acceptability should be included in the definition of forage, describing the portion of the forage production that is available and accessible for use by a specified kind or class of grazing animals. Cover is "the plants or plant parts, living or dead, on the surface of the ground or the area of ground cover by plants of one or more species", while density "the number of individuals per unit of area".

It is obvious that the estimation of the above vegetation parameters, especially forage production, are very useful to project stocking rates and feed requirements for specific time periods (i.e. annually, grazing season, rotation cycle, etc.). In addition, the degree to which a forage meets the nutritional requirements of a specific kind and class of animal (i.e. forage quality) is always important to be taken into account.

Clipping and Weighing

Weight is the most important expression of herbage production. Clipping is probably the most common method for determining herbage weight for pasturelands and grasslands where the vegetation is primarily herbaceous and is stratified into relatively homogeneous types. The herbage weight is determined in representative small plots (quadrats), which can be varied in size and shape (e.g. rectangular, quadrangular, and circular). Herbage weight from such quadrats is multiplied by a given factor to obtain kg/ha. Quadrangular quadrates with area of 250 cm^2 are the most common used for herbage determination. Several types of hand clippers and shears are used for plant material harvest of quadrats and the clipping is applied either on individual species or on the whole herbage indiscriminately. The clipping height is usually 5 cm above the ground and separation into live and dead components is frequently needed.

Browse production is based on current twig and leaf growth available to animals, ordinarily within 1.5 m of the ground. Measurements of browse are done on representative plants of each present woody species in the rangeland, which are cut at 10 cm height above ground and the collected material is hand-separated into browse material (leaves and twigs up to 2 mm diameter) and branches and all of them are weighed after oven drying at 70°C. Browse is expressed in kg DM per plant; whether the number of each woody species per unit area is known browse can be expressed in kg/ha, as well. However, in shrub lands composed of both woody plants and an undergrowth of herbaceous plants, measurements have to be done on both herbage and browse. This, can most of the time be achieved with clipped plots. The disadvantage of this method is that it is tedious and time consuming.

Estimation and Double Sampling

To overcome the disadvantage of the previous method, weight estimate method has been developed which is much faster than clipping and weighing and permits more intensive sampling. This method presupposes extensive training with actual clipping to adjust estimates and to improve accuracy. Disadvantages include the need to develop estimation skills, a high degree of concentration needed by the estimator, lack of accuracy and variation among observers. Estimators tend to underestimate quadrats with high herbage weights and overestimate those with low herbage weights. To improve the weight estimation technique some researchers used the double sampling, in which a small proportion of the estimated plots are independently clipped and weighed. From this small sample the relation between estimated and actual weight is calculated and used to adjust the estimate of the large sample. The double sampling method was widely used in shrub lands for the determination of their forage production.

Forage Intake

Intake is probably the most important variable determining animal performance and voluntary intake is generally correlated with the amount of nutrients that can be extracted from feed. The measurement of forage intake serves firstly as a means of explaining differences in animal performance between two species or cultivars, secondly as the basis for a hypothesis that attempts to provide a more general explanation for such differences between species or cultivars, and thirdly as an aid to explain variation in animal performance associated with different grazing regimes and management practices. Estimating the feed intake of a grazing animal must by definition rely upon techniques imposing minimal disturbance to the normal grazing activity of that animal.

Measuring Intake in Stall-Fed Animals

Ad libitum intake occurs when the feed supply is unlimited in quantity and quality so that animals are unconstrained in satisfying their hunger. Most determinations of forage intake have been conducted by feeding animals under controlled situations in which the feed is either harvested or manufactured (mixed) and provided to animals in dry lots or pens. Intake is measured as the difference between the feed offered and that refused when fed in excess of the amount the animal would consume. While these measurements may be quite accurate, variability among animals given the same diet, particularly a forage diet, may be quite high, ranging from 10 to 30% of the mean. This high variation may be caused by different nutrient demands among animals, by differences in the ability to consume and process feed, or the conditions under which the intake trials are conducted.

Measuring Intake in Grazing Animals

There is a huge dilemma for measuring intake under grazing because weighing the feed and refusal as in pen studies is not possible. Therefore, indirect measures or estimates have been used, and generally the accuracy and precision (measure of variability) are compromised. Several methods are available for estimating intake of grazing animals and all the methods have their own advantages as well as disadvantages (Table 1).

Table 1 : Some Advantages and Disadvantages of Different Methods used in Forage Evaluation in Grazing Animals

Parameter	Technique	Advantages	Disadvantages
Faecal output	Faecal bags	• Amount of faeces actually voided obtained • No chemicals required • Give rapid results	• Tedious • Problem with diet low in dry matter • Distortion of hind legs due to weight of faeces in bags • May affect animal behavior • Not practical with large animals
	Orally- dosed markers (Cr_2O_3)	• Amount of faeces actually voided obtained	• Rectal grab sampling required • Diurnal variation in faecal chromium concentration • Animal disturbance
	n-alkenes	• Easy to analyze	• Alkanes not completely recovered in faeces • Hydrocarbons could be absorbed in the small intestine

Contd...

Table 1. Contd...

Forage intake	Weighing animals	• No need for laboratory analyses (chemicals and specific apparatus) • An estimate of fresh forage consumed could be obtained	• Large investment (equipment and sophisticated computer software) • Still not used in practice • Need to account for weight loss (defection and urination)
	Differences in forage mass (HM)	• No need for laboratory analyses (chemicals and specific apparatus)	• Assumption that the decline in HM is entirely consumed • Not applicable for woody species • How to consider pasture growth and senescence?
	Bite counting	• Operator training for bite count • Observations is minimal and requires little specialist expertise	• Time consuming • Difficult at night • Unreliable (inter-observer variation)
	Esophageal fistulas	• Used in large and small size animals	• Incomplete extrusa recovery • May affect grazing behavior
Diet digestibility	48-h- *in vitro* and *in situ* techniques	• Less time required than *in vivo* digestion trial	• Not adequate for tannin-rich feeds • Forage consumed by penned animal is often not similar to that consumed by grazing animal • Same degree of digestion between penned and grazing animals
	Internal markers (lignin, silica, etc.)	• No use of external markers (cost) • Rapidity, accuracy	• Diurnal variation of lignin concentration • Soil contamination (silica) • Sophisticated equipment required • Database for calibration
Diet composition	Hand-plucking	• Sample free from - salivary contamination	Tractable animal which can be approached closely
	Esophageal fistulas	• Accurate representation of diets(selective grazing/browsing)	• Requirement for surgery • Biases due to short time of collection period
	Micro-histological analyses of rumen content (plant epidermal fragments)		• Laborious • Prone to error because of the differences in the digestibility of different component species and plant parts • Unequivocal identification of some species (dicots) is difficult

Contd...

Table 1. Contd...

	Micro-histological analyses of faecal samples	• Short term (single defaction) and long-term (series of faecal samples bulked over a number of days) estimates of diet composition could be made	• Forage species passed in the faeces are often not proportional to those consumed • Destruction of some plant species during preparation
	n- alkane in faeces	• Analysis easy and precise	• Plants (species or parts) low in alkanes could not be identified
	Bite size and count	• Requires little equipment	- Operator training in plant identification - Low precision of identification when forage species make up less than 2% of the diet
Sampling for intake, digestibility and diet composition determination	Hand sampling	• Little equipment and time	• This procedure is subjective • May not correspond to the actual plant parts consumed by the animal
	Esophageal fistula	• Little physiological disturbance compared to rumen fistula • Used in large and small size animals	• Salivary contamination of forage (interactions tannins -salivary proteins) • Chemical changes during mastication and salivation • Incomplete recovery

Fecal Excretion

If one knows the amount of feces voided (FO) each day and the digestibility (D) of the diet consumed, then intake can be calculated by the equation-

$$\text{Intake (kg/d)} = FO/(1-D) \qquad \text{Eqa.. 1}$$

However, collection of feces voided is quite difficult under grazing, although not impossible. Various kinds of collection bags have been used. A simple plastic bag can be clipped to the wool of small ruminants such as sheep. The bags are removed once or twice per day, feces samples are then dried and processed as in a typical digestion trial. However, with cattle and buffaloes, fecal bags are quite cumbersome, require moving the animal to an enclosure to remove and replace the bags, and often interferes with normal grazing, likely reducing intake. Hence, other methods have been devised to estimate fecal output. The use of external markers has probably been most widely used to estimate fecal output of grazing, and often stall-fed, cattle. The concept is that when a known amount of an inert (indigestible) marker is fed and equilibrium is obtained (~ 7 d), the same amount should be voided in the feces. If a small grab sample of daily fecal excretion is obtained, concentration of the marker in the feces can be used to calculate the amount voided. Assumptions include: 1) the marker is quantitatively recovered (completely indigestible or the digestibility is known and constant); 2) the marker is proportionately and evenly excreted (no pattern over hours or days); and 3) the marker does not interfere with ruminal activity. Markers that have been widely used include chromium sesquioxide (Cr_2O_3), rare earth elements such as Ytterbium (Yb), and long chain alkane waxes (C_{32} and C_{36}).

Diet Quality

Even if fecal output can be measured or estimated, digestibility of the diet is also required for equation 1. Most often a hand clipped sample of the herbage is evaluated by routine chemistry and in vitro organic matter digestibility (IVOMD) to represent nutritive value. However, IVOMD often fails to mimic in vivo digestibility determined with animals. To avoid this error, adjustment equations relating in vitro to in vivo digestion of similar forages may be used. A greater difficulty is obtaining a sample of the diet. Many studies have shown that animals graze selectively and that hand clipping a sample from the available herbage in a pasture does not adequately characterize the diet. An attempt was made to hand-pluck samples similar to that being grazed by cattle. However, esophageally-fistulated steers selected forage with more ash, protein, fat and less crude fiber than hand plucked samples. The degree of selectivity reported in the literature is quite variable due to season, forage species, and chemical anti-quality components, such as tannins. Because of animal selectivity, *most investigators in this field of research concluded that it is virtually impossible to sample a pasture by hand and have reliable representation of the animal diet.* The problems are exacerbated if the pasture contains a mixture of plant species or heterogeneity in canopy (e.g., with spot grazing or where there is mixture of dead and green material). With modern techniques for mapping (e.g., global positioning systems), the relative proportion of grazed to the total area available could be mapped and characterized. Because of the inadequacy of hand sampling, two methods have evolved to estimate diet quality-

i) The use of esophageally- (or occasionally ruminally-) fistulated animals, and

ii) The use of internal markers (as opposed to external markers mentioned above).

Critical Components of Diet Quality

Herbage nutrient analysis can be a good tool to determine diet quality. However, forage quality for grazing animals is more accurately determined by the following factors, which are affected by observation and adaptive management of the grazing resource-

- Herbage intake
- Forage diversity
- Forage quantity, availability, and density
- Appropriate supplementation (energy or protein), when necessary
- Appropriate minerals—offered free choice
- And clean, fresh water offered at all times

Cannulated Animals

The use of the esophageal fistula allowed the animal to collect the forage sample for the investigators. The fistula is simply an opening (~1.5 inches in diameter) through the neck into the esophagus through which the grazed forage can drop. When not being used for forage collection, the fistula is closed so that feed passes down the esophagus to the rumen. Later a cannula was developed for the fistula to facilitate opening and closing the fistula. Samples of forage collected by fistulated animals have consistently been shown to be different chemically from those collected by hand, whether by random selection of the forage to be clipped, or by seeking to harvest only that which the animal is eating.

While the use of animals fistulated at either the esophagus or rumen have been used for many years to collect forage samples representing the diet selected by the grazing animals, they are not without their problems. Esophageally fistulated animals are not easy to prepare and they require constant maintenance, because if the fistula opens accidentally then severe loss of electrolytes can occur. As a sampling method, questions have arisen over incomplete collection of the eaten forage and whether the fistulated animals ingest a diet similar to that of target animals (the ones the diet is supposed to represent). Little relationship ($r^2 = 0.13$) was found between the grass or legume content of diets selected by fistulated animals and diets from resident cattle (those that had been grazing the same pasture for the entire season) determined by the ä13C content of the forages and feces. Contamination by saliva can be problematic and variable. Only ash was found significantly different between hand-clipped and esophageal-fistula samples. However, later the differences were observed and adjustments were developed for each ash, protein and fiber components.

Incomplete recovery of the diet may also occur if some of the ingested material bypasses the opening and proceeds to the rumen. In order to alleviate the loss of ingesta, some have used animals with rumen fistulas fitted with a molded rubber cannula. Ruminally-fistulated cattle are not as likely to undergo stress due to loss of cannula or plug as those fistulated at the esophagus, and thus are easier to maintain. However, the same problems exist relating to grazing when samples are desired and whether the diet mimics resident animals.

Internal Markers

The use of internal markers to estimate digestibility of the diet has been widely investigated. Internal markers are natural plant constituents that are recovered quantitatively in the feces, that is either they are indigestible or the digestibility is known and constant. Diet digestibility (D) can be calculated from the ratio of the marker in the diet (Md) and feces (Mf)

$$D\ \% = 100 - (100 * Md/Mf) \quad \text{Eqa.. 2}$$

Markers that have been evaluated are silica, lignin, fecal N, chromogen, indigestible neutral detergent fiber, and acid insoluble ash. All have problems of consistent recovery. Chromogens are pigments associated with chlorophyll and xanthophylls and are specific for each forage. Thus they are more effective for lush growing forage and least effective for drought or stressed plant tissues. Calibration must occur for each set of conditions; therefore they have limited application. Lignin is probably the most widely used as internal marker, but suffers from variation in digestibility.

The most recent development of a marker system has been the use of the hydrocarbons (*n*-alkanes) of plant wax. By using dosed synthetic alkanes as external markers to estimate fecal output together with naturally occurring plant wax alkanes, diet selection, intake and digestibility can be determined. The technique has the advantage that determination of both internal (natural plant waxes) and external (dosed synthetic) markers are made with one process using gas chromatography. One must still collect samples of each plant part (leaf, stem, inflorescence) from each species that the animals might graze. Many plant parts and species have different proportions of alkanes that form a fingerprint for that particular tissue. By using the fingerprint information, diet composition can be determined by statistical evaluation of the pattern of waxes in the feces with that from all the available plant parts and species in the pasture. Problems with the alkane technique include variable recovery and low concentrations of natural waxes in some plants, especially tropical and subtropical species.

An ideal internal marker is currently not available. Moreover, because of the labour involved and the extreme number of laboratory assessments required, the use of markers have received little or no application by producers nor routinely in large experiments from which to develop a database of samples useful for equation development and prediction of diet quality from chemical or other characteristics of harvested forage. However, such a database would be very useful on which to base prediction equations to estimate diet quality and intake from more easily obtainable determinations.

Animal Production Strategies Under Grazing

The profit and sustainability of livestock enterprises rely on pasture to provide feed of suitable quantity and quality to meet livestock production targets at acceptable costs. We need to have a good knowledge of pasture species as well as an understanding of the objectives of the enterprise when planning for different strategies. Use of an appropriate strategy/ management is needed to provide forage in appropriate quantity and quality, and at the right time so that animals achieve their growth/ production potential. This is needed because pasture species are not ideally matched to-

- The climatic variability between and within years
- Varying soil types between and within paddocks
- Weed competition (although many weeds provide valuable feed!)
- Grazing habits of livestock, within and between species
- Grazing pressure
- And soil fertility, water logging, salinity, acidity, insect damage and so on.

Conditions on different farms even vary enormously, so it is impossible to outline a standard 'recipe' for grazing animals. But we do have a growing understanding of how to manipulate pastures through different strategies to provide for specific animal requirements.

Manipulating Species Composition

The botanical composition of a pasture affects the amount and quality of FOO (Feed on offer) which, in turn, is reflected in animal production. Botanical composition can be modified by grazing and a range of other farm practices (including the use of fertilizers, herbicides and insecticides).

Grazing Practices

The following grazing procedures affect the botanical composition of a pasture-

- Rotational grazing with adequate rest periods at three to five week intervals and high grazing pressures benefit desirable species
- Strategic grazing based on the stage of growth of the plant benefits desirable species
- Grazing strategies in conjunction with herbicides or fertilizers, spray-grazing, use of nitrogen etc affect composition of pasture
- Hard grazing over the flowering period reduces seed production in erect seeding annuals
- And heavy grazing pressure reduces seed set by annual grasses and reduces the quantity of annual grasses in the pasture the following year. This is also a consequence of conserving hay and silage.

Kind of Stock/ Animals

Different grazing habits and dietary preferences (Table 2) affect pasture composition. The effects of sheep or cattle are associated with grazing pressure and the ability of sheep to graze closer and more selectively than cattle. Sheep grazing nutritious or palatable perennial pasture plants often critically damage pasture plants. Cattle do less damage. Sheep and cattle do not need to be grazed together and this practice is likely to complicate management decisions. Run together, sheep will always benefit at the expense of cattle, but running cattle in front of sheep, on the same area, often increases animal production and provides simpler pasture control.

Other farm practices that affect composition by altering the balance of plant competition include use of herbicides and insecticides to control different weeds and pests & mites, respectively in a pasture.

Table 2: Grazing Preferences (as % of intake) from Range Vegetation in Animals

Species		Plant	
	Grass	Weeds	Browse
Horse	90	4	6
Cattle	70	20	10
Sheep	60	30	10
Goat	20	20	60

Applying Grazing Management

Different species and varieties of pasture plants vary greatly in their response to grazing. Knowledge of how each responds to grazing is essential to get the best animal performance and pasture persistence.

The objectives of grazing management are to-

- Optimize pasture growth rate
- Use feed efficiently and profitably
- Ensure quantity and quality are suited to satisfying the stock objectives
- Ensure persistence of desirable plant species
- Ensure that ground cover is adequate to prevent erosion and resist weed invasion
- Maintain stable pastures

Generally, management that assists the well-being of the pasture benefits the grazing stock. However in all systems compromises are likely. To employ the above objectives, we need to do confidently

- Read the pasture — assess its health, composition and productivity
- Assess the situation — livestock need, pasture requirements
- Apply appropriate strategies where they will be most beneficial and profitable for pasture and animals.

Grazing management should not be used as a strategy isolated from other management tools. Grazing strategies may well vary with different levels of other tools. Thus grazing management needs to be modified if poorly adapted species have been sown or plant nutrients are deficient.

Selective Grazing

Use the grazing action of animals to increase tillering and therefore the density of pastures. Keep pastures short (5–15 cm) so sunlight reaches most of the plant material and allows maximum production of new tillers or competing species. Given suitable moisture and temperature, pastures that are heavily grazed are slow to provide new growth and recovery is initially slow. Then, as new leaves are produced, they are able to trap more sunlight and the rate of growth increases. The lower a pasture is grazed and the longer it is grazed hard for, the greater will be that initial delay.

The critical factors for regrowth are

- To have plants with well developed roots to provide energy from the plant for regrowth and to supply plant needs from the soil, and
- To maintain adequate leaf area allowing plants to trap sunlight

To maximize growth rate after grazing, have plants with a good root system and sufficient residual leaf area. Pasture plants then regrow immediately. Regraze before the oldest leaves die. Do this in rotational grazing by adjusting grazing intensity and rotation speed. Do this in set stocked systems by adjusting stocking rate (and thus grazing intensity). Sheep and cattle have different feed preferences and different grazing behaviour. But both sheep and cattle select leaf in preference to stem, and stem in preference to dead material.

When FOO is high, sheep will

- Select green material
- Tend to graze closer to the ground in preferred spots (patch grazing)
- Select leaf material, and
- Have a greater impact on the botanical composition of the pasture than cattle

Cattle, on the other hand

- Also select green material
- Are less selective
- Cannot graze as closely as sheep, and
- Have a longer period of need for higher quality feed than sheep

Mature dry cattle can be used to 'condition' rank pastures by removing taller shading grass and encouraging legumes growth. They are also less selective than sheep and will remove more dry feed when mixed with green, even though it is of lower quality. However, performance of productive animals (lactating cows, growing yearlings) will fall if the overall quality of the pasture is poor. Mature dry cattle can also be forced to graze harder than the cow and calf herd without harming production. In a rotational grazing system the cow and calf herd is moved on when the FOO drops too low to maintain optimum production and a herd of dry cattle, such as heifers, may be used as followers. Use the heifers after the cows to graze the pasture lower and prevent patch grazing occurring.

Rotational grazing favours the persistence and productivity of perennial species compared to set stocking. When assessing pasture, do not simply look at total FOO, but keep an eye on the different species. Look at the legumes and see if they are being grazed heavily when grasses are not. Check whether some grass species are being grazed more than others. Animals continuously grazing a mixed pasture do not always select a particular species. Selection varies throughout the year depending on the nutritional status and palatability of the different components. In continuous grazing, grazing pressure and the pattern of the break of season are likely to have a greater effect on botanical composition than the grazing system itself. The invasion of weeds and undesirable species is a response to grazing management or other factors such as soil fertility. If desirable species are threatened, modify the grazing management or other factors involved.

Stocking Rate

It is important to distinguish between *stocking rate* and *grazing intensity*. Stocking rate is the number of defined stock units per unit area, usually expressed as ACU/ha (adult cattle unit per hectare). Grazing intensity includes a pasture component as well as stock numbers, for example, two paddocks stocked at the same rate but one with 1500 kg DM/ha FOO and the other with 2500 kg DM/ha FOO have different grazing intensities, the first being higher.

The aim of a grazing enterprise is to profit by-

- Optimum production from each animal; and
- Optimum animal production from each hectare of land

Increasing stocking rate to increase animal production per hectare can lead to lower production per head. As stocking rate increases, competition increases between animals for pasture and greater spoilage by dung, urine and trampling. Consequently, up to a certain stocking rate, animal production per hectare will rise as stocking rate increases, even though there is a decline in per head production. But increasing stocking rate past this point, results in a decline in both production per hectare and production per head. In meat animals that are being grown to a specific level of fatness the drop away is sudden because a point is reached where individual gains are too low to provide any finished product per hectare.

Fodder budgeting allows us to match pasture supply and animal production, thereby calculating the number of grazing days in a particular paddock. We need to achieve a compromise between per head and per hectare production. Use pasture benchmarks as a basis for varying stocking rates while ensuring the nutrient intake of livestock is maintained. Grazing management decisions aim to match feeding requirements with pasture production. That is, match sheep or cattle that have the lowest demand, like dry stock, with lower pasture quality. Match growing and lactating stock with the highest quality and quantity of pasture on offer.

Supplementation with Grazing

According to one estimate 406 million of livestock in India still depends on forages available from grazing lands which includes degraded range lands, forest lands and peripheries, tree grooves undegraded wastelands and agricultural fallows. But continuous shrinkage of range land, land under fodder production and forest has resulted in increased grazing pressure on such land and decreased the productivity of grazing land. Although, the production levels and their potential in different grass covers (Table 3) indicates that the possibilities are there for upgrading the production by technological interventions and management.

Table 3: Herbage Production (t/ha) from Different Grass Covers in India

Grass cover	Production level	
	Actual	Potential
Themeda-Arundinella (Sub-mountain region)	2.2	4.0
Phragmitis-Saccharum-Imperata (Sub-tropic, high humid)	4.0	5.0
Dicanthium-Cenchrus-Lasiurus (Dry and arid tract)	3.3	5.0
Sehima-Dicanthium (Central and Southern semi-arid tract)	3.5	6.0
Temperate Alpine type (Mountain region)	-	>6.0

Particularly, under semi-arid and arid agroclimatic ecology, characterized by short spell of monsoon, rapid vegetative growth and early withering with onset of dry season, the livestock productivity can not be sustained round the year. Hence, supplemental feed along with grazing is the only alternative for better livestock health and productivity. It has been observed that supplementation of concentrate feed/ top feed is beneficial as well as cost effective particularly for the small ruminant producers. However, to maximize the profit from livestock farming judicious supplementation of concentrate feed resources can be used as follows:

Protein Supplementation

During lean period when crude protein content of average range forages is less than 5-6%, the supplementation of one of the protein rich fed resources is particularly beneficial to increase the productivity of livestock. These types of supplementation are also cost effective and increase the intake and digestibility of forage. The protein sources like oil seed cakes (groundnut cake, mustard seed cake, and cotton seed cake) or legume leaf meals @ 0.5 to 0.9 kg for animal can be successfully supplemented to the large ruminants and in small ruminants the supplemental rate may be 0.1 to 0.2 kg. But to maximum cost effectiveness protein supplements can be used twice or thrice in a week.

Energy Supplementation

When livestock graze in forest land and or silvipastoral system, during lean period the critical nutrient which inhibit the livestock productivity may be the energy. Under such situation as well during drought conditions, one of the energy sources like barley grain, maize grain can be supplemented for better livestock productivity. These feeds usually depress the intake and digestibility of range forages and serve primarily as substitutes for range forages. In contrast to protein, energy supplements must be fed daily to obtain satisfactory animal performance. However, energy supplementation should be avoided during morning before grazing of livestock, which reduces the forage intake, and not satisfactory improves the livestock productivity.

Fig. 3. Growing Goats Supplemented with Energy Under Grazing

Mineral Supplementation

Minerals, the most critical nutrient, may affect the productivity of livestock even after the availability of protein and energy. Even supplementation of a pinch of salt regularly to the grazing livestock tremendously improves the productivity. Due to low cost, poorest of the poor farmer can also afford such type of supplementation. The recommended level of salt supplementation per month is 0.4 kg for growing cattle, 0.9 kg for lactating cows, 0.2 kg for small ruminants. As most of the range pasture deficient in phosphorus, which limit the livestock productivity in a major way, therefore, mixing of phosphorus sources with common salt give added advantage to the livestock farmers.

Conclusions

There will always be unanswered questions in the search to optimize livestock production in a grassland ecosystems. Grazing livestock producers and research scientists need to work together. Based upon our current understanding, the following issues are of critical importance-

- Practical decision making guidelines for economically and ecologically efficient grazing management practices
- Critical periods and intensity of grazing for individual species as well as plant communities
- Documenting seasonal shifts in plant species most preferred by livestock on native rangeland
- Critical forage allowances and/or remaining amounts of standing herbage for optimum animal performance and long-term sustainability of forage resources
- Comparative grazing studies of performance in different weight/age categories of livestock on rangeland
- Periodic summaries of practical information from research and long-term observations for major range ecosystems

References

Ahmed, J., and Bonham, C.D. 1982. Optimum allocation in multivariate double sampling for biomass estimation. Journal of Range Management 35: 777-779.

Allen, V.G. and Segarra, E. 2001. Anti-quality components in forage: Overview, significance, and economic impact. Journal of Range Management 54: 409-412.

Burns, J.C., Pond, K.R. and Fisher, D.S. 1994. Measurement of forage intake. *In* Fahey, Jr. G.C., Collins, M., Mertens, D.R., Moser, L.E. (eds.) Forage quality, evaluation, and utilization. ASA, CSSA, and SSSA. Madison, WI. pp. 494-526.

Coates, D.B., Schachenmann, P. and Jones, R.J. 1987. Reliability of extrusa samples collected from steers fistulated at the oesophagus to estimate the diet of resident animals in grazing experiments. Australian Journal of Experimental Agriculture. 27:739-745.

Dove, H. 1996. The ruminant, the rumen and the pasture resource: Nutrient interactions in the grazing animal. *In* Hodgson, J. and Illius, A.W (eds.) The ecology and management of grazing systems. CAB Intl., Wallingford, OX, UK. pp. 219-246.

Kundu, S.S., Singh, S., Mahanta, S.K. and Pailan, G.H. 2005. *Feeding Farm Animals*. Satish Serial Publishing House, Delhi. pp. 191-212.

Loredo, M.A., Simpson, G.D, Minson, D.J.and Orpin, C.J. 1991. The potential for using n-alkanes in tropical forages as a marker for the determination of dry matter intake by grazing ruminants. Journal of Agriculture Science, Cambridge 117:355-361.

Mahanta, S.K., Singh, B.B., Pailan, G.H., Verma, N.C. and Kumar, S. 2007. Effect of grazing management practices on nutritional status of grazing goats supplemented with barley grain during summer in Bundelkhand. Range Management & Agroforestry 28: 239-241

Moore, J.E. 1994. Forage quality indices: development and application. *In* Fahey, Jr. G.C., Collins, M., Mertens, D.R., Moser, L.E. (eds.) Forage quality, evaluation, and utilization. ASA, CSSA, and SSSA. Madison, WI. pp. 967-998.

NRC. 1996. *Nutrient Requirements of Beef Cattle*. 7th rev. ed. National Academy Press. Washington, D.C.

NRC. 2001. *Nutrient Requirements of Dairy Cattle*. 7th rev. ed. National Research Council, National Academy Press, Washington, DC

Pailan, G.H., Mahanta, S.K., Verma, N.C. and Kundu, S.S. 2007. Performance of sheep and gaots maintained on rotational grazing with different levels of supplementation. Indian Journal of Animal Science 77: 1161-1165.

Pathak, P.S. and Kumar, S. 2004. Forage and grazing resources in different agro-climatic regions of India. In: Kundu, S.S., Misra, A.K. and Pathak, P.S. (eds.) Buffalo production under different agro-climatic regions. International Book Distributing Co., Lucknow.

Platis, P.D. and Papanastasis, V.P. 2003. Relationship between shrub cover and available forage in Mediterranean shrublands. Agroforestry Systems. 57: 59-67.

Range Term Glossary Committee. 1989. A glossary of term used in range management. 3rd edition. Glossary Revision Committee, Publications Committee, P.W. Jacoby (chairman). Society for Range Management, Denver, Colorado.

Reese, G.A., Bayn, F.L. and West, N.E. 1980. Evaluation of double-sampling estimators of subalpine herbage production. Journal of Range Management 33: 300-306.

Senft, R.L., Coughenour, M.B., Bailey, D.W., Rittnehouse, L.R., Sala, O.E. and Swift, D.M. 1987. Large herbivore foraging and ecological hierarchies. Bioscience 37:789-799.

Torrell, D.T. 1954. An esophageal fistula for animal nutrition studies. Journal of Animal Science 13:878-884.

Van Soest, P.J. 1994. *Nutritional Ecology of the Ruminant*. Cornell University Press, Ithaca, NY, USA.

Verma, A. 1998. Feeding management strategies in different regions. Proceed. XXIX Dairy Industry Conference, Nov. 28-29, 1998 at NDRI, Karnal. pp. 84-95.

❑❑❑

Chapter 13

Plant Carbohydrate and Protein Fractions as Quality Determinants

A. K. MISRA, M. M. DAS, K. K. SINGH, ANIL KUMAR AND M.K. TRIPATHI*
Indian Grassland and Fodder Research Institute, Jhansi-284003 (U.P.), India
*Central Sheep and Wool Research Institute, Avikanagar (Rajasthan), India

Introduction

Farm livestock (cattle, buffalo, sheep and goat etc) belong to a class of animals called ruminants. Ruminants have a digestive system, which allows them to utilize roughages (e.g. hay, grass) as a major source of nutrients. These animals have a large (capacity up to 50 gal.) fluid filled digestive organ at the beginning of the digestive tract called the rumen. The rumen contains a large population of microbes (bacteria and protozoa). Microbes in the rumen do much of the initial digestion of feed. These microbes have the ability to break down cellulose and hemicellulose, which are main components of roughages. Rumen microbes also break down other components of the animal's diet such as protein and starch. The reticulum is a smaller organ, which acts as a holding area for feed after it passes down the esophagus. The omasum is an organ, which absorbs water from the digesta (mixture of feed and fluid) before it flows into the abomasum (true stomach). The animal's own digestive enzymes break down food in the abomasum and small intestine. Absorption of these nutrients occurs mainly through the small intestine. Monogastric (non-ruminant) animals (e.g. pigs, dogs, man) are not able to efficiently digest cellulose. When ruminants consume forages, they take fairly large bites and swallow the material with a minimum of chewing. After eating, they stand or lie down to "chew their cud". This involves regurgitating boluses (masses) of forage up the esophagus and into the mouth, where it is re-chewed and then swallowed. This reduces the size of the forage particles and greatly increases the surface area available for microbial digestion.

Efficiency of animal production is closely correlated with the nutrient value of the forage available. Feed quality has been defined as the amount of nutrient material that an animal can obtain from a feed in the shortest possible time. Among several factors, crude protein (CP), digestible dry matter (DDM) and metabolisable energy (ME) have been considered more appropriate attributes for evaluation of range forage quality. Forage plants are the basis of ruminant nutrition, and cell wall digestibility is the limiting factor of their feeding value. Cell wall digestibility is therefore "the" target for improving the feeding value of forage crops. Nutritionally balanced diets play a very important role in enhancing production performance of dairy animal. For exploitation of genetic potential of dairy animal synchronization of carbohydrates and protein in rumen is of great significance. Carbohydrates and proteins are major components of ruminant diet and play an important role in balancing of nutrients for preparation of efficient diet formulations. Carbohydrates (CHO) are the major source of energy for rumen microorganism and the single largest component (66-70%) of a buffalo's diet. They represent the major component of net energy for milk as precursors for lactose, fat and protein. Dietary protein content and its degradability is important in the rumen to maintain the ruminal ammonia pool as it serves as main nitrogen source to rumen microbes. In high yielder, microbial protein content arising due microbial fermentation is not sufficient to meet the requirement for milk synthesis, therefore, bypass protein which is not degraded in rumen and pass intact to the small intestine for tissue level absorption is provided. To explore the quality of feeds for efficient utilization in rumen different methods have been developed and used for evaluation of feeds and forages. These methods are helpful to examine the quality of particular feed or diet for efficient utilization of nutrients without waste for higher production. The objective of this chapter is to provide some information on available techniques for evaluation of carbohydrates and protein content in feed/diets for efficient utilization in ruminant digestive system mainly in rumen.

Importance of Carbohydrates in Ruminant's Ration

Carbohydrate is a main energy source for ruminants, and it is the nutrient that should be first considered along with protein. Carbohydrates are the largest component in the ration and contribute 60 to 70% of the net energy used for milk production. Carbohydrates may be divided into fiber and non-fiber components. The fiber or structural portion, commonly called neutral detergent fiber (NDF), includes cellulose, lignin, and hemicellulose. The NDF fraction represents the slower digestible fibrous portion of the ration. The nonstructural or non-fiber carbohydrates (NFC) include starch, sugars and pectin. Microorganisms ferment carbohydrates to volatile fatty acids (VFAs) in the rumen, where they are absorbed and utilized as energy sources for maintenance, synthesis of milk, and tissues. Also, the VFAs (acetic, propionic, and butyric) provide carbon units for synthesizing milk components such as milk fat. Glucose is needed by the mammary gland to synthesize lactose. The liver synthesizes glucose from propionic acid for use in other body tissues. Acetic acid is needed for about 50% of the milk fat produced in the mammary gland. The amount and ratio of volatile fatty acids (VFAs) produced in the rumen may alter milk composition. As an example, rapidly fermenting carbohydrates favor the production of propionic acid; slowly degrading fibrous or structural carbohydrates promote acetic acid production. Rations low in fiber may result in too little acetic acid being produced in the rumen, resulting in possible acidosis and reduced milk fat production. Adequate amounts of effective fiber are needed in the ration for the maintenance of health.

Carbohydrates Fractions

In 1864 in Germany at Weende experiment station, Henneberg and Stohmann proposed a system of feed analysis, which later became popularly known as "Proximate analysis. " In the 'Weende's analysis the feed component are divided into six parts. i.e. Moisture, Crude protein, Crude fiber, Ether extract, Ash and Nitrogen free extract. Crude fiber and nitrogen free extract represents the carbohydrate fraction. Crude fiber portion represents the cellulose, hemicelluloses and lignin fraction, which is poorly digestible in rumen. However nitrogen free extract represents the rapidly digestible portion, which includes sugars and etc. Proximate analysis values are used for about a century as the basis of predicting energy and protein availability form feedstuffs (Morrison, 1956).

In the year 1967, Van Soest proposed a most widely used method of classification for carbohydrate based on structural and non- structural carbohydrates. The structural CHO consist of elements mainly found in plant cell wall. The non- structural CHO are located inside the cells of plants and are usually more digestible than structural CHO. The structural components in the plants like cellulose, hemicellulose and lignin are measured as NDF. Acid detergent fiber measures lignin and cellulose. Even though pectins are a part of the cell wall, they are considered as a non-structural CHO because compared to hemicellulose the rumen microorganism completely and rapidly ferment the pectin. On the basis of detergent fractions system Goering & Van Soest (1970) classified the carbohydrates further into detergent fractions given in table 1. According to the classification fractions A and B1 are soluble in neutral detergent fraction A consist of sugar and organic acids whereas fraction B1 consist of starch, pactins and glucose (Sniffen *et al* 1992). Hemicellulose, cellulose and lignin which are insoluble in neutral detergent the model partitioned this fiber into B2 and unavailable fractions C is determined on the basis of lignin associated fibre. Unavailable fraction C is calculated as lignin x 2.4 (Mertens, 1973). Van Soest detergent system is certainly better than Weende's proximate analysis. Later on carbohydrates further partitioned in six fractions i.e. A1, A2, B1, B2, B3 and C presented in table 2 on the basis of ruminal degradability and intestinal digestibility (table2).

Table 1: Carbohydrates Fractions and their Ruminal Degradability and Intestinal Digestibility

Name	**Components**	**Ruminal degrada bility (% per Hour)**	**Intestinal digestibility (%)**
Fractions A	Sugars, Organic acids	200-350	100 (But traces of fraction A reaches to the Intestine)
Fractions B1	Starch, Pectin & Glucans	20-40	76
Fractions B2	Available Fibre, Cellulose, Hemicellulose	2-10	20
Fractions C	Lignin and fiber Associated, with lignin	Nil	Nil

In neutral detergent soluble (NDS) major parts are sugars, lipids, some ash, NPN and some protein. However variable amount of ash and protein can be remain with NDF and ash contamination can contribute up to four percent of NDF. Ideally, both ADF and NDF should be expressed on an ash free basis. Then several modifications have been suggested for analytical procedure of NDF. The first was the inclusion of heat stable amylase in the procedure to remove starch. The other was the use of sodium sulphite to minimize protein contamination. NRC 2001 included NDF values for dairy cattle of amylase - sulphite procedure. In using sulphite in the standard procedure reduces crude protein contamination but does not quantitatively remove all of it. Using of sodium sulphite in the NDF procedure is discouraged if the residues are to be assayed for neutral detergent insoluble protein. Sulphite also attacks lignin and should not be used in a sequential analysis of lignin or for subsequent *in vitro* digestion. Nonstructural CHO is readily digestible and is not recovered in the NDF. The non-fibre CHO (NFC) includes sugar, starch and other reserve CHO such as galactans and pectins. The NFC for feeds is calculated by the difference:

100-(%NDF + % CP + % Fat + % Ash) or

100-[(% NDF- NDFCP) + % CP +% Fat + % Ash)]

Table 2: Proposed Carbohydrates Fractions and their Ruminal Degradability and Intestinal Digestibility

Name	Components	Ruminal degrada bility (% per Hour)	Intestinal digestibility (%)
Fractions A	Sugars	200-350	100 (But traces of fraction A reaches to the Intestine)
Fractions B1	Organic acids	1-2	100
Fractions B2	Starch	20-40	75
Fractions B3	Soluble available fibre, Pectin, B-glucans	40-602-10	Small Intestine = 0 Large Intestine = 100
Fractions C	Lignin, Fibre associated with lignin	Nil	Nil

NDFCP is the neutral detergent insoluble crude protein. The first equation is preferred because it helps for correcting crude protein in the NDF. The nonstructural CHO or NSC is measured by enzymatic methods and includes only starch and sugars. When using the enzymatic procedure to NSC, Sucrose and fructans appear in the starch fraction. This applies mainly to forage particularly grasses as they contain little starch. The university of Florida has developed a system for partitioning the neutral detergent soluble carbohydrates (NDSC) or the CHO fractions excluding hemicellulose and cellulose (table 3). The system uses an extraction with 80% ethanol to separate low molecular weight and organic form the starch and non-starch polysaccharides. The sugars directly in the ethanol extract and starch on the ethanol insoluble residue. The organic acid is and non-starch polysaccharides, which can be the most divers fractions, are calculated by difference. The Cornell Net carbohydrates and protein system (CNCPS) has used three fractions to predict degradation in the rumen and the passage of un degraded feed to lower gut (Sniffen *et al* 1992). They estimated ruminal degradation of fraction A is 200 to 300 % /hr and less than 5 percentage reach the intestine. Some workers found that degradation rate of NDS are only 15 to 19% in forages. Neutral detergent soluble contains not only sugar but also including starch, protein and non-carbohydrates such as protein, ash and lipids also included. Ruminal degradation of fraction B1 is 20 to 40 %/hr only. Degradation of B1 fraction s depending upon the source of starch and methods of processing and storage of feed. About 70 to 90% starch is fermented n the rumen hemicellulose, cellulose and lignin are insoluble in the neutral detergent. This model partitions fibre into available (B2) and unavailable fraction on the basis of lignin and fiber. Unavailable fiber is lignin x 2.4 lignin simply creates problem in digestion by encrusting or covering nutrients. Degradation of fraction B2 is 2 to 10 % / hr. between 30 to 70 percentage available fibre B2 fraction is fermented in the rumen. C fraction is not fermented in the rumen nor digested in intestine.

Importance of Proteins in Ruminant's Ration

Protein is one of the main building blocks of the body. It is usually measured as %4 Crude Protein (CP). It is a major component of muscles, the nervous system and connective tissue. Protein is composed of chains of amino acids. Adequate dietary protein is essential for maintenance, growth, lactation and reproduction. Protein is composed of several fractions, which vary in their solubility in the rumen. Microbes in the rumen digest rumen soluble protein. Rumen insoluble protein passes intact through the rumen to the lower digestive tract. A portion of this bypass (or escape) protein is digested in the small intestine. Crude protein is estimated by measuring hay's nitrogen content and multiplying that by 6.25. Much of the protein in feeds for ruminants (cattle, sheep, and goats) is broken down by rumen bacteria and used by them in digestion of carbohydrates (cellulose, sugars, and starches) in the forage and supplemental grains in the ration. This produces more bacteria, which are digested in the true stomach of the animal. A portion of the feed N that enters the rumen will be degraded to compounds such as peptides, amino acids or ammonia. The primary mechanism for this breakdown in the rumen is microbial proteolysis. The solubility, structure, and particle size of the feed will all influence the amount of degradation that takes place. There will always be a portion of the feed N that enters the rumen that is not degraded (RUP).

All RDP does not breakdown and be converted to ammonia at the same rate. Van Soest (1994) provided an overview of a system to define N sub-fractions that would permit better characterization of feed N availability and use in the dairy cow. This system includes the following fractions:

A This is mainly NPN, amino acids, and peptides that are "instantly" available in the rumen.

B1 This fraction has a fast rate of degradation in the rumen.

B2 This fraction has a variable rate of degradation in the rumen.

B3 This fraction has a slow rate of degradation in the rumen.

The use of this approach assists in doing a better job of describing N utilization in the rumen and improving the efficiency of feed N use. The use of this approach does require additional feed analysis data and computer formulation programs designed to utilize this information. Microbial protein (MCP) is produced in the rumen by the rumen microorganisms. The key factors that determine the quantity of MCP synthesized is the quantity of ammonia available in the rumen and the supply of fermentable carbohydrates to provide an energy source. The availability of peptides may also stimulate the production of MCP by some rumen microorganisms. The NRC (2001) predicts MCP production as 13% of the discounted TDN (total digestible nutrients) available in the rumen.

Microbial protein can provide 50 – 80% of the amino acids required in the intestine by the dairy cow. Optimizing MCP production helps in increasing the efficiency of N use in the cow and controlling feed costs. The benefits of MCP are related to:

- MCP averages about 10% N (60-65% CP).
- MCP is a good source of RUP.
- MCP has a high digestibility in the intestine.
- The amino acid profile of MCP is fairly constant
- MCP has an excellent ratio of lysine to methionine.

Protein Systems

There are 2 systems used to evaluate and balance rations for dairy cows on a protein basis. These are the CP (crude protein) and MP (metabolizable) protein systems. The CP system has been the most commonly used system.

The CP system is easy to use and has tabular feed composition and animal requirement information. This system assumes that all N in different feeds is similar in use and value to the cow. The Dairy NRC (2001) indicated that CP was a poor predictor of milk production. Nutritionists have modified the CP system to better meet their needs. They have added SP, RDP and RUP as additional factors to consider when using CP as the base for formulating dairy rations on a protein basis.

The Dairy NRC (2001) has suggested moving to a MP system to better define and refine protein formulation and utilization. This system fits with the biology of the cow. The challenge is that this system is not tabular and requires the use of computer programs to calculate both MP requirements and the MP supplied by feeds and MCP. The industry is changing to an MP approach. This system should provide an opportunity to improve the efficiency of protein use in dairy cattle. The use of this system will also decrease N excretion to the environment and lower potential ammonia emissions.

Protein Fractions

In ruminants protein (nitrogen) nutrition is considered a complex and dynamic process. Nitrogen including dietary origin and arising due to degradation in rumen provides ammonia to ruminal pool for microbes and considered a critical nutrient in the ruminant as it is main component of protein. The ruminants cannot use nitrogen as a nutrient directly at the tissue level. But ruminal microbes converts protein from both true protein and non protein nitrogen (NPN) into microbial protein and supply needed amino acids for production of animal protein for synthesis or other productive functions. Generally protein is analyzed in Weende's system as crude proteins (N*6.25) which contains N from either sources i.e. true protein or NPN. However, CP value of feed does not give the true picture of N utilization in animal system. Later on, it was realized that the dietary protein consumed by ruminants is evaluated on the basis of degradability in the rumen and ARC (1980) has proposed the RDP and UDP system for expressing the protein quality of feedstuffs. Similarly in NRC in1989, endorsed and adopted the new system replacng the existing DCP system in use in the western countries.

CNCP System of Analysis

The above discussed methods of feed analysis (proximate and detergent system) applied for determination of nutritive value supply the information only about the concentration or quantity of feed nutrients and give little indication about their availability in animal system. Also there is considerable variability in the utilization pattern of different fractions of particular nutrient. For example different fractions of carbohydrate (soluble sugar, starch, non-starch polysaccharides) and protein (non-protein nitrogen, soluble peptides and soluble tree protein) are utilized by different rumen microbes and animals differently (Nocek and Russel, 1988; Sniffen *et al.*, 1992), indicating their importance of estimation in feedstuffs. Accurate prediction of forage biological value and performance of animals fed forages has been proposed by Russel *et al.*, 1992, using Cornell net carbohydrate and protein system (CNCP). In CNCP sysytem of analysis (sniffen et al 1992) protein fraction is divided into five further fractions on the basis of solubility. Mainly feed protein is partition into three broad fraction i.e non-protein nitrogen (NPN), true protein and unavailable nitrogen (Van Soest et al 1981). These fractions named A (Non protein nitrogen), B (True protein) and C (bound protein or unavailable protein) respectively. True protein is further divided into three sub fraction viz., B1, B2, B3 on the basis of their ruminal degradability (Krishnamurthy et al 1983). B1 is instantly degradable part of the protein in the rumen and contains mainly globulins and some albumins. B2 fraction of proteins contains mostly albumins and glutelins, whereas B3 fraction includes prolamins, extension protein and denatured protein. The C fraction is considered bound protein, which is unavaible when exposed to ruminal degradaton, having maillard products, and N bound of lignin. Fraction A and B1 are soluble in borate phosphate buffer. These can be partitioned further by extraction with TCA. Fraction B2, B3 and C are insoluble in borate-phosphate buffer. Extraction with neutral detergent without sodium sulphite isolates fractions A, B1 and B2 (soluble in neutral detergent) from fraction B3 and C (insoluble in neutral detergent). Acid detergent partitions protein in fraction C (insoluble in acid detergent) and fraction A, B1, B2, B3 (soluble in acid detergent). Fraction B2 is calculated as a difference between borate –phosphate buffer insoluble protein and neutral detergent insoluble protein. fraction B3 is the difference between neutral detergent insoluble protein and acid detergent insoluble protein. proteins insoluble in acid detergent are fraction C. Fraction A contains ammonia, nitrate, amino –acids and peptides and ruminal degradation of these are instantaneous and complete. B1 protein slightly is less degradable than A but almost part is degraded in rumen only small part is available in small intestine. B2 fraction degraded in rumen at the rate 5 to 15 %/hr with intestinal digestibility of 100%. B3 proteins prolamins and cell wall associated proteins are degraded only 0.1 to 1.5%/hr in rumen and digestibility in intestine is 80%. Fraction C is unavailable both in rumen and intestinal level of degradation (Table 4). Information on the nutrient components especially in term of carbohydrate and protein fractions along with their relationship of some of tree leaves, shrubs and grasses available in semi-arid region of India using CNCP system. Have been summarized in table 5 to 10.

Table 3: Partitioning of Neutral Detergent Soluble (NDS) Carbohydrates with 80% Ethanol by using Values from Direct Analysis and from Calculation

80% ethanol extract	80% insoluble ethanol residue
Organic acids (Calculated Value)	Starch (Analyzed value)
Sugars (Analyzed value)	Neutral Detergent Solubles (Calculated Value) NDF (Analyzed Value)

Table 4: Protein Fraction and their Ruminal Degradability and Intestinal Digestibly

Name	Components	Ruminal degradability (% per Hour)	Intestinal digestibility (%)
Fractions A	NH_3, NO_3, Amino acids, Peptides	Instantaneous	Not reaches to intestine
Fractions B1	Globulins, Some Albumins	200-300	100
Fractions B2	Most Albumins, Glutelins	5-15	100
Fractions B3	Prolamines, Extension proteins, Denatured proteins	0.1-1.5	80
Fractions C	Maillard products, tannin bound to lignin	Nil	Nil

Methods for Estimation of Ruminal Carbohydrate & Protein Degradability

In general in vitro and in sacoo methods are applied for estimating the degradability of nutrient. These methods are based on quantity of nutrient degraded in ruminal environment and their degradability values fitted in different model developed to calculate A, B and C values. Models developed by Orskov and McDonald (1979) and McDonald (19981) with lag phase are very common for these studies.

Table 5. Carbohydrate Fractions in Tree Leaves, Shrubs and Grasses

Name	Carbohydrates (%)	Non-structural carbohydrates (% of CHO)	Carbohydrate fractions (% of carbohydrates)			
			A	B_1	B_2	C
Tree leaves						
Albizia lebbek	64.89	28.71	13.35	15.36	30.57	40.72
Leucaena leucocephala	61.16	57.77	36.65	21.12	17.11	25.12
Hardwickia binata	72.99	52.06	31.42	20.65	19.92	28.02
Grewia optiva	65.62	49.38	26.14	23.24	36.23	14.38
Anogeissus pendula	74.59	50.22	30.10	20.12	27.38	22.40
Mean (Tree leaves)	**67.85**	**47.63**	**27.53**	**20.10**	**26.24**	**26.13**
Shrubs						
Acacia catechu	74.82	58.88	38.61	20.27	19.33	21.79
Securengia virosa	77.27	67.09	53.17	13.92	22.35	10.57
Zizyphus xylophyrus	77.68	67.09	47.61	19.48	17.48	15.43
Dichrostachys cinerea	69.44	43.44	28.57	14.87	10.38	46.18
Helictris isora	71.89	46.37	28.59	17.79	33.59	20.04
Mean (Shrubs)	***74.22***	***56.58***	***39.31***	***17.27***	***20.63***	***22.80***
Grasses						
Heteropogan contortus	87.80	44.38	20.57	23.81	42.90	12.72
Sehima nervosum	82.18	13.01	1.85	11.15	72.01	14.98
Chrysopogon fulvus	86.74	16.03	3.47	12.56	65.67	18.30
Cenchrus ciliaris	83.43	16.92	4.10	12.82	66.65	16.43
Dichianthium annulatum	83.62	15.09	1.86	13.23	70.76	12.15
Mean (Grasses)	***84.754***	***21.09***	***6.37***	***14.71***	***63.60***	***15.32***
P	**	**	**	**	**	**
SEM	1.12	0.97	1.18	0.47	0.78	0.79

P= Statistical significance; *SEM*= Standard error of means;
**=*P*<0.01 Source: Chaurasia et al., 2006

Table 6 : Protein Fractions and *in vitro* Crude Protein Digestibility (IVCPD) in Tree Leaves, Shrubs and Grasses

Name	In vitro	Per cent of CP			
	CP digestibility	Soluble protein	Non-protein nitrogen	Acid detergent insoluble protein	Neutral detergent insoluble protein
Tree leaves					
Albizia lebbek	53.80	35.08	58.81	18.27	34.22
Leucaena leucocephala	88.97	35.06	81.47	19.33	27.67
Hardwickia binata	54.43	50.39	76.32	18.66	39.14
Grewia optiva	66.53	39.73	66.82	18.51	41.54
Anogeissus pendula	78.37	18.84	59.49	20.95	30.50
Mean (Tree leaves)	**68.42**	**35.82**	**65.38**	**19.15**	**34.62**
Shrubs					
Acacia catechu	68.75	22.79	54.08	25.01	42.85
Securengia virosa	82.58	16.94	45.31	21.30	27.81
Zizyphus xylophyrus	69.29	18.17	36.12	35.42	54.01
Dichrostachys cinerea	52.50	35.05	16.46	31.10	33.20
Helictris isora	73.72	31.57	71.58	21.36	61.99
Mean (Shrubs)	***69.36***	***24.91***	***44.71***	***26.84***	***43.97***
Grasses					
Heteropogan contortus	60.97	31.31	78.75	38.21	54.98
Sehima nervosum	60.39	25.53	53.94	47.88	52.29
Chrysopogon fulvus	60.25	24.91	45.60	65.77	69.01
Cenchrus ciliaris	61.27	31.45	26.78	47.62	55.88
Dichianthium annulatum	73.47	38.75	17.97	30.64	40.77
Mean (grasses)	***63.27***	***30.39***	***44.61***	***46.02***	***54.59***
P	**	**	**	**	**
SEM	1.63	1.70	5.40	1.02	1.03

P= Statistical significance; *SEM*= Standard error of means; **=*P*<0.01
Source: Chaurasia et al., 2006

In Vitro Methods

These methods are based on the inoculation of small quantity of substrate with rumen microorganism under controlled conditions in the laboratory. These methods are easy, less expensive and not very laborious. In 20th century several workers attempted to develop and standardize in vitro techniques based on substrate disappearance (Marston 1948, Johnson) 1963). But in the year (1963) Tilley and Terry developed in vitro technique, which have been accepted by workers and widely used for feed evaluation. Later this in vitro technique was modified by Goering and Van Soest (1970). However, still the in vitro using Tilley and Terry procedure is method of first choice by large number of research workers and commonly used in different laboratories of the world. In Tilley and Terry in vitro procedure in first stage test substrate is incubated for 48 hour with rumen inoculum and then in 2nd stage followed by 48h incubation in a pepsin -hydrochloric acid solution. Then filtration and ashing is applied to quantify the undegraded substrate residue. In the (modified by the Goering and Van Soest) procedure of in vitro the second stage of pepsin digestion is avoided and residue obtained after 48h of incubation is treated with neutral detergent solution (NDS) to separate more completely adherent microbial biomass form the undegraded feed residue. Meneke et al (1979) developed an in vitro technique measuring gas production in a simple device consisting of a 100 ml calibrated syringe made up of glass fitted with a plunger. The gas arising due to the fermentation of substrate would push up the piston of syringe. In this technique gas production during 24h of incubation was one parameter and crude protein, crude fat and crude ash is used for prediction of organic matter digestibility and metabolizable energy (ME). No quantification is required in this method, which reduces potential of analytical error and increases the work efficiency.

Table 7. Nitrogen Fractions in Tree Leaves, Shrubs and Grasses

Name	Nitrogen fractions (%)				
	A	B1	B2	B3	C
Tree leaves					
Albizia lebbek	20.66	14.42	30.70	15.95	18.27
Leucaena leucocephala	28.55	6.51	37.27	8.34	19.33
Hardwickia binata	38.44	11.95	10.47	20.48	18.66
Grewia optiva	26.63	13.10	18.72	23.04	18.51
Anogeissus pendula	11.08	7.76	50.66	9.55	20.95
Mean (Tree leaves)	***25.07***	***10.75***	***29.57***	***15.47***	***19.15***
Shrubs					
Acacia catechu	12.22	10.56	34.37	17.84	25.01
Securengia virosa	7.66	9.29	55.24	6.50	21.30
Zizyphus xylophyrus	6.66	11.52	27.82	18.58	35.42
Dichrostachys cinerea	5.77	29.29	31.74	2.10	31.10
Helictris isora	22.47	9.10	6.44	40.63	21.36
Mean (Shrubs)	***10.95***	***13.95***	***31.12***	***17.14***	***26.84***
Grasses					
Heteropogan contortus	24.56	6.74	13.71	16.78	38.21
Sehima nervosum	14.04	11.50	22.17	4.41	47.88
Chrysopogon fulvus	11.11	13.80	6.09	3.23	65.77
Cenchrus ciliaris	8.26	23.19	12.67	8.26	47.62
Dichianthium annulatum	6.68	32.07	20.48	10.13	30.64
Mean (Grasses)	***12.93***	***17.46***	***15.02***	***8.57***	***46.02***
P	**	**	**	**	**
SEM	1.50	2.05	1.64	1.95	1.02

P= Statistical significance; *SEM*= Standard error of means; **=*P*<0.01 Source: Chaurasia et al., 2006

Table 8. Relationship of Cell Wall Fractions with *in vitro* DM Digestibility (IVDMD) of Tree Leaves, Shrubs and Grasses

Attributes	NDF	ADF	Cellulose	Hemi cellulose	ADL	IVDMD
NDF	-	0.92**	0.89**	0.87**	0.04	-0.33*
ADF	0.92**	-	0.96**	0.61**	-0.01	-0.45**
Cellulose	0.89**	0.96**	-	0.60**	-0.26	-0.32
Hemicellulose	0.87**	0.61**	0.60**	-	-0.08	0.48*
ADL	-0.04	-0.01	-0.26	-0.07	-	-0.48**
IVDMD	-0.33*	-0.45**	-0.32	0.48*	-0.48**	-

* and **= Correlation is significant at the P<0.05 and P<0.01 level, respectively;
Source: Chaurasia et al., 2006

Table 9. Relationship of Carbohydrate Ractions with *in vitro* DM Digestibility (IVDMD) of Tree Leaves, Shrubs and Grasses

Attributes	Fractions of carbohydrate				In vitro DM digestibility (%)
	A	B1	B2	C	
Fraction-A	-	0.499**	-0.869**	0.005	0.443
Fraction B1	0.499	-	-0.562	-0.035	0.014
Fraction B2	-0.869**	-0.562**	-	-0.471**	0.089
Fraction C	0.005	-0.035	-0.471**	-	-0.364*
In vitro DM digestibility (%)	0.443	0.014	0.089	-0.364*	-

* and **= Correlation is significant at the P<0.05 and P<0.01 level, respectively
Source: Chaurasia et al., 2006

In Sacco Methods

In sacco technique is based on nylon bag method which measures the disappearance of feed components form bag containing the test diet/feed after incubation for a period in the rumen of an animal fitted with a rumen cannula. It is based on the silk bags technique used by Quin et al (1939). Silk bag were later replaced by nylon bags by Erwin and Elliston 1959, Johnson

1966 and Demarquilly and Chenost 1969). The method is easy and dismisses a number of criticisms related to microbial synthesis. Kinetics of degradation for longer period of incubation also provides the information of feed degradation characteristics in rumen when data is analyzed and fitted in Orskov and Mac Donald model (1979). The only problem associated with this technique is low repeatability as suggested by the diversity of values obtained by different researcher for a similar feeds. Differences also found from one laboratory to another in the application of this method. There are several factors associated with differences in values of nylon bag degradability. These mainly characteristics, influx of microbial populations, undegraded particle, feed retention of solubilised products, sample size to bag surface ratio etc. By above techniques different workers estimated the NDF digestibility of diet / feeds and find out the Kinetics of particular feed using different model. Similarly protein degradability was also estimated in different feeds by in situ or in sacco methods. The effective protein or DM degradability is then calculated by the expression $p= a+bc/(c+k)$ where

a = intercepts

b = potentially degradable fraction

a+c = potentially degradability

c = degradation constant

k = rumen outflow rate

Other Techniques

A. Degradation by Commercial Proteases of Microbial Origin

The rumen inoculums used in in vitro studies has been a major source of variation, therefore different workers used bacterial proteases that are commercially available (Kirshnamurthy et al., 1983, Poes et al., 1985 and Assoumani et al., 1992). Although this approach is correct but not practiced due to lake of commercially available source of proteolytic enzymes from rumen bacteria.

B. *In vitro* Ammonia Release and *in vitro* Inhibitor Method

In vitro NH3 production has been used for obtaining the estimates of protein degradation for the ARC system of protein evaluation and has been also been used in other countries. The method is used to measures the amounts of NH3 produced in vitro from a feed incubated with rumen fluid (Broderick, 1978). The chef advantage of this method is its simplicity and result for protein degradability agree fairly with those obtained with other e.g. Solubility (Crawford et al., 1980). One of the difficulty of some in vitro system based on

Table 10. Relationship of Protein and Nitrogen Fractions with *in vitro* CP Digestibility (IVCPD) of Tree Leaves, Shrubs and Grasses

Attributes	In vitro	Protein fractions			Nitrogen fractions				
	CP digestibility (IVCPD)	Non protein nitrogen (NPN)	Acid detergent insoluble protein (ADIP)	Neutral detergent insoluble protein (NDIP)	A	B1	B2	B3	C
ADIP	-.345*	-.353*	-	.730**	-.417**	.184	-.447**	-.463**	-
NDIP	.324*	-.042	.730**	-	-.090	-.029	-.780**	.269	.730**
N-A	.063	.844**	-.417**	-.090	-	-.482**	-.326*	.471**	-.417**
N-B1	.327*	-.833**	.184	-.029	-.482**	-	-.206	-.297*	.184
N-B2	.502**	-.068	-.447**	-.780**	-.326**	-.206	-	-.381**	-.447**
N-B3	.065	.443**	-.463**	.269	.471**	-.297*	-.381**	-	-.463**
N-C	-.345**	-.353**	-	.730**	-.417**	.184	-.447**	-.463**	-
IVCPD	-	.188	-.345**	.324**	.063	.327*	.502**	.065	-.345**

NH3 release has been incomplete recovery of degradation products. Ammonia concentrations in the fomenters have been found to decrease after reaching their peak values. The addition of energy to the fermentation medium causes more efficient utilization of NH3 (Bergner *et al.,* 1972). Attempts have been made to overcome these problems. Brodesick (1987) reported a method that incorporated inhibitors of microbial N metabolism thus allowing the quantitative recovery of protein degradation products. Hydrazine and its derivations are used inhibition of amino acid deamination.

Table 11. Relationship of Total Digestible Nutrients (TDNX) with *in vitro* DM (IVDMD) and CP Digestibility (IVCPD) of Tree Leaves, Shrubs and Grasses

Attributes	IVDMD	IVCPD	TDNIX
IVDMD	-	.699**	.415**
IVCPD	.699**	-	.592**
TDNX	.415**	.592**	-

** Correlation is significant at the P< 0.01 level, Source: Chaurasia et al., 2006

Table 12: Increasing NDF Content Affecting Forage Quality

Forage Quality	% Forage NDF Dry Matter Basis	Dry Matter Intake as a % of Body Weight
Excellent	38	3.16
*	40	3.00
*	42	2.86
*	44	2.73
*	46	2.61
*	48	2.50
*	50	2.4
*	52	2.37
Poor	54	2.22

Source: Mertins (1985)

C. Near Infrared Reluctance Spectroscopy

NIRS is a physical analytical method, which is based on the absorbance of light at wavelength regions that relate to chemical components within feeds.

Compared to other methods. HIRS is unique because it is non-destructive. It requires no reagents and is therefore non-polluting and characteristics that entire feed rather than specific components of interest. NIIRS has been successfully used to predict a wide of forge quality parameters. In several studies gives superior productions of nutritive value than chemical analysis or bioassay (Barber et al 1990, Adesogan et al 1998). However NIRS equipment needs calibration with feed characteristics. Reeves II(2000) noted that NIRS estimates of forage quality are also affected by particle size, temperature, Sample homogeneity, packing and type of feed.

Conclusions

Various techniques or methods are available for estimating ruminal protein or carbohydrate degradation of feed stuffs. Every technique has its own merits and limitaions. In sacco methods are widely used for evaluation of feed nutrients degradation characteristics in rumen but number of factors associated techniques affects the reproducibility. Similar is the case with in vitro methods. However methods adopted by CNCPS model are simple, easy and based on different fractions and their degradability in the rumen have potential for use in our conditions. In our conditions, using CNCPS various feeds of different origin should be screened for developing a primary database of carbohydrate and protein fractions. Later these values may be utilized for developing practical and balanced rations for livestock.

References

Adesogan, A.T., Owen, E. and Givens, D.I.(1998). Animal feed Science and techonology, 74:259:272.

ARC (1980) The nutrient requirements of ruminant livestock. Commonwealth Agricultural Research Council, Slough, U.K

Assoumani, M.B., Vedeau, R., Jacquot, L. and Sniffen, C.J.(1992). Anim. Feed Sci. Tech., 39:357

Barber, G.D., Givens, D., Krids, M.S., Offer, N.W. and Murray, I. 1990. Animal feed science and techonology, 28:115-128.

Broderick, G.A(1978). J. Nutr., 108:181

Chaurasia, M, Kundu, S.S., Singh, Sultan and Misra, A.K. 2006. Cornell net carbohydrate and protein system for nutritional valuation of tree leaves, shrubs and grasses. Indian Journal of Animal sciences 76(01): 79-85.

Crawford, R.J.Jr., Hoover, W.H. and Junkins, L.L(1980)J. Anim. Sci., 51:986

D'Mello (Editors), Farm animal metabolism and nutrition. CAB International, Wallingford, pp. 185-207.

Demarquilly, C. and Chenost, M., 1969. Ann. Zootech., 18:419-436.

Ervin, E.S. and Elliston, N.G.,(1959). J. Anim. Sci., 18:1518(abstract).

Goering,H.K and Van Soest, P.J(1970). Forage fibber analysis(apparatus, reagents, procedures and some application) Agric. Handbook 370, ARS, USDA. Washington. D.C.

Johnson, R.R.(1963). Journal of Animal Science, 22:792-800.

Johnson, R.R., 1966. J.Anim. Sci., 25:855-875.

Krishnamoorthy, U., Sniffen, C.J.., Stern, M.D. and Van Soest, P.J.(1983) Br. J. Nutr. 50:555-568.

Krishnamorthy, U., Snffen,C.J., Srern, M.D. and Van Soest, P.V., (1983). Br.J. Nutr., 50:555-568.

Marston, H.R.(1948). Biochemcal Journal, 42:564-574.

McDonald, H.R., (1981). J. Agric. Sci. Cambridge, 96:251-252.

Menke, K.H., Raab, L., Salewski, A., Steingass, H., Fritz, Dand Schneider, W. (1979). J. Agric. Sci.(Camb.), 92:499.

Merterns, D.R (1973). Application of theoretical mathematical models to cell wall digestion and forage intake in ruminants. PhD Dissertation, Cornell Univ., Ithaca, NY

Morrison, F.B(1959) Feeds and Feeding. Clinton, lowa: Morrison publishing Co. Van Soest, P.J (1967) J.Anim. Sci.,26:119-128

NRC (1989). Nutrient requirements of Cattle. Nat Res. Council, Washington, D.C

Orskov, E.R. and McDonald, I.,(1979). J. Agric. Sci., Cambridge, 92:499-503.

Poos-Floyd, M., Klopfenstein, T. and Britton, R.A., (1985). J. Dairy Sci., 68:829-839.

Quin.J.I., Van der Wath., J.G. and Myburgh. S.(1939). Journal of Veterinary Science and Animal Industry 11, 341-360.

Reeves III, J.B.(2000). Use of near inflectance spectroscopy. In: J.P.F.

Sniffen, C.J.,O'Connor, J.D. Van Soest, D.G. and Russell, J.B(1992) J. Anim. Sci. 70:3562

Tilley, J.M.A and Terry, R.A.(1963). J.Br. Grassl. Soc., 18:104.

❑❑❑

Chapter **14**

In vivo Techniques for Measurement of Forage Digestibility

SULTAN SINGH
Indian Grassland and Fodder Research Institute, Jhansi-284003 (U.P.), India

Introduction

The chemical composition of a feed/fodder determines its nutritive value. The difference in energy value of feeds is due to their differing digestibility. Digestibility can be defined as the proportion/fraction of a feedstuff or dietary constituent/ingredient that is lost on its passage through the digestive tract. The digestibility of a feed determines the amount that is actually absorbed by an animal and therefore the availability of nutrients for growth, reproduction etc. Digestibility may be determined by measuring the quantity of feed consumed and quantity of feces voided after an animal has had sufficient time to become accustomed to a feed/fodder/diet. Conventional digestibility measurements have contributed substantially to develop systems that describe nutritive value of feed/fodder. The net energy system of feed evaluation is still used to represents the most useful approach to describe nutrient requirement and nutritive value. As per the energy partitioning scheme that provides net energy system, the sole maximum energy loss during fees/fodder utilization by ruminants is that voided by the animal as feces. Though, the determination of the dry matter or organic matter digestibility is not strictly equivalent to the fecal energy loss but closely related to digestible energy. Indeed conventional digestion trial is an integral part of digestible energy determination. Therefore, because of the magnitude of feed/fodder energy loss and its importance in measuring/defining the nutritive value, the measurement of digestibility in vivo is a fundamental technique to study the nutrition of ruminants consuming forage based diets. Undoubtedly, the *in vivo* feeding trial procedure is the one that most accurately determines the nutritive value of a given feed, since it measures the actual amount of feed ingested and digested. *In vivo* digestibility is essentially measured for 2 purposes: (i) for some standard feed evaluation methods *in vivo* digestibility must be measured at maintenance; and (ii) digestibility measured as part of *ad lib* consumption trials.

There are several factors of errors in digestion experiments but the procedural variability in the conduct of the digestion trials is undoubtedly an important factor. Information on procedures used to conduct digestion trials, general experimental protocol to determine the forage digestibility in ruminants' are described here.

Requirements of Digestion Trials

Housing

While planning a digestion experiment, initial consideration should be given to the housing facilities and collection materials as they affect the intake and in many cases digestibility. Environmental factors (temperature, photoperiod, muddy pens etc) are known to influence the intake and in many instances digestibility. A suitable house for conducting digestion trial should consist of a fully enclosed building where environmental variables are controlled. Temperature is the most critical environment factors as intake and passage rate increases and digestibility decreases as temperature falls below the thermo neutral zone of a species (5-20 C) under experiment/trial.

As photoperiod effect the intake, a consistent light pattern should be maintained during the course of digestion trial/experiment. Intake is usually more with longer photoperiods. In case when digestion experiments are conducted without environmental control, environmental information on mean temperature and its variability, season, light/dark cycles etc) should be mentioned while reporting digestibility results.

Animal Restraint

Once the housing requirement is met, the method to restrain the animals should be considered. During digestion trial, individual animal is used as experimental unit for quantitative collection of each animal feces/excreta and measurement of daily intake. For achieving animal restraint individual stalls should be to provide free movement to animal, tie stalls or metabolic stalls may also be used depending up on the method of excreta collection. Metabolism stalls are the most restrictive approach to restraint the animal that tend to decrease feed intake and may be sometimes uncomfortable to animal. Selection of docile animals and exposure to metabolism stalls before the experiment minimizes the intake problems.

Fig 1: Digestion Stall for Individual Animal Feeding

Initially tie stalls and labours were used to manually collect the excreta/ feces during digestion trial. As a result of tedious nature of work and labour cost metabolism stalls designed. The desirable design of metabolic stalls allows easy adjustment of length and width to restrict the movement and position of animal to ensure quantitative, uncontaminated collection of feces. With loose stalls or tie stalls separate collection devices are required for feces and urine. Urine collection is not required for a digestion trial, for measurement of N balance and determination of metabolizable energy urine collection is required.

Collection Materials

Fecal collection bags were originally developed to direct feces into a pan behind the metabolism stall. Subsequently, faecal bags were found suitable to collect whole feces from grazing animal. Faecal bag with specific deigns are required to collect faeces from female animals to avoid faecal contamination with urine. This requires simple deflection of urine or collection of urine via receptacles designed for use in conjunction with a faecal collection bag. Urine collection bags are available for both male and female animals. As the metabolism stalls are usually elevated they allow urine collection via gravity flow.

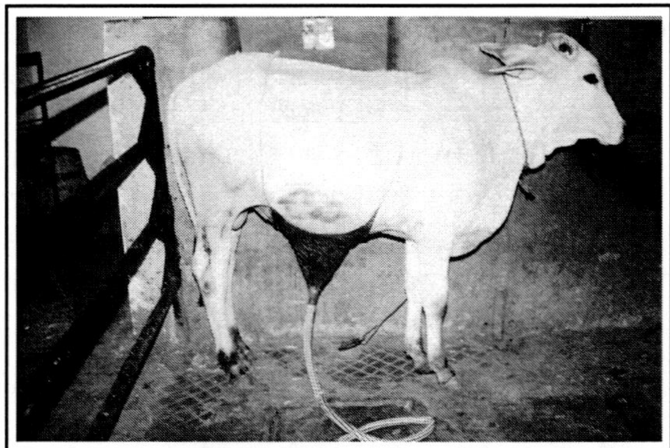

Fig 2: Male Animal Urine Collection Bag

The animals are confined in metabolism cages, to which an arrangement of sieves is attached to separate faeces and urine. The cages have to be wide enough to accommodate the animals, which (especially goats) should be chained so that they can move freely without being able to jump out.

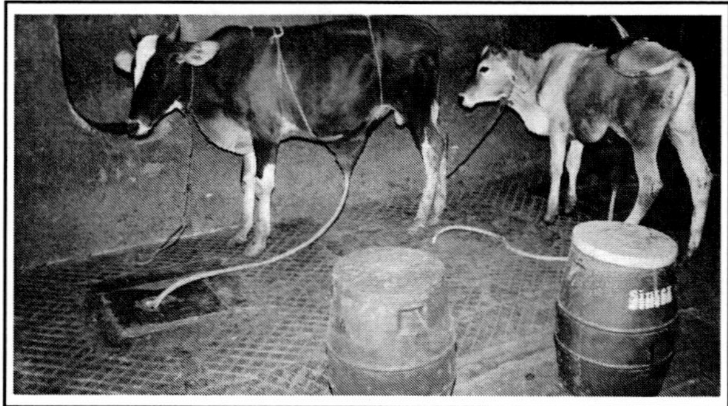

Fig 3: Animals in Metabolic Stalls with Provision of Urine & Faeces Collection

Animal Considerations

Animal Selection and Training

Selection of animals is the most critical element to ensure the success for conducting digestibility experiments. Selected animals should be trained for sufficient time prior the start of experiment to avoid the disturbance during the digestion trial. Initially 25-30 % more animals should be selected than to be used in experiment. During training animals are exposed to considerable handling by human beings (grooming, hand feeding etc) and animals that are excessively nervous or not adapted well to training are culled. Animals selected for digestion trials should have similar genetic makeup, physiological status, age, sex, weight and previous nutritional history.

In order to minimize variability, mature animals of similar body weight and age of the same sex and breed are selected. If females are used, they should not be pregnant, to avoid unnecessary stress and variability due to differing physiological requirements. Animals that show signs of behavioral stress (e.g., nervousness, low intake, weight loss) due to confinement in metabolism cages should be replaced with similar calm ones before proceeding with the digestion trial. Results obtained from aberrant animals usually deviate greatly from the normal. It. is important also to make sure that the animals are healthy and free from internal parasites. To eradicate internal parasites, each animal is dosed with albendazole 600, 3.8 mg/kg body weight before the experimentation.

Number of Animals

It is important to have enough animals in a digestion trial. Most experiments use 4-10 animals per treatment depending on the available resources. The higher the numbers of animals the more accurate and reliable are the results. The best approach to determine the number of replicates is based on decisions such as; 1) variability of the trait under evaluation Assuming such data are available); 2) size of the treatment difference that one desires to detect and 3) how willing one is to accept (i.e. the probability) that a significant treatment may go undetected.

Methods for Conducting Digestion Trial

Period Structure and Length

After animals have been trained to acquaint with general surroundings (feeding, watering, animal attendant and collection device) they are ready for the actual experiment. Digestion trial usually consists 2 phases i.e. preliminary period and collection period. The purpose of preliminary period is to establish the intake level, to ensure that all the residue of the previous diet are voided from the gastrointestinal tract and to allow the ruminal microbial population time to adapt to the diet under evaluation. Preliminary periods between 24-30 days have been recommended by many workers, while most recommendations fall in the 10-14 days range. Preliminary period of less than 1o days is not advised as 10-14 days are required for adaptation to a new diet.

The collection period is the period when feed, orts and feces/excreta are collected quantitatively. Early animal nutritionists suggested a preliminary period between 10-14 days but the subsequent research suggested that there is a little advantage of extending the collection period beyond 6-d days compared to 10 days collection period. Recent work recommends a collection period between 5-10 days with ideal period of 7 days. A collection period even in limited time and facility should not be less than 4 days. In case of use of markers for grazing animals, total number of sample collected is more important than total length of collection period.

Feed Intake/Consumed Determination

To establish feed intake is one of the objectives of the preliminary period. Level of intake influences rate of passage and subsequently digestibility. Therefore to avoid the compounding effect of intake level on digestibility level of intake should be maintained same across treatments where digestion coefficients are to me measured. Treatments should be such that all animals receive same amount to their body weight as there is potential for variability in the relationship between body weight/size and ruminal capacity.

It is important to ensure that the amount of feed to be used is sufficient to cover the whole duration of the experiment. The food used for the trial should be as homogeneous as possible. With conventional feeds, two meals per day are offered and the amount of feed offered is approximately 10% more than the maximum predetermined voluntary intake. In the case of shrub feeding, the feed is offered continuously throughout the day. In order to simulate natural grazing where feed selection exists, the feed should be offered as branches in quantities of about 3 times the estimated daily intake. To assess the dry matter intake, the percentage of weight lost through evaporation in the course of the day must be calculated by setting aside a batch of feed similar to those offered to the animals and weighing it at the start and at the end of daily feeding. Intake is calculated from the difference between the feed offered and the feed rejected, taking into account evaporation loss.

Frequency of feeding should also be considered when conducting a digestion trial. However ad libitum feeding allows consumption of many meals throughout a day, few animals are offered feed more than 2 or 3 times daily under production conditions. To save the labour feed is offered either once or twice daily during the digestion trial as there is little effect of frequency of feeding on digestibility.

Make sure that the conditions of the feed do not change during the trial. If possible in small ruminant work, make up the feeds for the whole experiment at one time.

If the roughage in question contains less than 2% crude protein for every 10% estimated digestibility (approximately) then a flat rate of high protein cake should be given to the animal twice daily to make up the deficit. Soya, cottonseed, mustard seed cake, ground nut cake or sunflower or any other available protein sources can be used. We must be sure that intake is not limited by deficiencies in the diet. If *a* few diets are tested in the same design, the same flat rate should be used for all, based on the diet with the lowest CP.

Animals should also have continuous access to mineral licks and clean water throughout both the preliminary and the test period. Ideally, dry matter of the feed offered should be determined every time the feed is given to the animal. This is necessary when silages or fresh green material are fed to the animals. Silages must if possible be corrected for volatile compounds e.g., acetic lactic, NH3. The animal should be allowed to refuse about 10-15% for cattle and 20-25% for sheep and goats for shrubs and browse; other conditions need to be introduced such as the woody material not eaten and could be much more than 20-25%. The dry matter (DM) content of the refusals must be measured. For sheep and goats, the degradability measurements must be done on the refusals. In the case of shrub feeding it is necessary to feed maybe 3 times more than the animals will consume as often only the leaves are consumed. It is especially important that the dry matter concentrations and weight, both the feed offered and refused are measured. The feed must be introduced twice daily, but the refusals need only be recorded once daily.

Feed intake is the amount of feed an animal consumes. It can be estimated by using either digestibility data or markers.

When digestibility data are available, intake can be estimated by multiplying the dry matter (DM) weight of faces by a digestibility factor. The factor is known as the feed: faeces ratio and is expressed as:

DM intake (g/day) = 100/ (100 – digestibility) × DM weight of faeces

where digestibility is in per cent.

For example if the dry-matter weight of faeces of an animal is 820 g/day and the percentage digestibility of the feed consumed is 60%, then the amount of dry feed consumed would be:

DM intake = 100/ (100 – 60) × 820 = (100/40) × 820 = 2050 g/day

Digestibility and intake data can be derived from the indigestible components of a diet, known as 'markers'. Internal markers are dietary constituent ordinarily present in the diet (e.g. lignin), and external markers are not present in feed/fodder/diet and they are added to the diet (e.g. chromic oxide, iron oxide and barium sulphate). These markers are ued in conditions where feed intake and faecal output is difficult.

The formula used to estimate faecal output is:

Faecal output (g) = grams of marker in feed per day/concentration of marker in faeces

For example an animal is dosed with 40 g of chromic oxide per day to determine its daily faecal output. The concentration of marker in the dry faeces sample is 5.00% and the dry-matter weight of the faecal output is:

Faecal output = 40/0.05 = 800 g

If the proportion of marker in the diet is 4.0%, then we calculate dry-matter intake as follows:

DM intake = 800 × (5.00/4.0) = 1000 g/day

Fecal Output Measurement

To determine total tract DM or OM "apparent" digestion, there is need to measure fecal DM or OM output. To estimate ME and N balance there is need to measure the urine output also. Urine collections are easily accomplished either in metabolism stalls or via collection devices strapped to the animal. To measure fecal output, one can either physically collect and weigh the total fecal excretion, or output can be estimated via use of inert markers. Alternatively, total tract digestion can be determined directly via internal marker rations in situations where diet intake is not measured directly as with grazing animals.

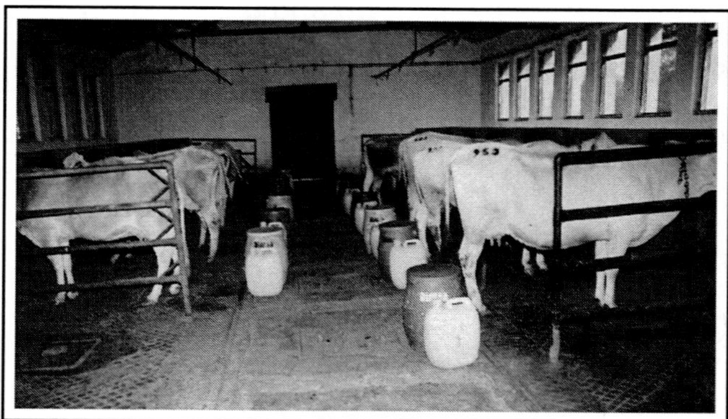

Fig 4: Animals in Digestion Stall with Feacal Collection Pots

Weighing and sampling of feces is usually done at the end of each 24-h cycle as there is no significant changes in chemical composition or heat of combustion of feces allowed to stand at room temperature for either 7 or 24

h. Emptying the bags once daily is also quite common, particularly with sheep or younger cattle. It may be desirable to empty fecal collection bags several times daily with large cattle; use of at least two bags per animal facilitates this procedure. One disadvantage of using fecal bags for total collection, particularly with large cattle, is that extended collection periods (> 7 d) may result in significant soreness under the tail and over the withers.

After weighing, feces must be sub sampled for subsequent chemical analysis. Two approaches may be used for sub sampling of faeces. First, a fixed percentage of each animal's daily fecal output may be saved and used to form a composite for that animal. The percentage retained depends on the amount of feces excreted and the size of sample needed for subsequent analysis. Amount of sample retained vary from complete collection with some small ruminants like sheep, to collecting only4 to 20 % of daily fecal excretion of cattle and buffaloes. The most commonly used approach for proportional sampling is to collect 5 % of daily excretion. The second approach to sub sampling feces is to collect a fixed amount of feces from each animal daily. Following drying and grinding, each animal's ground subsample can be composited in proportion to each daily's fecal output. Hence, each day's fecal output is expressed as a percentage of total output during the collection period. Then, the ground subsample from each day is added to each animal's final composite in the same proportion as that day's fecal output relative to the entire collection period.

Total collection of feces can be tedious and, in some cases, labor-intensive, it should usually be the method of choice whenever possible. However, in some experiments, total fecal collections are either not feasible or are logistically difficult like grazing conditions. Under such conditions, the use of external or internal markers affords an opportunity to estimate digestibility without conducting total fecal collections. In grazing trials where intake is not known, internal markers are used in conjunction with either external markers or total fecal collection to determine digestibility and intake.

Markers use for Intake and Digestibility Measurement

Intake and digestibility of a feed/fodder can be derived from the indigestible component of a diet, known a "markers". Markers are of two types, 1) Internal markers: present in the feed/fodder/diet (lignin) and 2) External markers: are chemical substances not present in feed/fodder/diet and added in diet (chromic oxide, iron oxide and barium sulphate). These markers are used when measurement of intake and fecal out is difficult as in grazing animals.

External Markers

To estimate fecal output, markers may be administered continuously, frequently, or in a single dose. When delivered in a single dose, fecal markers excretion over time can be described mathematically and used to predict fecal output. Single dose cause considerable variability in the accuracy of fecal output predictions, besides it extensive sampling are required following dosing (multiple samples at different times of the day over several days). Continuous or frequent (at least daily) administration of an external marker and collection of fecal "grab" samples once marker concentration has reached equilibrium in the feces is more precise procedure. Average marker concentration in the grab sample is divided in to the dose of marker to estimate fecal output. Several days are required for marker concentrations to achieve equilibrium in the feces. Once the marker concentration has reached equilibrium, "grab" samples of feces can be collected to determine fecal output. A common problem encountered when frequent marker administration is used to estimate fecal output is diurnal variation in the fecal marker concentration and this is particularly true when markers are administered only once daily. Continuous or frequent administration of marker (> twice daily) has reduced diurnal variation. Diurnal variation may exist, at times, even with frequent administration of marker. Many workers have described the characteristics of an ideal marker, but an ideal marker should have following characteristics: 1) must not be absorbed, 2) must not affect or be affected by the digestive tract or its microbial population, 3) must flow parallel with or be physically similar to or intimately associated with the material it is to mark, and 4) must have a specific method of estimation. A marker must be fully recoverable at sites distant from the site of administration (e.g., small intestine, feces) and complete in order to generate accurate digestibility data. Obviously, the most desirable approach is to work with a marker that is recovered quantitatively in the feces. Requirements regarding length of preliminary or collection periods and animal numbers are not lessened simply because one is interested in determining recovery of a marker. Thus, care should be taken to avoid the use of recovery estimates from one or two animals to adjust fecal recoveries of the animals in an experiment. Adjustments based on such an approach are as likely to bias digestion coefficients as they are to improve their accuracy. If one desire to estimate fecal output using markers for a large number of animals and has concern regarding the recovery and (or) release rate of a marker, a full-scale total collection trail should be conducted for determination of marker recovery (with adequate number of animals). In such cases, the conditions of the digestion trail must match the conditions of the larger trial as such as possible in order to ensure applicability of marker recovery data to the population of interest.

Internal Markers

Internal markers are inherent dietary constituents that, because of their resistance to digestion, may be used to monitor digestibility of other constituents in the diet. When intake is known internal marker can be used in same manner as external marker. In case intake is not known, an internal marker may be used to estimate the digestibility directly and this can be used in subsequent estimation of intake. Along with concentration of marker other nutrients of interest (DM, OM and N) are determined in representative samples of diet/ feed/fodder consumed and feces excreted. By calculating the differences in nutrient (DM, OM, N etc) concentration of feed/fodder/diet and feces relative to the internal marker, one can determine the digestibility of these nutrients.

Feed and Fecal Samples Processing

After collection of fecal material, it should be appropriately processed for its storage and subsequent analysis for chemical and biochemical parameters. Drying and grinding are the most common method of processing feces after sampling. Nutrients losses on drying occur but appear to be very small. Nutrients loss is more for both feed and feces when dried at low temperature (less than 50 °C). On the other hand fresh forages dried at higher temperature (near 100 °C) promote artifact production. Feces are less subject to artifact production than forages. Normally drying temperature near 50 °C is employed in digestion experiments. In case of forages collected from oesophageally or ruminal fistulated animals, freeze drying is preferred over oven drying.

Faeces can be collected either in faecal collection bags fitted to a harness or in metabolism crates or stalls allowing for collection of faeces and separation of urine and faeces. Faeces can be stored at 0°C for the collection period after which it must be mixed and sub sampled for appropriate analysis. Generally faeces should be dried at 60°C for approximately 48 hours. Feed refusal if any must be recorded and analyzed for each individual animal. If urine is collected it must be collected in dilute sulphuric acid to ensure that PH is less than 3 when urine is collected.

Samples of feed/fodder and feces are normally ground through 1 or 2 mm sieve using electrically operated mill depending on the parameter to be estimated.

Faeces and urine are collected at the same time once a day and stored at composite sample in separate containers in a refrigerator. In order to avoid loss of ammonia from urine due to evaporation, 5.0 ml of 25% H_2SO_4 is added to the urine collectors daily. One sub-sample of feed and of faeces is stored in a deep freeze for possible later use. Part of the other is oven-dried at 105°C to determine the DM content. A 10 ml sample of urine is weighed and used to determine the weight of the total urine output. The rest of the urine sample is freeze-dried and used for chemical analyses. The animals are weighed two days before collection starts and at the end of the collection period. Before weighing they are deprived of feed and water for 12 hours.

Calculations for Digestibility

When the experiment is over, the total amount of daily feed intake (kg DM/$BW^{0.75}$) and daily excretion of feces (kg DM) and urine (liter per day) are calculated. Once the intake is measured along with fecal dry matter output and other nutrients (OM, N, ADF, NDF, cellulose etc) in feces, urine and orts/residue, the calculations are made for digestibility coefficients. Following are the formulas used to calculate digestibility either from total fecal collection (digestion trial) or using markers.

$$\%\ \text{DM Digestion} = \frac{\text{DM consumed-kg (Feed DM offered-ort/refusal DM)-DM in feces-kg x100}}{\text{DM consumed-kg}}$$

When intake is known and external marker is frequently fed, then fecal output is calculated as below.

$$\text{Fecal DM output (g/d)} = \frac{\text{marker dose (g/d)}}{\text{Concentration of marker in feces (g/g of DM)}}$$

Once the fecal output is measured/determined with external marker, the digestion calculation can be made using above mentioned formula. In case of internal marker where intake is known, the internal marker can be used to estimate the fecal out as described above by determining the amount of marker consumed by each animal (concentration of marker in diet x amount of diet consumed). If the intake is known, then digestibility coefficient of DM and other nutrients (OM, N, NDF, ADF and cellulose) can be calculated using following formula.

$$\%\ \text{nutrient digestion} = 100\text{-}100 \times \frac{\%\ \text{marker in feed x}\ \%\ \text{nutrient in feces}}{\%\ \text{marker in feces x}\ \%\ \text{nutrient in feed}}$$

The digestibility of a feed (DM or OM) or any constituent of the feed is calculated as follows:

$$\text{Digestibility \%} = \frac{\text{Feed intake} - \text{Faeces} \quad \%}{\text{Feed intake}}$$

Nitrogen and energy balance can be calculated in the same way

Nitrogen balance = (% N in dry feed **X** daily dry wt of feed) - (% N in dry faeces **X** daily dry wt of faeces) - Daily urine N.

Energy balance = (Cal content of feed **X** daily wt of feed) - (Cal content of faeces **X** daily wt of faeces) - Daily energy content of urine

Sources of Errors in Digestion Experiments

Consideration of error sources is important to the design of any experiment. In digestibility experiments analytical and technical errors along with variation due to animals are the major sources of error.

Sampling procedures for diets, digesta, feces, orts/refusal etc should ensure that representative aliquots are collected for analysis. Hence feces, digesta, urine, orts/refusal samples should be mixed thoroughly before sub sampling and composite samples should reflect weighing for mass/volume of each collection that comprises the composite. All personnel involved in digestibility experiment should be aware/known to routines of feeding, sample collection and sample analysis. Processing (drying, grinding, mixing etc) should be applied uniformly across the samples of feed/fodder, feces, urine, ort/refusal etc to reduce the errors. Analytical instruments/equipments to be used in subsequent analysis of samples (feed/fodder, feces, ort, urine etc) should be checked routinely for accuracy and precision. Use of known standards to test equipments and procedures reduce the chance of errors.

Animal to animal variation is typically the single largest source of variation in digestion experiment. Careful selection of animals for genetic and physiological uniformity is important.

References

AOAC .1960. *Oficial methods of analysis* (gth ed.). Association of Official Agricultural Chemists, Washington, D.C., p. 832.

Maynard, L.P. and Loosely, J.K. (1969). *Animal Nutrition* (6th ed.). McGraw-Hill, Inc.

Donald, P., Edwards, R.A. and Greenhalgh, J.F.D. 1981. *Animal Nutrition* (3rd ed.). Longman Publ., New York.

Nastis, A.S. and Meuret, M. 1987. *Methods for esfimation of the nutritive value and grazing intake by goats in the Mediterranean area.* Agriculture, EUR 11893 FR-ENp, p. 142-158.

Osuji, P. O., Sibanda, S. and Nsahlai, I.V. 1993b. Supplementation of maize stover for Ethiopian Menz sheep: Effects of cottonseed, noug *(Guizotia abyssinica)* or sunflower cake with or without maize on intake, growth, apparent digestibility, nitrogen balance and excretion of purine derivatives. Animal Production 57:429–436.

Forage Quality, Evaluation and Utilization. 1994. (Ed. George C. Fahey, Jr.) A Publication of American Society of Agronomy, Crop Science Society of America and Soil Science Society of America, Madison, Wisconsin, USA.

❑❑❑

Chapter **15**

Estimation of Forage Degradability in Ruminants by in situ Technique

M. M. DAS, K. K. SINGH, S. K. MAHANTA, ANIL KUMAR AND A. K. MISRA
Indian Grassland and Fodder Research Institute, Jhansi-284003 (U.P.), India

Introduction

Feeding accounts for up to 70% of the cost of livestock production and forage continue to represent the single most important feed resource. Thus forage quality play an important role in growth and milk production of livestock. Forage quality can be defined as a function of both forage intake and digestibility. Measurement of plant cell wall and crude protein content are generally used as most important indicator of forage quality and ultimate utilization for animal productivity. Cell wall content is considered to be the most important factor affecting the forage utilization because it is composed of major fraction of forage dry matter and is correlated with forage intake and digestibility. The cell wall composed mainly of the structural polysaccharides viz. cellulose and hemicelluloses and is degraded by the microbes in the rumen. The microflora's ability to degrade and ferment the structural polysaccharides determines the availability of digestible energy from the forages.

Therefore, the rumen degradation characteristics of a feed is a guide to it's nutritive value for ruminants. The potential degradability is indicative of overall digestibility and the rate of degradation reflects the extent that digestible dietary nutrients escape the rumen fermentation. Forage degradation in the rumen depends on many factors like stage of maturity, presence of tannins, microbial proteolytic activity and rumen retention time.

In situ testing offers the opportunity to evaluate digestibility of a feedstuff which is fed to the animal. Reducing forage particle size by grinding reduces the digestion time, so it is recommended to estimate digestion rates of feed particles in a form similar to that entering the rumen. Furthermore, the test feedstuff should be exposed to the normal diurnal fluctuations in ruminal pH and any potential associative effects resulting from normal protein and energy supplementation. *In situ* digestibility data incorporated into a kinetic model resulted in close approximations of *in vivo* digestion (Moore *et al.*, 1989) that suggests a distinct advantage in using *in situ* digestibility data for ration balancing.

In situ Method

The test feed is weighed into a cloth (dacron or nylon) bag with a defined porosity to minimize loss of feed dry matter while allowing adequate microbial entry to produce normal digestive activity. The test feed materials in to the rumen allows immediate contact with rumen environment (temperature, pH, buffer substrate and enzyme). Extent of digestion can be measured using a single incubation time, which is equal to the expected ruminal retention time of the feedstuff while multiple incubation times can be used to develop rates of digestion. In this method the feed under study is put in a artificial fiber bags made of generally dacron or nylon which then are suspended in the rumen. The bags are then withdrawn from the rumen at 6, 12, 24, 48, 72, 96 and 120 h. They are washed under running tap water and gently squeezed until clear water come out of the bags. The zero time loss of DM is determined by soaking four weighed nylon bags containing the samples of forages in cold water for 1 hour, followed by washing of each bag under running tap water. The bags are then dried in an oven for 48 h at 60°C to a constant weight. DM and N loss from the bags is used to estimate rate and extent of degradation (Table 3, 4 and 5). The potential DM disappearance and the effective degradability of the forages using the nylon bags can be described by the model of Ørskov and McDonald (1979) as follows:

$$Y = a + b\,(1 - e^{-ct})$$
$$ED = a + bc\,/(k+c)$$

where

Y = the potential disappearance of DM at time (t),

a = the rapid soluble fraction,

b = potentially degradable fraction and

c = rate of degradation of b.

k = passage rate at outflow rate of 3% /hour.

The potential degradability PD = a + b

Feeds entering in the rumen go through one of two possible routes: they can be (1) degraded or (2) passed to the lower tract. In some sense, the two processes compete against one another: An increase in passage leads to a decrease in degradation. Since digestion of a nutrient is proportional to degradation rate and inversely proportional to passage rate. These relationships form the foundation of the equation used to predict nutrient digestibility:

Digestibility coefficient = $k_d/(k_d + k_p)$,

where k_d is rate of digestion and k_p is rate of passage

Rate of digestion (k_d,) is determined by measuring nutrient disappearance *in situ* at a number of time points. In the estimation of k_p, the rate of passage, is done through an independent procedure using markers. Due to the cost and difficulty involved in direct estimation of passage rate, standard values are often used. The rate of passage is an important factor in the calculation of the digestibility coefficient and similarly the rate of degradation also.

Kinetics of Forage Protein Degradation in the Rumen

The most widely used model for protein degradation is similar to that used for fiber digestion. The model describes total CP in feed into fractions A, B, and C, which sum to unity. Fraction A is the proportion of total CP already degraded at 0-time (i.e. mostly as NPN), B is the fraction of total CP which is potentially digestible or degradable, and C fraction of total CP is considered to be completely indigestible. The fraction of total CP degraded in the rumen is estimated by the fractional rates of passage and degradation. Degradation and passage both when account for removal, the rate of disappearance of fraction B, with respect to time, is expressed by the differential equation: dB / dt = $-k_dB- k_pB$, where k_d and k_p are fractional rates for ruminal degradation and passage, respectively (Broderick, 1994). When only ruminal degradation is considered as disappearance, the equation is $dB / dt = - k_dB$. Integration from zero to infinity yields, respectively, the amounts disappearing from the rumen due to degradation plus passage, $k_d + k_p$, and due to degradation only, k_d (Hungate, 1966). The proportion of total protein disappearance resulting from degradation in the rumen is given by the ratio: $k_d / (k_d + k_p)$ (Mertens, 1987). Total protein degradation in this model is represented by the equation:

$$\text{Degraded Protein} = A + B \times [k_d / (k_d + k_p)]$$

The proportion of total protein leaving rumen as undegraded is given by the complementary equation:

$$\text{Escaped Protein} = B \times [k_p / (k_d + k_p)] + C$$

This model for protein degradation is widely accepted and used (Ørskov and McDonald, 1979).

Bag Characteristics

The type of cloth, uniformity of pores and bag size influence the *in situ* degradation. Woven fabrics have little effect on *in situ* results. Pore size has significant effect on degradability results of feeds. Pore size must be large enough to allow free exchange of fluid and microorganisms between the bags and the ruminal milieu but it should be small enough to prevent the loss of indigestible particles or the entry of feed particles. The appropriate porosity is a compromise between limiting influx of rumen contents not associated with the test feed and allowing influx of microbial populations to degrade the test feed while at the same time limiting the efflux of undegradable feed particles (Fahey,1994).

When the woven fabric bags have pore size 10 µm or less, the number of total and cellulytic bacterial counts, protozoal counts within the bags was depressed (Lindberg *et al.*, 1984). When the large pore size bags were used (>100 µm), the influx of material in to the bag cause reduction in apparent disappearance and some times even resulting in negative digestibilities. The soluble and mechanical particle loss can account for a considerable proportion of nutrients particularly of N in fermented forages. Therefore the nylon bag made of woven material should have a pore size of 40-60 µm. The size of the bag has minimal interference on the *in situ* degradability results but the ratio of sample size to bag surface area has a large impact on *in situ* results. The optimum sample size is that which provides enough residue at the end of rumen incubation for chemical analysis without over filling the bag. It delays bacterial attachment, increase lag time, and underestimate digestion rates.

The increasing sample size to fixed surface area of bags may result in reduction in rate and extent of disappearance. The woven fabric bags having pore size between 40-60 μm should have surface area not greater than 10 mg of DM/ cm^2. Therefore, the amount of sample (bag size) to be incubated is dependent on the potential extent of ruminal digestion of a given ingredient or forage in relation to extended ruminal incubation times and the number of chemical analyses that is to be conducted on the residue.

Particle Characteristics

Generally, longer and coarser materials are associated with slower rates of digestion and grater variation. However, finely ground materials are subject to greater mechanical losses from the bags (resulting sometimes in unrealistically rapid rates of digestion) but variation is more controlled.

Homogeneity of the sample (initial particle size of the forage) should be given consideration in *in situ* studies. Forage samples grinded through screens ranging from 0.3- 5mm had minimal effect on *in situ* disappearance when extent of digestion is for 48 hrs (Nocek and Kohn, 1988). Increasing particle size may increase lag time and decrease the extent of disappearance for short-term incubations. Wide variation in the particle size may decrease the precision of *in situ* results. Therefore sample should be milled through a relatively small screen sieve (<5mm) to increase precision. Drying of samples at temperatures above 60°C should be avoided as higher temperature may alter the chemical composition of feed.

Dietary Effects

Diet is a major factor that determines the quantity and types of microbes and therefore the rate and extent of digestion of dietary nutrients and subsequently has great interference on the *in situ* disappearance measures. The most important effect is caused by the ratio of forage to concentrate (more specifically the amount of starch) fed to the recipient animal. The increasing amount of grain fed to the recipient animal decreases extent and rate of *in situ* DM disappearance of forages The effect of increasing grain fed to the recipient animal on in situ measures is greater for lower quality forages and roughages. The type of forage fed to the recipient animal may also affect the *in situ* disappearance of forages. *In situ* disappearance of alfalfa, bermuda grass, cottonseed hulls, corncobs, soyabean straw were greater when incubated in animals fed bermuda grass than when incubated in animals fed alfalfa (Neathery, 1969). Therefore, the type of forage fed to the animal should be similar to the type of forage being used for evaluation. Basal diet should contain a small amount of a wide range of ingredients to establish a diverse microbial population.

Analytical Effects

Most commonly species used are cattle and sheep. Sheep had higher ruminal ammonia N than steers, lower VFA and similar ruminal pH and rumen fluid dilution rate. Although protein degrading activity are not different between species. Within species, differences that relate to sex and physiological status are also sources of potential variation. Differences in rumen parameters also associated with age, pregnancy and stage of lactation. Many of these differences may be related to the specific type of diet associated with a specific physiological state and may influence ruminal factors other than digestion . In any event, caution must be used when comparing digestion rates of unlike animals.

The lack of standardization is the major source of variation in the *in situ* results. The method of inserting bags in to the rumen, the location of the bags within the rumen and rinsing technique may affect the *in situ* disappearance values. All the bags should be placed in the rumen at the same time and then the bags should be removed at various times for kinetic studies. Alternatively, bags can be placed in to the rumen at various times and all bags can be removed at the same time (Complete exchange method). Complete exchange method increase the labour use efficiency and reduce the variation in the rinsing technique. The standard method are more appropriate when *in situ* results are extrapolated to once a day feeding situation. When animals are fed once daily and all bags are placed in the rumen at the time of feeding, the diurnal pattern may exert its effect on all bags. When the bags are put in to the rumen in different times, the diurnal pattern may exerts effects differently on the different bags. Therefore, animals should be fed a total mixed diet more frequently which will help to reduce the diurnal variation.

Bags should be placed in the ventral sac of the rumen so that the bags must be in contact to the rumen contents. The bags are anchored with about 25 cm of nylon cord to cannula top in sheep and with about 50 cm or more in cattle. The length allows the bags to move freely within the digesta both in liquid and solid phase. Another variable component of the *in situ* assay is the method of washing adopted for the bags after incubation. A common method of rinsing bags is to soak or agitate bags in warm water until the rinsing solution is clear. Rinsing of bags can be done by hand or using washing machine.

Microbial Contamination

Potential contamination with microbial constituents is an inherent obstacle and source of variation associated with estimating true nutrient digestibility of feeds by the *in situ* technique. Microbial attachment specially the bacterial attachment to test feed particles within bags may increase the amount of DM present and may subsequently reduce the apparent extent and rate of DM digestion. The rate of *in situ* DM digestion showed 29 to 42% higher when residue weights were corrected for bacterial attachment than when uncorrected weights were used (Olubobokun *et al.*, 1990). Corrections for bacterial attachment also essentially eliminate lag time. Microbial correction is important for lower quality forages as well as for higher quality forages. The degree of bacterial contamination is not constant among the substrates. Forage species and N content of forages may affect the bacterial contamination. It may lead to faulty interpretation of treatment differences for uncorrected DM disappearance. Correction for microbial attachment imposes an additional variable to the *in situ* assay (table 1).

Table 1. Effect of Correction for Bacterial Nitrogen Contamination on the Percent error Associated with Determination of Residual Nitrogen at Specific Ruminal Incubation Times.

Ingredients	Ruminal incubation time (h)		
	5-6	12	24
Corn	0	4.8	3.6
Rapeseed meal	1.8	3.9	0.9
Barley straw	164.5	146.3	204.6
Alfalfa hay	25.0	22.0	44.0
Canary grass	-	33.8	27.8
Timothy	3.0	1.0	30.6

Source: Nocek, 1998

Other Measures

The duration of the incubation period can be varied depending on the desired use of the data. When digestion rate and extent are required then a series of bags are needed. These kinetic parameters then are incorporated into the models to estimate the effective degradability. Effective degradability is an estimate of the extent of ruminal digestion that occurs with specific rates of passage. When a single time point is used to estimate the digestibility, the length of incubation should be chosen based on estimated passage rate. The proper passage rate will depend on the DM intake of the animal being fed, type of feedstuff and composition of the diet. *In situ* DM disappearance values (72h) for forages were higher than digestibilities determined in *in vivo* in sheep and cattle fed at or above maintenance requirements. When *in situ* DM disappearance was measured at 48 h, the values were fairly similar to *in vivo* DM digestibility (Lindberg, 1983). For straw, hay and other fibrous materials, longer incubation intervals are generally required and for succulent feeds, the intervals should be shorter.

Besides there are three sources of variations also exists e.g. a) Variation among identical samples incubated in the same animal at the same time. b) Variation among identical samples incubated in the rumen of different animals at the same time. c)Variation among identical samples incubated in the same animal at different times.

Therefore, incubations should be replicated over animals, time, and within animals and time (table 2).

Table 2. Recommendations for Ruminal *in situ* Digestion Technique

A. Bag porosity	40 to 60 μm
B. Particle size	Protein and energy supplements, 2mm Whole cereal grains, fibrous by products, 5mm Hay (>80% DM), 5mm Silages (60-70% DM after drying), 5mm
C. Sample size to bag surface area	10-20mg/cm^2
D. Microbial contamination	Regardless of marker (DAPA, N^{15} etc) a correction should be used especially for low quality forages.
E. Diet	Feed to meet nutrient requirement in basal ration. Feed ingredient to be tested in the basal ration Feed a total mixed ration, *ad libitum*
F. Animal/period	Use the type of animal for which the digestion rate determinations will be used. At least two time period replications should be used if only using one animal Insert bags at the same time in relation to feeding for each animal and period.
G. Standard ingredients	One bag per each incubation time, suspended with test feeds.
H. Preruminal incubation	Soak bags in water or buffer prier to ruminal incubation.
I. Postruminal washing	Rinse in tap water until rinse water is clear (about 90 s / bag) with moderate manipulation.
J. Incubation times	0 to 6 h: 3 to 6 time points 6 to 24 h: 3 to 6 times > 25 h: 6 to 12 h intervals

1.*In situ* forage digestibility determinations have a higher variability than do *in vitro* tests. Coefficients of variation for *in situ* forage digestibility evaluations will normally be in the range of 3-5%.

2. The animal used in the *in situ* test must be similar both with regards to physiology and feeding program of the animal for which the ration is to be formulated.
3. *In situ* digestibility results are sensitive to forage: grain ratio and stage of lactation of the animal.
4. Restriction of microbial access to the protein
5. Microbial N contamination with bags
6. Losses of soluble and particulate N, which have been degraded.

Factors Affecting Cell Wall Degradability of Forages

Plant Maturity

Concentrations of cell wall constituents and lignin increases and soluble carbohydrate content decreases during forage maturation. The lower digestibility of the mature grass is also associated with a shift in site of cell wall digestion toward the hindgut, which accounts for 25% of total tract cell wall digestion (Hogan *et al.*, 1969). The reduction in cell wall degradability during maturity is due to progressively greater lignification of plant cell walls (Jung and Deetz, 1993). However, the degree of lignification varies with the different plant parts and tissues. The tissue-specific differences in lignification thus may affect the limited negative correlation of lignin concentration with cell wall degradability with forage samples of similar maturity. As grass stems develop, sclerenchyma and the mid-region parenchyma tissues develop thick, lignified cell walls. These lignified grass tissues are degraded slowly, but after extended periods of degradation only a very thin residual wall layer remains undegraded. This pattern of extensive degradation of thick-walled, lignified tissues is markedly different than what has been observed in legumes. Xylem tissue of alfalfa (*Medicago sativa* L.) stems has thick, heavily lignified walls and is virtually nondegradable. The reason for this difference among these forages in degradability of lignified tissues is unclear. Legumes generally have higher concentrations of lignin than do grasses of similar maturity; although some of this reported difference results from the poor recovery of grass lignin in the acid detergent lignin method. Both types of forages undergo a shift in composition from primarily guaiacyl-type lignin to mixed syringyl-guaiacyl lignin during maturation (Jung and Deetz, 1993). While lignified legume and grass cell walls are both rich in cellulose and xylans, grasses contain very little pectin and legumes do not contain appreciable amounts of ferulates. Early studies with a synthetic lignin model (polyeugenol) complex with cellulose indicated that cross linking was necessary for lignin to inhibit cellulose degradation by both isolated cellulase and rumen microbes (Jung and Ralph,

1990). Both ferulate monomers and dimers have been identified as being cross-linking agents in grasses (Ralph *et al.*, 1994). Ferulate esters in perennial ryegrass (*Lolium perenne* L.) act as the initiation site where lignin polymerization begins. Therefore, ferulate-mediated cross-links, formed during cell wall development, have the potential to influence cell wall degradability. Cross-linking of arabinoxylan to lignin by ferulate monomers and dimers reduced polysaccharide degradability by isolated enzymes (Grabber *et al.*, 1995). Neutral detergent fiber degradability of smooth bromegrass (*Bromus inermis* Leyss) plants, selected for divergent ferulate ether and Klason lignin concentrations, was negatively impacted by both cell wall components. However, these components were at least partially independent of one another in their effects, and ferulate ethers had a larger relative impact on fiber degradation than did Klason lignin (Casler and Jung, 1999). The nondegradable cell wall layers would prevent access by rumen microbes to potentially degradable cell walls in adjacent plant cells that had not been ruptured by chewing or grinding. Therefore lignin's impact on degradability of forage cell walls is not directly related to concentration. However, strong quantitative negative correlations exists between cell wall degradability and lignin concentration (Jung and Deetz, 1993). Each unit of lignin depressed cell wall degradability by by two units (Grabber, 2005).

Processing Effects

Usually conservation (dehydration and ensiling) has little effect on structural carbohydrate composition of forages, although the soluble carbohydrate fraction of forages is reduced considerably by ensiling (McDonald and Edwards, 1976). Changes in structural carbohydrate composition of hays are basically due to the consequence of leaf loss during harvest. Drying of forages has no apparent effect on the ruminal or total tract digestibility's of soluble or structural carbohydrates (Thomson and Beever, 1980). The extent and site of digestion of structural carbohydrates does not differ much between hays and wilted silages (Merchen and Satter, 1983).

In chemically (e.g., sodium hydroxide, ammonia, ozone, sulfur dioxide) treated forages and crop residues, considerable alteration occurs in cell wall structure and composition. Hydrolytic treatments may partially solubilize hemicelluloses and lignin, whereas oxidative treatments may attack and degrade lignin and can result in extensive solubilization of cell wall polysaccharides. Chemical treatments increase extent of digestion of cell wall polysaccharides by ruminants but generally do not affect site of cell wall digestion.

Associative effects

Digestibility of components in mixed diets may be additive at maintenance levels of intake, while it is negative when mixed forage-concentrate diets are fed at high intakes. Site of digestion of structural carbohydrates often is shifted toward the lower tract when ruminants consume increasing proportions of grain in mixed forage-concentrate diets (Bourquin *et al.*, 1990). Therefore,the negative effects of dietary non-structural carbohydrates on fiber digestion are: 1) The rumen microorganisms preferentially choose non -structural carbohydrates rather than structural carbohydrate; 2) Rapid non-structural carbohydrate digestion may lead to a decrease in ruminal pH and 3) More proliferation of non-structural carbohydrate-digesting microorganisms and competition for essential nutrients. Ruminal digestion of cell wall carbohydrates was inhibited both by low ruminal pH and by the availability of a readily fermentable energy source. Structural carbohydrate digestion generally is markedly depressed when ruminal pH is depressed below 6.0 (Hoover, 1986).

Effect of Anti-Nutritional Factors

Forage containing various anti-nutritional factors like tannins, saponins, mimosine etc have anti microbial activity, which indirectly affect the degradation of forages.

a) Tannin

Tannins are phenolic compounds with molecular weight ranging from 500 to 3000 Da. Tannins are usually secondary metabolites of plants and are broadly classified into two groups e.g. hydrolysable and condensed tannin. Tannins are most effective against the fiber degrading bacteria. Animals fed tannin rich *Calliandra calothyrsus*, the population of *Ruminococcus* spp and *Fibrobacter* spp. was reduced considerably, but fungi, protozoa population were less affected (McSweeney *et al.*, 1999).The number of total bacteria in the rumen of goats decreased significantly when the animals were fed tannin rich plant (*Acacia Nilotica*) and the decrease in the numbers was directly proportional the level of this feed in the diet. The ruminants if fed continuously tannin rich diet, the microflora will develop tannin tolerance.

b) Saponins

Saponins are steroidal sapogenin covalently linked to oligosaccharide moiety. The saponins of lucern origin had a detrimental effect on rumen fermentation and caused a reduction in total volatile fatty acids production and acetate to propionate ratio from 1.93 to 1.37 in the presence of 1% saponin in the medium (Lu *et al.*, 1987). The rumen microbes are able to deglycosilate the saponins to release the steroid moiety, which affects rumen fermentation.

Factors Influencing Ruminal Degradation of Forage Protein

Plant Maturity

Enhancing forage maturity is often linked with a reduced extent of degradation of forage protein in the rumen. Decreased CP content of forage plants at later maturities and decreased production of microbial protein resulting from lower intakes of digestible OM may result in reduced net flows of N into the intestine despite decreased degradability of forage protein in the rumen (Kawas *et al.*, 1990). Some of the plant N, while protected from ruminal degradation, is also more refractory to postruminal digestion.

Method of Conservation

The method of forage conservation may influence protein degradation in the rumen due to changes in the composition of the CP fraction of the forage. During cutting and drying of forage in the field, plant proteases transforms N compounds resulting in increases in soluble and non-protein N as well as an altered distribution of amino acids in the non-protein N fraction in comparison to the parent protein (Sullivan, 1973). Utilization of forage CP may thus be inversely related to changes in N solubility that occur during harvest and storage. Therefore, changes in N solubility due to conservation might have modest effects on extent of degradation of forage protein.

Physical Processing

Physical processing (e.g., chopping, fine grinding/pelleting) may affect digestion of forage components. Processing of hay by coarse chopping has little effect on site or extent of digestion. Fine grinding and pelleting of forages may increase ruminal dilution rate of particulate digesta, which are associated with increased escape of feed protein from ruminal degradation. Heat generated during the process of pelleting may provide protection of feed protein in the rumen. So, grinding and pelleting enhances the flow of N and amino acids to the small intestine (Thomson and Beever, 1980).

Legumes Versus Grasses

Legume protein is degraded more extensively than that in grasses. Net losses of N often occur during ruminal digestion of good-quality legumes (particularly alfalfa) such that duodenal N flows in animals fed these forages are lower than intake (Kennedy *et al.*, 1986). Similar losses also occur when grasses are fed but it is less as grasses contain less N than legumes (at similar maturities). The protein in grasses is usually less extensively degraded in the rumen. Therefore, the efficiency of utilization of legume protein is compromised more greatly than that of grasses due to attendant losses of N that results from extensive degradation of protein in the rumen.

Heat Damage in Conserved Forages

Excessive heating in conserved forages may lead to formation of acid detergent insoluble nitrogen(ADIN). The relationships between extent of heating, (ADIN) formation, and apparent N digestibility strongly imply that ADIN content of a feedstuff is reflective of the formation of indigestible N-containing complexes (Maillard products). ADIN is thus frequently been used as an indicator of heat damage in forages and generally suggested for use in defining the unavailable protein fraction of feedstuffs in systems designed to model ruminant N metabolism (NRC, 1985).

Therefore, excessive heating and ADIN formation in forages should be avoided. Moderate heating usually enhances protein value by reducing ruminal degradability. ADIN at modest levels in forages do not signal any undesirable effects and rather it reflects desirable effects of heating.

Table 3. Dry Matter Degradability of Different Forage Resources (at K=0.05% per hour)

Feed	**DM loss after 48 hrs**	**a (%)**	**b (%)**	**C (fraction per hr) %**	**Potential degradability (a+b) %**	**Effective degradability (%)**
C. ciliaris	43.34	1.50	46.40	0.035	47.90	19.11
Sorghum stover	37.43	2.10	41.55	0.043	43.65	21.31
Grass	36.09	1.00	42.18	0.027	43.18	15.79
Black gram straw	46.25	12.50	34.80	0.033	47.30	29.91
Groundmut haulm	50.04	14.00	36.30	0.042	50.30	30.57
Subabul leaves	63.55	15.00	49.23	0.044	64.23	38.22
S. seaban	75. 62	32.20	46.90	0.076	82.10	68.83
L. purpureus	60.25	26.40	41.10	0.056	70.60	55.40

Source: Samanta *et al.* (2004)

Table 4. *In Sacco* Digestibility of Nutrients (%) in Tree Leaves and Forages

Species	DM	CP
A. lebbeck	51.56	60.11
A. nilotica	71.57	79.33
L. leucocephala	75.98	82.92
S. aegyptica	77.11	80.12
S. hamata	64.28	70.11
A. scabroidus	68.82	72.99
L. purpureus	71.33	77.40
Wheat straw	48.47	-
Paddy straw	57.60	-
Oat hay	51.01	62.86

Source: Singh *et al.* (1995)

Table 5. Crude Protein Degradability of Different Forage Resources (at K=0.05%per hour)

Feed	DM loss after 48 hrs	a (%)	b (%)	C (fraction per hr) %	Potential degradability (a+b) %	Effective degradability (%)
C. ciliaris	37.53	-1.00	41.49	0.011	40.49	12.48
Sorghum stover	33.57	0.63	36.69	0.027	36.75	13.49
Grass	29.47	-1.50	25.70	0.019	31.20	12.32
Black gram straw	57.82	1.12	58.62	0.049	59.74	30.67
Groundmut haulm	55.68	4.25	55.78	0.045	60.03	30.67
Subabul leaves	62.42	8.50	55.15	0.063	63.65	39.25
Concentrate mixture	80.98	21.00	59.98	0.126	80.98	63.94

Source: Samanta *et al.* (2004)

References

Broderick,G.A.1994. Quantifying forage protein quality. In: Forage Quality Evaluation and Utilization. (Eds.)Fahey Jr, G.C., Collins,M., Mertens,D, R. and Moser,L.E. American Society of Agronomy, Inc. Crop Science Society of America, Inc.Soil Science Society of America, Inc. Madison, Wisconsin, USA.

Bourquin, L.D.K.A., Garleb, N.R.M and Fahey, G.C.1990. Effects of intake and forage level on site and extent of digestion of plant cell wall monometric components by sheep. *Journal of Animal Science* 68, 2479-2495

Casler, M.D. and Jung, H.J.G.1999. Selection and evaluation of smooth bromegrass clones with divergent lignin or etherified ferulic acid concentration. *Crop Science* 39,1866 - 1873.

Fahey, G. C., Collins, M., Mertens, D, R. and Moser, L.E. 1994. *Forage quality, evaluation and utilization*. American Society of Agronomy, Inc. Crop Science Society of America, Inc.Soil Science Society of America, Inc. Madison, Wisconsin, USA.

Grabber, J.H. 2005. How do lignin composition, structure, and cross-linking affect degradability: A review of cell wall model studies. *Crop Science* 45,820-831

Grabber,J.H., Hatfield, R.D., Ralph, J., Zon, J. and Amrhein, N. 1995. Ferulate cross-linking in cell walls isolated from maize cell suspensions. *Phytochemistry* 40, 1077-1082.

Hogon, J. P., Weston, R.H. and Linsay,J.R.1969. The digestion of pature plants by sheep. IV. The digestion of *Phalaris tuberosa* at different stages of maturity. *Australian Journal of Agriculture Research* 20,925-40.

Hoover, W. H. 1986. Chemical factors involved in ruminal fiber digestion. *Journal of Dairy Science* 69,2755-66.

Hungate,R.E.1966. *The rumen and its microbes*. p.65-70. Academic Press, New York.

Jung, H.G. and.Deetz, D.A. 1993. Cell wall lignification and degradability. P. 315 –346. In H.G.Jung et al (ed.) Forage cell wall structure and digestibility. ASA, CSSA, and SSSA. Maadison, WL.

Jung, H.G., and Ralph, J. 1990. Phenolic-carbohydrate complex in plant cell walls and their effect on lignocellulose utilization . p. 173 –182. In: D.E. Akin et al (ed.) Microbial and plant opportunities to improve lignocllulose utilization by ruminants. Elsevier, New York.

Kawas, J. R., Jorgensen, N. A. and Lu, C. D. 1990. Influence of alfalfa maturity on feed intake and site of nutrient digestion in sheep. *Journal of Animal Science* 68,4376-4386.

Kennedy. P. M., Christopherson, R. J. and Milligan, L P. 1982. Effects of cold exposure on feed protein degradation, microbial protein synthesis and transfer of plasma urea to the rumen of sheep. *British Journal of Nutrition* 47,521-535.

Kennedy, P.A. and Black, J.L.1986. Effect of simulatedsward structure on the rate intake of subterranean clover by sheep. Proc. *Australian Society of Animal. Production* 16,251-254.

Lindberg, J.E., Kaspersson, A. and Ciszuk, P. 1984. Studies on pH, number of protozoa and microbial ATP concentrations in rumen incubated nylon bags with different pore sizes. Journal of Agriculture Science (Camb.) 102,501- 504.

Lindberg, J.E.1983. Factors affecting predictions of rumen degradability using the nylon bag technique and a comparison between in vivo and in *in sacco* degradability measurements. PhD. Diss. Swedish University of Agriculture Science, Uppsala.

Lu,C.D.,Tsai, L.S., Schaefer, D.M. and Jorgensen, N.A.1987. Alteration of fermentation in continuos culture of mixed rumen bacteria by isolated alfalfa saponins. *Journal of Dairy Science* 70, 799-805.

McDonald,P and Edwards, R.A.1976. The influence of conservation methods on digestion and utilization of forages by ruminants. *Proceeding of Nutrition Society* 35,201-211

McSweeney, C.S., Palmer, B., Krause, D.O. and Brooker, J.D. 1999. Rumen microbial ecology and physiology in sheep and goats fed a tannin containing diet. In Tannin in Livestock and Human Nutrition. Proc. International Workshop, Adelaide, Australia, 31st May-2 nd June, pp140-145.

Merchen,N.R. and Satter, L.D.1983. Changes in nitrogeneous compounds and sites of digestion of alfalfa harvested at different moisture content. *Journal of Dairy Science* 66,789-801.

Mehrez, A.Z. and Ærskov, E.R.1977. A study of the artificial fibre bag technique for determining the digestibility of feeds in the rumen. *Journal of Agriculture Science* (Camb.) 88, 645-650.

Mertens, D. R.1987. Predicting intake and digestibility using mathematical madels of ruminal function. *Journal of Animal Science* 64, 1548-1558.

Moore, J.A., poore, M.H. and Swingle, R.S.1989. Calculated ruminal extent of digestion as influenced by in situ particle size, forage dosing time, and asmpling site in Hotstein cows. *Journal of Animal Science* 67(suppl 1) , 554.

Neathery, M.W.1969. Dry matter disappearance of roughages in nylon bags suspended in the rumen. *Journal of Dairy Science* 52,74-78.

Nocek, J. E. 1998. In situ and other methods to estimate ruminal protein and energy digestibility: A Review. *Journal of Dairy Science* 71,2051-69.

Nocek, J.E. and Kohn, R.A. 1988. *In situ* particle size reduction of alfalfa and timothy hay as influenced by form and particle size. *Journal of Dairy Science* 71,932-945.

NRC. 1985. Ruminant nitrogen usage. National Academy Press, Washington, DC.

Olubobokun, J.A., Craig, W.M. and Pond, K.R.1990. Effects of mastication and microbial contamination on ruminal in situ forage disappearance. *Journal of Animal Science* 68,3371-3381.

Ørskov, E.R., and McDonald, I. 1979. The estimation of protein degradability in the rumen from incubation measurements weighted according to rate of passage. *Journal of Agriculture Science* (Camb) 92, 499-503.

Ralph, J., Hatfield, R.D.,Quideau, S. and Helm, R.F., Grabber,, J.H. and.Jung, H.J.G. 1994. Pathway of p-coumaric acid incorporation into maize lignin as reveased by NMR. *Journal of American Chemical Society* 116, 9448 –9456.

Samanta, A.K., Singh, K. K., Das, M. M., Kundu,S.S.2004. Complete Feed Block- A balance ration for livestock (Technical Bulletin),IGFRI, Jhansi. (U.P.), India.

Singh, K. K., Samanta, A. K. and Srinivas,B.2001. Nutritional evaluation of tree leaves and range legumes. *Indian Journal of Animal Nutrition* 12,234-236.

Sullivan, J. T. 1973. Drying and storing herbage as hay. p. 1-31. In: G. W. Butler and R. W. Bailey (ed.) *Chemistry and biochemistry of herbage*. Vol. 3. Academic Press, Inc. London.

Thomas, D. J., and Beever, D. E. 1980. The effect of conservation and processing on the digestion of forages by ruminants. P. 291-308. In Y. Ruckebusch and P. Thivend (ed.) Digestive physiology and metabolism in ruminants. AVI Publishing Co., Westport, CT.

❑❑❑

Chapter **16**

Recent Developments in Physico-Chemical Techniques for Forage Quality Determinations

SUBIR KUMAR NAG
Indian Grassland and Fodder Research Institute, Jhansi -284003 (U.P.), India

Introduction

Animal performance is the best indicator of forage quality that depend upon intake, digestibility and efficiency of nutrient utilization. Variation in intake may account for 60-90% difference in digestibility. So it is imperative to measure forage characteristics, which actually govern, forage intake and digestibility. The constituents such as fibre, protein, lignin, dry matter (DM) etc. have been found to control intake and digestion. There are various analytical methods available for the analyses. These includes Proximate principle, Van Soest (detergent method) and Cornel Net Carbohydrate and Protein system which give valuable information about the chemical composition and constituents of the forages. Component chemical analyses also provide detail biochemical information leading to better understanding about nutrient utilization and animal performance. But there are some limitations, responsible for lack of animal response to forage analysis, which may be recognized like poor relationship between fibre and protein estimations and digestibilities, lack of uniform chemical rest, variation in energy prediction equations from fiber content etc. Many of these problems can be overcome by using the physico-chemical techniques that are being increasingly used of late in many places. Among various physico-chemical techniques involving spectroscopic instruments presently available and being used, the near infra red reflectance spectroscopy (NIRS) is used for routine testing of forage samples for many parameters like DM, fibre content, protein content, prediction of digestibility etc. Other techniques involving more sophisticated instrumentation like NMR, mass spectrometer, Raman spectroscopy, microspectroscopy and imaging etc. are used for very specific purposes and in few occasions.

Near Infrared Reflectance Spectroscopy (NIRS)

The discovery of infrared region in 1800 is credited to William F. Hershel's famous work "Experiments of the Refrangibility of the Invisible Rays of the Sun" read at the Royal Society on April 24, 1800. Infrared energy is the electromagnetic energy of molecular vibration. Spectroscopy is based on the interaction of electromagentic radiation with the matter to be analysed. The energy band is defined for convenience as the near infrared (0.78 to 2.50 microns), the infrared (or mid-infrared) 2.50 to 40.0 microns and the far infrared (40.0 to 1000 microns). The NIR region of the electromagnetic spectrum extends from the upper wavelength end of the visible region at about 770nm to about 2500 nm (13000-4000 cm^{-1}). However, most analytical work use NIR in between 1100-2500nm. Most valuable absorption bands that are overtones or combinations of fundamental stretching vibrational bands involving C-H, N-H and O-H characterize this region. The absorption bands seen in the near infrared region are the weaker overtones (1000-1900nm) and combinations (1900-2500 nm) of the much stronger fundamental absorption that occur in mid infrared region. The relative intensities of C-H stretch bands for infrared and the various NIR overtone regions are given in Table 1.

Table 1. Relative Intensities of C-H stretch Bands for Infrared and the Various NIR Overtone Regions

Band	Wavelength region	Relative intensity	Pathlength	Optical Density
Fundamental (n)	3380-3510 nm	100	0.01 mm	2
1st overtone (2n)	1690-1755 nm	1	1.0 mm	2
2nd overtone (3n)	1127-1170 nm	0.1	1.0 cm	2
3rd overtone (4n)	845-878 nm	0.01	10.0 cm	2
4th overtone (5n)	690-780 nm	0.005	10.0 cm	1

NIRS is a physical non-destructive method which depends on the measurement of light absorption by the surface of a sample using wavelength in the infrared region. The absorption spectrum produced depends on the chemical bonds within the constituents of the feed sample. The predominant near-infrared spectral features include: the methyl C-H stretching vibrations, methylene C-H stretching vibrations, aromatic C-H stretching vibrations, and O-H stretching vibrations. Minor but still important spectral features include: methoxy C-H stretching, carbonyl associated C-H stretching; N-H from primary amides, secondary amides (both alkyl, and aryl group associations), N-H from primary, secondary and tertiary amines and N-H from amine salts.

A Brief History about Evaluation of Forages by NIR Spectroscopy

Near-infrared reflectance spectroscopy is a unique technique in which the entire sample of a forage is characterized in terms of its absorption properties in the NIR region rather than separate sub samples being treated with various chemicals to isolate specific components. It is a widely used technique for feed and forage analysis. Feed samples are exposed to infrared radiation, and the reflectance of the radiation by the sample in measured. The instrument in calibrated by mathematically calibrating reflectance values with known chemical values for a series of samples of a particular ingredient. Thus the technique is of most value when a large number of similar samples are to be analysed.

In 1973 P. Williams reported the use of commercial NIR grain analyser for analysis of cereal products. Forage quality measurement using NIR was first reported by Norris *et al.* (1976). They measured a diverse range of forage samples and found standard errors 0.95, 3.1, 2.5, 2.1 and 3.5% respectively for CP, NDF, ADF, lignin and DMD in vitro. After this work NIR was adopted as an official method by the AOAC for measurement of CP and ADF in forages. Forage analysis using NIR measurement has been a major application of the technique largely due to the work of J.S. Shenk, M. Westerhaus, W. Barton, G. Marten, N. Martin, and a host of others who improved upon the technique and worked toward its widespread use and acceptance among scientists as a valid analytical technique. The handbook edited by G.C. Marten, J.S. Shenk and F.E. Barton (Marten *et al.*, 1985) has become the most used handbook for NIR forage analysis. NIR's use in forage testing was reviewed by Brown *et al.* (1990).

The successful use of NIR to measure the major organic components of forages has been demonstrated by many workers, and was reviewed extensively by Murray (1993). Protein, fibre fractions, water soluble carbohydrates have been measured successfully by NIR in perennial rye grass (Parnell and White, 1983) and total non-structural carbohydrates in lucerne roots (Brink and Marten, 1986) and tropical grasses (Brown *et al.*,1987). Introduction of alternative sampling devices for NIR instruments has allowed the analysis of fresh (undried) and unmilled forages (eg., silage and fresh grass) by NIR spectroscopy (Shank,1992). Later Williams and Norris (2001) edited a comprehensive text on NIR analysis for commercially important biological materials. A more recent comprehensive information source for NIR analysis of forages is a book edited by C. Roberts (Roberts *et al.* 2004).

NIR Spectra

Qualitative and quantitative near infrared (NIR) spectroscopic methods typically require the application of multivariate calibration algorithms and statistical methods (i.e. chemometrics) to model NIR spectral response to chemical or physical properties of the samples used for calibration . The NIR method relies on the spectra-structure correlations existing between a measured spectral response caused by harmonics of the fundamental vibrations occurring at infrared frequencies. These harmonic vibrations occur at unique frequencies depending upon the quantity of absorber (analyte), type of absorbing molecules present within the sample, and the sample thickness. Quantitative methods are possible where changes in the response of the near infrared spectrometer are proportional to changes in the concentration of chemical components, or in the physical characteristics (scattering/absorptive properties) of samples undergoing analysis.

NIR spectra are essentially plots of reciprocal $\log_{10}$ reflectance (log 1/R) against wavelength. The spectra appear as smooth rolling lines with few well defined features but they consist of many overlapped bands, since reflectance spectra of a forage sample is the summation of the spectra its major chemical components. The challenge is to extract useful information on nutritive value of forages from the reflectance data.

Prominent absorption spectra in NIR spectra of forages include water with two bands at 1940 and 1450nm, aliphatic carbon (C)-hydrogen (H) bands (lipids) at 2310, 1725 and 1400 and 1210nm, and oxygen (O)-H bands (carbohydrates) around 2100 and 1600nm. The nitrogen (N)-H-absorption due to amide structures present in protein occur predominantly at 2180 and 2055nm (Fig.1). It is the shape of the spectra line, or the rate of change in slope with wavelength, that conveys chemical information. Hence first and second derivative plots of log 1/R are useful as they can resolve overlapping bands into component absorptions, which may appear as shoulders on larger peaks, as well as minimizing the effect of particle size (Fig.2). The second derivative is particularly useful, as it produces a negative peak.

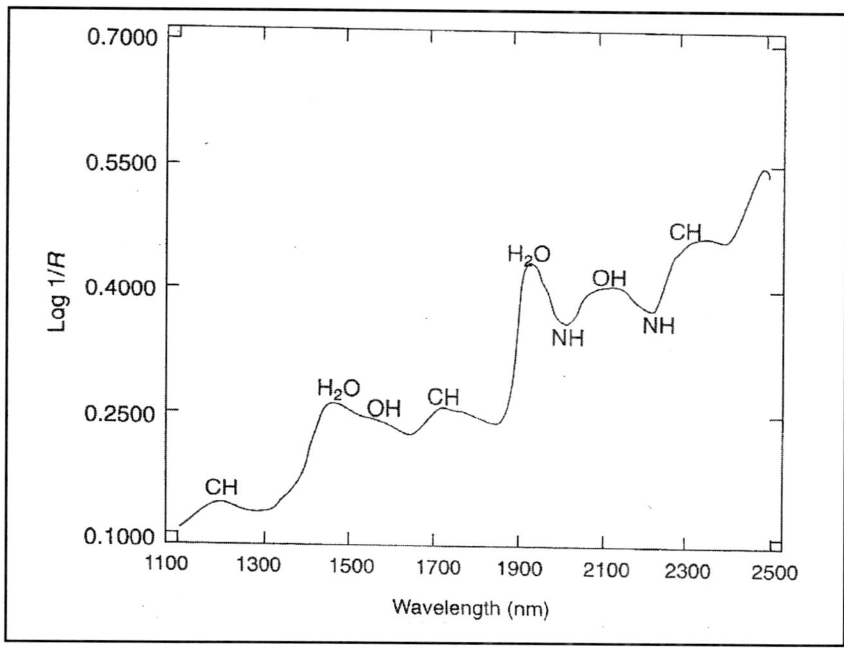

Fig. 1. NIR log 1/R spectrum of Hay Sample (Deaville and Flinn, 2000).

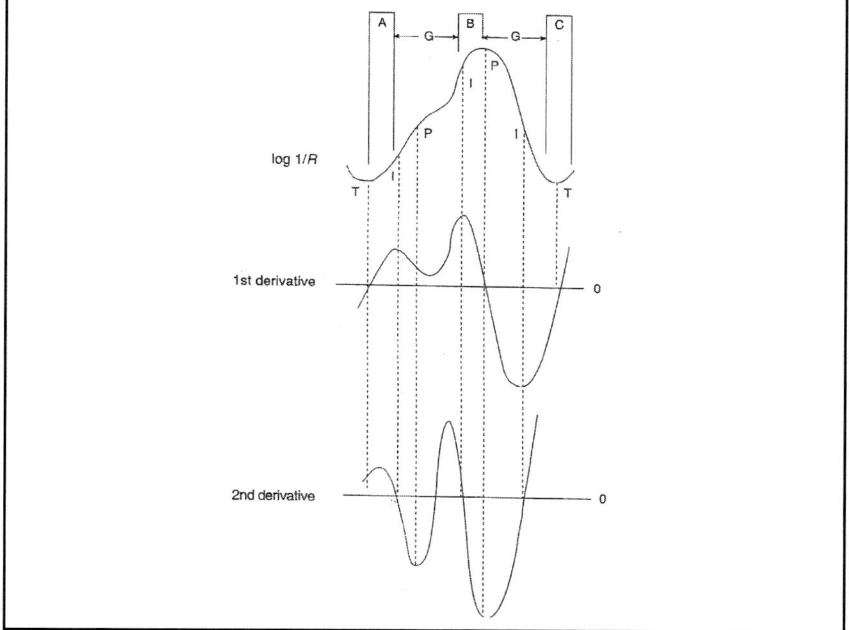

Fig. 2. The Relationship Between log 1/R and First & Second Derivative log 1/R Spectra (Deaville and Flinn, 2000).

NIR Instrumentation

The major components of NIR instrument consists of a source of irradiation usually tungsten filament lamp, optical system, detector and sample holder. There are many instrument types available for NIR analysis. The three major types of devices for wavelength selection are monochromators, filters and diode arrays. Monochromators scan the whole wavelength range by using a grating or prism and they provide maximum versatility. Instead of monochromator, radiation can also be analysed by means of a interferometer. Tilting filter instruments have there or six filters, which in effect, allow segments of spectra to be scanned. Fixed filter instruments measure log 1/R at discrete wavelengths only, but are quite satisfactory for applications where the constituents are well defined and the appropriate wavelength well known. Diode array instruments use no moving parts for wavelength selection and have the advantage of being able to scan at high speed. Silicon diode array or InGaAs (Indium, Gallium, Arsenic) peltier cooled detectors are used now a days.

There are basically two alternative procedures for presenting a forage sample to an NIR instrument. Small sample cups are used for finely ground, uniform materials and large rectangular cells are used to pack unground and often high moisture heterogeneous sample. Until recently most NIR forage analysis has been done using dried, ground samples and the recommended procedures for drying, grinding mixing and packing are well established. However, there are advantages in analysing forages in their natural, unground stage, due to reduced sample preparation time and avoidance of error due to losses of volatile compounds by oven drying.

Calibration of NIR Data

The NIR instrument is first to be calibrated before its use for forage quality testing. Calibration in the mathematical process required to relate NIR optical measurements to the desired constituent or property used to define the nutritional quality of forages. Calibration is the key for the successful use of any analytical technique. There are number of basic principles and many approaches that have been found to be vital for the success of NIR calibration. The major steps that are essential to obtain a satisfactory working calibration are sample selection, acquisition of spectra and reference data, pretreatment of spectra data, derivation of the regression model and validation of the model.

Selection of appropriate sample is very critical for a successful calibration. The calibration population must include all sources of variation likely to be found in future unknown samples of similar material, much on chemical, physical and botanical characteristics. The traditional method of sample selection involved either structured or random sampling from a population, using information already available on chemical or botanical composition. A more convenient procedure is to choose representative samples from a large population on the basis of their spectral characteristics. Software like CENTER and SELECT are available for this purpose. Murray (1986b, 1988) suggested that the calibration set should have "WEPT" criteria i.e., a WIDE range of composition, an EVEN (not normal) distribution, PRECISE analytical data from some reference methods and with TYPICAL properties of those encountered in future routine analysis.

The number of samples needs for calibration may vary. For agricultural products calibration based on fewer than 50 samples are seldom satisfactory. Windham et al. (1989) concluded that 50-100 samples were adequate for narrow based or closed population whereas 150 or more samples were necessary for broad based or open population. With recent advances in software and computer power, it is now possible to establish calibration population or 'Libraries' containing hundreds or thousands of samples, which are necessary for multi product forage calibration.

The quality of reference method analysis has a crucial effect on the accuracy of NIR calibration. Poor calibration accuracy has often been incorrectly blamed on NIR instruments when in most cases the laboratory reference method and other associated factors were at fault. There have been major improvements in NIR technology in the last few years. Standardization of instruments has been one of them.

Treatment of Spectral Data

Because NIR spectra are affected by particle size, light scatter and path length variation, so treatment of the spectral data prior to calibration can often improve calibration accuracy. Commonly used corrections are:

- Standard Normal Variate - Detrending (SNV-D)
- Multiplicative Scatter Correction - (MSC)
- Derivatisation

Detrending removes the linear and quadratic curvature of each spectrum. SNV scales each spectrum to have a standard deviation of 1.0 and MSC expands or contracts each spectrum and sifts at up or down to look most like the target spectrum. First and second derivative mathematical treatments are most frequently employed (Deaville and Flinn, 2000).

Calibration Methods

NIRS analysis relies on statistical procedures to obtain accurate predicted value. The statistical technique used to predict concentration of a component is usually multiple linear regression (MLR). The use of multiple regression to relate changes in apparent absorption at selected wavelengths in the near infra red spectrum to the concentration of a number of constituents has been the basis of almost all NIR results. Multiple regression analysis was the statistical technique that enabled Norris to develop NIR spectroscopy as a new technique. Alternate calibration methods include principal component regression (PCR), partial least square regression (PLS) and artificial neural networks. In PCR the original spectral data is compressed into a small number of independent components, or scores, which are used in the regression equation instead of spectra themselves. This overcomes the problem of inter correlation between the spectral data points. Typically, a model containing between five and ten principal components give satisfactory predictions. PLS is similar to PCR except that PLS uses the laboratory reference data as well as the spectral data in computing the factors using the regression. This often results in better prediction. A variation of this procedure called modified PLS (MPLS) is frequently used in forage calibration and involves standardizing the variables in the regression after each factor in calculated.

There is, however, no best technique to use in all situations. MPLS has been shown to be superior to multiple linear regression.

The validation of calibration using MLR is done by randomly selecting samples from the calibration population, exclude them from the calibration and use them as a validation set to test the calibration equation. The error thus occurred is measured as standard error of prediction (SEP). This test is satisfactory for closed population but inadequate in case of open population for which independent set of samples should be used for validation. In case of PCR or PLS/MPLS regression cross validation is used where calibration set is split into several groups and calibration progressively performed on each group until every sample has been tested for both calibration and validation. The validation errors are combined to a standard error of cross validation (SECV).

It has now been widely accepted that broad based NIR calibration for forage analysis are more useful than those based on separate species. Whilst the latter can be slightly more accurate, they lack robustness and involve more work and expense. Studies have also found that multiforage calibrations (hay, fresh forage, haylage and silage) can be nearly as accurate as those for single products provided enough samples are present in the calibration population.

Stepwise NIR Scanning and Calibration

1. Scan samples
2. Check scan visually
3. Add calibration data
4. Create PCA (Principal Component) and lib (library) files
5. Check population distribution and examine correlations to calibration data
6. Create equations
7. Check equations against independent samples of known composition
8. If equations work, use with unknown samples

Evaluation of Different Parameters of Forages by NIR Spectroscopy

Digestibility

The use of NIR spectroscopy to predict digestibility in vivo has been widely reported. In most cases NIR spectroscopy provided a better prediction of digestibility than traditional laboratory procedures. Barber *et al.* (1990) showed that for grass silage NIR spectroscopy was able to predict OMD *in vivo* more accurately than a range of commonly used laboratory methods. They used a stepwise MLR (multiple linear regression) of eight first and second derivatized spectral terms.

Table 2: Comparison of NIR Spectroscopy with Several Laboratory Methods for Predicting Organic Matter Digestibility (OMD) in vivo of Grass Silage

Method	Validation Statistics		
	R^2	SEP	Slope
NIR – 8th term	0.76	2.6	0.93
In vitro OMD	0.64	3.6	0.89
Pepsin-cellulase	0.40	4.7	0.71
Acetyl bromide lignin	0.14	5.3	0.48
Modified lignin	0.20	5.1	0.52

SEP = Standard error of prediction (%)

De Boever *et al.* (1996) have also shown for grass silage made in Belgium that NIR spectroscopy provided the best calibration relationship for OMD in vivo with R^2 value being 0.79 compared to 0.68, 0.64 and 0.53 for cellulase digestibility, rumen fluid digestibility and acid detergent lignin determination respectively.

NIR calibration has also been successfully developed for prediction of the digestibility and metabolic energy (ME) content of other forages including grasses and straws. Given *et al* (1992b) showed that the ME content of mixed growth grasses could be predicted by a five term NIR calibration equation and this technique was equal to the best traditional laboratory method (pepsin-cellulase).

Voluntary Intake

NIR spectroscopy has been successfully used to predict the intake of mainly non-fermented forages. The following table summarizes some of the work done earlier.

Table 3: Prediction of Voluntary DM Intake of Forages by NIR

Forage type	No. of Measurement	Intake		NIR prediction		Reference
		Species	Range	R^2	SEP	
Mixed	76	Sheep	39.7-114.3	0.62	7.8	Norris et al., 1976
Grazed pasture	21	Cattle	52.6-112.3	0.72	9.6	Ward et al., 1982
Grazed pasture	80	Sheep	430-1458	0.80	140	Flinn et al., 1992

Later on studies have also shown that voluntary intake of grass silage by cattle can also be accurately predicted by NIR spectroscopy. Park *et al.* (1997) measured the DM intake of 136 grass silage and showed that an MPLS spectra relationship could achieve a prediction accuracy (SECV) of 5.05 DM Kg^{-1} $W^{0.75}$. Agricultural and Food Research Council (AFRC,1991) proposed that NIR spectroscopy could be one of the promising areas for predicting the voluntary intake of grass silage, since the spectra contain information on the entire chemistry of the feed. Steen et al. (1998) showed NIR spectroscopy to be superior to some 35 other measurements of silage quality, digestibility *in vivo* and rumen DM degradability in situ.

Prediction of Fermentable Energy of Forage

Very often it is important to distinguish between the energy in the forage that is available for rumen microbial growth (fermentable energy) and that which is directly available to the host animal. Fermentable energy represents the portion of dietary energy available for microbial protein synthesis. NIR spectroscopy could be used to predict the concentration of individual or total fermentation acids (Sinnaeve *et al.,* 1994). An alternative approach which showed that fermentation acids in sorghum silage could be adequately predicted by NIR spectroscopy if the acids were first converted to their corresponding salts before microwave drying. Deaville and Givens (1996a) showed that an MPLS model of the NIR spectra obtained from 800 fresh silages could be used to predict lactic acid and VFA. The prediction of silage fermentation acids represents an approach to estimate fermentable energy content by difference. The integrated studies on forage using both gas production techniques and NIR spectroscopy by Herrero *et al.* (1996) showed that it was possible to derive reasonable NIR predictive relationship for cumulative gas volumes produced between 8-48h of incubation, although rates of gas production could not be predicted.

Prediction of Soluble and Insoluble Protein

NIR spectroscopy is used to predict the CP content of forages and other feeds. However, protein quality is more accurately described by estimating the immediately soluble fraction, the insoluble but degradable fraction and the rate of degradation of degradable fraction. These measurements are normally carried out using in situ technique which is expensive and slow and requires modified animals. NIR spectroscopy can be used to predict nitrogen solubility. Abrams *et al.* (1988) showed that with various fresh silages the solubility and insolubility of nitrogen in buffered solution of NaCl could be estimated by a sixterm multiple stepwise regression (MSR) calibration of second order derivatized log1/R spectral data.

Table 4: Estimation of Nitrogenous Constituents in Silage by NIR Spectroscopy

Nitrogen fraction	Mean value (g/kg fresh wt)	Calibration Statistics		
		SD	R^2	SEC
Total N	9.55	4.49	0.98	0.70
Soluble N	3.48	1.99	0.92	0.58
Insoluble N	5.97	2.7	0.98	0.35

Use of NIR Data to Monitor Forage Digestion

Several wavelength region of NIR spectra which appear as being most important in relation to forage digestibility is mentioned in the following table (Table 5). Spectral regions around 1650-1670 and 2260-2280 nm appear as most important regions associated with digestibility. The primary wavelength for gas production is in the same region of digestibility located between 1664-1696 nm. Givens et al. (1992a) demonstrated that in cereal straws the 1650 and 2254nm regions related to indigestible fractions of the forage. Other findings indicate that these regions relate to lignin bonding.

Table 5: Primary and Secondary NIR Spectroscopy Wavelength Segments used to Predict Forage Digestibility

Forage	Predicted term	Wavelength (nm)	
		Primary	Secondary
Mixed	DMD in vitro	2266	1662
Grass	ME	1668	1364
Grass silage	Lignin	1658	2286
Hays	DMD in vitro	1326	2266
Hays/Grass	ME	1666	1690
Grass silage	OMD in vitro	2266	1662
Grass	ME	1658	2280

Minerals and Antinutritional Factors

Minerals do not absorb in the NIR region. But correlations between minerals and other organic components allow reasonable NIR calibrations to be obtained in some cases. Clark *et al.* (1987) concluded that accurate NIR analysis was limited to only Ca, P, K and Mg which were found to have coefficient of variation (CV) ranging from 10-20%, for other minerals the CV exceeded 20%. Smith *et al.* (1991) obtained a calibration for Mg in perennial rye grass with a CV of 16% and concluded that NIR could be a useful tool for preliminary screening of rye grass lines for Mg in a breeding programme.

NIR can also be used to screen various forages for antinutritional factors/ secondary plant products, which are normally present in small quantities but may be detectable if they affect NIR spectra by means of secondary correlations (Deaville and Flinn, 2000).

Advantages of NIRS

1. It can scan samples in a wavelength range where it can actually see the organic compounds like cellulose, protein, starch etc.
2. It can detect physical differences of the forage where chemical tests can not.
3. It is a versatile technique for simultaneous measurement of several parameters.
4. It is very fast - measurement results obtained within seconds.
5. It is safe and green technique as no chemical is used.
6. NIR equations can also directly predict animal performance.

7. NIR equation has been developed for estimation rumen undegraded protein (RUP) for which no current widely accepted method is available for accurately estimating RUP differences in forages.
8. New tests for additional feed factors like particle size of the forage, grass percentage in a legume or hay, leaf to stem ratio in alfalfa, onset of diseases and pest are also being developed by NIR.

Disadvantages of NIRS

1. The main disadvantage of NIR spectroscopy is that of high initial cost of the instrumentation and its reliance on chemometrics.
2. The prediction depends on calibration and calibration data come from chemical analysis. So the performance depends on the chemical tests of the samples.
3. A major limitation in the use of NIR spectroscopy for forge analysis is the inadequacy of many conventional laboratory reference methods.

Other Physico-Chemical Techniques

The use of high resolution (HR) spectroscopic techniques like NMR (Nuclear Magnetic Resonance), mid infrared (MIR), Raman spectroscopy and pyrolysis mass spectroscopy (PyMS) are finding increased usage in forage assessment. These techniques, as opposed to low resolution (LR) spectroscopic techniques like NIR, UV-VIS spectroscopy, generally provide more specific and detailed information of a primary nature. The employment of HR techniques becomes absolutely necessary where LR techniques are not sufficient to give the desired information.

The UV-VIS spectroscopy can give excellent primary information if the information sought is for a single or isolated chromophore. However, forage samples may contain multitude of chromophores, making such analysis difficult. NIR spectroscopy is generally not a source of direct information but an indirect or secondary technique, requiring independent calibration by a primary technique, using chemometric approaches. The primary technique is normally some form of gravimetric analysis of moisture, lignin, fibre etc. Although this approach is sufficient and cost efficient for many needs, the dependence on proximate analysis limits the effectiveness of NIR spectroscopy.

NMR Spectroscopy

NMR has been a powerful tool for examination of finer details of forages. Both solution and solid state NMR is employed for the purpose. In true solution

NMR studies the extractable components, which are soluble in deuteriated solvents, are subjected to wide array of single and multidimensional experiments. If the material is pure, sufficient information is obtained to assign the molecular structure. Solid state NMR is well suited for examining the intact structure of forage materials, without requiring extraction. New experiments, which lie somewhere in the realm between the solution and solid methods, is now in vogue. These are experiments in which all or part of the sample is made mobile by hydration.

Several hydroxyl cinnamic oligosaccharide esters (Fig.3) have been isolated from barley, bamboo, Bermuda grass, maize etc. and their structures have been subjected to detailed characterization by solution NMR with the aim to determine the specific involvement of hydroxycinnamic ester moieties in restricting the availability of cell wall polysaccarides of forage grasses to ruminant digestion. First the proton NMR spectrum of the sample was taken. The assignment was straight forward for the phenolic acid portion of the molecule, but required a more sophisticated approach for the carbohydrate portion due to many overlapping signals. The assignment of the carbohydrate portion of the proton spectrum could be accomplished by the use of one dimensional (1-D) analogue of the 2-D (^{1}H, ^{1}H) Homonuclear Hartmann-Halm (HOHAHA) or total correlation spectroscopy (TOCSY), which involved selective irradiation of a known position, such as the anomeric portion of each molecule. Knowing the proton assignments the corresponding ^{13}C spectrum could be assigned by employing inverse detected (^{1}H) one bond (^{1}H, ^{13}C) heteronuclear multiquantum correlation (HMQC) experiment. After both the ^{1}H, ^{13}C spectra have been assigned for individual subunits, then the linkages between them could be found by employing inverse detected long range (^{1}H, ^{13}C) hetero nuclear multiple bond connectivity (HMBC) experiment. The actual solution conformation of the molecule in the NMR tube can be established by 2-D experiments the phase sensitive (^{1}H, ^{1}H) rotating frame Overhauser enhancement spectroscopy (ROSEY) and compared with a (^{1}H,^{1}H) double quantum filtered correlation spectroscopy (DFQ-COSY) experiment. In addition, the angle of rotation of the hydrocinnamic acid in relation to arabinofuranocide was estimated by analysis of coupling constants. The final conformation then could be expressed as population of rotomers.

Solution NMR has also been used to provide the insight as to how the hydroxycinnamic acids linked to hemicellulose polysaccharides are incorporated into the cell wall. It has been used to compare the isolated forage lignins and show that the warm season grass forage differ from cool season in that they incorporate p-coumaryl units.

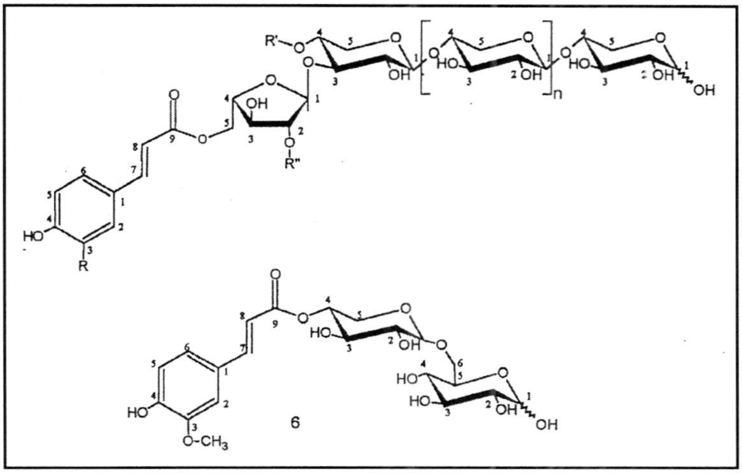

Fig. 3. Hydroxycinnamic oligosaccharide esters.

Solid State NMR Spectroscopy

Solid state NMR spectroscopy can overcome many of the limitations of solution NMR spectroscopy and is ideally suited to the study of intact forages or their residues. Unlike solution NMR spectroscopy a higher field is not always better. This is especially true for carbonaceous material, if one is only interested in obtaining isotrophic chemical shift information. The most popular technique for obtaining solid state ^{13}C spectra has been that of cross polarization/magic angle spinning (CP/MAS). Modification of the basic solid state ^{13}C CP/MAS NMR can often be used more effectively to study forages. Interrupted or delayed decoupling is one such modification which has been used to detect difference in normal versus brown mid rib mutant lines in sorghum (Akin *et al.*, 1986) and pearl millet (Morrison *et al.*, 1993). The mutant lines are characterized by the brown coloration in the mid vein of their leaves or stems and are generally more digestible by ruminants than the normal lines. By employing the delayed decoupling technique change in the lignin structure can be directly detected in the intact forage since the loss of signal intensity at 154 ppm due to syringyl units could be observed in mutant lines which would not have been possible by using standard CP/MAS experiment. Thus isolation of lignin was not required for getting the desired information.

Solid state ^{13}C CP/MAS NMR spectroscopy was also used in the same study to show that lignin polymerization had not occurred to a large extent in the mutant lines of pearl millet, where phenolics were mostly ester linked than in the normal lines where they were ether linked i.e. polymerized and this accounted for greater digestibility of the mutant lines.

Solid state NMR was used to investigate changes in the composition of residues from red clover and pea nut hulls obtained by four commonly used gravimetric methods for lignin in forages (Reeves and Schmidt, 1994). The samples had previously been analysed by IR and PyMS procedures. Solid state ^{13}C NMR enabled the confirmation of a structural interpretation of why the different gravimetric procedures were not chemically equivalent. The three instrumental techniques together enabled a more accurate evaluation of the chemical composition of lignin obtained by each of the four different gravimetric procedures. The NMR spectral data of dimers and oligomers of hydroxycinnamaldehyde, coniferaldehyde and sinapyldehyde have allowed a more substantive assignment of aldehyde components in lignins (Kim *et al.*, 2003).

Other Spectroscopic Techniques

Diffuse reflectance infrared Fourier transform spectroscopy (DRIFTS) in the mid infrared region, reveal more clearly the digestion of carbohydrates which is shown by decrease in O-H stretch bands (~3000-3600 cm^{-1}) and C-O mixed mode heavy atom vibrations (~900-1200 cm^{-1}). Like the NMR the MIR spectra show that there is no digestion of lipids or lignin. The indications for lignin or phenolics actually slightly increase as digestion proceeds.

Fourier transformed Raman spectra is another spectroscopic technique which is highly responsive to the aromatic components because the Raman active vibrations must be accompanied by a change in polarisability and aromatic compounds contain highly polarisable bonds. In Raman spectra small difference in the phenolics content of the residue are revealed as large differences at ~1600 cm^{-1}, which is barely detectable in solid state NMR or MIR spectra. However, in contrast, the loss of carbohydrates is not as evident in Raman spectra as in MIR or NMR.

Pyrolysis mass spectrometry (PyMS) has also been employed effectively to examine the compositional difference in forages. When electron impact (EI) ionization is used, phenolic components can be examined. Even with softer ionization like chemical ionization (CI) carbohydrates can be emphasized. Coastal Bermuda grass and Kentucky-31 tall fescue were subjected to EI MS and specific markers were found for p-coumaric and ferulic acid esters at m/z 120 and 150 respectively. The marker for p-coumaric acid showed greater abundance in coastal Bermuda grass, which was in agreement, with the NMR spectroscopic results on lignin (Himmelsbach, 2000).

Microspectroscopy and Imaging

These are the techniques which combine spectroscopy with microscopy employed to study the chemical factors along with anatomical compartments at specific morphological stages of plants.

UV spectroscopy in conjunction with microscopy, called UV micro-spectroscopy or microspectrophotometry is one such attempt in this direction. By restricting the spectra to specific cell types the interference between compounds that give similar spectral response can also be minimized. UV microspectroscopy was used in the detection of phenolic components in specific cells of forages that showed different degradability (Akin and Rigsby, 1992).

The concept of the use of microspectroscopy in the MIR region is also helpful. The technique got resurgence after the development of FT-IR. The technique has been reviewed by Reffner (1988). *MIR microspectroscopy* has been used to compare the level of phenolics found in the vascular bundle and pith of normal and brown midrib mutant tissue. In MIR imaging a synchrotron light source is used to enhance the signal noise ratio and a spatial resolution of ~6 mm is obtained. The technique was used on a leaf vascular bundle of coastal Bermuda grass (Wetzel *et al.*, 1998). The phenolics, lipid and protein of the vascular bundle were imaged at 1509, 1550 and 1469 cm^{-1}, respectively. The greatest concentration of phenolics was found in the sclerenchyma and parenchyma bundle sheath. Lipids appeared to be located in the epidermal tissue and protein nearby. MIR microspectroscopy is capable of locating more chemical constituents than UV microspectroscopy but it is limited by its spatial resolution in discriminating between anatomical structures.

Raman microspectroscopy imaging has a high spatial resolution of the order 1mm and spectral resolution is equivalent to MIR. But the technique is not very applicable to forage materials that contain highly fluorescent molecules such as chlorophyll. But it may be applicable to residues, especially useful in mapping the location of lignin in digested residues.

NIR imaging, started to be explored extensively, represents a compromise between MIR and Raman imaging with regard to spatial and spectral resolution. Most of the application in NIR imaging is restricted to medical arena.

Conclusions

The developments in NIR spectroscopy has revolutionized the analysis and nutritional characterization of forages by allowing rapid evaluation thus helping in quick, strategic and economic decisions to be made regarding appropriate supplementation or forage upgrading. Therefore, prediction of forage quality by NIR spectroscopy should be used as a tool for further better estimation of nutrient supply to animals. Among other physico-chemical techniques NMR is exceptionally specific. Each technique has its own advantages and problems. Purchase of these latest state of the art equipment, their operation and maintenance, that too for forages, is in itself very difficult job particularly where we have different priorities. Nevertheless, to keep pace with advancement of science and technology worldwide we must attempt to introduce some of the techniques for greater understanding of the forage value in terms of digestibility and animal performance.

References

Abrams, S.M., Shenk, J.S. and Harpster, H.W. 1988. Potential of near infrared reflectance spectroscopy for analysis of silage composition. *Journal of Dairy Science* 71, 1955-1959.

AFRC, 1991. AFRC technical committee on responses to nutrients, Report No.8, Voluntary intake of cattle. *Nutrition Abstracts and Reviews* (Series B) 61, 815-823.

Akin, D.E. and Rigsby, L.L. 1992. Scanning electron microscopy and ultraviolet absorption microspectrophotometry of plant cell types for different biodegradabilities. *Food Structure* 11, 259-271.

Akin, D.E., Hanna, W.W., Snook, M.E., Himmelsbach, D.S., Barton, F.E. and Windham, W.R. 1986. Normal 12 and brown mid rib 12 sorghum. Chemical variations and digestibility. *Agronomy Journal* 78, 832-837.

Barber, G.D., Givens, D.I., Kridis, M.S., Offer, N.W. and Murray, I. 1990. Prediction of the organic matter digestibility of grass silage. *Animal Feed Science and Technology* 28, 115-128.

Brink, G.E. and Marten, G.C. 1986. Analysis of alfalfa root carbohydrate concentration by near infrared reflectance spectroscopy. *Crop Science* 26, 159-161.

Brown,W.F., Moore, J.E., Kunkle, W.E., Chambliss, C.G. and Porter K.M. 1990. Forage testing using near infrared reflectance spectroscopy. *Journal of Animal Science* 68, 1416-1427.

Clark, D.H., Mayland, H.F. and Lamb, R.C.1987. Mineral analysis of forages with near infrared reflectance spectroscopy. *Agronomy Journal* 79, 485-490.

De Boever, J.L., Cottyn, B.G., De Brabander, D.L., Vanacker, J.M. and Boucque, C.V. 1996. Prediction of the feeding value of grass silages by chemical parameters, in vitro digestibility and near infrared reflectance spectroscopy. *Animal Feed Science and Technology* 60, 103-115.

Deaville, E.R. and Flinn, P.C. 2000. Near Infrared Spectroscopy: an Alternative Approach for the Estimation of Forage Quality and voluntary Intake. In Givens, D.I., Owen, E., Axford, R.F.E. and Omed, H.M. (eds.) Forage evaluation in ruminant nutrition. CABI, UK. pp. 301-320.

Deaville, E.R. and Givens, D.I. 1996a. Prediction of the non fermentable energy fraction of grass silage using near infrared reflectance spectroscopy. *Animal Science* 60, 681 (Abstract).

Flinn, P.C., Windham, W.R. and Dove, H. 1992. Pasture intake by grazed sheep estimated using natural and dosed n-alkanes – a place for NIR? In: Hildrum, K.L., Isaksson, T, Naes, T. and Tandberg, A. (eds.) Near Infrared Spectroscopy. Ellis Harwood, Chicester, pp. 175-178.

Givens, D.I., Baker, C.W. and Zamime, B. 1992a. Regions of normalized near infrared reflectance difference spectra related to the rumen digestion of straws. *Animal Feed Science and Technology* 36, 1-12.

Givens, D.I., Moss, A.R., Baker, C.W. and Adamson, A.H. 1992b. Influence of growth type and season on the prediction of the metabolisable energy content of herbage by near infrared reflectance spectroscopy. *Animal Feed Science and Technology* 37, 281-296.

Herrero, M., Murray, I., Fawcett, R.H. and Dent, J.B. 1996. Prediction of the in vitro gas production and chemical composition of kikuyu grass by near infrared reflectance spectroscopy. *Animal Feed Science and Technology* 60, 51-67.

Himmelsbach, D.S. (2000). NMR and other physiological techniques for forage assessment. In Givens, D.I., Owen, E., Axford, R.F.E. and Omed, H.M. (eds.) Forage evaluation in ruminant nutrition. CABI, UK. Pp. 321-344.

Kim, H., Ralph, J., Lu, F., Ralph, S.A., Boudet, A.M., Mackay, J.J., Sederoff, R.R., Ito, T., Kawai, S., Ohasi, H. and Higuchi, T. 2003. NMR analysis of lignin in CAD deficient plants. Part 1. Incorporation of hydroxycinnamaldehydes & hydroxybenzaldehydes into lignins. *Organic and Biomolecular Chemistry* 1, 268-281.

Marten, G.C., Shenk, J.S. and Barton, F.E. 1985. Near infrared reflectance spectroscopy-Analysis of forage quality-*USDA Agric. Handbook No. 643*, p 96.

Morrison, W.H., Akin, D.E., Himmelsbach, D.S. and Gamble, G.R. 1993. Investigation of the ester and ether linked phenolic constituents of cell wall types of normal and brown mid rib pearl millet using chemical isolation, microspectrophotometry and ^{13C} NMR spectroscopy. *Journal of the Science of Food and Agriculture* 63, 329-337.

Murray, I. 1986b. Near infrared analysis of forages. In: Haresign, W. and Cole, D.J.A. (eds.) Studies in the Agricultural and Food Sciences, Recent advances in Animal Nutrition. Butterworths, 141-156.

Murray, I. 1988. Aspects of the interpretation of NIR spectra. In: Creaser, C.S. and Davies, A.M.C. (eds.). Analytical applications of spectroscopy. RSC, pp 9-20.

Murray, I. 1993. Forage analysis by near infrared reflectance spectroscopy. In: Davies, A., Baker, R.D., Grant, S.A. and Laidlaw, A.S. (eds.) Sward management handbook, British Grassland Society, Reading, UK, pp. 285-312.

Norris, K.H., Barnes, R.F., Moore, J.E. and Shenk, J.S. 1976. Predicting forage quality by infrared reflectance spectroscopy. *Journal of Animal Science* 43, 889-897.

Park, R.S., Gordon, F.J., Agnew, R.E., Barnes, R.J. and Steen, R.W.J. 1997. The use of near infrared reflectance spectroscopy on dried samples to predict biological parameters of grass silage. *Animal Feed Science and Technology* 74, 2664-2673.

Parnell, A. and White, J. 1983. The use of near infrared reflectance spectroscopy in predicting the digestibility and the protein and water soluble carbohydrate contents of perennial rye grass. *Journal of the National Institute of Agricultural Botany* 16, 221-227.

Reeves, J.B. and Schmidt, W.F. 1994. Solid state ^{13C} NMR analysis of forage and byproduct derived fiber and lignin residues. Resolution of some discrepancies among chemical, infrared and pyrolysis-gas chromatography-mass spectroscopic analysis. *Journal of Agricultural and Food Chemistry* 42, 1462-1468.

Reffner, J.A. 1998. Instrumental factors in infrared microspectroscopy. *Cellular and Molecular Biology* 44, 1-7.

Roberts, C., Workman, J. and Reeves, J. (eds.) *Near-infrared Spectroscopy in Agriculture*, ASA-CSSA-SSSA, Madison, WI, 2004.

Shenk, J.S. 1992. NIRS analysis of natural agricultural products. In: Hildrum, K.I., Isakson, T., Naes, T. and Tandberg, A. (eds.). Near infrared reflectance spectroscopy: Bridging the gap between data analysis and NIR applications. Ellis Horwood, Chichester, UK, pp. 235-240.

Sinnaeve, G., Dardennne, P. and Agneessens, R. 1994. Global or local? A choice of NIR calibrations in analyses of forage quality. *Journal of Near Infrared Spectroscopy* 2, 163-176.

Smith, K.F., Wills, S.E. and Flinn, P.C. 1991. Measurement of the magnesium concentration in perennial rye grass (*Lolium perenne*) using near infrared reflectance spectroscopy. *Australian Journal of Agricultural Research* 42, 1399 – 1404.

Steen, R.W.J., Gordon, F.J., Dawson, L.E.R., Park, R.S., Mayne, C.S., Agnew, R.E., Kilpatrick, D.J. and Porter, M.J. 1998. Factors affecting the intake of grass silage by cattle and prediction of silage intake. *Animal Science* 66, 115-127.

Ward, R.G., Smith, G.S., Wallace, J.D., Urquhart, N.S. and Shenk, J.S. 1982. Estimates of intake and quality of grazed range forage by near infrared reflectance spectroscopy. *Journal of Animal Science* 54, 399-402.

Wetzel, D.L., Eilert, A.J., Pietrazk, L.N., Miller, S.S. and Swear, J.A. 1998. Ultraspatially resolved synchrotron infrared microspectroscopy of plant tissue in situ. *Cellular and Molecular Biology* 44, 145-167.

Williams, P. and Norris, K. (eds.) *Near-Infrared Technology in the Agricultural and Food Industries 2nd Edition*, American Association of Cereal Chemists, St. Paul, Minnesota, USA, 2001.

Windham, W.R., Mertens, D.R. and Barton, F.E. 1989. Supplement 1. Protocol for NIRS calibration: sample selection and equation development and validation. In: Marten, G.C., Shenk, J.S. and Barton, F.E. (eds.) Near infrared reflectance spectroscopy : Analysis of forage quality. USDA Agriculture handbook No.643, Washington, DC, pp 96-103.

□□□

Chapter 17

Gas Production Technique for Forage Evaluation

SULTAN SINGH
Indian Grassland and Fodder Research Institute, Jhansi-284003 (U.P.), India

Introduction

Evaluation of feeds and forages provide essential information on their nutritive value to formulate economic diets desirable for animal physiology to achieve optimum production. The need for precise nutritional evaluation of feeds/forages will be more in future due to advancement in molecular biology (development of transgenic plants), crop improvement programmes, environmental and economic constraint related to use/disposal of crop byproducts and residues(Theodorou *et al*, 1994). As early as 1725 the ruminant's feeds were evaluated as Straw Units (Blaxter, 1986). Initial techniques of feed/fodder evaluation were mainly developed to characterize their nutritive value rather than to predict the animal performance. Increase in knowledge of plant chemistry, animal physiology and rumen microbiology had led to development of more precise evaluation methods for feeds and forages.

Gas production has been widely used by the nutritionists for the evaluation of feed and fodders since its development by Menke and Steingass (1988). This technique not only measures the possible extent of fermentation of feeds and forages but also the rate of fermentation. This technique measures the animal performance in respect of intake and digestibility. Many gas measuring techniques and in vitro gas production methods are in use with little variations in different laboratories.

Here methodology/procedure of pressure transducer gas production technique along with its application in evaluation of feeds and fodders for different nutritional attributes are delineated.

Gas Production (Pressure Transducer Technique)

In principal the pressure transducer technique is similar to Menke *et al.* (1979) where gas production incubations of ground substrate samples (feed/fodder/diet) are carried out in gas tight culture bottles containing anaerobic medium inoculated with rumen fluid at 39 ^{0}C. With the progress of fermentation process gases accumulated in the head space and are measured using pressure transducer connected to a 3 way gas tight syringe assembly. The syringe is used to measure and release the gas volumes.

The method was developed to get information on the fermentation kinetics of ruminant feed and fodder based on long term end point fermentation (166 hours). This technique is equally applicable for for short period fermentations (48 and 72 hours) to evaluate the tropical forages and rank them according to their fermentability. This technique alike the Menke gas production has its applicability in the biological evaluation of the effect of phenols on fermentation by adding chemical agents like polyethylene glycol (PEG), polyvinylpyrrolidone (PVP) and polyvinyl polypyrrolidone (PVPP).

Preparation of Feed/Fodder Sample

The dried samples of feed/fodder are ground through 1 mm sieve using electrically operated mill. Weigh 1.0 gm sample of feed/fodder in triplicate and put in the serum bottles (160 ml). Use 3 serum bottles as control.

Reagents

A. Composition of Digestion Media

(a) Buffer Solution (g/l)

Ammonium hydrogen carbonate (NH_4HCO_3) = 4
Sodium hydrogen carbonate ($NaHCO_3$) = 35

(b) Macro Mineral Solution g/l)

Disodium hydrogen ortho phosphate 12hydrate ($Na_2HPO_4.12H_2O$) = 9.45

Potassium dihydrogen ortho phosphate (anhydrous) KH_2PO_4 = 6.2

Magnesium sulphate 7-hydrate ($MgSO_4.7H_2O$) = 0.60

(c) Micro Mineral Solution (gm/100ml)

Calcium chloride 2-hydrate ($CaCl_2.2H_2O$) = 13.2

Manganese chloride 4-hydrate ($MnCl_2.4H_2O$) = 10.2

Cobalt chloride 6-hydrate ($COCl_2$. $6H_2O$) =1.0

Ferric chloride 6-hydrate ($FeCl_2.6H_2O$) =8.0

(d) Resazurin Solution (gm/100ml)

Rresazurin (redox indicator) =0.1

(e) Reducing Agent (gm/100ml)

Cystein-HCl = 0.625

Distilled water =95 ml

1M NaOH= 4ml

Sodium sulphide = 0.625

Digestion Media

Distilled water: 500 ml

Micro solution: 0.1 ml

Buffer solution: 200 ml

Macro mineral solution: 200 ml

Resazurin solution: 1 ml

Prepare digestion medium by adding above ingredients except reducing agent in said mentioned order and allow each ingredient to get dissolved after each addition. Pass the carbon dioxide gas through the above prepared digestion media/solution till it turns pink/purple colour.

Pour 85 ml of digestion media in serum bottles having sample and used a control. Now add 4 ml of reducing agent in each bottle and gas the bottles with CO_2 till the pink colour fads. Now put the rubber stopper on bottles and crimp with aluminum seal. Chill the bottles at 4 ^{0}C in refrigerator/fridge. Put the bottles in incubator at 39 ^{0}C for 2-3 hours before the start of gas production experiment (before inoculation).

Preparation of Microbial Inoculum

Collect digesta from rumen fistulated animal (sheep/cattle/buffalo) in a thermos and transfer it in laboratory to a pre heated flask. Strain the rumen liquor through a double layer of muslin cloth and collect in a CO_2 filled beaker. Stir the rumen liquor manually or using a magnetic stirrer and bar and pass the CO_2 while using for inoculation.

Inoculation

The bottles should be degassed before inoculation. This is achieved by using the pressure transducer to achieve the zero reading on the LED display. Inoculate the serum bottles with 10 ml rumen fluid using a 10 ml syringe fitted with needle (21 gauge x1.5 inches). If there is frequent blockage of needle, refilter the inoculum through muslin cloth. Once inoculation is over adjust the bottle head space pressure zero.

In case of use of incubator, remove 15-20 bottles at one time for inoculation and put them back into incubator before removing the other 15-20 bottles.

Fermentation/Incubation Period

Pressure and volume are recorded at 3-24 hour interval during the 120 hour fermentation period with more frequent measurements during early 40 hours. Typical times of gas measurements (recording readings) are 3,6,9,12,16,24,28,33,38,45,60,70,94,118 and 145 hours. After measurement bottles should be shaken between readings. After gas measurement bottles should be kept back in incubator quickly to avoid their cooling.

Application of Technique for Estimation of Different Nutritional Parameters

The volume of gas produced from the incubation of substrate (feeds/ fodder) is closely associated to their digestibility and energetic value. The gas production data can be used to predict the organic matter degradation. The primary application of the gas production technique is to evaluate the energetic value of feeds/fodder. With the improvement in technique it has also been used to assess the presence of anti nutritional compounds in different feed stuff particularly in the unconventional feeds and browse species. The use of tannin binding complex agents' e.g. insoluble polyvinylpolypyrrolidones was first proposed by Andersen and Sowers (1968) as an indirect method to determine biologically active tannins in plant materials. Makkar *et al* (1995) first work on the use of tannins binding complex agents in the in vitro gas production to evaluate anti nutritive factors in feeds. The gas production technique is quite useful in prediction of the short chain fatty acids (SCFA) as the gas produced on fermentation/incubation of a substrate (feeds/fodder) is mainly from acetate and butyrate and gas production from propionate is less. Further the composition of SCFA will help to predict the relative production of methane from a feed/fodder on its fermentation. Gas production technique is quite useful in determination of microbial mass as it is better index to assess the nutritive value of a feed/fodder since microbial protein is ultimately used by the animal. Thus this technique is of quite significance particularly in developing countries where most of the laboratories are not equipped with modern facilities to estimate microbial mass and SCFA.

Measurements

Gas Volume

Subtract the mean of gas produced from set of three blanks from the volume of gas produced from each sample. This will represent the observed value of gas per unit amount of substrate (sample). After knowing the DM of each sample, the volume of gas per unit weight of sample can be calculated. First 9-12 hours of gas production are generally not used in calculations

This data of gas production is fitted to the following exponential equation given by Ørskov and Mc Donald (1979) to estimate the gas volume at a particular time.

P=a+b (1-exp^{ct}) where p represent gas production at time t, (a+b) the potential gas production, c the rate of gas production and a, b and c are the constant in the exponential gas production.

Dry Matter Disappearance

At the end of the incubation/fermentation period filter the contents of bottle through pre weighed Gooch crucibles. Wash the bottles with water to remove residue or any feed particle attached to the bottle walls followed by washing residue in crucible. After filtration oven dry the crucibles at 65-70 ^{0}C till a constant weight is attained. The loss in original sample weight is expressed as dry matter degradation.

Pressure Transducer to Measure the in vitro Gas Production

Metabolizable Energy and Organic Matter Digestibility

The gas production technique determines the energy value of feeds/fodder in terms of metabolizable energy and organic matter digestibility using gas production and chemical constituents of feeds and fodder into account. For roughages, the relationships between ME and OMD with gas production and chemical composition are as below:

ME (MJ / Kg DM) = 2.20 + 0.136 Gp + 0.057 CP

OMD (%) = 14.88 + 0.889 Gp + 0.45 CP + 0.0651 XA

For compound feeds (cereal and byproducts)

ME (MJ/kg)=1.06+0.1570GP+0.0084CP +0.0022EE-0.0081 ash

DOM %=9+0.9991GP+0.0595CP+0.0181 ash

Where, ME is the metabolisable energy; OMD organic matter digestibility; CP, crude protein in %; XA, ash in %; and Gp, the net gas production in ml from sample after 24 h of incubation.

Short Chain Fatty Acids (SCFA)

The level of SFCA in a feed is the indicator of energy availability to the animal. Measurement of SFCA is important for relating the feed composition to production parameters and net energy value of diets. Short chain fatty acid production can be predicted from the gas production values using the following relationship.

SFCA (without PEG) = -0.0601 + 0.0239 x Gas

SFCA (in presence of PEG) = 0.0521 +0.0207 x Gas

Over all SFCA = -0.00425 + 0.0222 x Gas

If molar proportions and amount of SCFA is known CH_4 and CO_2 expected from rumen fermentation could be calculated.

Microbial Mass

With the advancement in nutritional concepts to evaluate the ruminant feeds, the current concept aimed at high microbial efficiency that cannot be achieved by merely measuring the gas production. The in vitro microbial mass production can be calculated as below.

Microbial mass = Substrate truly degraded – (gas volume x stoichiometrical factor). For roughages the recommended stoichiometric factor is 2.20.

Microbial Nitrogen

Microbial N (MN) could also be measured after incubation by following nitrogen balance approach.

MN = TN – (NDF-N + Ammonia-N),

Where, TN is total nitrogen i.e., feed N + N in buffered rumen fluid in the syringe/bottle before incubation (at 0 time), NDF-N is the N bound to NDF fraction following incubation and ammonia-N is the ammonia-N in the supernatant following the incubation. In a closed system, the total N present at the start of the incubation can be in microbial mass, NDF-N, ammonia-N and amino acids during any time of the incubation. Negligible amounts of amino acids and peptides are present in the supernatant during fermentation and therefore these are ignored in calculation of microbial N.

For determination of NDF fraction for N bound to NDF after incubation, the syringe/bottle contents are transferred into a 600 ml beaker and the syringes/bottles are washed twice with a total of 50 ml neutral detergent solution (NDS) and emptied into the beaker. The contents are refluxed for 1 h, and then filtered through pre-weighed gooch crucibles. The crucibles are dried overnight at 100 ^{0}C and weighed. True degradability can be calculated as the weight of substrate incubated minus the weight of the residue after NDS treatment. The residue after NDS treatment (neutral detergent residue, NDF) is subjected to micro-Kjeldahl digestion for determination of NDF-N.

Advantages of Technique

1. This technique can be used to screen a large number of feed resources like in breeding programmes for developing nutritionally superior cultivars of forages.
2. It could be of great use in developing supplementary strategies with locally available conventional and unconventional feed constituents to attain higher microbial efficiency and ultimately higher microbial mass and nitrogen.
3. This technique provides a better insight into nutrient-anti nutrient and anti nutrient-anti nutrient interactions.
4. It has significant role in rumen modulators for increasing microbial protein synthesis efficiency and decreasing in methane emissions.

Limitations of the Technique

(i) To achieve the reliable and precise results it is a constant supply of electricity is must in order to maintain a constant temperature during incubation period.

(ii) Similar to other *in vitro* techniques, it is a closed system where end products accumulate and can inhibit fermentation or create an environment unfavorable to the rumen.

(iii) The technique will underestimate the nutritive value of feeds that are high in protein. This is due to the fact that protein fermentation contributes little to the total volume of gas production.

Future Research

The future research area of feed evaluation systems should include new information on the relationship between specific end-products of digestion and performance of the animals, as well as information on animal and microbial metabolism, feed composition, and effects of various factors on feed utilization (Flatt, 1988). An adequate feed/fodder (dietary) analysis should have the requirement that methods employed are relevant to a nutritional classification of the dietary chemical components (Van Soest and Robertson, 1985). The *in sacco* technique is probably the most widely used method for feed studies, however some drawbacks have been pointed out in this technique by some workers. Future research areas should also consider the inoculums sources other than use of fitulated animals to make the technique more applicable under farm conditions. Further the new methods should involve animal intake and performance attributes to make the evaluation system more close to animal conditions.

References

Andersen, R. A. and Sowers, J. A. 1968. Optimum condition for bonding of plant phenol to insoluble polyvinylpyrrolidone. *Phytochemistry* 7,293.

Blaxter, K. L. 1986. An historical perspective: the development of methods for assessing nutrient requirements. *Proceedings of the Nutrition Society* 45, 177.

Flatt, W. P. 1968. Feed evaluation Systems. Historical Background. World Animal Science. Disciplinary Approach B4. (Ed. Ørskov, E. R.). Elsevier Science Publications. Pp: 1-22.

Makkar, H. P. S., Blummel, M. and Becker, K. 1995. Formation of complexes between polyvinyl pyrrolidone and polyethyl glycol with tannins and their implications in gas production and true digestibility in in vitro techniques. *British Journal of Nutrition* 73, 897

Menke, K. H., Raab, L., Salewski, A., Steingass, H., Fritz, D. and Schneider, W. 1979. The estimation of the digestibility and metabolizable energy content of ruminant feedingstuffs from the gas production when they are incubated with rumen liquor *in vitro*. *Journal of Agriculture Science* 93, 217.

Menke, K. H. and Steingass, H. 1988. Estimation of the energetic feed value obtained from chemical analysis and *in vitro* gas production using rumen fluid. *Animal Research and Development* 28,7.

Ørskov, E . R. and Mc Donald, P. 1979. The estimation of protein degradability in the rumen from incubation measurements weighed according to rate of passage. *Journal of. Agriculture Science* 92,499.

Theodorou, M. K., Willians, B. A., Dhaona, M. S., McAilan, A. B. and France, J. 1994. A simple gas production method using a pressure transducer to determine the fermentation kinetics of ruminants feeds. *Animal Feed Science and Technology* 48, 185.

Van Soet, P. J. and Robertson, J. B. 1985. *Analysis of forages and fibrous foods*. A laboratory Manual for Animal Science 613. Report of Research of the Cornell University Agricultural Experimental Station.pp.2,3,35.

❑❑❑

Chapter **18**

Insect and Pathogen Infestation : Impact on Forage Yield and Quality

SHARMILA ROY AND PRADEEP SAXENA
Indian Grassland and Fodder Research Institute, Jhansi-284003 (U.P.), India

Introduction

Plants are exposed to insects and pathogens during their life cycle. The diversity of these organisms and the intensity of their damage may vary from plant species to season and region. This biotic pressure affects the plant performance in short term and the evolution of plant species in the long term. In nature the plants that can tolerate or compensate in terms of recover equivalent yield or fitness (quality of production) for pest attack have oblivious selective advantage that lead to genotype maintenance. In agriculture system, successful crop production depends on managing the insect pests and pathogens. In this processes a cornucopia of chemicals, naturally produced secondary chemicals and plant genetic engineering have been used to regulate the pest population and recover losses in yield and quality of the yield. In nature, the plants have their own compensation processes that help them to recuperate the losses. In fodder production systems (perennial tree-grass based and annual cultivated crops) both the natural and the human interventions are used for the optimum qualitative and quantitative production. In this chapter the effect of insect herbivory and the pathogens on plant in general and especially on the forage plants have been discussed.

Effect of Insects on Plant Performance

Insects feeding on plants are the most diverse group of animal kingdom. They can be classified based on their feeding mechanisms (guilds) and at least one group of insects can be assigned to every guild of feeding mechanism (Table 1). The insects may further be classified as monophagous, feed exclusively on a particular plant type or polyphagous feed on diverse group of plants.

Table 1 : Feeding Guilds of Insects

Feeding Mechanisms	**Insect Examples**	
	Common Name	**Insect Order**
Cambium bark or sap wood borer		
Main stem	Bark beetles	Coleoptera
Roots	Long horned beetles	Coleoptera
Sap Feeders: leaves, twigs, seeds	Aphids, Jassids, bugs	Homoptera
Leaf Feeders		
Young leaves	Shoot fly	
Mature/ any age leaves	Semiloopers	Lepidoptera
Shoot Feeders	Stem bore larva	Lepidoptera
Root Feeders	White grubs	Coleoptera
Root Sap Feeders	Cicada nymphs	Homoptera
Feeding pith and wood	Stem borers	Lepidoptera
	Saw flies	Hymenoptera
	Long horned beetles	Coleoptera
Feeding on Leaf parenchyma	Leaf miner flies	Diptera
	Leaf miner beetles	Coleoptera
Seed eating	Bruchids	Coleoptera
	Pod borers	Lepidoptera
Seed harvesters	Ants	Hymenoptera
Induces galls on stem, leaves	Gall midges	Diptera
	Eriophyids	Mites
	Gall wasps	Hymenoptera
Consume nectars, pollens	Bees	Hymenoptera

In general insect herbivory has a negative effect on individual plant growth and productivity. The intensity of these negative impacts depends on six factors (Orcutt and Nilsen, 2000). The factors are as follows:

i. *Life stage of plant* : Insect bites of the leaves of a mature plant has little effect on the production aspects while similar extent and type of damage on a seedling may affect the survival or growth of a plant.

ii. *Type of tissue damage* : Damage of inactive tissues like bark, cork, senescent leaves has less effect on plant performance than the damage of active tissues like growing root/shoot tips.

iii. *Time/ plant stages of damage* : Insect defoliation at germination or at the early seedling stage affect the crops severely than the late season defoliation.

iv. *Plant genotype* : With in a plant population a particular plant type may be vulnerable to insect herbivory than the others or vice versa. This variation in genetic resistance has a greater impact on plant evolution.

v. *Climate and environmental stresses* : Under various climate conditions and stresses like drought, excess water or temperature the impact of herbivory on plants may be of greater than the normal climatic conditions.

vi. *Plant species variation* : Different plant responds differently to the herbivory. A particular plant species may be susceptible /resistant to particular insects while others may not be.

All these characters indicate that a simple measure of mass loss (yield) is not enough to describe the losses in production but the plant fitness is also need to be considered for the quality production.

The influence of herbivory on plant performance can be estimated experimentally by excluding the pests either by use of pesticides or by exclosure methods and comparing the growth and production. However, the plants have their own mechanism of defence against the insect feeding and compensate the losses. This compensation is also affected by endogenous and exogenous factors. The endogenous factors are the mechanisms that influence allocation and reallocation of resources within plants (Trumble *et al.*, 1993). These include regrowth patterns, photosynthetic activities, senescence and leaf morphology. The endogenous factors include nutrient availability, intensity and timing of defoliation, herbivore distribution *etc.* Insect herbivory may influence plant community dynamics because of successional loss of nitrogen and carbon by attacked plants. The influence of insect attack on crops affect the plants at physiological and as well on ecological levels. These responses may be summarised in following points:

A. Influence of Insect Herbivory on Plant Morphology and Physiology

1. In general insect attack effect in net productivity if the attack continues. In some instances insect damage causes an increase in plant productivity.
2. When insects are excluded from crops, the plant showed enhanced productivity and fecundity.
3. Plants show compensatory growth in response to herbivory and can regain their competitive advantage.
4. Herbivory induces production of secondary plant protection chemicals that can help defend against insect damage.
5. Herbivores can enhance seed production or reduce yield.
6. Herbivores feeding on seeds and flowers can reduce seed yield and pollination.
7. Herbivores feeding on leaves can reduce seed yield by reducing leaf area, which reduces photosynthetic activities required for flowering and seed yield.
8. Reallocation of nutrient occurs in the plants to reduce the yield.
9. Insect feeding can influence the sex ratio of a plant population as well as the architecture of plants.
10. Seed eating insects may affect germination and seed spreading mechanism.
11. Insect damage may become responsible for secondary infestation of pathogens to the plants.
12. Insects may act as vectors of plant pathogens.
13. Herbivory can influence plant's capacity to tolerate environmental stresses.

B. Influence of Insect Herbivory on Plant Ecology

1. In an ecosystem insects may determine system productivity.
2. Insect feeding may responsible for plant species diversity of ecosystems.
3. Herbivory may reduce the competitive advantage of potentially dominant species or vice versa.
4. Moderate feeding by insects may directly encourage the predator population in an ecosystem and/or indirectly stimulating effectiveness of the predators by producing secondary chemicals.

Effect of Pathogens on Plant Performance

A. The Process of Disease in Plants

Disease develops in individual plants by a series of sequential steps beginning with the arrival of inoculums at the plant surface and ending with the terminal stages of pathogens. The innumerable stages, in between, are unique to each pathogen, host (crop) and variation of the environment.

To cause disease by a pathogen, penetration to its hosts is first essential stage. Recognition and compatibility phenomena are next stage of pathogenesis. In the third stage of pathogenesis, invasion begins in earnest. Biotic pathogens spread from cell to cell within the host tissue living at the cost of its host. To obtain energy and substance, for their own nourishment, many pathogens secrete enzymes and other metabolites that breakdown the cell wall, cell membranes and other complex organelles and constituents of the host cytoplasm. Some living pathogens, secrete toxins with disturb the normal metabolic machinery of the host cells. Other produce, growth regulatory substances with alter the normal patterns of cellular development and differentiation. Viruses take over the synthetic machinery of host cells and divert it to synthesis of viral nucleic acids and proteins instead of host nucleic acids and proteins. In all stages of pathogenesis the pathogen force the host to alter its vital functions. The alterations constitute invisible symptoms of disease. They also give rise to the characteristic visible symptoms by which a person can recognize a disease (Fig 1).

B. How Plants Suffers to Plant Pathogens

The effect of a disease involving chlorosis or necrosis of infects at tissues is usually an ultimate reduction in photosynthetic activities of the plant, a reduction contributed to by the reduced photosynthesis capacity of infected plant and a reduced plant growth rate. In disease caused by necrotrophs, the production of tissue necrosis, also results in increased oxidative enzyme activity and rapid enhancement of O_2 uptake. In these cases, potential respiration occurs during both the early and later stages of pathogenesis.

The viral pathogens do not penetrate plant cell wall directly; they are able to cause the plant cell to respond to their presence by callus deposition (wall thickening), lignifications and aromatization of the cell wall. The formation of paramural bodies and vesicles depicts a process for callus deposition between the plasmalemma and the plant cell wall.

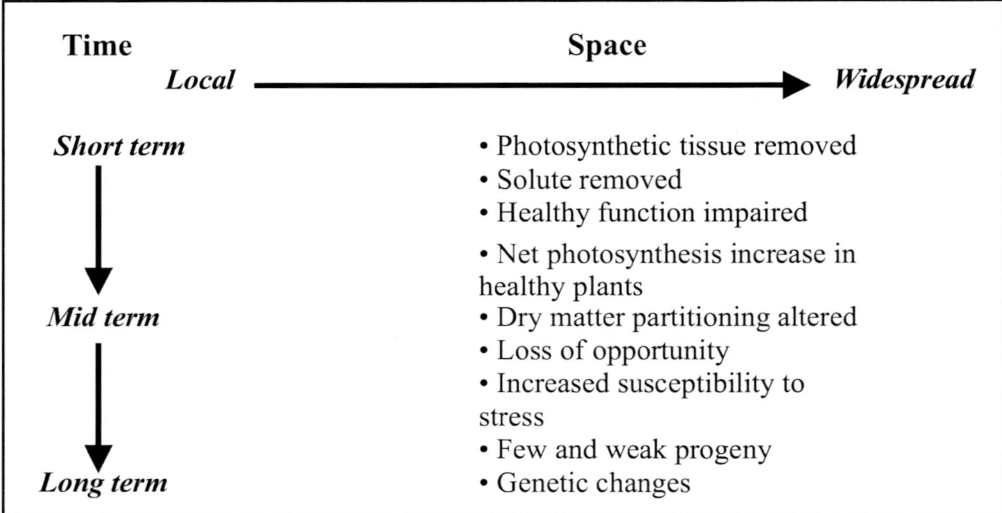

Fig. 1 A Generalized Scheme to Demonstrate the Spread of Impacts of Foliar Diseases in Space and Time, After Establishing the Infection.

It is apparent that both bacteria and fungi can rapidly degrade the plant cell wall in appropriate pathogen-hot - combination. Degradation proceeds in a sequential fashion, and the order generally followed in chitin, pectin, hemicelllulose and cellulose. The appropriate enzymes appear to be synthesized as greater amount of substrate are exposed. In addition, each in turn becomes crucial to the continuing process of wall degradation and host cell death. The essentiality of cutinase was clearly documented that indicated that blockage of the enzyme precluded infection. The micro-environment is also crucial in determining the rate at which plant cell wall degrades.

Infection by pathogens is also accompanied by the biosynthesis of new pathogen proteins, which may be few in number in case of the simple viruses or several hundred in the case of bacteria and fungi. Additionally, there may be both qualitative and quantitative changes in the pattern of host protein biosynthesis or degradation which may change as infection processes. The early stages of infection particularly in fungi may be characterized by stimulation in host protein synthesis. In some cases, there may be a redirection of host protein synthesis to permit extensive syntheses of pathogen proteins.

The activity of several host enzymes includes PAL and peroxidase, may increase markedly, particularly when tissue responds hypersensitivity. In the later stages of infection a decrease in over all rate of protein synthesis and enhancement of proteolytic enzyme activity may be observed as host cells senescence. The effect of pathogens on major processes of host plants is summarized in Table 2.

Table 2 Principal Process in Plants which can be Affected by Pests

Process	***Growth Reducing Factors***
A. Carbon and energy economy	Weeds, Necrotrophic fungi, leaf eating insects and viruses
(i) Photosynthesis	
CO_2 diffusion	
CO_2 fixation	
Other chloroplast process	
(ii) Respiration	Many fungi & viral pathogens, mites, phloem feeder,
Maintenance	
New synthesis (Growth)	Leaf diseases
Photo respiration	Biotrophic fungi, phloem feeder, viruses, CO_2
(iii) Allocations	Phloem feeding insects
(iv) Transport	
B. Water economy	
(i) Uptake	Root pathogens, nematodes
(ii) Transport	vascular wilt diseases
(iii) Transpiration, stomatal regulation	leaf diseases
C. Nutrient economy	
(i) Uptake	Root pathogens; nematodes
(ii) Transport	vascular wilts
(iii) Redistribution	leaf diseases
D. morphogenesis	Mycoplasmas; Phytoplasmas
(i) Organ initiation	
(ii) Organ growth	

Major Pest and Diseases of Forage Crops/Plants

Forage crops are subjected to the depredation of pests and diseases in the same manner as in other agricultural produce (Table 3). These pests and diseases hamper crop establishment, impair forage quality and reduce green fodder and seed yield. Besides this, they also cause indirect losses, like knot formation in legumes ultimately resulting to reduction of nitrogen fixation capacity. Nematodes, aphids, jassids and other sucking pests inflict direct losses to the crop, yet are also indirectly responsible for the transmission of a number of viral and other pathogens. Further disease and pest interactions in the plant produce toxins, adversely affecting the animal health, which sometimes can be fatal.

Influence of Insects and Pathogens on Production and Quality of Forage

The plant growth is usually supposed to be equal to yield and leaf area is related with the carbohydrate production. In forage crops the green leaf and the seed productions are directly affected by various pests. These losses depend on many factors such as incidence and intensity of pests, crop type, yield potential of variety, availability of nutrients and climatic conditions. Yield losses by various pests in different fodder crops are presented in Table 4. The data presented in the table is based on the survey and field experiments (Saxena *et al.*, 2002).

Table 3. Forage Crops Pest Spectrum

Crop	Insect pests	Pathogens
Sorghum (*Sorghum bicolor*)	Shootfly (*Atherigona varia soccata*), Stem borer (*Chilo partellus*), Aphids (*Schizaphis graminum*), Sorghum midge (*Contarinia sorghicola*), Pyrilla (*Pyrilla perpusilla*), Pollen eating beetle (*Chiloba acuta*), Army worm (*Mythimna separata, Spodoptera exigua*), Cut worms (*Agrotis* sp.), Leaf roller (*Marsemia trapezalis*), Ear head bug (*Caocoris angustatus*), Ear head caterpillar (*Helicoverpa armigera*)	Anthracnose (*Colletotrichum graminicola*), Sooty stripe (*Ramulispora sorghi*), Zonate leaf spot (*Gloeocercospora sorghi*), Downy mildew (*Sclerospora sorghi*), Gray leaf spot (*Cercospora sorghi*), Rough leaf spot (*Asochyta sorghina*), Turcicum leaf blight (*Helminthosporium turcicum*), Maydis leaf blight (*H. maydis*), Charcoal rot (*Macrophomina phaseolina*), Rust (*Puccinia purpurea*), Ergot (*Claviceps fusiformis*), Leaf spots (*Ramulispora sorghi, Pyricularia* sp.)
Oats (*Avena sativa*)	Bird cherry aphid (*Rhapalosiphum maidis*), Thrips (*Halothrips ganglbaueri*)	Leaf blotch (*Helminthosporium avenae*), Crown rust (*Puccinia coronata avenae*), Stem rust (*Puccinia graminis avenae*), Sclerotial wilt (*Sclerotium rolfsii*)
Maize (*Zea mays*)	Shootfly (*Atherigona varia soccata*), Stem borer (*Chilo partellus*), Aphids (*Schizaphis graminum*), Sorghum midge (*Contarinia sorghicola*), Pyrilla (*Pyrilla perpusilla*), Pollen eating beetle (*Chiloba acuta*), Army worm (*Mythimna separata, Spodoptera exigua*), Cut worms (*Agrotis* sp.), Leaf roller (*Marsemia trapezalis*), Ear head bug (*Caocoris angustatus*), Ear head caterpillar (*Helicoverpa armigera*)	Brown stripe downy mildews (*Scleropthora rayssi* var. *zeae*), Bacterial stalk rot (*Erwinia carotovora* var. *zeae*), Turcicum leaf blight (*Helminthosporium turcicum*), Maydis leaf blight (*Dreschslera maydis*), Common Rust (*Puccinia sorghi*), Leaf spots (*Helminthosporium nodulosum, H. tetramera*), Seedling blight (*Fusarium moniliforme*), Stalk rot and seed rot (*Cephalosporium acrimonium, Diplodia maydis, Erwinia carotovora*)

Contd...

Table 3. Contd...

Pearl Millet (*Pennisetum glaucum*)	Shootfly (*Atherigona varia soccata*), Stem borer (*Chilo partellus*), Aphids (*Schizaphis graminum*),Pollen eating beetle (*Chiloba acuta*), Blister beetle (*Zonabris pustulata*), Army worm (*Mythimna separata, Spodoptera exigua*), Cut worms (*Agrotis* spp.)	Downy mildew (*Scleropthora graminicola*), Ergot (*Claviceps fusiformis*), Smut (*Tolyposporium penicillariae*), Rust (*Puccinia penniseti*), Leaf spots (*Cercospora fusimaculans, Curvularia penniseti, Helminthosporium australiense, H. rostratum, Pyricularia pennisetii*), Leaf blotch (*Xanthomonas annamalaiensis*), Twisted top (*Fusarium moniliforme*)
Berseem (*Trifolium alexandrinum*)	Gram caterpillar (*Helicoverpa armigera*), Bihar hairy caterpillar (*Spilosoma obliqua*), Hairy caterpillars (*Euproctis virguncola* and *E. lunata*), Green semilooper (*Plusia nigrisigna, Thysanoplusia orichalcea*), Aphid (*Aphis craccivora*), Dusky bug (*Oxycaraenus* sp.), Leaf miner (*Phytomyza* sp.)	Root rot complex (*Rhizoctonia solani, Fusarium semitactum* and *Tylenchorhynchus vulgaris*), Stem rot (*Sclerotinia trifoliorum*), Leaf spots (*Alternaria, Myrothecium* and *Epicoccum* spp.)
Cowpea (*Vigna unguiculata)*	Flea beetles (*Monolepta signata*), Hairy cater pillars (*Spilosoma obliqua*), semilooper (*Plusia nigrisigna*), Leaf hopper (Empoasca kerri), Blue butterfly (Limpidness booties), Pod borer (*Maruca testulalis*), Gram caterpillar (*Helicoverpa armigera*), Black aphid (*Aphis craccivora*), Blister beetle *(Mylabris phalerata*)	Root rot (*Macrophomina phaseolina anamorph Rhizoctonia bataticola*), cowpea mosaic virus (*CPMV*), Leaf spots (*Colletotrichum lendmuthianum*), Bacterial blight (*Xanthomonas compestris* pv. *vignicola*), Seedling rot (*Rhizoctonia, Pythium, Sclerotium, Phytopthora* sp.), Top necrosis (*Fusarium equiseti*), Wilt (*Fusarium* spp.), Powdery mildew (*Oidium* spp.)
Lucerne (*Medicago sativa*)	Lucerne weevil (*Hypera postica*), Hairy caterpillar (*Spilosoma obliqua*), Aphid (*Acyrthosiphon pisum* and *Therioaphis tifolii f. maculata*), Leaf hoppers *(Exitianus indicus, Empoascanara* sp., *Austroagalia* sp., *Macrosteles* sp. and *Empoasca* sp.), Alfalfa caterpillar (*Spodoptera exigua*), Tobacco caterpillar (*S. litura*), Lucerne butterfly (*Terias hecabe*), Semilooper (*Thysanoplusia nigrisigna*), Thrips (*Halothrips tenuipennis*), Seed chalcid (*Bruchophagus roddi*)	Downy mildew (*Peronospora trifoliorum*), Rust (*Uromyces striatus*), Common leaf spot (*Pseudopeziza medicagenis*), Anthracnose (*Colletotrichum trifolii*), Crown warts (*Physoderma alfalfae*), Leaf spots (*Cercospera mediagenis, Stemphylium botryosum, Alternaria medicagenis, Xanthomonas alfafae*), Powdery mildew (*Erysiphe polygoni*), Mosaic (*Alfalfa mosaic virus*), Crown and root rot (*Sclerotinia trifoliorum*), Damping off or Root rot (*Phytopthora megasperma*), Root rot (*Rhizoctonia* and *Fusarium* sp., *Corynobacterium insidiosum*)

Contd...

Table 3. Contd...

Guar (*Cymopsis tetragonaloba*)	Leaf hoppers (*Empoasca* sp., *Exitianus indicus*), Yellow mite (*Polyphagotarsonemus latur* and *P. datus*)	Bacterial blight (Xanthomonas compestris pv. Cyamopsidis), Leaf spots (Alternaria cyamopsidis), powdery mildew (Erysiphe polygoni), Root rot (Rhizoctonia solani) and Anthracnose (Colletotrichum lindmuthianum)
Stylo (*Stylosanthes* spp.)	Pod borer (*Helicoverpa armigera*), Hairy caterpillars (*Spilosoma obliqua*) Leaf hopper (*Empoasca kerri*), Black aphid (*Aphis craccivora*)	Anthracnose (Colletotrichum gloeosporiodes and C. dematium f. truncata), Leaf spot (Cercospora spp.), Leaf and stem spot (C. truncatum), Seedling rot and blight (Corticium rolfsii), Foliar blight (Rhizoctonia solani), Inflorescence blight (Botrytis cinerea, Rhizopus stplonifer)
Leucaena (*Leucaena leucocephala*)	Psyllids (*Heteropsylla cubana*)	Root rot (Ganoderma lircidium), Gumosis (Fusarium solani), Pink disease (Corticum salamanicolor), Necrotic lesions (Botryodiplodia theobromus), Seedling rot (Colletotrichum spp.), Leaf spot (Camptomeris leucaenae), Black spot(Colletotrichum gloeosporiodes), Rust (Uredo leucaenae- glaucae), Damping off (Pythium solani, Rhizoctonia solani and Fusarium oxysporum), Seedling rot (Colletotrichum spp.)
Grasses	Grasshopper (*Acrida crenulata, Thisoicetrus pulcher, Cyrtocanthacris tatarica, Hieroglyphus nigrorepletus, Catantops pinguis, Acrida exaltata, Oedaleus abruptus, Chrotogonus trachypterus, Aelopus tamulus, Colemania* sp., *Oxya* sp., *Locusta migratoria, Melanoplus spretus, Attractomorpha crenulata, Acrida exaltata),* Yellow hairy caterpillars *(Psalms pentacle*), Hairy caterpillar (*Euproctis* sp., *Creatonotus gangis*), Red hairy caterpillar (*Amsacta* spp.), Army worms *(Cirphis unipuncta, C. albistigma, Spodoptera mauritia, Euxoa spinifera*), Termites (*Microtermes* sp., *Odontotermis* sp.), Aphid (*Rhapalosiphum maidis*), Leaf hoppers (*Nephotattix* sp., *Empoasca* sp., *Empoascanara* sp., *Exitianus indicus*)	Rust (Puccinia and Uromyces spp.), Leaf spots (Pyricularia, Phylochora, Dreschslera, Rhynchosporium, Sorosporium, Colletotrichum and Helminthosporium spp.), Ergot (Claviceps purpurea), Smut (Ustilago sp.), Blight (Rhizoctonia solani)

Table 4. Losses in Forage Crops Due to Various Diseases and Pests.

Crop	Pests	% loss in fodder/seed yield
Sorghum	Diseases (General)	58
	Charcoal Rot	15-55
	Rust	50
	Anthracnose	48
	Leaf blight	45
	Shootfly	15
	Nematodes	7
Pearl-millet	Diseases (General)	30
	Downy mildew	10-45
Maize	Diseases (General)	13.2
	Maydis leaf blight	30
	Rust	32
	Nematodes	10
	Stem borer	27-30
	Shoot fly	20
Oats	Diseases	55
	Nematodes	5
Cowpea	Yellow mosaic	10-100
	Leaf spot	47
	Insect	38-41
	Nematodes	15
Field beans	Nematodes	10.9
Guar	Diseases	75
	Nematodes	12
Lucerne	Diseases	72
	Nematodes	7
	Insects	30
Grasses	Insects	18-20

The pest and pathogens not only hamper green and dry fodder yield but also affect the qualitative aspects of the fodder by altering the plant physiology, thus affecting the nutritional quality and the recovery of plant growth (Table 5, 6 & 7). Indirect effects of insect feeding on crops reduce the vigour and affect the competitive ability.

Table 5 Quantitative Impacts in Forage Crops Due to Insect Pests

Insect herbivory	Crop	Examples	Degree of infestation	Plant compensation/Qualitative loss	References
Foliage Feeding	Sorghum	Shoot fly, Shoot Borer	Low - Moderate	Tillering enhanced by reallocation of available nutrients	Sharma and Nwanze, 1997
			Heavy	Plant dies	
	Maize	Corn ear worm	Low - Moderate	Size of new leaves increases due to mesophyll elongation	Heilman et al., 1981
	Cowpea, Soybeans	Leaf beetles, leaf miners	Low - Moderate	Mild decrease in leaf area, restitutive cell growths with new chloroplasts, increase in photosynthetic rate	Martens and Trumble, 1987
		Semiloopers	Heavy	Leaf area reduced, photosynthesis rate decreased, Carbohydrate and protein production enhanced	Ingram *et al.*, 1981; Thorne and Koller, 1974
	Lucerne	Leaf beetles, leaf miners	Low - moderate	Increasing photosynthetic activity in the remaining leaf area, reduces intracellular resistance to CO_2 transport	Hodgkinson, 1974; Daley and Mcneil, 1987
			heavy	6-10 days delay in regrowth, Loss in crude protein (0.3-1.4t/ha)	Saxena *et al.*, 2002
	Grasses	Grasshoppers	Low - moderate	Tillering enhanced by reallocation of available nutrients Water availability in remaining leaves increases, induces enhanced protein synthesis	Sosebee and Wiebe, 1971; Ryle and Powell, 1975; Dyer and Bokhari, 1976; Deregibus and Trlica, 1990
Sap suckers (foliage, branch, shoot)	Cowpea, Lucerne, Berseem,	Aphids, Jassids	Heavy	Plant/Seed vigour, viral disease carriers	Pandey *et al.*,1995
	Leucaena	Psyllids	Heavy	Plant vigour, reduction in photosynthesis	Saxena *et al.*, 2002
Root feeders	Sorghum, maize, Grasses	Nematodes, White grubs, termites,	Moderate - Heavy	Loss in plant growth and water absorption capacity	Hou *et al.*, 1997: Blossey and Hunt-Joshi, 2003
	Alfa alfa	Mealy bugs	Moderate - Heavy	Chlorosis, stunting	Godfrey and Pickel, 1998
	Clover, Lucerne	Root curculio	Moderate - Heavy	Survival, stunting (20-40%, yield loss (30%), shows signs of water stress, dormancy induced	Goldson *et al.*, 1985, 1987
	Forage rape	Root fly	Moderate - Heavy	Reduction in root, shoot, leaf biomass	Griffiths *et al.*, 1994
Flower feeding	Cowpea	Mylabris beetle	Moderate - heavy	Reproductive loss	Pandey *et al.*,1995
Flower bud sap suckers	Sorghum	Head Bugs, Midges	Moderate - heavy	Seed yield and seed vigour loss	Sharma and Nwanze, 1997
Seed feeding	Cowpea	Pulse beetle	Heavy	Germination loss	Saxena *et al.*, 2002
Seed harvesters	Grasses	Ants	Moderate - High	Seed removal from the sites affects the plant population	Brown *et al.*, 1979

Table 6. Qualitative and Quantitative Effect of Powdery Mildew on Forage Legumes

Characters	Cowpea Degree of Infestation					Methi Degree of Infestation				
	20%	40%	60%	80%	100%	20%	40%	60%	80%	100%
Plant Height (cm)	-1.75	-8.22	-14.54	-15.79	-21.67	-0.53	-7.81	-9.37	-10.42	-15.62
Branches/plant (No.)	-0.22	-13.34	-24.66	-32.35	-49.09	-12.57	-24.95	-31.33	-56.28	-81.23
Pods/plant (No.)	-9.44	-17.55	-27.00	-32.44	-35.11	-17.89	-32.15	-35.69	-41.47	-48.12
Seeds/pod (No.)	-17.82	-4.62	-43.25	-45.12	-46.25	-3.33	-17.86	-33.33	-34.33	-51.00
DMY (%)	+5.43	+6.08	+8.04	+9.78	+23.04	+0.62	+1.52	+2.14	+6.20	+9.36
GFY (%)	-5.34	-13.60	-23.13	-23.13	-27.42	-10.00	-34.00	-40.00	-54.00	-64.00
Crude Protein (%)	-1.96	-15.57	-16.49	-16.73	-19.49	-11.97	-13.50	-14.00	-24.00	-24.50
NDF (%)	+9.15	+17.67	+17.96	+24.25	+25.00	+8.15	+15.00	+25.00	+34.00	+35.00
ADF (%)	+13.29	+14.58	+16.11	+25.82	+27.29	+27.29	+12.78	+20.14	+37.82	+48.33
Cell Contents (%)	-2.56	-5.70	-6.00	-8.10	-8.60	-0.50	-1.98	-2.40	-4.10	-4.30
Hemicellulose (%)	-25.00	-29.01	-31.45	-35.80	-40.45	-32.35	-33.19	-47.59	-54.62	-55.46

(- = Reduction over Healthy plant; + = Increase over healthy plant) (Source: Saxena, 1982; Saxena *et al.*, 1992)

Table 7 : Affect of Diseases on Qualitative Aspects of Forage Legumes

Characters	Host Pathogen Interaction			
	Oat Rust (Puccinia coronata)	Guar Mildew (Leviellula taurica)	Cowpea Mildew (Oidium sp.)	Methi Mildew (Leviellula taurica)
Moisture content (%)	(-)	-1.34	-1.90	-1.50
GFY/DMY	(-)	-4.69	-8.18	-4.50
Total Chlorophyll (mg/l)	-37.15	-83.03	-87.82	-71.85
Chlorophyll a (mg/l)	-34.16	-73.73	-87.27	-72.46
Chlorophyll b (mg/l)	-49.42	-78.68	-84.26	-71.82
N (%)	(-)	+27.85	+18.50	+27.13
P (%)	(-)	+127.53	+88.80	+81.00
K (%)	(-)	+39.24	+13.33	+36.50
Ca (%)	(-)	+77.27	+38.88	+23.88
Mg (%)	(-)	-49.16	-75.06	-44.06
Ash (%)	(-)	(-)	+5.53	+11.36
Total Phenol (%)	(-)	(-)	-26.80	-20.00
Reducing Sugars (%)	(-)	(-)	+33.33	+55.17
Non Reducing Sugars (%)	(-)	(-)	-56.84	-80.26
Total Sugars (%)	(-)	(-)	+25.34	+34.28
Starch (%)	(-)	(-)	-34.59	-39.65

(- = Reduction over Healthy plant; + = Increase over healthy plant; (-) = data not available)
(Source: Ahmad and Yadav, 1978; Ahemad and Saxena, 1985; Saxena, 1982)

Summary

Insects and pathogens affect various aspects of plant performance. The moderate damage incurred by the insects boost the plant physiology and many times plants compensate the losses in terms of yield and the quality. For example, moderately damaged plants show high tillering, branching, larger leaves, and efficient nutrients reallocation to other plant parts etc. However, the repeated and heavy damages kill the plant and affect the yield. The severe attacks by the insects alter the plant population dominance pattern and dynamics like increase in weed population in a crop field.

Unlike the insect damage pathogens affect the plant physiology more severely even at moderate levels of infestations. The toxins secreted by the pathogens and the interruptions of pathogens at cellular level initiate necrosis and deleterious biochemical processes, reducing the quality and yields.

A massive literature on the plant, insects and pathogen interactions are available. However, specific studies on the various forage crops is needed for integrated pest management as well for animal nutrition managements view point. Most of the work in this area was old requires new information under the changing climate and human need to enhance the fodder as well the milk/meat production.

References

Ahmad, S. T. and Yadav, R. B. R. 1978. Effect of crown rust on the chlorophyll contents of *Avena sativa*. *Forage Research* 4 (2), 177-179.

Ahmad, S T and Saxena, P. 1985. Studies on some of the induced changes in leaves of cluster bean (Cyamposis teragonloba) by powdery mildew infection. *International Journal of Tropical Plant Diseases* 3, 165-169.

Blossey, B and Hunt-Joshi T. R. 2003. Belowground herbivory by insects: Influence on plants and aboveground herbivores. *Annual Review of Entomology* 48, 521-548.

Brown, J H; Reichman, O. J. and Davidson, D. W. 1979. Granivory in desert ecosystems. *Annual review of Ecology and Systematics* 10, 201-227.

Daley, P. F. and Mcneil, J. N. 1987. Canopy photosynthesis and dry matter partitioning of alfa alfa infested by the alfa alfa bloch leaf miner (*Agromyza frontella* Rondani). *Canadian Journal of Plant Sciences* 67, 433-443.

Deregibus, V A and Trlica, M J. 1990. Influence of defoliation upon tiller structure and demography in two warm season grasses. *Acta Oecologia* 11, 693-699.

Dyer, M. I. and Bokhari,U. G. 1976. Plant- animal interactions: Studies of the effects of grasshopper grazing on blue gamma grass. *Ecology* 57, 762-772.

Godfrey, L. D. and Pickel, C. 1998. Seasonal dynamics and management scheme for a subterannean mealy bug (*Rhizoecus Kondonis* Kuwana), pest of Alfa alfa. *Southwest Entomology* 23, 343-350.

Goldson, S. L., Dyson, C. B., Proffit, J. R., Frampton, E. R. and Logan, J. A. 1985. The effect of *Sitona discoideus* Gyllenhal (Coleoptera: Curculionidae) on lucerne yields in New Zealand. *Bulletin of Entomology Research* 75, 429-442.

Goldson, S. L., Bourdot, G. W. and Proffit, J. R. 1987. A study of the effects of *Sitona discoideus* (Coleoptera: Curculionidae) larval feeding on the growth and development of Lucerne (*Madicago sativa*). *Journal of Applied Ecology* 24, 153-161.

Griffiths, D. W, Birch, A. N. E. and MacFarlane-Smith, W. H. 1994. Induced changes in the indole glucosinolate content of oil seeds and forage rape (*Brassica napus*) plants in response to either turnip root fly (*Delia floralis*) larval feeding or artificial root damage. *Journal of Science, Food and Agriculture* 65, 171-178.

Heilman, M. D., Manken, L. N. and Dilday, R. H. 1981. Tobacco bud worm effect of early season terminal damage on cotton lint yieldsand earliness. *Journal of Economic Entomology* 74, 732-735.

Hodgkinson, K.C.1974. Influence of partial defoliation on photosynthesis, photorespiration and transpiration by Lucerne leaves of different ages. *Australian Journal of Plant Physiology* 1, 561-578.

Hou, X., Meinke, L. J. and Arkebauer, T. J. 1997. Soil moisture and larval western corn root worm injury: influence on gas exchange parameters in corn. *Agriculture Journal* 89, 709-717.

Ingram, K. Y., Herzog, D. C., Boote, K. J., Jones, J. W. and Barfield, C. S. 1981. Effects of defoliating pests on soybean CO_2 exchange and reproductive growth. *Crop Sciences* 21, 961-968.

Martens, B. and Trumble, J. 1987. Structural and photosynthetic compensation for leaf miner injury in lima beans. *Environmental Entomology* 16, 374-378.

Orcutt, D. M. and Nilsen, E. T. 2000 Herbivory and plant stress. *In: Physiology of Plants Under Stress : Soil and Biotic Factors.* John Wiley & Sons, Inc. Verginia. US. 264-328.

Pandey, K. C., Hasan, N., Bhaskar, R. B. and Hazra, C. R. 1995. *Pests and diseases of major forage crops: Problems and prospects* Technical Bulletin. AICRP Forage Crops. IGFRI, Jhansi. India. 40pp.

Ryle, G. C. and Powell, C. E. 1975. Defoliation and regrowth in the graminaceous plant: the role of current assimilate. *Annals of Botany* 39, 297-310.

Saxena,P. 1982. Studies on certain powdery mildews of fodder legumes. *Ph D Thesis Giwaji University*, Gwalior India 237pp.

Saxena, P., Chauhan, S. and Ahmad, S. T. 1992. Astimation of losses in fodder cowpea due to powdery mildew. *Indian Phytopathology* 44, 376-379.

Saxena, P., Pandey, K. C., Faruqui, S. A., Hasan, N., Bhaskar, R. B., Roy, S., Azmi, M. I., Shah, N. K. and Ch Padmavati. 2002. *Forage Plant Protection.* Technical Bulletin. IGFRI, Jhansi, India. 38pp.

Sharma, H. C. and Nwanze, K. F. 1997. *Mechanism of resistance to insects in sorghum and their usefulness in crop improvement.* Information Bulletin 45. ICRISAT Patancheru. AP, India 51pp.

Sosebee, R. E. and Wiebe, H. H. 1971. Effect of water stress and clipping on photosynthate translocation in two grasses. *Agronomy Journal* 63, 14-19.

Thorne, J. H. and Koller, H. R. 1974. Influence of assimilate demand on photosynthesis, diffusive resistance, translocation and carbohydrate levels of soybean leaves. *Plant Physiology* 54, 201-207.

Trumble, J. T., Kolondy-Hirsch, D. M. and Ting, I. P. 1993. Plant compensation for arthropod herbivory. *Annual Review of Entomology* 38, 93-119.

❑❑❑

Chapter 19

Pesticide Residues in Animal Feeds and Fodder and their Management for Quality Livestock Production

SUBIR KUMAR NAG
Indian Grassland and Fodder Research Institute, Jhansi-284003 (U.P.), India

Introduction

Pesticides are very essential and invaluable input in modern day agriculture due to the fact that pests, diseases, and weeds destroying 1/3rd of the crop or commodity during growth, harvesting and storage. India is loosing about Rs 90,000 crores worth of agricultural produce annually due to pests. Among different plant protection measures like cultural, mechanical, biological and chemicals, the last one i.e. use of pesticides is still the most convenient, easiest and effective means for pest management. Beside, these chemicals are inseparable for successful control of various vectors of human and animal diseases and ectoparasites on animal body. The use of pesticides is all likely to continue in the near foreseeable future despite the fact that various alternative methods of pest control are being promoted.

Pesticides, therefore, are an integral part of the agricultural production systems. Even in the IPM programme pesticides are kept as the last resort particularly when the other options fail to deliver the desired result. India consumes less than 300g active ingredient (a.i.) per hectare of pesticides, the total consumption being 2% of total world consumption of around 20,00,000 million tonnes (MT). Although there is a steady decline in the consumption of pesticides in India from 73652 (MT) in 1995-96 to 39773 MT in 2005-06, the use of these substances would be continued in future to protect crops, and also the human and animal health. One of the main reasons for the success of green revolution in India was the use of synthetic pesticides as plant protection measures. Dr. Norman Borlaug, the eminent agricultural scientist and Nobel Laureate once said that "People of the world must decide that we either use agricultural chemicals and use them wisely, in the right amount and the right kind to produce the food or WE ALL STARVE".

However, because of ignorance and malpractices, continuous, prolonged, non-judicious and excessive use of the plant protection chemicals has posed lot of problems to environment and its inhabitants. Most of the pesticide molecules and/or their metabolites are toxic xenobiotics and are reported to have toxicity related health problems like carcinogencity, teratogenicity, mutagenicity, genotoxicity, neurotoxicity and many other adverse effects. Over the years there has been a mounting fear and concern that indiscriminate and disproportionate use of pesticides may lead to harmful and ill effects in human beings and animals and may have deleterious effect on the environment. The term 'Pesticide Residues' has thus become much talked about environment related issue infamously popularized in the society and a matter of dispute in trade and commerce both nationally and internationally. The term 'Pesticide residue' may be defined as any specific substance in food, agricultural commodity, or animal feed resulting from the use of a pesticide. The term includes any derivative of a pesticide such as conversion product, metabolite, reaction product and impurities considered to be toxicological significance. Whenever pesticides are applied on a crop or commodity or on any substrate to control pests, diseases, weeds etc. they invariably leave toxic residues on the substrate on which they are applied. The amount of residues left depends on the nature of the chemical, dose and volume of application, nature of substrate and prevailing weather condition. The occurrence of pesticide residue in ecosystem may have a threefold impact on man, by reducing the reproductive capacity of his environment or by contamination of either his food or work environment. Man and animals are exposed to low level of pesticide residues in air, water and food chain for a prolonged period of time, which pose problem to all animals, birds and other organisms and also human being. Man is the ultimate consumer of pesticide. As we see in the pesticide cycle, man ultimately consumes residues of pesticides in the grains, pulses, oil and oilseeds, milk, ghee, coffee, fish, meat etc. Through fodder, water, air and other foodstuffs pesticide residues reach animals and then through milk and/or meat products it is accumulated in the body of human being. This accumulation is five times higher in man in comparison to crops. The bioaccumulation and biomagnifications of certain pesticide residues in human food chain is an important aspect of pesticide induced environmental pollution, which needs attention.

The presence of pesticide residues has been detected in various items and in food chain. Among the different group of pesticides organochlorines are most persistent in nature. Being lipophilic they are deposited in body fat of man and animal. Such pesticides have an affinity for lipid material and are not easily biodegradable. So they cause contamination of animal products like milk, meat, fish, egg, ghee, butter etc. Animal products contribute 70-90% of the total oragnochlorinated pesticides in the diet of an Indian, particularly milk and milk products (Kannan et al., 1992). The residues of these pesticides in animal products also decline at a very slow rate even after the sources of contamination are eliminated. All the feed resources like cereal grains, oilcakes, brans, legume and nonlegume forage crops, crop residues are found to be contaminated with organochlorinated and other pesticides.

How the Pesticides Enter into the Animal Body?

Pesticides can enter into the animal body in three possible ways (Fig. 1).

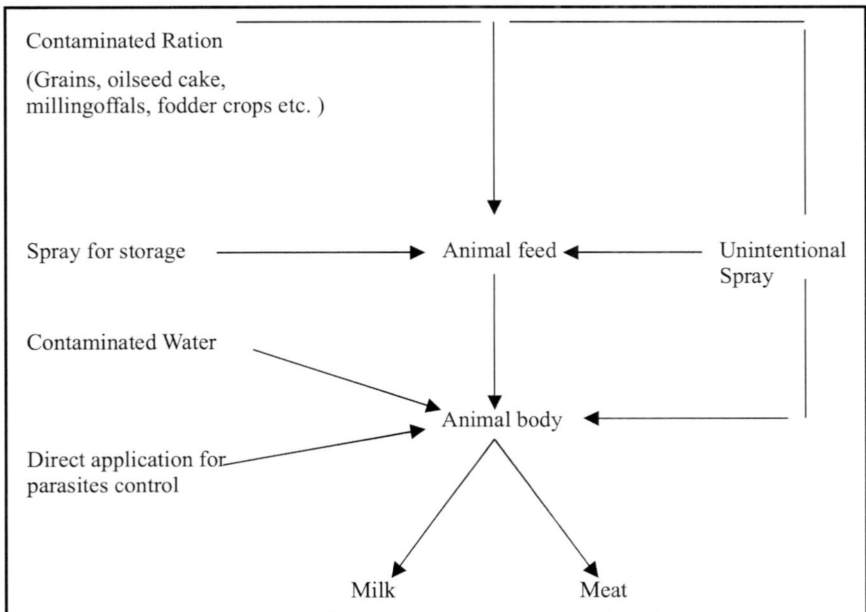

Fig. 1: Possible Sources of Contamination of Animal Body and Animal Products with Pesticides.

i) Through contaminated feed and fodder
ii) Through contaminated water
iii) Through skin pores when sprayed or drenched to animal body surface for killing ecto parasites

The main source of entry is the contaminated feed and fodder offered to animals for consumption. The feed materials may be contaminated by pesticides either at the time of cultivation of the crops in the field, from which they have been prepared like oilseed crops (mustard, linseed, sesamum, cotton etc.) in case of oilseed cakes, cereals (paddy, wheat, maize, etc.) in case of straw, chuni and hay. During the storage of the feed materials, grains, seeds etc. in the warehouse, godowns and other places pesticides are used to keep away the storage pests and diseases. Green fodders and/or dual purpose crops like sorghum, maize, bajra, oat, berseem, cowpea, different grasses grown in the field are often sprayed with pesticides to save them from the onslaughts of diseases and pests. This may act as direct source of contamination because many a times fodder crops after being treated with pesticides are fed to animals straight way without waiting sufficient time for the residues to go down below maximum residue limit. Pesticides are also used in a large scale for their useful properties to kill ectoparasites like ticks, mites, insects on the animal body. As a result they may enter into the animal body through skin pores when applied by drenching. Water sometimes may also be a source of contamination although concentration may not be such as to worry about. After entry into the animal body pesticides are distributed in different tissues and organs. Since many of them particularly organochlorines and synthetic pyrethroids are lipophilic in nature, they have a tendency to accumulate in the adipose tissues and usually they do not get converted into water soluble metabolites. As the concentration in the fatty tissues exceeds a threshold level, the excess residues are translocated into the lactating glands of the animals from where they are secreted in the milk. Because of special position of milk in the diet of infants and children the presence of significant amount of residues in milk is undesirable. That is why the residue limit for milk and dairy products tend to be more severe than those for other products.

Pesticide residues in feed: Feed is main source for the contamination of animal body with pesticide residues. As discussed earlier, feeds coming as by products of agricultural crops such as straw, oilseed cakes, chunnies, brans are contaminated in the field in many ways directly or indirectly by pesticides. Apart from applying pesticide to the growing crops in the field, spraying of pesticide on feed concentrate bags in godowns for controlling storage pest is also a common practice which can also lead to contamination. The readymade feed concentrate sold in the market is actually a mixture of many grains and jaggery/molasses which many contain pesticide residues. Rice and wheat straw are one of the important feed materials used in India for cattle and buffaloes. Rice consumes second largest share of total pesticide (17%) consumed in India. Rice bran, a rich source of HCH and other residues is used as an ingredient in the readymade feed concentrate. Although wheat crop as such does not consume much pesticide but during the storage of wheat straw pesticides are used to prevent to attack of insects, termites etc.

Numerous example of the contamination of animals through animal feed have been reported in the literature (Back, 1970, 1975; Von Houweling et al. 1977). But, public awareness of this did not become acute until 1974 when polybrominated biphenyls (PBD) were associated with an accident in Michigan in USA. The direct impact of this incident resulted in the destruction of 30,000 cattle and 6000 swine as well as sheep, poultry and dairy products. In USA screening of random samples of feed comprising of grains, cereals, by products and miscellaneous for organochlorine insecticides like lindane, β-BHC, aldrin, dieldrin, pp^1-DDE, op^1-DDT, pp^1-DDD, pp^1-DDT, heptachlor, and its epoxide revealed average contamination levels of less than 10ppb (Pierson et al.1982).

Status of Pesticide Residues in Feed and Fodder in Indian Scenario

Information regarding the status of pesticide residues in feed and fodder crops in India are very scanty unlike those for food and horticultural crops for which plenty of work has been done. Only little information is available about the status of residues in feed and fodder crops, that too mostly concentrated on organochlorines like DDT and HCH, which are summarized below.

Contamination of cattle feeds by DDT and HCH in some species of fodder grass was reported by Kaphalia and Seth (1982). Shastry (1983) depicted results of monitoring of different categories of feed samples like poultry and cattle feed, different feed ingredients, hay, grains, cereals, pulses, oilseeds and some green fodders. Out of 1241 samples tested 747 (60.19%) were found to contain residues of two the then most important insecticides DDT and BHC (HCH), along with endrin and lindane in some samples.

Punjab as a leading state in the era of green revolution benefited maximum by implementing the technologies and boosting the production. But the state has also suffered maximum due to the negative impacts of green revolution like degradation of soil health, loss of soil fertility and productivity, lowering of ground water table, contamination of soil and water, and adverse effects of large scale use of pesticides on man and environment. Widespread and high level contamination of different food and agricultural commodities with pesticide residues has been reported many times from different sources. Singh et al. (1988) determined insecticide residues in wheat grains and straw arising from their storage in premises treated with BHC and DDT under malaria control programme in Punjab. All the 120 samples of wheat grains collected from BHC as well as DDT sprayed districts contained residues of both HCH and DDT at varying concentration of 0.01-5.18mg/kg and 0.01-2.14mg/kg respectively. Analysis of 40 wheat straw samples also showed presence of BHC (mean concentration 0.24 mg/kg) and DDT (1.35mg/kg) residues in all the samples. Again, Battu et al. (1996) on monitoring of 105 samples of different feeds collected form different places of Punjab found 100 and 80% contamination with residues of HCH and DDT respectively. Feed concentrates, straw and green fodder showed total HCH residues comprising α, β, γ and δ isomers up to 0.26, 2.28 and 0.69mg/kg and DDT residues up to 0.28, 6.70 and 0.35mg/kg respectively. The tolerable content of 0.02 mg/kg β-HCH and 0.10mg/kg ΣDDT in the feed samples (dry weight basis) were exceeded in about 35% samples, while those of α, β and δ HCH exceeded in about 3% samples. In addition to HCH and DDT, residues of endosulfan and malathion were also detected in a limited number of feed concentrate samples out of a total of 17 samples tested by Kang et al. (2002) at Ludhiana (Punjab). Dicofol, an acricide organochlorine group, was also detected in two feed concentrate samples. Limiting value of β, γ and δ HCH and total DDT were exceeded in about 13, 6.5, 65 and 9.7% of animal feed samples on dry weight basis. However these results showed substantial decline in pesticide residues in feed samples in Punjab with respect to β-HCH and total DDT as compared to earlier report of 35% samples exceeding the limiting values of two insecticides (Battu et al. 1996).

Analysis of 32 cattle feed samples collected from local market of Lucknow revealed presence of α, β and γ isomer of HCH in all the samples and the mean concentration of total HCH was 0.393 ppm (Dikshit et al. 1989b). Residues of PP[1] DDE, OP[1] DDT, PP[1] DDD and PP[1] DDT were also present in all the samples and the total DDT content comprising of all these isomers was 0.236ppm. The level of aldrin was found to be 0.156ppm. Endrin, one of the most toxic organohlorines, was also present (0.02ppm) in all the feed samples along with α-endosulfan (0.029ppm). Dikshith, et al. (1989a) in their attempt to quantify residues of different organochlorine pesticides in oilseeds used in the country found that mustard, groundnut and sesamum seeds obtained from Lucknow were contaminated with DDT to the extent of 130-3200ppb and HCH to the extent of 91-250ppb. From down South Unnikrishnan et al. (1998) also reported that BHC, DDT and endosulfan were detected in samples of cattle feed and feed ingredients collected from Bangalore. Total BHC content of the compounded feeds ranged from 0.0076-0.1197mg/kg, DDT ranged from traces to 0.01 mg/kg. The active isomer γ–BHC constituted only a minor portion of total BHC content in most of these samples. In groundnut cake samples the predominant contaminant was DDT, while BHC was the major contaminant in compounded feed.

In a multilocational study conducted by All India Coordinated Research Project on Pesticide Residues (Agnihotri, 1999) in 15 locations on straw to determine HCH and DDT residues it was revealed that out of 377 samples analysed, most samples were found to contain residues of both HCH and DDT. The mean value was from <0.01-0.60 mg/kg for DDT and <0.01-2.27mg/kg for HCH. The presence of residues of HCH and DDT in straw and feed in almost all samples of different locations in amounts above the prescribed limit proved that they were a potential source of contamination in milk (Agnihotri, 1999). HCH and DDT residues were also monitored in certain dry fodder (pearl millet, wheat and barley) and feed concentrates (sesamum, mustard, and cottonseed cake) collected from Jaipur in Rajasthan (Gupta et al. 2000). Although no DDT residue was detected in any of the samples but HCH residues was present in the range of 0.3-1.73mg/kg in feed concentrates. In case of dry fodder wheat and barley had no residues of HCH but pearl millet samples were contaminated with HCH in the range of 0.06-0.11mg/kg.

Monitoring of HCH residues in feed samples from Hisar, Haryana showed that all four isomers of HCH i.e. α, β, γ and δ were present in mustard cake, cottonseed cake, churi and wheat straw. The mean concentration of total HCH varied 0.2 to 0.76 mg/kg (Singh et al., 1997). In another study conducted at Karnal, Haryana (Prasad and Chhabra, 2001) it was found that contamination in feed concentrates and roughages was caused by BHC (73.3%), endosulfan (8%), heptachlor (8%) DDT (4.2%) and aldrin (2.6%). The content of BCH residues was highest in concentrate mixture followed by maize, groundnut cake, wheat bran, rice bran, cotton seed cake, barley, and mustard cake. Among the roughages it was highest in maize fodder followed by oat, berseem, wheat straw, lucerne and paddy straw in the decreasing order. Among the different isomers of HCH, α and γ-isomers were more than β and δ. The two isomers of DDT (OP^1 and PP^1) were present in all the forages and in most of the concentrate feeds as compared to fodders. Again, Sharma et al. (2005) in a multiresidue monitoring programme of feed concentrates done at Karnal, found DDT contamination in oilseed cakes, groundnut husk, wheat straw and fodder to the extent of 0.107, 0.228, 0.37 and 0.05mg/kg respectively.

Feed samples collected from Jorhat district of Assam shown presence of lindane or γ-BHC (0.025-0.041mg/kg), β-endosulfan (0.0028mg/kg) and endosulfan sulfate (0.045-0.049mg/kg) residues in them (Deka et al. 2004).

Monitoring of feed and fodder samples for multiresidues of different groups of pesticides like organochlorines, organophosphates & carbamate and synthetic pyrethroids have been carried out in 533 feed samples (oilseed cakes, straw, chunnies. compound feed etc.) collected from different places of Bundelkhand region (Nag and Raikwar, 2006). Out of them 301 (56.47%) samples were found positive containing residues of different organochlorinated pesticides like HCH isomers, DDT complex, endosulfan. The mean total HCH concentration varied from 0.0099 to 0.3055mg/kg in different samples. Among different isomers of HCH the concentration of b-HCH was highest. Endosulfan (α, β and sulfate) was detected in the range of 0.0095-0.237mg/kg. Total DDT varied from 0.0161 to 0.1179mg/kg. Dicofol, an acaricide and DDT analogue, could be detected in very few samples. In samples positive with HCH isomers and DDT complex the concentration of residues was found to exceed the respective limiting values in very limited samples. Otherwise they were mostly within safe limits. Contamination of feed samples with organophosphate (OP) pesticide residues was lesser as out of 533 samples 274 (51.4%) were contaminated. Out of ten selected OPs, residues of mainly chlorpyriphos, malathion and quinalphos were detected. Others like dichlorvos, phorate and dimethoate were also found but less frequently. The concentration of different OP compounds in the contaminated samples was very less in most of the cases and the range was found to be from traces to 0.8mg/kg. In case of synthetic pyrethroids (SP) contamination was still lower as 264 (49.53%) samples were contaminated with SP residues, mainly contributed by cypermethrin, fenpropathrin, cyhalothrin, fenvalerate and deltamethrin. The concentration of cypermethrin in some mustard and cotton seed cake samples was quite high (about 2mg/kg). Otherwise, it was mostly low as envisaged from the range of concentration in the positive samples which varied from traces to 0.5mg/kg. Among the different type of feed samples mustard, linseed, cottonseed cake, gram and lentil chunni samples were found to be more contaminated than others.

Green fodder: Work on green fodder crops with respect to pesticide residues, their dissipation pattern and safety evaluation has been very limited. Endosulfan, the recommended and commonly used insecticide in forage crops, was evaluated for persistence of residues and safe-waiting period following application at the recommended dose in fodder crops like sorghum, cowpea, lucerne and maize. The residual half-life around 3-4 days and safe-waiting period around 9-14 days have been recommended (Nag and Shah, 1999, 2001,

2002; Raikwar et al., 2003). Carbendezim, the systemic fungicide recommended and used in fodder crops was also evaluated for persistency in sorghum (Raikwar and Nag, 2004). Residual half-life of 6days was observed and safe-waiting period around 9-10days was recommended.

The status of pesticide residues in green fodder can be assessed only from the very few monitoring reports available on the subject. Gupta et al. (2000) monitored HCH and DDT residues in certain green fodder crops like pearl millet, guinea grass, mustard, fenugreek, oat, chinese cabbage and lucerne collected from Jaipur (Rajasthan) and they concluded that all the samples contained HCH residues above MRL. However, DDT residue was not detected in any sample. Prasad and Chhabra, (2001) analysed few green fodders such as maize, oat, sorghum, berseem and lucerne collected Karnal, Haryana for organochlorine pesticide residues. According to their results, total HCH content was abnormally high ranging from 15-44 mg/kg DM basis, lowest in lucerne and highest in maize. Total DDT and endosulfan residues ranged from 1.3-2.8mg/kg and 1.9-5.9mg/kg respectively. Fourteen green fodder samples collected from Ludhiana district of Punjab contained HCH residues (range 0.03-0.82mg/kg) but in only two and three samples total DDT (BDL-0.18mg/kg) and endsulfan (BDL-1.58 mg/kg) residues were detected. Among the OPs, malathion was found in eight samples (Kang *et al.* 2002).

Sorghum, maize, bajra, oat, berseem and sugarcane top collected from different places and locations of Bundelkhand region were monitored for pesticide residues (Nag and Raikwar, 2006). About 50% samples were contaminated with different organochlorine pesticides like DDT, HCH and endosulfan ranging from Tr-0.8310mg/kg, 0.0067-0.0290mg/kg and Tr-0.6985mg/kg respectively. In case of organophosphate group of insecticides, the contamination level was 45%. Malathion, chlorpyriphos and quinalphos were mainly detected. Synthetic pyrethroids particularly cypermethrin, cyfluthrin and fenvalerate were found present in some samples and overall contamination was 42%. However in grasses like setaria, guinea, dinanath, sirsa, paragrass, Bothriochloa, hybrid napier and fodder cowpea collected from Jorhat district of Assam no detectable residues of any pesticide was found (Deka et al. 2004).

Strategies for Overcoming the Problems

As stated earlier that the main problem faced with the pesticide residues present in feed and fodder is that ultimately they would be transferred to dairy and animal products meant for human consumption. So efforts should be made to keep the residues at minimum level, if not totally unavoidable, so that contamination, if at all, present in food of animal origin, can be below the safe or allowable limit.

Limiting values are the concentrations of pesticides in feed and fodder if fed to lactating animals daily, the likely residues in milk will be less than their MRL values and safe for human consumption. The limiting value of a pesticide in feed can be derived on the basis of its legal permissible limits in milk and its rate of transference from feed to milk. On the basis of short term feeding experiment on buffaloes the transfer coefficient (The percentage of daily intake of pesticide which is excreted in the milk each day after equilibrium conditions have reached) of HCH isomers and DDT complex have been recommended as 12-15% for α-HCH, 31-36% for β-HCH, 2.7% γ-HCH, 8.5% for δ-HCH and ~5% for most of the DDT isomers and its analogous (Kalra *et al.*, 1986; Kapoor and Kalra, 1997). The coefficient was higher in buffalo milk than in cow milk. Similarly Nag *et al.* (2007) recommended a transfer coefficient of 0.23-0.33% for endosulfan on the basis a short term feeding study on milking goat. Transfer coefficients of carbaryl (0.16%), fenvalerate (0.44-0.64%) and deltamethrin (0.27%) in bovine milk have been mentioned by Kalra and Chawla, 1985) but based on unpublished literature. Residue level of a particular pesticide found in milk is in direct proportion of the transfer coefficient following the ingestion of contaminated ration by lactating animals. The limiting value of DDT complex, α, β, γ and δ isomers of HCH in feed and fodder have thus been worked out to be 0.1, 0.12, 0.02, 0.12, and 0.07 mg/kg (dry wt. basis) (Kang *et al.*, 2002). If the feed containing residues above the limiting values are fed to the lactating animals, it is likely to be passed to the milk. The limiting value of remaining organochlorinated and different other groups of pesticides are yet to be worked out.

Safety Measures

If people are careful from the very beginning right from the sowing of the crops, then most of the problems related to pesticide residues can be avoided. Few points in this regard may be mentioned here.

- Synthetic pesticides should be used when absolutely necessary.
- Proper pesticide should be used in proper condition as per the recommendation.
- They should be applied at the exact dose.
- Farmers/users should not be guided by the dealers/retailers in choosing the chemical.
- Application equipment should be perfect and there should be minimum possible drift
- Application should be avoided immediately before harvest.

- Safe waiting period or pre harvest interval (the prescribed minimum time gap between the last application of a pesticide to a crop/ commodity/animal and harvesting or grazing/feeding the livestock or milking/slaughtering the animal for human consumption) should be strictly maintained.
- In case of feed/fodder the maintenance of safe waiting period becomes all the more important and should be strictly adhered to because the fodder is normally straightway fed to animals and thus there is hardly any chance for decontamination. Freshly treated material should never be fed to animals particularly to those which are in milking stage.

However, even in spite of all efforts, if the contamination of feed and fodder with pesticides can not be checked, then there are some processes which help in decontaminating the treated materials.

- Sun drying of the material for few days can remove some of the residues.
- In case of green fodders simple washing or dipping in salt/dilute acid/ lime solution can be done to remove a certain percentage of residues. But for concentrated feed this may not be feasible. Silage making process may also lower the concentration of residues in fresh green fodder.
- If the animals somehow consume feed materials contaminated with pesticide residues, then charcoal supplementation @ 1-2g/kg body wt. can prove to be effective antidote.
- For animals consuming pesticides at a high dose closer to toxic level, the treatment would be symptomatic.

Conclusions

Presence of varying types of toxic substances in feed material is undesirable because from there they enter into the animal body system and redistribute themselves in different tissues, organs and deposited there for a long time due to their persistence nature. They directly affect the growth of animals, incite many diseases, affect and reduce the immune system, reduce the reproductive efficiency which ultimately leads to loss of production efficiency. They generally combine with one or more bioactive ligands such as -OH, $-COO^-$, OPO_3H, >C=O, -SH, -S-S-, $-NH_2$ and –NH that is essential for normal physiological function and activation of several enzymes. Most importantly they contaminate human food chain like milk, meat etc. and thus enter human body and pose serious health problems.

Now-a-days there is much consciousness about the food safety. Consumers worldwide expect protection from hazards occurring along the entire food chain. It is, therefore, the need of the hour to emphasis greatly on food safety as well as food quality in respect of pesticide residues, antibiotic residues, heavy metals, aflatoxins, pathogens and other contaminants. Access of countries to the food export markets depends on their capacity to meet the regulatory requirement (non tariff trade barriers) of the importing countries.

There is a direct link between the safety of feed and the safety of food of animal origin because feed and food are interrelated as one compliments the other with the supply of feed for production of milk, meat, egg etc. Feed production must, therefore, be subjected to the same way as food production, quality assurance including safety systems based on principal of HACCP (Hazard Analysis and Critical Control Point) endorsed by the Codex Alimentarius Commission (CAC), an inter governmental body, which has formulated international standards for a wide range of food products and specific requirements concerning pesticide residues, food additives, veterinary drug residues, hygiene, food contaminants *etc.*

References

Agnihotri, N.P.1999. In: Pesticide safety evalution and monitoring. All India Coordinated Research Project on Pesticide Residues, Indian Agricultural Research Institute, New Delhi, pp. 104-111.

Battu, R.S., Singh, B., Chahal, K.K. and Kalra,R.L. 1996. Contamination of animal feed with residues of HCH and DDT. *Pesticide Research Journal* 8, 172-175.

Buck, W.B. 1970.. *Journal American Medical Association* 156, 1434

Buck, W.B.1975. *Journal American Veterinary and Medical Association* 166, 222.

Deka, S.C., Barman, N. and Baruah, A.A.L.H. 2004. Monitoring of pesticide residues in feed, fodder and butter in Assam. *Pesticide Research Journal* 16, 86-89.

Dikshit, T.S.S., Kumar, S.N., Raizada, R.B. and Srivastava, M.K. 1989b. Organochlorine insecticide residues in cattle feed. *Bulletin of Environmental Contamination and Toxicology* 43, 691-696.

Dikshit, T. S. S., Kumar, S. N., Tandon, G.S., Raizada, R. B. and Ray, P. K. 1989. Pesticide residues in edible oils and oilseeds. *Bulletin of Environmental Contamination and Toxicology* 42, 50-56.

Gupta, A., Singh, B., Parihar, N.S. and Bhatnagar, A. 2000. Monitoring of HCH and DDT residues in certain animal feed and feed concentrates. *Pestology* 24, 47-49.

Kalra, R.L. and Chawla, R.P. 1985. Pesticidal contamination of foods in the year 2000 A.D. In : *Proceedings of Indian National Science Academy* B52, 188-204.

Kalra, R.L., Chawla, R.P., Joia, B.S. and Tiwana, M.S. 1986. Excretion of DDT residues into milk of Indian buffalo, *Bubalus bubalis* (L.) after oral and dermal exposure. *Pesticide Science* 17, 128-134.

Kang, B. K., Singh, B., Chahal, K.K. and Battu, R.S. 2002. Contamination of feed concentrates and green fodder with pesticide residues. *Pesticide Research Journal* 4, 308-312.

Kannan, K., Tahabe, S., Romesh, A., Subramanian, A. and Tatsukawa, R. 1992. Persistent organochlorine residues in food stuff from India and their implication on human dietary exposure. *Journal of Agricultural and Food Chemistry* 40, 518-524.

Kaphalia, B.S. and Seth, T.D. 1982. Organochlorine pesticides contamination in some species of fodder grasses. *Environmental Pollution Series* B.3, 231-237.

Kapoor, S.K. and Kalra, R.L. 1997. Transfer of HCH isomers from feed into milk of Indian Buffalo (*Bubalus bubalis* L.). *Pesticide Research Journal* 9, 72-78.

Nag, S.K. and Raikwar, M.K. 2006. Final Report of Lal Bahadur Shastri Young Scientist Scheme "Monitoring of pesticide residues in animal feed, fodder and milk with special reference to Bundelkhand region" submitted to ICAR, New Delhi.

Nag, S.K. and Shah, N.K. 1999. Residual persistence of endosulfan in fodder sorghum. *Pestology* 23, 52 - 55.

Nag, S.K. and Shah, N.K. 2001. Endosulfan residue and its dissipation in Lucerne (*Medicago sativa* L.). *Pesticide Research Journal* 13,109-111.

Nag, S.K. and Shah, N.K. 2002. Residues and efficacy of endosulfan against pest complex of fodder cowpea. *Journal of Entomological Resesrach* 26, 133-139.

Nag, S.K., Mahanta, S.K., Raikwar, M.K. and Bhadoria, B.K. 2007. Residues in milk and production performance of goats following the intake of a pesticide (endosulfan). *Small Ruminant Research 67*, 235-242.

Pierson, D.A., Hoffman, J.S., Nord, P.J., Gebhart, J.E. and Frank, C.N. 1982. Distribution of chlorinated pesticides in animal feed components and finished feeds. *Journal of Agricultural and Food Chemistry* 30, 187-189.

Prasad, K.S.N. and Chabara, A. 2001. Organochlorine pesticide residues in animal feeds and fodders. *Indian Journal of Animal Sciences* 71, 1178-1180.

Raikwar, M.K. and Nag, S.K. 2004. Dissipation of carbendazim residues in fodder sorghum. *Indian Journal of Agricultural Sciences* 74, 227-229.

Raikwar, M.K., Nag, S.K., Banerjee, T. and Shah, N.K. 2003. Persistence behaviour of endosulfan in fodder maize. *Pesticide Research Journal.* 15, 186-190.

Sharma, V., Wadhwa, B.K., Stan, H.J. 2005. Multiresidue analysis of pesticide in animal feed concentrate. *Bulletin of Environmental Contamination and Toxicology* 72, 242-245.

Shastry, M. S. 1983. Monitoring of pesticide residues in animal feeds and animal products. *Pesticide*, December issue. pp. 36-38.

Singh, P.P., Battu, R.S. and Kalra, R. L. 1988. Insecticide residues in wheat grain and straw arising from their storage in premises treated with BHC and DDT under malaria control programme. *Bulletin of Environmental Contamination and Toxicology* 40, 696-702.

Singh, R., Kumari, B., Madan, V.K., Kumar, R. and Kathpal, T.S. 1997. Monitoring of HCH residues in animal feeds. *Indian Journal of Animal Sciences* 67, 250-252.

Unnikrishanan, V., Surendara Nath, B., Gayathri, V., Sampath. K.T. and Rama Murthy, M.K. 1998. Organochlorine pesticide residues content in feed and feedstuff. *Indian Journal of Dairy & Biosciences* 9, 59-64.

Van Houweling, C.D., Bixler, W. B. and Mc Dowell, J.R. *Journal of American Veterinary Medical Association* 171, 1153.

❑❑❑

Chapter 20

Microbial Protein Synthesis in Ruminants Fed Forage

S. K. MAHANTA, K. K. SINGH, M. M. DAS AND A. K. MISRA
Indian Grassland and Fodder Research Institute, Jhansi-284003 (U.P.), India

Introduction

Unlike non-ruminants, formulating rations for precise nutrient requirements in ruminants is a challenge. Both ruminants and non-ruminants utilize nutrients in tissues but ruminants have another metabolic system i.e. microbial metabolism in the rumen. This posses a nutritional challenge because nutrient requirements of the two systems are distinctly different. Maximizing ruminant productivity involves meeting the requirements for both systems. Inherent losses in energy and nitrogen associated with pre-gastric fermentation by microbes in the rumen make the digestion and subsequent absorption of dietary nutrients less complete in ruminants. The rumen is a complex microbial ecosystem and not easy to manipulate like an industrial fermentation. The primary research focus, from an evolutionary standpoint has been on making rumen microbes degrade forages quicker and more extensively. In that respect, attempts have been made to modify feeds to make the cell wall more accessible to microbial enzymes. Rumen microorganisms are interdependent for supplies of carbon and nitrogen; improving the availability of carbon sources to increase digestibility would provide maximum benefit to overall fermentation if it were synchronized with the availability of nitrogen. New strains of genetically engineered bacteria that are more effective fiber digesters have been introduced into the rumen. Although temporal variations occur, the 'dynamic steady-state' of the flora seems highly resistant to changes. As a result the introduction and subsequent growth of exogenous bacterial strains with desirable biochemical properties is difficult if not impossible to sustain over long periods in the hostile rumen environment. Perhaps the greatest potential to manipulate ruminal fermentation has been realized with the use of dietary additives that alter microbial physiology to either enhance beneficial processes like fiber digestion, lactate fermentation and non-protein nitrogen use or minimize inefficient processes like methane production and ammonia absorption. Biochemical reactions in the rumen are highly interactive. Interfering with a reaction almost always results in a cascade of interrelated reactions, some of which may not necessarily be beneficial.

In India, most of the livestock population is reared under subsistence oriented mixed farming systems and the livestock owners have to raise their animals on fibrous forages/ crop residues, thus, any strategy for improvement in livestock production requires efforts to maximize the efficiency of utilization of available feed resources in the rumen by appropriate supplementation to meet the requirement. The inefficient utilization of nutrients, result in wastage, particularly of nitrogen, adding to the cost of production and ultimately leading to environmental pollution. Thus judicious use of protein supplements has become important and necessitates estimation of protein that is digested and absorbed from intestine.

The Rumen

The four compartments of the ruminant stomach are the rumen, reticulum, omasum, and abomasum (Fig 1). The rumen is the largest compartment, and it contains billions of bacteria, protozoa, molds, and yeasts. These microorganisms live in a symbiotic manner with the animal, and they are the reason ruminant can eat and digest large amounts of forages. The bacteria and protozoa do most of the digestion of feeds for the animal. This is a tremendous factory. There are 25 to 50 billion bacteria and 200 to 500 thousand protozoa in every milliliter of rumen fluid. The microorganisms digest the plant fiber and produce volatile fatty acids. These fatty acids are absorbed directly through the rumen wall and supply 60 to 80 % of the energy needed by the animal. In addition to energy, the microorganisms produce protein including essential amino acids from the protein and nitrogen the animal ingests. Because the microbes can use nitrogen to make protein, animals can eat urea and other sources of non-protein nitrogen that would kill non-ruminants. The microbes also make vitamins B and C.

The conditions in the rumen are not only complex but they are intermittent. The rate at which feed enters the rumen will be very different during grazing or stall feeding. Salivary flow rate is not steady and rumination activity that is not continuous will depend on the type of diet. The flow of substances into and out of the rumen may involve more than one pathway. Volatile fatty acids can leave the rumen via passage into the lower tract or they can be absorbed and partly metabolized in the epithelium. Urea may enter the rumen via saliva or directly from blood through the rumen epithelium. Despite all these variations there are certain generalizations that we can make regarding the rumen-

- Temperature is usually maintained within the range of 38-41°C with 39°C used as a common mean temperature
- Rumen pH can range from around 7.0 on forage diets to as low as 4.6 on high-grain diets
- Mean redox potential is -350mv reflecting the strong reducing environment and the absence of oxygen
- Carbon dioxide and methane are the major gases present in the rumen
- The solid and liquid digesta leave the rumen at different rates

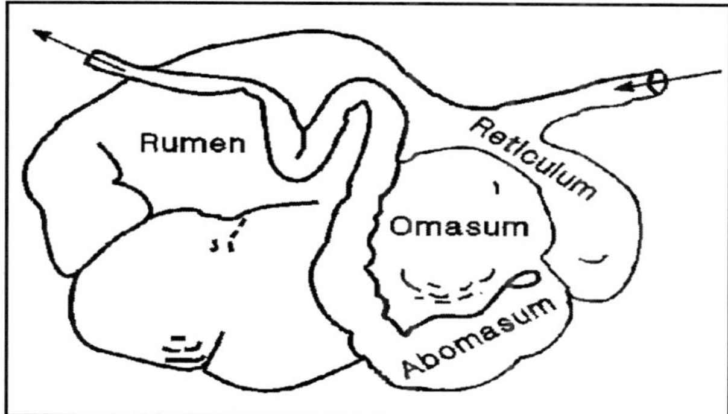

Fig 1: Schematic Diagram of Four Compartmented Stomach of Ruminant

Microbial Ecology

The rumen contains one of the most diverse and dense microbial ecologies known in nature. A possible explanation for the diversity is the complex nature of the feed, which contains carbohydrate, proteins, fats, other organic compounds and minerals. In order to utilize these compounds organisms are either highly specialized (narrowly adapted) to compete for a few of the feeds or become widely adapted and are capable of using many nutrients (Table 1 & 2). In addition to these categories there are microbes in the rumen that alter their metabolism depending upon the availability of nutrients. Another factor that selects for diversity is the ability of organisms to accomplish the most growth. Since growth is limited by the quantity of available food, efficiency at which food is transformed into cells will dictate survival. The conversion of carbohydrate to acetate, propionate, butyrate, carbon dioxide and methane results in a cell yield that is greater than other theoretical biochemical pathways. However, when nature or amount of carbohydrate is altered, heterolactic fermentation replaces the acetic-butyric type by microbes that are more efficient in the production of new cells. Given the various possible biochemical interactions, certain combinations of reactions produce more cells than do others and the conditions necessary for each may vary.

Table 1: Rumen Bacteria and their Substrates

Kind of substrates utilized/ fermented	Bacteria
A. Cellulose degrading	*Bacteroides succinogenes* *Ruminococcus flavefaciens* *Ruminococcus albus* *Butyrivibrio fibrisolvens*
B. Hemi cellulose degrading	*Butyrivibrio fibrisolvens* *Bacteroides ruminicola* *Ruminococcus sp*
C. Pectin degrading	*Butyrivibrio fibrisolvens* *Bacteroides ruminicola* *Lachnospira multiparus* *Succinivibrio dextrinosolvens* *Treponema bryantii* *Streptococcus bovis*
D. Amylose (starch) degrading	*Bacteroides amylophilus* *Succinimonas amylolytica* *Bacteroides ruminicola* *Streptococcus bovis*
E. Urea degrading	*Succinivibrio dextrinosolvens* *Bacteroides ruminicola* *Selenomonas sp.* *Ruminicoccus bromii* *Butyrivibrio sp.* *Treponema sp.*
F. Methane producing	*Methanobrevibacter ruminantium* *Methanobacterium formicicum* *Methanomicrobium mobile*

Contd...

Table 1. Contd...

G. Sugar utilizing	*Treponema bryantii*
	Lactobacillus vitulinus
	Lactobacillus ruminus
H. Acid utilizing	*Megasphaera elsdenii*
	Selenomonas ruminantium
I. Protein degrading	*Bacteroides amylophilus*
	Bacteroides ruminicola
	Streptococcus bovis
	Butyrivibrio fibrisolvens
J. Ammonia producing	*Bacteroides ruminicola*
	Megasphaera elsdenii
	Selenomonas ruminantium
K. Lipid utilizing	*Anaerovibrio lipolytica*
	Butyrivibrio fibrisolvens
	Treponema bryantii
	Eubacterium sp.

Table 2: Rumen Ciliate Protozoa/ Fungi and their Substrates

Genera	**Kind of substrates utilized/ fermented**
A. Protozoa	
Isotricha	
intestinalis	Starch, sucrose, glucose, pectin
prostoma	Starch, sucrose, glucose, pectin
Dasytricha	
ruminantium	Starch, maltose, cellobiose, glucose
Entodinium	
bursa	Starch, hemicellulose
caudatum	Starch, cellobiose, glucose, maltose, sucrose
furca bilobum	----
simplex	Starch
Diplodinium	
Polyplastron	Cellulose, glucose, starch, sucrose
Diplodinium	Cellulose, hemicellulose, starch
Diploplastron	Cellulose, hemicellulose, starch
Eudiplodinium	Cellulose, hemicellulose, starch
Ostracodinium	Cellulose, hemicellulose, starch
Eremoplastron	Cellulose, hemicellulose, starch
Epidinium	
Ecaudatum caudatum	Cellulose, hemicellulose, starch, sucrose, maltose
Ophryoscolex	
caudatus	Cellulose, hemicellulose, starch
B. Fungi	
Neocallimastix frontalis	Degrading plant (fibrous) particle, cellulose, xylan
Sphaeromonas communis	Degrading plant (fibrous) particle, cellulose, xylan
Piromonas communis	Degrading plant (fibrous) particle, cellulose, xylan

The contents in the rumen are heterogeneous and consist primarily of a microbial suspension in free liquid, a solid mass of digesta, and a gas phase. Microbial populations of the rumen are characterized into those organisms free in the rumen fluid, those associated with feed particles and those associated with the rumen wall. Those associated with the solid particles have been further subdivided into those organisms firmly associated with particles and those more reversibly associated. In a fully functioning rumen, there is a dynamic equilibrium, as ruminal microbes adhering to and detaching from feed particles are constantly leaving or re-entering the fluid compartment. The rumen harbors a large number of different species of bacteria, fungi and protozoa. Metabolic interactions between these different populations are critical for their collective survival. Moreover, products of the metabolism of some species of microorganisms are sources of energy for the other species. These interactions regulate in a large part the concentrations and activities of individual species as well as the nature of the fermentation products.

Energetics of Rumen Fermentation

Rumen microorganisms require strict anaerobic conditions for normal function. Consequently, a key feature of fermentation is the partial oxidation of substrates within the rumen. The energetics of microbial metabolism can be characterized into three main processes: (1) amount of organic matter fermented, (2) concentration of relative proportions of fermentation products produced and (3) amount and efficiency of microbial protein synthesis.

Anaerobes conserve ATP in the form of a trans-membrane electrochemical gradient commonly referred to as the Proton Motive Force (PMF). However bacterial cell growth depends largely upon a membrane bound ATPase for the transfer of ATP from PMF. The rumen is a highly reduced environment and energy is often limiting. The survival of rumen microorganisms is dependent upon the efficiency with which ATP is produced, transferred and utilized during bacterial growth. Maintenance of normal fermentation within the rumen requires that the large amounts of reducing equivalents produced in the form of NADH must be re-oxidized. Microbial populations have evolved fermentation pathways that effectively lower the concentration of reducing equivalents in the rumen. The main products of these fermentation reactions are SCFA (short chain fatty acids), CO_2 and CH_4. Acetate is the predominant SCFA found in the rumen and its formation is largely a function of the production of hydrogen from reduced cofactors. However, a high concentration of hydrogen gas is thermodynamically unfavorable and will inhibit further fermentation. Low hydrogen levels are maintained by methanogens resulting in greater hydrogen and consequently acetate production. Anaerobic conditions within the rumen result in most of the energy (ATP) in the fermented organic matter being retained in the products (SCFA and microbes) with some losses occurring in the form of CH_4 and heat. The route with which reducing equivalents are disposed determines the availability of ATP. If the removal of hydrogen is coupled to acetate production, ATP yield is highest compared to ATP yields for butyrate and lactate. It is now believed that ATP yield from propionate can be comparable to that from acetate. If ATP production is uncoupled with microbial growth, excess SCFA production will be inversely related to microbial cell synthesis. This will have a major impact on the supply of the two most important sources of nutrients i.e. SCFA (energy source) and microbial biomass (protein supply).

Factors Affecting Microbial Growth and Yield

At any given time the rumen will contain variable levels of peptides, amino acids, ammonia, carbohydrates, polymers, isoacids, lipids, vitamins, and minerals. The levels of these substrates will vary with time after feeding and with the source and quantity of feed consumed. In order to optimize microbial growth a constant supply of nutrients is required. We know the major nutrients that are needed by microbial species in the rumen but we know little about the amounts and combinations needed to optimize microbial growth. Ammonia is the most important source of nitrogen for protein synthesis in the rumen. Ammonia concentration in the rumen can fluctuate between <1 mM (low protein roughage) and as high as 40 mM following feeding of rapidly degraded protein. Many microbes that use ammonia can also use amino acids or peptides. The use of ammonia is reduced in the presence of amino acids and peptides explaining the variability in ammonia concentrations in the rumen. Some studies suggest that larger peptides may be metabolized more rapidly than smaller peptides and that hydrophobicity reduces the rate of peptide hydrolysis. However, the combination of amino acids and peptides stimulate total microbial growth to a greater extent than either protein by itself. Microbial nitrogen requirements are dependent upon the type of carbohydrate being fermented in the rumen. Bacteria that ferment NSC prefer amino acids and peptides and those that ferment fiber rely primarily on ammonia as their nitrogen source.

Carbohydrates are the major source of energy for rumen microbes. Studies suggest that following the uptake and hydrolysis of polysaccharides the resultant hexoses and pentoses are readily fermented and support microbial growth with similar efficiencies. However, due to differences in the rate of hydrolysis of different oligosaccharides total protein yield per unit of time may vary. Some bacteria, particularly the cellulolytic species, may require the isoacids in order to meet their amino acid requirements. Isoacids are produced via the deamination of dietary amino acids as well as from the growth and lysis of other microbes.

Ruminal pH has a major impact on fermentation of microbial growth. When sources of readily fermentable carbohydrates are added to forage diets fiber digestion may be impaired. Depression in digestibility due to a drop in pH is not equal for all nutrients. While fibrolytic and proteolytic activities are severely depressed, fermentation of starches and sugars remain very high. Liquid and solid turnover rates in the rumen are not the same and vary with intake. Since microbes are present in all compartments within the rumen their survival will depend on their ability to reproduce at a rate equal to or greater than the turnover of that compartment. Attachment to feed particles is necessary for survival of microbes that have a slow rate of growth, such as protozoa and fiber digesters. Both total yield and efficiency of microbial yield are influenced by liquid and solid turnover in the rumen. Microbial yield (Fig 2) is important because this is an index of the amount of microbial protein made available to the animal each day. Since microbial protein can supply well over 50% of the requirements of a lactating cow, the practical significance of maximizing microbial yield is obvious. Microbial efficiency is important because it is part of the calculation of yield. Intake, rumen turnover and microbial efficiency are all positively related. An increase in intake will increase ruminal turnover, resulting in an increase in microbial growth. If organic matter digestion in the rumen is reduced it may have a negative impact on the efficiency of microbial growth.

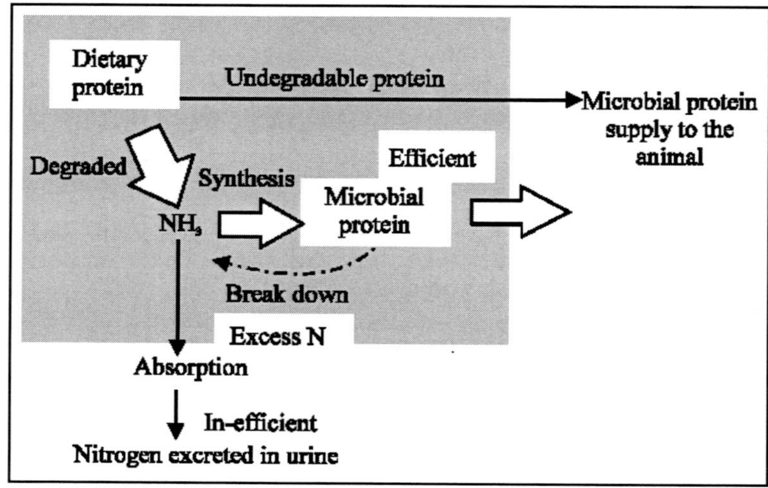

Fig 2: Inter-Conversion of Dietary Nitrogen into Microbial Protein

Nitrogen Metabolism in Rumen

Microbial cells and dietary nitrogen that escapes ruminal degradation are the major sources of protein and amino acid requirements of ruminants. Although plant materials that comprise the bulk of ruminant feeds are composed of a vast array of nitrogenous compounds, most of the nitrogen contained in the forages and cereals fed to ruminants is protein. Compounds that are not true protein, but contain nitrogen are non-protein nitrogen (NPN) and include nucleic acids, nitrates and supplemental urea. Enzymatic activity in the rumen converts dietary protein into amino acids, which are in turn deaminated to ammonia and various carbon skeleton compounds (organic acids). This affects the composition of dietary protein that escapes the rumen as well as the microbial protein fraction. This process appears to be wasteful since the animal requires some amino acids for its own use. However, some plant proteins may be very indigestible by the host enzymes and these proteins often have a low content of the essential amino acids. Since the microbes synthesize protein from ammonia and other suitable carbon skeletons they contribute to the nutritional metabolism of the animal. Nucleic acids (5 to 10% of dietary N) are fermented rapidly and the nitrates are converted to nitrite, which is fermented to ammonia. The most common strategy is to increase the escape of high quality dietary protein by minimizing proteolysis, peptidolysis and deamination. It is well accepted however, that increasing the undegradable intake protein fraction must not be at the expense of lowering the degradable intake protein in the diet. Optimal protein supply to the animal depends on adequate degradable protein to maximize capture of organic matter in microbial biomass. Synchronizing the rate of nitrogen hydrolysis with the rate of energy release will increase the rate of assimilation of ammonia by microbes and maximize nitrogen use by the animal (Fig 3).

Estimation of Microbial Protein Synthesis

In ruminants, microbial cells are major source of protein, formed as a result of feed digestion under anaerobic conditions. They provide greater part of amino acids that the host animal requires for tissue maintenance, growth and production. In forage fed animals, microorganisms are virtually the only source of protein. Therefore, knowledge of microbial contribution to the nutrition of the host animal is vital for improving animal production.

Several methods of protein evaluation have been employed in animal nutrition. Majority of existing methods that are used to measure microbial protein supply are microbial markers such as RNA (ribonucleic acid) and DAPA (diaminopimilic acid) or isotopes ^{35}S, ^{15}N, ^{32}P etc. These techniques require ruminally and post ruminally canulated animals. Moreover, these processes are complex, tedious and difficult to practice extensively, especially under field conditions.

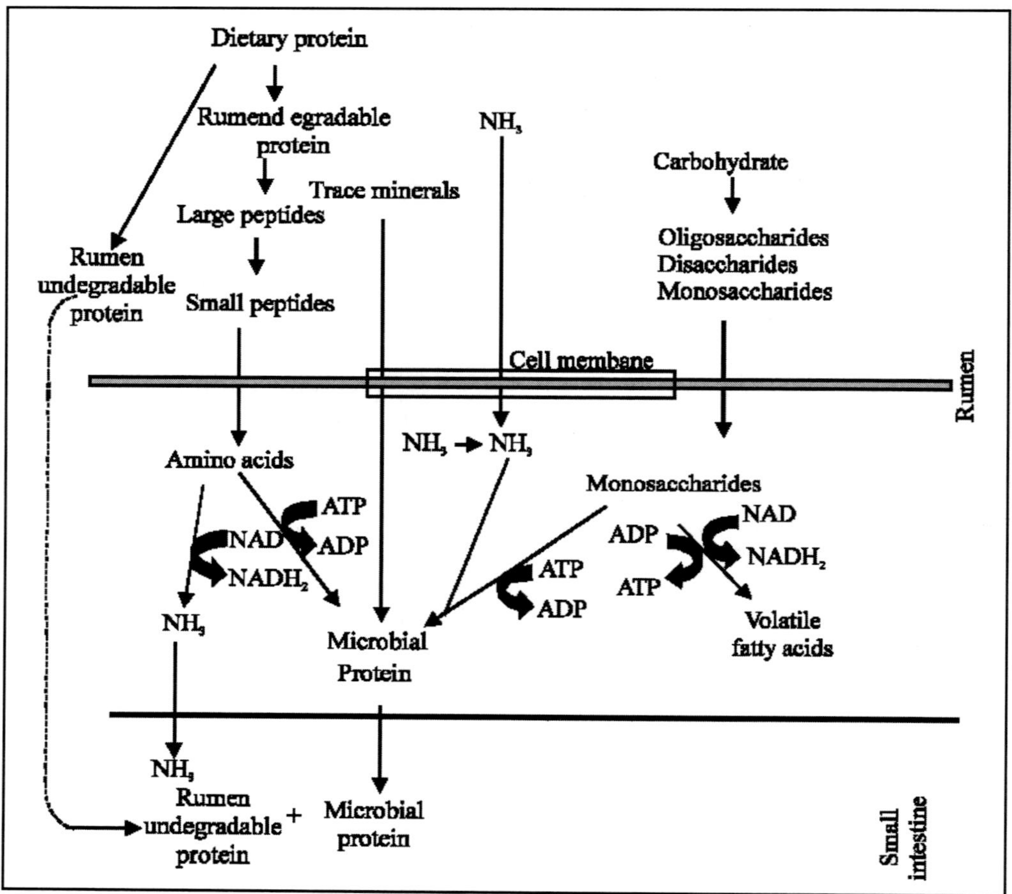

Fig 3: Nitrogen Metabolism and Microbial Protein Synthesis in Ruminant Animals

The method of using microbial purine compounds, as a marker for the rumen microbial biomass was first suggested by McDonald. Since micro-organisms have high concentration of purine containing compounds such as RNA and DNA relative to their concentrations in plants and animal cells. Therefore, any purines present in digesta can be expected to be only of microbial origin. Recently, it has been suggested that the estimation of purine derivatives that include xanthine, hypoxanthine, allantoin and uric acid, in urine could serve as a simple method of quantifying the intestine flow of microbial protein to the animal (Sumsel *et al.*, 1994).

In fact, the purine excretion method is simple, non-invasive and does not require surgical preparation of the animal. It is being used in many laboratories, e.g. for comparisons between different dietary regimes, and shows good agreement with other methods used for measuring microbial yield, including the methods based on the determination of digesta flow markers and microbial markers such as ^{15}N and ^{35}S. However, the relationships between microbial yield of purines from the rumen and urinary excretion of purine derivatives may differ between different breeds and species of ruminants.

Nucleic Acids and Degradation of Purines in Rumen and Small Intestines

Nucleic acids are high molecular weight compounds that yield a mixture of basic substances called purine and pyrimidines on acid hydrolysis. Purines have heterocylic ring structures with varying functional groups. The purine bases, adenine and guanine, are found in DNA and RNA. Rumen microorganisms have high concentrations of purine containing compounds (RNA and DNA) relative to concentrations in plants and animal cells. Of the total N present in rumen bacteria, nucleic acids and proteins comprise 13-19% and 75-85%, respectively.

The term purine derivatives (PD) refer to the sum of allantoin, uric acid, xanthine and hypoxanthine (Fig 4). All these compounds are excreted in the urine of sheep, goats, and camels but xanthine and hypoxanthine are practically absent from the urine of cattle and buffaloes (Chen and Gomes, 1992). Usually, the free nucleic acid and derivatives from diet are quickly degraded in the rumen. Thus, the majority of purine nucleosides and free bases present in the small intestine in ruminants are microbial in origin. The purine content of bacteria can vary widely due to different species and types of bacteria, energy sources and the time post feeding. There is little change in the flow of nucleic acids in the digesta. Microbial nucleic acids entering the small intestine (of which about 60-70% is RNA) are extensively degraded to mononucleotides and enzymes capable of removing the phosphatase groups from the mononucleotides to form the nucleosides. The end products of these nucleosides are adenine, guanine and ribose are absorbed in the small intestine. The digestibility of nucleic acids in the small intestine ranges from 80-90% for RNA and 75-85% for DNA in ruminants. However, the relationship between purine absorption and urinary purine excretion are curvilinear (Balcells *et al.*, 1992), probably reflecting the degree of biochemical feedback on *de novo* synthesis, which is a source of endogenous purines and the balance between the utilization of absorbed purines by tissues, and the degree of salvage of PD released within tissue cells.

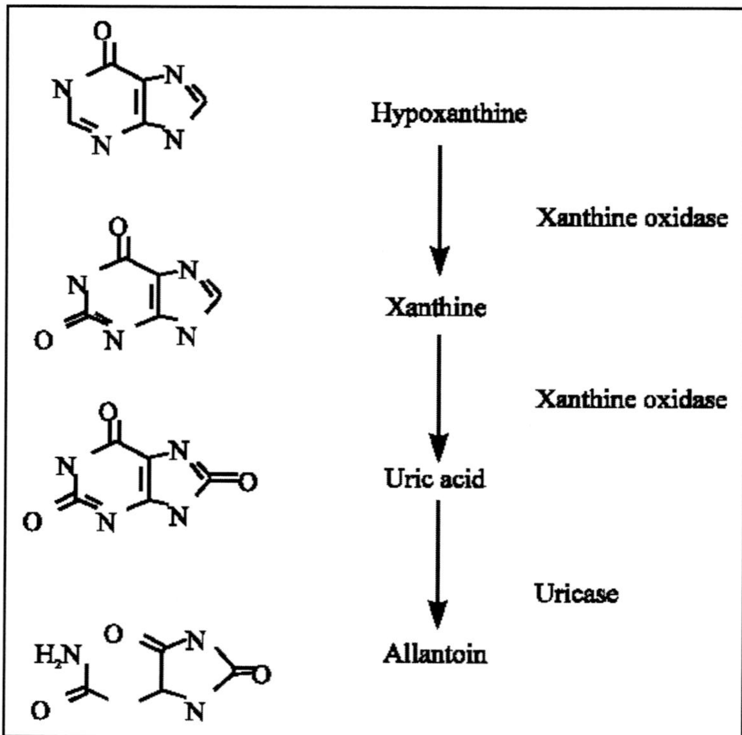

Fig 4: Chemical Structures and the Enzymes Involved in the Conversion of the Purine Derivatives

Excretion of Creatinine and Purine Derivatives

Creatinine (C) excretion is almost constant in the urine of different animal species with varying digestible organic matter intake (DOMI), when expressed on metabolic weight basis, it could be used as an internal marker, relating the ratio of PD to C concentrations to the daily output of PD (IAEA, 2000). The indicative values for C excretion (mmol/kg $W^{0.75}$/d) for cattle and buffalo are 0.8 and for sheep and camel are 0.4.

For the PD technique to be applied under farm conditions, it would be ideal if spot measurements such as spot urine, plasma or milk could be used instead of total urinary excretion. A linear correlation was observed between the ratio of allantoin-N to creatinine-N and ME intake in sheep, suggesting the possibility of using the allantoin: creatinine ratio in spot urine as the parameter. Subsequent studies were carried out in steers, dairy cows and sheep to examine the diurnal variations in spot urine measurements. More recent studies were carried out under the FAO/IAEA research programme.

A review on spot urine measurements was given by Chen *et al.* (2003). The conclusion from the studies conducted are-

- PD to creatinine ratio correlates to feed intake and intestinal flow of microbial purines, and thus can be used as an indicator of microbial protein supply
- Sufficient number of measurements should be made in order to reduce errors
- The variability of spot measurements would be greater than that based on total urine collection

Many studies have used the direct (PD/Creatinine) concentration ratio in their presentation of results (Chen and Orskov, 2003). However, data of the direct ({PD}/{creatinine}) ratio can only be compared within the same animal or among animals of the same body weight, since the daily creatinine excretion is a function of body weight. Chen *et al.* (2003) proposed to use a term " PDC index"

PDC index = $[PD]/[Creatinine]*W^{0.75}$ (1)

Where W is the body weight (kg), [PD] and [Creatinine] are PD and creatinine concentrations in urine, in mmol/L. Differences in PDC index should now be due to urinary PD output. Data of PDC index can be compared among animals of different body weights (but the same breed). By definition, the PDC index and PD excretion has following linear relationship:

PD excretion = PDC index x C (2)

Where, C is the daily creatinine excretion (mmol/kg $W^{0.75}$) for a specific breed of animals, and should have been previously measured using complete urine collection. From this equation, total daily excretion of PD can be calculated from the PDC index, yet without the need for total urine collection. The possibility of using plasma PD concentration or PD out put in milk were not considered as appropriate index of microbial protein supply (Shingfield, 2000).

Again using a modelling approach, the urinary PD excretion can be predicted in a quantitative manner that may be related to microbial N supply in cattle with reasonable confidence. Further research is however needed to understand better the source of variability in PD excretion between animal species and between animal individuals of the same species, so that more precision can be attached to the estimation of microbial protein supply based on the PD excretion.

Methods for Determination of Purine Derivatives in Urine

For sheep urine, purine derivative determination includes analysis of xanthine, hypoxanthine, uric acid and allantoin. However, for cattle urine, only the analyses of uric acid and allantoin are required since xanthine and hypoxanthine are present in trace quantities. The urine samples which have been collected previously and diluted before storage need further dilution. The next dilution should be such that the concentrations of purine derivatives in the final sample would fall within the range of the standards used in the assays (5-50 mg/L for both uric acid and allantoin). The dilution factor needed therefore depends on the feed intake, and thus the microbial protein supply of the animal. An estimate of the dilution factor required can be calculated based on the following:

- Feed intake (fresh or DM)
- Daily urine output (volume)
- Body weight (not so necessary for sheep urine)

(a) Calculation of Microbial-N (MN) Yield

MN = 32 g/ kg DOMR

Where, DOMR= Feed intake × DM content × OM content × OM digestibility x 0.65

(b) Calculation of the Equivalent Amounts of Punne Absorbed (Pa) by the Animal

Pa (mmol/ d) = MN (g N/ d) - 0 727

(c) Calculation of Total Purine Derivative Excretion (PDe mmol/d)

For sheep: PDe = 0.84 Pa + 2 (assume the endogenous contribution = 2 mmol/d)

For cattle: PDe = 0.85 Pa + 0.385 × $W^{\circ 0.75}$

where $W^{0.75}$ = metabolic body weight in kg

On average, allantoin and uric acid account for 80% and 15% of the total PD. Therefore, their daily outputs can be calculated.

(d) Calculation of the Dilution Factors

Concentration of allantoin and uric acid in the original urine can be calculated as:

= daily output ÷ volume of urine

The required dilution factor can then be calculated as:

= original concentration ÷ target concentration.

The target concentration is between 10-40 mg/L for both allantoin and uric acid.

As an example to illustrate the above calculation, for a steer of 300 kg body weight, consuming 8 kg fresh feed and excreting 10 L/d of urine, the urine requires to be diluted between 10-30 times for uric acid assay and between 40-160 times for allantoin assay. Based on the values obtained for the individual samples, a common dilution factor could be used for all urine samples. Again animals in tropics tend to have a lower excretion of purine derivatives than European animals at a given feed intake. In this case, a dilution factor of half the calculated value may be used for dilution. For example, if the above calculation indicates that a dilution of 40 time is required, then a 20 time dilution may be adopted for urine from tropical animals. If precipitates are visible in the sample of urine, place the sample in an ultrasonic bath for 20 min to break up the particles before dilution. It is strongly advised not to filter the urine using filter paper since the particles may be uric acid precipitates.

Determination of Allantoin (by colorimetric method)

Principle

This is based on the colorimetric method described by Young and Conway (1942). In this procedure, allantoin is first hydrolysed under a weak condition at 100 ± 2^{0}C, to allantoic acid which is hydrolysed to urea and glyoxylic acid in weak acid condition. The glyoxylic acid reacts with phenylhydrazine hydrochloride to produce a phenylhydrazone derivative of the acid. The product forms an unstable chromophore with potassium ferricyanide. The colour is read at 522 nm.

Reagents

- 0.05 M NaOH
- 0.01 M NaOH
- 0.05 M HCl
- 0.023 M Phenylhydrazine hydrochloride (freshly prepared before use).
- 0.05 M Potassium ferricyanide (freshly prepared before use).
- Concentrated hydrochloric acid (11.4 N) cooled at –20^{0}C for at least 20 min before use.

- Alcohol bath, 40% (v/v) alcohol, kept at –20⁰C (40% NaCl solution instead of alcohol solution may be used).
- Allantoin (*from Sigma or BDH*).

Equipments

- Spectrophotometer.
- Boiling water bath.

Standards

Prepare a 100 mg/L stock solution of allantoin. It is diluted to give working concentrations of 10, 20, 30, 40, 50 and 60 mg/L-

i) Weigh 50 mg of allantoin and transfer it into a 500 ml volumetric flask. It is dissolved in about 100 ml of 0.01 M NaOH, and topped up to volume with distilled water. The addition of NaOH is only to dissolve allantoin.

ii) 5, 10, 15, 20, 25 and 30 ml of stock solution is pipetted into 50 ml volumetric flask and volume is made up with distilled water to prepare 50 ml of the working standards containing 10, 20, 30, 40, 50 and 60 mg allantoin/L, respectively. For routine analysis, larger volume of stock solution may be prepared.

iii) Each working standard is stored as small aliquots in the freezer. Only the necessary quantities are thawed and any left over discarded. This is to ensure that fresh standards are used for each analysis run.

Preparation Before the Analysis

a) Put the alcohol bath into the freezer overnight

b) Put the concentrated HCl into freezer just before the beginning of the analysis

c) Switched on the water bath

d) Prepare fresh solution of Phenylhydrazine hydrochloride and Potassium ferricyanide on the day of analysis and kept in freezer before use. Fifty ml solution is enough for 10 samples in duplicate. Each set of 50 ml solution of Phenylhydrazine hydrochloride and Potassium ferricyanide is prepared by dissolving 0.1663g and 0.835g, respectively in 50-ml volumetric flasks. Volume is made up with distilled water.

Procedure

This procedure required critical timing of the reactions. The reading of OD of standards and samples must be made within a shortest possible time span, since OD decreased with time. Therefore, no more than 10 samples in duplicate are processed in each run. A set of standards and a blank (using distilled water) in duplicate are processed.

a) Pipette 1 ml of sample, standard or distilled water (blank) into 15 ml tube

b) Add 5 ml of distilled water

c) Add 1 ml of 0.05M NaOH and mixed well using a vortex mixer

d) Place the tubes in the boiling water bath for 7 min. Remove from the boiling water bath and cool the tubes in cold water.

e) Add 1 ml of 0.05 M HCl to each tube. The pH after addition of HCl must be in the range 2-3 (more HCl may be added if necessary and the required amount should be followed in later runs)

f) Add 1 ml of Phenylhydrazine hydrochloride. Mixed and transfer the tubes in the boiling water bath for exactly 7 min. Care must be taken not to interrupt steps 6-10.

g) Remove tubes from the boiling water and place immediately in the icy alcohol/ NaCl solution bath for several minutes. The idea is to slow down the reaction by reducing the temperature. The use of alcohol and water mixture is to achieve a temperature below zero for the liquid.

h) Pipette 3 ml of conc. HCl (operated in a fume cupboard) and 1 ml of Potassium ferricyanide. This is performed for all samples within the shortest possible time span. Start the timer for 20 minutes.

i) Mix the contents thoroughly and transfer to 4.5 ml cuvettes at room temperature

j) Read the absorbance at 522 nm after exactly 20 minutes after starting the timer. It is important that the same duration is used consistently for all samples and runs. Once started, reading as OD is done as quickly as possible (otherwise the colour will fade gradually). It is important that OD for samples and standards are read within the shortest possible time span.

Calculations

The standard curve is linear. Therefore, we can fit linear regression between the known allantoin concentrations (standards; mg/L) (X) and the corresponding OD (Y) to determine the concentration of test samples (mg/L).

Allantoin (mmol/d) = {(Allantoin mg/L x Total urine Lit) / Molecular weight of Allantoin 158}

Determination of Xanthine and Hypxanthine (by enzymatic method)

Principle

In this method, xanthine and hypoxanthine are enzymatically converted to uric acid and thus determined as uric acid which is monitored by its absorbance at 293 nm The OD at 293 nm increases after the enzyme treatment. The net increase in OD is then used for the calculation of the amount of uric acid formed based on the uric acid standard curve.

Reagents

- KH_2PO_4 buffer, 0 2M pH 7. 35, adjust pH with either H_3PO_4 or KOH
- L-histidine 4.3 mM
- Xanthine oxidase, add 25 µl of the solution to 3 ml of the buffer
- Uric acid

Equipment

- Spectrophotometer
- Water bath
- Ultrasonic bath (optional)

Standards

Prepare a 100 mg/L stock solution of uric acid (add about 100 µl of 0.6 N NaOH to 1 litre of the solution to help dissolve the uric acid). Dilute it to give working concentrations of 20, 40, 60, 80 and 100 mg/L. Store each working standard as small aliquots at -20°C.

Procedure

a) Pipette 1 ml of urine, standard or blank (distilled water) into test tubes All samples and standards are done in duplicates. Distilled water is used as the blank. Prepare two sets.

b) Add 2 5 ml phosphate buffer

c) Add 0 35 ml L-histidine solution. Mix well

d) To one set, add 150 µl of buffer. To the other set add 150 µl of the XO solution. Mix well and incubate at 37°C for 60 mm*

e) Read OD at 293 nm

**The conversion of xanthine and hypoxanthine to uric acid is complete when the OD of the samples remains constant.*

Standard Curve and Calculations

a) Use OD of the standards without XO added for the construction of uric acid standard curve. Transform both X and Y into natural logarithmic function. Fit the Ln (Y) into a linear function of Ln (X).

b) Calculate the AOD for the samples, i.e. the difference between two sets with and without XO; addition: AOD = OD with XO - OD without XO.

c) Calculate the corresponding concentration of uric acid from AOD based on the above standard curve. This increment in uric acid concentration corresponds, on a molar basis, to the sum of xanthine and hypoxanthine present in the samples.

d) Estimate the approximate contribution of OD reading from the xanthine (ODX) in the set without XO based on a pre-determined xanthine standard curve*. For practical purposes, use the concentration of xanthine plus hypoxanthine calculated in Step c to calculate ODX

e) Re-adjust the AOD (i.e. AOD2 = AOD + ODX), and repeat from step c once.

Note: The activity of xanthine oxidase can be inhibited by excess amount of substrate in the sample. The addition of L-histidine reduces this inhibition.

* Xanthine can also absorb UV at 293 nm although the absórbame is 10 times lower than uric acid at the same concentration. Hypoxanthine does not absorb UV at this wavelength. To correct for the absorbance due to xanthine in the sample set without addition of XO, we need to generate a standard UV absorbance curve for xanthine. This can be done the following way:

Use xanthine solutions of concentrations ranging from 10 to 50 mg/L, and go through the above procedure (without XO added). Fit the OD of xanthine (ODx) into a linear function of xanthine concentration. A xanthine standard may not be needed for every run.

Determination of Uric Acid (by uricase method)

Principle

Uric acid absorbs UV at 293 nm, although other compounds may also absorb at this wavelength. When samples are treated with uricase, uric acid is converted to allantoin and other compounds that do not absorb UV at 293 nm. Therefore, the reduction in OD reading after treatment with uricase is correlated with the concentration of uric acid in the sample. After treatment, the OD of the standards should be zero if the conversion is complete.

Reagents

- KH_2P0_4 buffer, 0.67 M, pH 9.4. Adjust the pH with KOH
- Uricase from porcine liver (e.g. SIGMA). Prepare an enzyme solution of 0.12 unit/ml buffer
- Uric acid

Equipment

- Spectrophotometer
- Water bath
- Ultrasonic bath (optional)

Standards

See preparation of uric acid standards. In this case, prepare standard working concentrations of 5, 10, 20, 30 and 40 mg/L.

Procedure

a) Pipette 1 ml of urine or standard or blank (distilled water) into 10 ml tubes. Mix with 2.5 ml phosphate buffer. Prepare two sets of tubes.

b) To one set, add 150 µl buffer and to the other add 150 µl of uricase solution. Mix well

c) Incubate in the water bath at 37°C for 90 mm

d) Remove from water bath, mix and transfer the solutions to cuvettes and read the OD at 293 nm. If the enzymatic conversion is complete, the OD of the standards with uricase added should be zero If not, incubate in water bath for an additional 30 min and read again.

Standard Curve and Calculation

a) Standard curve is curvilinear. When both X and Y are transformed to Ln functions, Ln (Y) is linearly correlated to Ln (X). Use the OD reading of the set without addition of uricase for the construction of standard curve. Please refer to the earlier section for construction of standard curve.

b) Calculate the net reduction in OD (AOD) for the samples due to uricase treatment AOD = OD without enzyme - OD with enzyme

c) Calculate the uric acid concentration from AOD based on the established standard equation (as in step a)

Determination of Creatinine (by colorimetric method)

Principle

The method is based on the Jaffe reaction. Creatinine reacts with picrate ion formed in alkaline medium to develop a red orange colour. The colour produced from the sample is then compared in a colorimeter with that produced by a known amount of creatinine under the same condition.

Reagents

- *Stock standard (1 mg/ ml):* 1 g of creatinine is made upto 1000 ml with 0.1 M HCl. This solution is stable at room temperature. The working standards of 10, 20, 30, 40, 50, 60, 70, 80, 90, 100 and 120 mg/L are prepared from stock solution.
- Stock reagent

a) *Stock reagent A (0.5 M NaOH):* 20 g NaOH is dissolved in distilled water and made up to 1 litre. This solution is stable for 1 year at room temperature.

b) *Stock reagent B (saturated picric acid*): 7-8 g of reagent grade picric acid is weighed and dissolved in 500 ml distilled water. The excess picric acid is allowed to remain in contact with water and bottle is shaked occasionally.

- *Working reagent (alkaline picrate):* This solution is prepared fresh daily before assay.

Procedure

a) Take 0.5 ml of sample (distilled water for blank, creatinine standard solution from 10 to 120 mg/L and diluted urine) to the test tubes in duplicate.

b) Add 6 ml of distilled water to each tube and mixed the solution thoroughly.

c) Add 1 ml of the freshly prepared working reagent (alkaline picrate) and mixed well.

d) Tubes containing solution are left at room temperature for 45 min.

e) Read the OD at 505 nm using cuvettes.

Calculation

a) Subtracted the value of the blank from the observed OD of the standards and samples to obtain corrected values.

b) Create a standard curve from observed OD of standards. The curve is linear. The linear regression analysis is done on the data to obtain *a* and *b* value of the linear equation:

OD = a + b concentration (mg/L)

c) The amount of creatinine in urine (mg/L) is calculated and corrected for dilution factor to calculate the total amount of creatinine excreted in urine.

Creatinine (mmol/d) = {(Creatinine mg/L x Total urine Lit) / (Molecular weight of creatinine 113)

Daily Excretion of Purine Derivatives

Calculate the excretion of allantoin, uric acid, and xanthine plus hypoxanthine, using SI units (mmol/ d). The proportions of individual components of the total PD are normally-

- *Sheep urine* allantoin 60-80%, uric acid 30-10%, xanthine plus hypoxanthine 10-5%. As the total excretion increases, the proportion of allantoin increases.

- *Cattle urine* allantoin 80-85%, uric acid 20-15%. Within the same animal, the proportions are relatively constant, but there seems to be variation between animals

PD (mmol/d) = Allantoin (mmol/d) + Uric acid (mmol/d)

PDC index = [PD] / [Creatinine] x $W^{0.75}$

Where $W^{0.75}$ represents the metabolic body weight (kg) of the animal. [PD] and [Creatinine] were PD and creatinine concentration in mmol/L in urine.

Microbial N Supply

Microbial nitrogen supply is calculated using endogenous PD values obtained and compared with the values derived from the model developed for European Cattle, by using equations given below:

European Cattle Y = 0.85X + (endogenous PD $W^{0.75}$)

Crossbred Cattle Y = 0.83X + (endogenous PD $W^{0.75}$)

Where $W^{0.75}$ represents the metabolic body weight (kg) of the animal. The slope of 0.85 and 0.83 in equation represents the recovery of absorbed purines as PD in urine. The component within parenthesis represents the net endogenous contribution of PD to total excretion after correction for the utilization of microbial purine by the animal. In European cattle, the endogenous contribution is taken as a constant at 0.385 mmol/kg $W^{0.75}$ per day (Verbic *et al.,* 1990).

Daily Purine Absorption

The calculation daily purine absorption (X, mmol/d) is done by using equation

$$X = (Y - 0.385 \times W^{0.75}) \div 0.85$$

where Y represents daily urinary PD (mmol/d) excretion.

Intestinal Flow of Microbial N

The following factors are used for the calculation of intestinal flow of microbial N (g N/d) from the microbial purine absorbed (X, mmol/d) as described by Chen and Gomes (1992).

i) Digestibility of microbial purines is assumed to be 0.83. This is taken as the mean digestibility value for microbial nucleic acids on observations reported in the literature.

ii) The N content of purines is 70 mg N/mmol.

iii) The ratio of purine-N: total N in mixed rumen microbes is taken as 11.6:100.

$$\text{Microbial N (g N/d)} = X\ (\text{mmol/d})/0.116 \times 0.83 \times 1000$$
$$= 0.727\ X$$

The above method for the calculation of protein supply from purine absorption presumes that the purine: protein ratio in mixed rumen microbes remains unchanged by dietary treatments (IAEA-TECDOC-1093, 1999).

However, it is also important to recognize that there are some potential sources of error. There is a relatively small loss of purine derivatives via non-renal routes that may not be a constant fraction of the total loss from the plasma pool. If the renal and non-renal losses from the body purine pools differ between different types of ruminants the accuracy of prediction of microbial flow from the rumen from urinary purines may be affected. For example, the slope of the line relating urinary purine excretion to digestible dry matter (DM) intake, which reflects this loss-partition ratio, has been shown to differ between sheep, cattle and buffaloes. This ratio may vary with type of diet and other factors (e.g. glomerular filtration rate (GFR), kidney function, digesta flow rate through the gut). The purine excretion method depends on an assumption that purines reaching the intestine are derived exclusively from rumen micro-organisms. There are reports that dietary purines can escape rumen breakdown in certain dietary components, e.g. fish meal. If materials known to escape degradation in the rumen form a major part of the diet, then results obtained with the method should be interpreted with caution.

References

Aghazadeh, A.M., Aghsaghali, A.M., Sis, N.M., Dizaji, A.A. and Shaddel, A. 2007. Estimation of microbial protein synthesis and urinary excretion of purine derivatives in sheep offered alfalfa. *Journal of Animal and Veterinary Advances* 6, 1140-1145.

Balcells, J., Parker, D.S. and Seal, C.J. 1992. Purine metabolite concentrations in portal and peripheral blood of steers, sheep and rats. *Comparative Biochemistry and Physiology* 101B, 633-636.

Chen, X. B. and Gomes, M.J. 1992. Estimation of microbial protein supply to sheep and cattle based on urinary excretion of purine derivatives- *An overview of the technical details.* Rowett Research Institue.University of Aberdeen U. K.

Chen, X.B. and Orskov, E.R. 2003. Research on urinary excretion of purine derivatives in ruminants : past and future. In: Makkar H.P.S. (eds) Development, standardization and validation of nuclear-based technologies for estimating microbial protein supply in ruminant livestock for improving productivity. IAEA-2003, Viena.

Chen, X.B., Jayasuriya, M. C. N. and Makkar, H. P. S. 2003. Measurement and application of PD : creatinine ratio in spot urine samples of ruminants. In: Makkar H.P.S. (eds) Development, standardization and validation of nuclear-based technologies for estimating microbial protein supply in ruminant livestock for improving productivity. IAEA-2003, Viena.

IAEA. (2000). *Report of IIIrd Research Co-ordination committee Meeting.* International Atomic Energy Agency, March 20-24, 2000, Malaysia.

IAEA. 1997. Estimation of rumen microbial protein production from purine derivatives in urine. International Atomic Energy Agency, *IAEA-TECDOC-945,* Vienna.

IAEA. 1999. Nuclear based technologies for estimating microbial protein supply in ruminant livestock. International Atomic Energy Agency, *IAEA-TECDOC-1093,* Vienna.

Karsli, M.A. and Russel, J.R. 2001. Effects of some dietary factors on ruminal microbial protein synthesis. *Turkish Journal of Veterinary and Animal Sciences* 25 : 681-686.

Kundu, S.S., Singh, S., Mahanta, S.K., and Pailan, G.H. 2005. *Feeding Farm Animals.* Satish Serial Publishing House, Delhi.

McDonald, P., Edwards, R.A., Greenhalgh, J.F.D. and Morghan, C.A. 1995. *Animal Nutrition.* 5th edn. Longman Scientific and Technical, New York.

❑❑❑

Chapter 21

Fodder Availability and Livestock Management in North-West Himalaya

SUDESH RADOTRA, INDER DEV AND JP SINGH
Indian Grassland and Fodder Research Institute, Jhansi-284003 (U.P.), India

Introduction

The entire Himalayan region is a vast and its diverse geographical grouping is unique and have complex ecosystem. The entire Himalayan region occupies about 18% of the total geographical area of India. The northwestern Himalaya generally represents the Himalayan tract from Kumaon in Uttar Pradesh to Chitral in Gilgit. The Himalayan ranges situated in this region exhibit a diverse climate, topography, vegetation, ecology and land use pattern. The annual average rainfall varies from 8 cm in Ladakh to over 200 cm in some parts of Himachal Pradesh and Uttar Pradesh. The vegetation may vary from the scrub in the lower hills to arctic in the Greater Himalaya.

The concept of fodder production is rather a recent introduction in the north western Himalayan region. In this region the cultivation of fodders is restricted to the lower hills. Beyond this zone, the cultivation of fodder is not practical because of adverse climatic and the land constraints. Almost all the available cultivable land, which is around 10% of the total geographical area, is used for the cultivation of cereals, pulses and oil seeds. Rangelands and forests, comprising 50 to 70% of the total geographical area are the major sources of forage for sheep, goats, rabbits, cattle and other animals. Traditionally, the pastures in sub- alpine and alpine regions have been the only source of fodder availability. Nomadic glaziers (*Khadwals, Gaddis, Gujjars, khas*) use sub-alpine and alpine pastures for rearing their livestock. Due to ever-increasing demand of animal products, the livestock population has increased manifold, thereby increasing the grazing pressure on these pastures. This has resulted in the deterioration of the grass cover as well as valuable forest species.

In the lower Himalayan region also the natural grasslands are under heavy grazing pressure and have been badly degraded. Insignificant land area is under fodder cultivation and there is least scope of increasing this.

Availability of Fodder

A) Jammu & Kashmir

In Jammu & Kashmir the net area sown is about seven lakh ha and the area under permanent pastures and grazing lands is about 1 lakh ha. The area under fodder crops has gradually increased from 9000 ha during 1955-56 to 46,000 ha during 1998-99. Most of the increase in area under fodder cultivation has taken place in Jammu division. However, the area under Oat cultivation has substantially increased in Kashmir division.

Forage Resources

The Kashmir valley always faces acute shortage of fodder. Though grazing is the major feed resource but during winter the entire landscape gets covered with snow and the animals can not graze. This crucial period lasts from November- March when paddy straw is the mainstay for survival of animals. From April- October besides grazing following forages resources are used.

1. *Aquatic weeds*: Aquatic weeds mostly *Nelumbium* and *Limneria* leaves and stems are collected from lakes and other water bodies and fed to the animals. These weeds are devoid of any nutrition and are used for maintenance.

2. *Fodder trees*: Use of fodder trees is not very common. Even the numbers of species available for fodder are few. *Celtis australis, Celix alba, Robinia pseudoacaia* are the major fodder trees. Out of these the leaves of *Robinia* and *Celix* are not relished by the livestock because of their bitter taste.

3. *Crop residues*: Rice and maize are the major crops of the area and paddy straw and maize stover is extensively used to feed the animals. Crop residue of minor crop like soybean, lentils and other pulses are also used for feeding the animals.

4. *Natural grasses*: All the pastures and grazing lands are never protected; hence there are no hay fields in the valley. However, the field bunds and interspaces of orchards produce enough quantities of hay which is conserved on trees and attics of the houses for use during lean period.

Pastures: Pastures are the major resources of animals feed in the valley. Most of these are located in sub alpine and alpine areas. All the pastures are the property of state departments of Forest and Revenue. There are well defined migratory routes to various pastures from different areas. The *Gujjars* and *Bakarwals* have grazing rights in specified areas. Grazing areas other than the pastures are lands of revenue department and village common properties. Foothills where forests exist are also used for grazing.

Due to an ever increasing demand of livestock products the grazing pressure on the alpine and sub- alpine pastures is increasing unabated.

Table1: Biomass Availability from Representative Alpine Pastures

Site	Altitudes (m)	Green herbage (t/ha)
Duksum	2500	29.1
Drobmarg	3200	10.8
Dachigam	1900	20.5
Dagwan	3400	14.0
Daomarg	3000	4.7
Gaobal	2000	10.5
Kralpathri	2500	14.6

Fodder requirement: The age old tradition of livestock rearing has influenced the sedentary system of livestock rearing of *Gujjars* who possess an unparallel wisdom in this pursuit. Their standard daily feeding regime consists of feeding green fodder @ 40, 25, 5, 8 and 30 kg per head of buffalo, cow, sheep, goat and horse, respectively. Though the green fodder is not available round the year, the dry matter fed to the animals is mostly in the form of wheat straw and is taken as an equivalent of 20% of the green weight.

Table 2: Fodder Imported in J&K State

Year	Quantity (t)
1973-74	1070
1977-78	2480
1980-81	3670
1985-86	4310
1990-91	5980
1992-93	7800
1993-94	6580
1994-95	4800
1995-96	4670
1996-97	7150
1997-98	660
1998-99	5960
1999-2000	4890

Source: Digest of Statistics, 1999-2000 Directorate of Economics & Statistics, J&K Govt.

Table: 3 Fodder Production J&K (lac tones)

Year	Green Fodder	Dry Fodder	Total
1977-78	58.95	26.52	85.47
2000-2001	60.83	23.65	84.48
2002-2003	61.42	25.10	86.52

During 2002-2003 the green fodder availability was 1.62 kg/head/day whereas dry fodder availability was 0.66 kg/head/day. Thus the total availability is much less than the National average of 5.15 kg/head/day of which 2.94 kg/head/day is green.

Source: Singh Rajiv, A.K. Srivastava & Nagendar Sharma Status of Livestock and Livestock Products in S.K.U.A.S.T., Jammu

Table 4: Native Grasses of Jammu

Species	Vernacular Name	Distribution
Themeda anaethera	Kangaroo grass	Western Himalaya extending to Afghanistan between 1000-3000 m altitude
Chrysopogon echinulatus	Love grass	North- Western Himalaya
Heteropogon contortus	Spear grass	Widely distributed in the tropics ascending in the Himalaya upto 2000m altitude
Bothriochloa intermedia	Forest Blue grass	Widely distributed upto 200m altitude in Himalaya
Dichanthium annulatum	Marvel grass	Distributed from plains upto 2000m altitude in Himalaya.
Imperata cylindrica	Cotton wool grass	From Himalaya extends to Afghanistan, Iraq and Iran. May be found upto 2500m altitude.
Pennisetum orientale	Thin Napier	Western Himalaya upto 2500m altitude
Panicum antidotale	Blue Panic grass	Distributed from plains upto 1000m altitude in Himalaya
Setaria verticillata	-	Distributed upto 2000m altitude in Himalaya
Arthraxon lancifolius	-	Hills of India, Burma, Ceylon, China. Distributed from 500m upto 2500m altitude in Himalaya

B) Himachal Pradesh

In Himachal Pradesh pastures and grasslands are the major forage resources of the state. About 12 lakh ha of land in the state is covered under pastures and grasslands. This area form 35.44% of the total area. Besides barren and unculturable land (3.30% of the area), culturable waste lands (2.58% of the area) are also put under grazing of the animals. Field bunds and terrace risers comprising 1.64% of the total area are also a source of herbage. The pasture and grasslands are at a critical stage of denudation and degradation. These areas have never been fertilized or improved. The pastures and grasslands are socially fenced and the entry of animals is stopped in June at the onset of Monsoon. The closure remains till first fortnight of October when the herbage is harvested and conserved as hay for use in winter. Immediately after the grass harvest the pastures and grasslands are again opened for indiscriminate grazing under heavy stocking till next June. Other important forage resources are weeds from the cultivated land, fodder trees and agro-wastes. The total green fodder availability from all these resources is about 3145 thousand tones green fodder while, the agro- wastes provide 2841 th. tones dry fodder. The estimates made also indicate that Chamba district has the highest fodder availability followed by Kangra and Mandi. The lowest availability of fodder has been recorded in district Hamirpur.

The gap in the fodder availability is met with large scale import of wheat straw from adjoining states of Punjab and Haryana.

Grazing Land: The total geographical area of 55,67,300 ha have been classified as permanent pastures and grazing lands in Himachal Pradesh. However it is very difficult to ascertain the actual area available for grazing. The percentage composition, herbage production and nutritive value of grazing areas from three zones are presented in the tables 5, 6a, 6b and 7.

Table 5: Botanical Composition of *Ghasnis* (Grassland) of the three Hill Zones in Himachal Pradesh.

Zone	Species	Percentage composition
Lower Hills	*Arundinella nepalensis*	07
	Bothriochloa pertusa	11
	Cynodon dectylon	06
	Chrysopogon gryllus	13
	Dichanthium annulatum	07
	Eragrostis spp	08
	Imperata cylindrical	18
	Saccharum spontaneum	22
	Other grasses like themeda	08
Middle Hills	*Agrostis stolonifera and A gigantea*	13
	Alopecurus myosuroides	7.6
	Chyrsopogon echinulatus	39.4
	Dactylis glomerata	05
	Dichanthium annulatum	04
	Eragrostis spp	03
	Festuca rubra	01
	Imprerata cylindrica	10
	Pennisetum orientale	09
	Poa pretense	02
	Trifolium repens	02
	Lotus corniculatus	01
High Hills	*Agropyron spp.*	06.3
	Agrostis stolonifera	10.5
	Andropogon ischaemum	10.5
	Alopecurus myosuroides	04.2
	Dactylis glomerata	15
	Festuca arundinacea	06.5
	Festuca rubra	09
	Lotus corniculatus	08
	Pennisetum flaccidum	13
	Poa alpinum	07
	Phleum alpinum	06
	Trifolium repens	04

Table: 6a. Herbage Production of Grasslands (t fresh weight per ha)

Month	Lower Hill	Middle Hills	High Hills
July	4.9	5.02	7.06
August	3.44	3.68	4..51
September	1.67	2.40	2.58

Table 6b. Biomass Production on Lower and Middle Hills Areas in March – June (t/ha)

Month	Fresh Weight	Dry Weight
Lower Hills		
March	3.24	0.83
April	3.10	0.83
May	2.75	0.66
June	1.76	0.44
Middle Hills		
March	1.59	0.30
April	3.08	0.47
May	6.62	2.20
June	1.36	0.35

Table 7. Nutritive Value of Pasture Herbage (% age in dry matter)

Parameter	July	August	September	Mean
		Lower Hills		
Crude protein	4.32	4.87	3.91	4.36
NDF	69.25	70.29	75.62	71.72
ADF	43.12	46.22	48.12	45.82
Calcium	1.32	1.12	1.31	1.25
Phosphorus	0.13	0.16	0.14	0.14
		Middle Hills		
Crude protein	8.27	10.22	10.13	9.54
NDF	71.32	71.76	74.27	72.45
ADF	37.33	38.46	41.12	38.97
Calcium	1.29	1.47	1.63	1.46
Phosphorus	0.11	0.17	0.16	0.15
		High Hills		
Crude protein	10.10	9.78	10.32	10.04
NDF	28.32	62.51	67.11	62.65
ADF	35.10	34.70	42.12	37.31
Calcium	0.83	0.91	1.13	0.96
Phosphorus	0.19	0.13	0.17	0.16

Fodder Trees: Trees provide a major proportion of livestock's feed in the middle hills with 84 major fodder trees and 40 shrubs. Migratory grazier do not own trees however, they use tree leaves wherever available. Trees in open forests and road sides are regularly lopped; illicit lopping is done in reserved forest as well. At times tree leaf fodder is purchased from adjoining villages during migration. The tree fodder is either fed at stopovers or during the winter stay in the outer hills. The leaf biomass production of some important trees is presented in Table 8.

Table 8. Leaf Yields of Important Fodder Trees.

Tree	Age at lopping (year)	Fresh yields (kg/ tree)
Bauhinia variegata	8-10	15-20
Dendrocalamus hamilitonii	8-10	30-40
Grewia optiva	8-10	8-10
Quercus incana	10-12	8-10
Robinia pseudoacacia	6-8	10-15
Terminalia arjuna	8-10	40-50

Forage Demand and Supply in Western Himalaya: A Balance Sheet for Himachal Pradesh

Forage Resources A0vailability

The forage available from different land resources and fodder availability is presented in Table 9 and 10. The grazing resources of the state comprise of permanent pastures/grasslands (24.45%), forest area (11.20%), barren & uncultivable land (3.19%), cultivable wastelands (2.12%) and barren lands (1.0%). These resources account for 41.0% of the total geographical area. District-wise analysis shows that Chamba has maximum availability of area (21.8%) followed by Kangra (18.0%), Kullu (15.8%) and Shimla (13.9%). Area under forest was observed maximum in Chamba and lowest in Bilaspur. The maximum area under culturable wastelands and barren lands (33.8%) was found in Kangra and lowest in Lahaul & Spiti districts (0.6%). The Kangra district has maximum area under field bunds (11.89 th. ha) followed by Shimla and Mandi.

Projected demand for fodder: The projected demand for green and dry fodder has been presented in Table 11. A perusal of the table reveals that total demand for green and dry fodder for 2002 is 10,144 and 7,890 thousand tones, respectively and for 2010 is estimated at 11,205 and 8,715 thousand tones, respectively. The total livestock population in terms of ACU have been projected about 3088 thousand and 3411 thousand by 2002 and 2010.

Table 9. Forage Resources ('000 ha) in Himachal Pradesh

Districts	**Geographical area (1)**	**Total cultivated area (2)**	**Area under fodder cultivation (3)**	**Forest area (50% of reported) (4)**	**Permanent pastures and grasslands (5)**	**Cultivable wasteland (6)**	**Barren and uncultivable land (7)**	**Area under bunds (8)**	**Total area available for fodder/ grazing (4+5+6+7+8)**	***Per cent* of total reported area**
Bilaspur	116.7	32.4	16.2	5.7	42.6	4.7	6.2	4.24	63.44	54.36
Chamba	652.8	42.4	21.2	135.8	353.3	6.0	5.2	3.24	503.54	77.14
Hamirpur	111.8	36.9	18.45	10.0	0.3	4.9	17.6	3.69	36.49	32.64
Kangra	573.9	118.9	59.45	112.2	97.1	39.9	28.1	11.89	289.19	50.39
Kinnaur	640.1	7.6	3.8	13.6	160.6	6.3	34.1	0.76	215.36	33.64
Kullu	550.3	36.1	18.05	98.7	167.6	3.7	4.7	3.61	278.31	50.57
Lahaul& Spiti	1383.5	3.1	1.55	55.0	95.4	0.7	4.4	0.31	155.81	11.26
Shimla	395.0	94.8	47.4	86.6	97.2	6.1	12.0	9.48	211.38	53.51
Mandi	513.1	73.3	36.65	56.9	200.8	9.9	16.6	7.33	291.53	56.82
Sirmaur	282.5	41.4	20.7	24.33	60.7	13.0	7.8	4.14	109.97	38.93
Solan	193.6	39.3	19.65	10.1	78.7	12.6	11.1	3.93	116.43	60.14
Una	154.0	42.1	21.05	14.4	9.6	10.3	29.7	4.21	68.21	44.29
Total	5567.3	568.3	284.15	623.3	1363.9	118.1	177.5	56.83	2339.63	42.02

Source: Statistical outline of Himachal Pradesh (2000)

Table 10. Fodder Availability in Himachal Pradesh

Districts	Green fodder Production (000 t/annum)					Dry fodder production (000 t/annum)		
	30% of grazing lands @ 5.0 t/ha	Weeds from cultivated area @ 2.5 t/ha	Cultivated fodder (30 t/ha)	Fodder trees	Total green fodder	Agrowaste	70% of grazing lands @ 1.5 t/ha	Total dry fodder
Bilaspur	76.13	6.42	9.63	2.00	94.18	277.90	66.61	344.51
Chamba	604.25	6.06	9.09	11.24	630.64	171.38	528.72	700.10
Hamirpur	43.79	7.21	10.82	1.92	63.73	232.25	38.31	270.57
Kangra	347.03	21.45	32.18	9.88	410.53	492.10	303.65	795.75
Kinnaur	258.43	0.93	1.40	11.04	271.80	9.60	226.13	235.73
Kullu	333.97	5.87	8.81	9.52	358.17	175.93	292.23	468.16
Lahaul & Spiti	186.97	0.33	0.50	23.92	211.72	3.30	163.60	166.90
Shimla	253.66	16.43	24.64	6.80	301.53	595.43	221.95	817.38
Mandi	349.84	10.50	15.75	8.72	384.81	187.90	306.11	494.01
Sirmaur	131.96	7.82	11.73	4.88	156.39	276.93	115.47	392.40
Solan	139.72	6.60	9.90	3.36	159.58	191.68	122.25	313.93
Una	81.85	7.18	10.77	2.72	102.52	227.03	71.62	298.65
Total	2807.56	96.8	145.20	96.00	3145.56	2841.43	2456.64	5298.07

Table 11. Projected Demand of Fodder (000 t/annum) for Livestock in Himachal Pradesh

Animals	2002			2010		
	ACU ('000)	Green fodder requirement	Dry fodder requirement	ACU ('000)	Green fodder requirement	Dry fodder requirement
Cattle	1663.4	5464.3	4250.0	1916.1	6294.4	4895.6
Buffaloes	914.5	3004.1	2336.5	969.2	3183.8	2476.3
Sheep	219.9	722.4	561.8	223.2	733.2	570.3
Goats	235.7	774.3	602.2	245.4	806.2	627.0
Others	54.6	179.4	139.5	57.1	187.6	145.9
Total	3088.1	10144.5	7890.0	3411.0	11205.2	8715.1

Livestock Enterprises in The North Westeren Himalaya

Animal husbandry practices vary with the agroclimatic conditions and socio-economic dimension of different regions of India. Livestock acquires special importance in mountain farming system on both ecological and socio economic point of view. They are an integral part of the farming system and a "bridge" connecting two types of land viz. forest and cultivated.

A) *Jammu & Kashmir*: The state of Jammu and Kashmir is agrarian and about 75% population is engaged in various agriculture and livestock related pursuits. Being hilly in physiographic the net sown area is very less. The livestock reared in the state comprise of sheep, goat, horses & ponies, donkeys, mules camels, pigs, yaks, cattle and buffalo. The livestock population is presented in Table 12 and 13A& 13 B.

Table 12: Livestock, Population (in Lakhs)

S. NO	Year	Sheep	Goats	Horses & Ponies	Donkey/ Mules	Camels	Pigs	Yaks	Total livestock population
1	Anantnag	3.600	0.361	0.168	0.025	-	-	0.002	4.675
2	Baramula	3.228	0.804	0.145	0.002	-	-	-	7.682
3	Budgam	1.439	0.248	0.145	0.001	-	-	-	4.170
4	Doda	3.910	1.499	0.091	0.064	-	-	0.126	9.765
5	Jammu	1.030	2.499	0.145	0.063	0.026	0.105	-	11.030
6	Kargil	1.821	0.912	0.061	0.054	-	-	0.107	3.630
7	Kathua	3.066	3.899	0.072	0.032	0.008	1.005	-	10.003
8	Kupwara	1.603	0.616	0.122	0.004	-	-	-	4.806
9	Leh	0.805	1.103	0.042	0.076	0.001	-	0.093	2.527
10	Poonch	1.944	1.075	0.044	0.00	-	-	-	4.906
11	Pulwama	1.970	0.335	0.141	0.005	-	-	-	7.569
12	Rajourai	2.409	2.448	0.101	0.051	-	0.003	-	7.514
13	Srinagr	1.171	0.206	0.141	0.001	-	-	0.002	2.726
14	Udhampur	3.699	2.090	0.087	0.045	0.002	0.005	-	10.748
	Total	31.695	18.095	1.505	0.433	0.037	1.118	0.330	91.751

Source: Digest of Statistics, 1999-2000 Directorate of Economics & Statistics, J&K Govt.

Table 13 a: Cattle and Buffalo Population of J&K State

S. No.	District	Cattle				Buffaloes			
		Male	Female	Young Stock	Total	Male	Female	Young Stock	Total
1	Anantnag	0.850	1.703	0.761	3.314	0.010	0.070	0.017	0.097
2	Baramula	1.120	1.506	0.755	3.381	0.041	0.055	0.026	0.122
3	Budgam	0.767	1.174	0.386	2.327	0.002	0.006	0.003	0.011
4	Doda	1.326	1.783	0.500	3.609	0.032	0.0351	0.081	0.464
5	Jammu	2.461	1.554	0.554	4.569	0.390	1.750	0.454	2.594
6	Kargil	0.034	0.310	0.061	0.405	0.093	0.138	0.038	0.269
7	Kathua	0.852	0.940	0.342	2.134	0.148	0.483	0.157	0.788
8	Kupwara	0.807	1.196	0.407	2.410	0.005	0.039	0.003	0.050
9	Leh	0.034	0.205	0.040	0.279	0.056	0.066	0.005	0.127
10	Poonch	0.516	0.390	0.120	1.026	0.021	0.662	0.124	0.807
11	Pulwama	0.512	1.102	0.507	2.121	0.017	0.043	0.043	0.103
12	Rajourai	0.708	0.551	0.187	1.446	0.053	0.813	0.189	1.055
13	Srinagr	0.408	0.562	0.316	1.286	0.004	0.011	0.004	0.019
14	Udhampur	1.703	1.273	0.471	3.447	0.163	0.910	0.299	1.372
	Total	12.098	14.249	5.407	31.754	1.035	5.397	1.446	7.878

Table 13 b. : No. of Sheep and Goat imported in J&K State

Year	No's
1973-74	547139
1977-78	675001
1980-81	932772
1985-86	1149855
1988-89	845156
1989-90	795548
1990-91	724839
1991-92	905860
1992-93	915713
1993-94	1075496
1994-95	964150
1995-96	929797
1996-97	943062
1997-98	1044794
1998-99	1263703
1999-2000	1242634

Source: Digest of Statistics, 1999-2000 Directorate of Economics & Statistics, J&K Govt.

Livestock rearing and management: The livestock is reared under three well recognized systems: Sedentary, Semi- migratory and migratory. Earlier the sedentary system was confined to the outer Shiwaliks only but now it is becoming common on upper reaches as well. Under this system the animals are stationed near individual households. Under semi- migratory system the animals are taken to the upper reaches, sometimes upto sub-alpine pastures for grazing during summer and these are brought back during winters. Farmers in a village hire a few persons who carry the animals for summer grazing and bring them back. These people are paid in cash or kind. Some of the well recognized tribes also practice this system. The migratory system of animal rearing is an ancient system and is very well established in the Himalaya. *Gujjars* and *Bakarwal* of Jammu and Kashmir and *Gaddis* of Himachal Pradesh are dominant tribes who rear their animals, mostly sheep and goats under this system. Large flocks of sheep and goat stay for winter months in the outer Shiwaliks and the adjoining plains and river valleys. During the month of March their upward migration starts through the middle and higher Shiwaliks and finally they reach sub- alpine pastures in the month of May. During their downward movement, the migration starts from November-December. The most preferred migratory routes adopted by *Gujjars* of Jammu and Udhampur hereafter are:

1. Lowermunda- Duru- Pahalgam- Aru- Liderwat- Kolahi
2. Lowermunda- Kulgam- Shopyan- Sedav- Aharbal- Kongwattan-Konsernag
3. Lower munda- Srinagar- kangan- Sonmarg.

Grazing in highland pastures: The migratory flocks are accompanied by both male and female members of the family but the females do not go beyond the sub alpine areas where most of the *Gujjars* have constructed their summer houses (mud huts). The women stay here till the return of flocks from alpine areas. Buffaloe and cows are not taken for grazing in alpine area and they remain here during summer. The milk collected is converted into butter, ghee (clarified butter), cheese and milk bread (Kalari) and these are regularly sent down to the villages for sale. The pastures which have larger concentration of *Gujjars* are frequently visited by the traders who purchase these products and take these down to the cities.

Every flock of sheep has 2-3 goats who act as guide for grazing. The sheep always follows the goat for foraging .Besides the goats every flock has 2-3 ferocious dogs who guard the sheep during night. In most of the pastures the graziers earmarks certain grazing areas and do not use these for grazing.

These preserved areas are opened for grazing before downward migration. The *Gujjars* believe that grazing in these areas fattens the sheep and goat and it helps in withstanding the arduous journey to the lower hills. The *Gujjars* are well aware of the traditional herbal systems of medication for animals. In case of wounds they fumigate it with burning a common grass *Stipa sibirica.* Similarly *Thymus, Aconitum, Podophylum, Plantago, Rhodoendron* etc, are used for various ailments of livestock.

The animals reared by the *Zamindars* who own cultivable lands but still possess a few buffaloes and cows. They retain sufficient milk for their consumption and the surplus is sold to *Dodhi Gujjars.* The Dodhi *Gujjars* on the contrary, are exclusively engaged in milk production. They, generally do not own large land holding and buy the fodder from *Zamindars.* Their herd strength is much larger than the *Zamindars* and they also collect milk from other producers in the area and sell this in the neighboring towns and cities

1. Dodhi (Producing and trading milk and milk products)
2. *Zamindar* (practicing arable agriculture along with livestock rearing)
3. Others (nomadic grazier and other professionals)

However, the nomadic livestock rearing is still practiced extensively by the *Bakarwals* a tribe identical to the *Gujjars.* The *Gujjars* in the districts of the J&K state, particularly in Kashmir, have also preferred to be sedentary but migratory livestock rearing is still a significant occupation there.

Herd Strength: The herd strength of the *Gujjars* who sustain on sedentary livestock rearing only prefer to possess maximum numbers of buffaloes which could provide commercially viable quantities of milk. The average number of the animals owned by a Dodhi *Gujjar* is presented in Table 14. On an average, Dodhi *Gujjar* in the area owns 10 buffaloe, 7 cows and 3 horses. The horses are used to pull the cart loaded with milk to the nearby urban areas.

Table 14. Average Herd Strength of a *Gujjar* Family

District	Occupation	Buffaloes	Cows	Sheep	Goat	Horses	Total
Udhampur	Dodhi	8	6	-	-	2	16
	Zamindar	4	2	-	-	-	6
	Others	5	6	55	45	-	111
Jammu	Dodhi	12	8	-	-	4	24
	Zamindar	6	4	-	-	-	10
	Others	3	5	30	35	-	73

On the contrary the *Zamindar Gujjars* whose basic occupation is the arable agriculture possess at an average, 5 buffaloes and 3 cows only. The *Gujjars* who are settled in professions and have to lead a sedentary life process 2-3 buffaloes and cows and the entire milk produced is used by the family. The nomadic people who form a very insignificant proportion possess on an average 4 buffaloes, 5 cows, 42 sheep and 40 goats per family. These *Gujjars* still practice the migratory system of animal rearing and their herd strength remains almost static.

C) Livestock Population of HP

Himachal Pradesh is having 5,221,586 livestock population. Among total livestock population cattle constitutes the major share of about 42%. The goat population is second highest and constitutes about 22% of the population followed by sheep (21%). Buffalo constitutes about 14.0%, while other animals constitute only 1% of the total livestock population. The district wise analysis shows that cattle population is distributed throughout the state with maximum numbers in Mandi and Kangra and lowest in Kinnaur and Lahul and Spiti District. Buffalo population decreased with the increase in elevation, while reverse trend was noted in sheep and goats.

Feeding Calenders: The information on feeding calendars of three zones of the state has been collected through Focussed Group Discussion (FGD) and are described here.

Table 15: District-wise Livestock Population in Himachal Pradesh

District	Livestock					
	Cattle	Buffaloes	Sheep	Goats	Others	Total
Bilaspur	58,825	93,637	20,319	64,163	1,322	238,266
Chamba	242,304	36,292	257,355	196,146	5,196	737,293
Hamirpur	56,397	101,310	45,861	26,746	1,855	232,169
Kangra	392,914	152,685	189,737	236,502	10,583	982,421
Kinnour	20,465	2	63,213	33,033	5,679	122,392
Kullu	162,575	351	110,144	53,478	1,353	327,901
Lahul & Spiti	8,463	-	37,377	11,156	5,367	62,901
Shimla	324,511	24,448	113,397	92,621	5,317	560,294
Mandi	463,211	118,003	199,731	209,595	5,901	996,441
Sirmour	239,946	42,950	22,453	119,645	4,836	429,830
Solan	142,550	80,155	17,005	86,831	3,080	329,621
Una	61,414	98,413	3,886	38,070	834	202,617
Total	2,173,575	748,246	1,080,456	1,167,986	51,323	5,221,586

Source: Integrated Sample Survey: Directorate of Animal Husbandry (Himachal Pradesh) 2001-2002

Feeding Calendar (ZONE-1)

January	*Grewia,Artocarpus, Dendrocalamus,Bauhinia, Albizia, Terminalia,Ficus* and dry grass	*Ficus religiosa*	*Grazing*
February			
March			
April	*Acacia, Morus, Celtis* and dry grass		
May			
June			
July	Green fodder from haylands and stall feeding		
August			
September			
October	*Grewia,Artocarpus, Dendrocalamus,Bauhinia,ta Albizia, Terminalia,Ficus* and dry grass		
November			
			Grazing
December			

Feeding calender of the zone 1 reveals that the fodder trees alongwith dry grasses are fed to the animals from October to March; however Acacia, *Morus* and *Celtis* are fed from April to June. Stall feeding is done during July to September. *Ficus religiosa* is fed to the animals through out the year. Grazing is done during November to May.

Feeding calendar (Zone-11)

January	*Grewia,Artocarpus, Dendrocalamus, Bauhinia, Terminalia,Ficus* and dry grass	*Ficus religio sa*	*Grazing*
February			
March			
April	*Morus, Celtis* and dry grass		
May			
June			
July	Green fodder from haylands and stall feeding		
August			
September			
October	*Grewia,Artocarpus, Dendrocalamus, Bauhinia, Terminalia,Ficus* and dry grass and dry grass		
November			
			Grazing
December			

By and large feeding calendar of the zone 11 is same as that of zone 1.

Feeding calendar (Zone-111)

January	*Quercus* and dry fodder	*Grazing*
February		
March		
April		
May	*Morus, Robinia* and *Celtis* and dry fodder	
June		
July	Green fodder from haylands and stall feeding	
August		
September		
October		
November		
		Grazing
December	Quercus and dry fodder	

In zone 111 *Quercus* is the most abundant fodder tree used for animal feeding during October to March. In this zone also grazing forms one of the most important component of feeding.The herders do not own the land but by way of usage they have the grazing rights, which pass on by inheritance in the family. Graziers have to renew their permits annually by paying a grazing fee of Rs. 1/- for a sheep and 1.25/- goat.

Feeding and livestock management practices: The availability of green fodder is the most important single factor on which the success of animal husbandry will depend. With the intensification of livestock development, the importance of fodder production is fully recognized. Quantitative and qualitative insufficiency of feed and fodder is one of the main reasons for poor yield of milk and other animal products in hilly areas. The various feed and fodder resources, their time of use and quantity offered per adult animal recorded in villages of Chamba Disstt. is as under:

About 92% animals received sufficient or more than sufficient amount of dry matter to meet their requirements for dry matter whereas, 8% animals were under fed as far as dry matter requirements were concerned. About 62% animals did not receive any green fodder. Tree leaves formed the source of green fodder in the study area. The important tree species used for feeding were *Bauhinia variegata, Grewia optiva, Bambusa arundinacea, Melia azedarachta,*

Acacia catechu, Pyrus, Robinia pseudoacacia, Morus alba, Quercus leucotrichophora, Ficus species, and *Celtris australis.*

Table 16 :

Feed resources	Time of use	Quantity/adult animal (kg)
Grass from grassland	July to Oct/Nov	10 – 15
Maize stover	Nov to March	8 – 15
Wheat straw	June – July and Nov to Feb	2 – 5
Hay	Nov to June	4 – 9
Weeds from wheat crop	March to May	8 – 12
Tree leaves	Nov to Feb	3 – 10
Concentrate feed	Working and lactating animals	0 – 1
Shortage of feed	Nov to June	-

References

A.O.A.C (1990). Official methods of analysis. Association of Official Analytical Chemists, 15th edition, Arlington, Virginia

Anonymous (1996) Statistical outline – HP Economics and Statistics Development Govt. of HP Shimla.

Anonymous, 2001-2002. Integrated sample survey, Directorate of Animal Husbandry (H.P), India.

Inder Dev, Bimal Misri and M.S Pathania (2006) Forage demand and supply in western Himalaya: A balance sheet for Himachal Pradesh. *Ind Jour of Animal Science* 76(a) : 720-726.

Katoch, D.C. (1998). Present status of forage resources of Himachal Pradesh. Lecture to Development Officers. Directorate of Extension Education CSK HPKV, Palampur (India).

Misri, B. and Inder Dev (1997). Traditional use of fodder trees in the Himalaya. *IGFRI, Newsletter* 4(1).

Negi, G.C. (1990). Livestock Development in Himachal Pradesh in retrospect and prospect. Report submitted to HPKV, Palampur, India.

Singh, Punjab 1996. Status of Himalayan Rangelands in India and their sustainable management 13-22. In Proc. *Rangeland and Pastoral Development in Hindu Kush-Hiamalayas* (November 5-7, 1996), Kathmandu, Nepal. (Ed. By Daniel J. Miller and Sienna R. Craig).

Technical Bulletin. 2007. Research Initiatives for Fodder Resource Development in North-Western Himalaya.

Vashist, G.D and Pathania, M.S. (2001). Requirement and availability of feed and other fodder resources for bovine population in Himachal Pradesh. *Indian Journal of Animal Sciences.* 71 (4): 367-374.

❑❑❑

Chapter 22

Use of Behaviour Knowledge to Enhance Feed Intake in Livestock

S. B. MAITY AND N. DAS
Indian Grassland and Fodder Research Institute, Jhansi-284003 (U.P.), India

Introduction

Voluntary food intake is the weight of food eaten by an animal or group of animals during a given period of time during which they have free access to food. Generally livestock are fed *ad libitum* i.e. they have free access to food all of the time and they control the feeding in such a way that it satisfies their energy requirements including fat deposition in the adult. But many situations arise when they are either over or under-fed. Cattle, sheep, pigs and broiler are prone to obesity (under *ad libitum* feeding) than other species, possibly as a result of selection by man for fast growth even if much of the weight gain is in the form of fat. Generally high level of feed intake leads to excessive fat deposition. Recently consumer preference has been shifted to lean meat, hence it is in the interest of the producer to reduce fat deposit in animal body because of its sale value and because the high energy content and low water content of adipose tissue make it very expensive to produce. In farm situation under eating may also occur in ruminants when they are offered high fibrous bulky food because such foods are digested slowly and its rate of passage in the rumen sets a limit to the rate at which more food can be eaten. When voluntary feed intake is too low, a larger portion of metabolizable energy in the food utilized for body maintenance and this causes a poor efficiency of food conversion. The problem of under eating is severe when other abdominal organs (gravid uterus) or changes (deposition of fat) compete for space or when the energy requirement is very high, as in early lactation. Food intake is depressed when food is deficient in an essential component such as protein, mineral vitamin etc. Further, when the herbage available for grazing is sparsely scattered in the pasture, the animal may not get sufficient time to eat the grasses during day time grazing.

The physiological factors and behaviour pattern which control and influence feed intake are of great importance to animal owners. Farmers often face the problem of balancing food intake with optimum nutrient requirement that is needed for maximization of animal production. The aim must be to balance food consumption with the desired level of production. The optimum level of production depend on I) relative cost of different types of feed and their nutritive value and ii) production response curve to changes to feed quantity and quality. Animal owners use behaviour knowledge while handling animal or identifying their problems. Their purposes for watching the animal behaviors are to increase productivity. No matter how nutritious the diet, it would not be converted into livestock products (milk, meat, egg) unless the animal consumes the proper quantities. If we intend to influence the eating pattern of animals, we must have acquired knowledge of ingestive behaviour and social behaviour. Optimal use of behaviour knowledge is made while handling the animal particularly under intensive management conditions where animals have no choice or preference to select food.

Ingestive Behaviour

When an animal starts to eat it is said to do so because it is in a state of hunger; when it stop eating it does so because it is satiated. These two terms have no precise physiological meaning. Animal looses energy continuously in the form of heat and metabolites and in the absence of any intake of food; it will enter into an energy deficit state. When the deficiency level crosses the threshold limit, internal sensors of the animal are activated and send signals, which in turn activate various body systems are needed for searching food. Due to initial time delay between searching and feeding, there will be negative overshoot; body energy level depleted further. With the initiation of feeding, body energy level is gradually restored to normal level and at this point internal sensing system detect surplus energy and send signals to stop eating. However some positive overshoot may occur if there is time delay in stopping intake. Food intake is continuous unless turned off by satiety signals. In animal, time lag can be considerable between the end of eating and the actual return of energy balance. However feeding is often terminated will in advance of the animal's restoration of energy to a steady state of optimum zone. In some way the amount of food or calories necessary to make up an incurred deficit is anticipated so that the animal avoids going from a state of deficiency to one of excess. The various control systems involved in the control of feeding are mentioned below.

1. *Glucoprivic control*: When blood glucose level declines, its utilization by certain cell declines; when the threshold is crossed, the negative feedback signal are originated and the animal start feeding. Peripheral receptor system for glucose is found in liver and gut, which sends moment-to-moment information to brain. Center sensor is situated in lateral hypothalamus where special receptors detect levels of blood-linked clues of nutrition such as glucose, lipids, satiety hormone and metabolic heat.

2. *Lipolytic control*: When glucose level in blood declines, growth hormone release which initiates breakdown of fat and also initiate feed intake (plasma level of growth hormone rises immediately before eating but nearly absent after feeding). On the other hand relatively high level of insulin associated with lipogenesis following eating sends stimuli to central satiety detector.

Ventromedial nucleus of the hypothalamus (Vem cells) contains the cell which constitute the CNS detector of satiation with respect to caloric balance. When satiety signals reach Vem cells, it inhibits the cells of lateral hypothalamus which then turn off the ongoing feeding behavioiur. Gastrointestinal signals do not constitute major signals - nor does a glucose repletion detection system play a major role in suppression of behaviour.

Other Factors Influencing Feed Intake

1. *Defense of body weight*: Constancy of adult body weight is the result of an innate set point for weight or perhaps a set point for total body fat store. Starved animal (for few days) eat more rapidly when offered feed. When animal under forced feeding gains weight, voluntary feed intake decreases until body weight comes back to the normal level. Set point of young animal is constantly being shifted upward.

2. *Stimulants / depressants of CNS*: CNS stimulant drugs decrease intake whereas CNS depressing drugs increase intake.

3. *Estrogen level*: During estrus feed intake decreases.

4. *Rumen fill:* Neural signals from distended stomach travel through vagus nerve to brain. Intake falls in pregnant or fat animal because of pressure on rumen wall.
5. *Stress*: Such as crowding, heat, polluted air continued over extended time spans may results in lessened feeding sufficient to produce body weight loss.
6. *Temperature effect*: High ambient temperature drastically reduce food intake and also shifts in the types of foods animal select or prefer. At low temperature food intake is increased.
7. *Sex*: Male animals grow more faster and consume more feed compared to females.
8. *General vs. specific needs*: When animals are offered balanced but insufficient diet, only energy deficiency may occur and this can be rectified by intake of feeds that are compatible with its enzyme systems and or intestinal flora. However when animals offered imbalance diet, which is deficient of particular nutrients, feeding behaviour is changed to 'search behaviour for selective diet' which will continue till the deficiencies are rectified. Search includes tasting of or sniffing at various potential edibles in the animals surrounding. The ability of animal to correct a specific mineral deficiency ,even when given free access to minerals, is poor. It is believed that nutritional deficiencies in general do not have homeostatic method of self-correction although some can affect behaviour through creation of depraved appetite (PICA)
9. *Frequency of feeding*: Body weight regulation is better accomplished by eating small but frequent meals than by eating large meals less frequently. Large meals help energy storage, usually as fat and liver glycogen.
10 *Dietary composition*: There will be high intake if feed is bulky with non-nutritive materials; animals eat more frequently rather than eating larger meals.
11. *Palatable food* : The total intake increases when foods are palatable. The word palatability has often been used to describe the overall sensory impression the animal receives from its food. However, there is no precise way of measuring palatability.
12. *Texture*: More intake is observed with pelleted food rather than pulverized food. Finely divided diets absorbs quickly from gut, which sends negative feedback to shut off feeding.

13. *Motivation*: Physiological facts alone cannot explain all about feeding behaviour - motivation plays significant role; individual animal feed less compared to group animals.

14. *Circadian rhythm*: Circadian rhythm influences feeding. Unless the lights are very long poultry do not eat during the hours of darkness. In pigs there are two peaks of feeding, one in the morning and one in the early afternoon. Some feeding also takes place at night. In cattle, a high level of feeding activity is usually observed after fresh food is offered. Peaks of feeding are observed in the morning before sunset and after midnight. Dairy cows eat few meals from around 1.00 - 6.00 hrs. Sheep shows similar circadian rhythm of feeding to cattle.

Feeding Pattern

Cattle graze forages in a characteristic way, it wraps (roll up) its tongue around herbage and tear rather biting clearly. While grazing, cattle walk slowly forward, moving their head from side to side. Animal keeps its muzzle close to the ground as it bites and tears off mouthful of grass, which it swallows without much chewing. As the animals consume the preferred herbage immediately under its muzzle it shifts slightly to the right, left or forward to select the next bite. Onset of grazing is mostly related with sunrise and most of the daylight hours occupied with grazing. Major periods of grazing are just after the dawn, mid-morning, early afternoon and near sunset. Night grazing is also seen. Day: Night Grazing ratio depends on climate - more night grazing in summer and more day grazing in winter. Area grazed by the animal depend on i) topography, ii) availability of water source, iii) breed (Zebu in tropical graze more than European cattle) and iv) inclement weather like strong winds, rain grazing time declines. Group structure during grazing also depends on weather, close grazing is practiced during winter whereas in summer distance between animals' increases. During grazing cattle constantly sniff the herbage. The odour of an exotic grass species or an area contaminated with faeces may influence grazing activity. Cattle like certain plant species compared to other species (creaming). Moreover within the same plant species, the stage of growth also influences the animal preference and it adopts progressive defoliation. Cattle react more negatively to bitter taste than to salt or sour solutions. Some plants develop chemical materials with a strong taste at late stage of growth and are then rejected by cattle. Sheep eat by holding the grasses with their lips and tongue, pulling it back into the mouth for extensive chewing. Sheep can select food well as they have narrow bite. When they are selecting actively (e.g. green material from a predominantly dead sward) the rate of eating is greatly reduced.

In group-housing environment, the feeding behaviour of an individual cow is mostly dependent on the availability of feeding space (i.e. length of trough per animal). If the space is sufficient for all animals to stand before the feeding manger, there will be no competition for eating. All the animals eat simultaneously when fresh food is offered or when they return from milking. At other times a proportion will be eating leisurely. But when the manger is insufficient for all animals to eat simultaneously, then competition for food starts among the animals, dominants animals displace submissive ones, and meals are taken in succession of brief episodes interspersed with movement to other feeding positions. Cows lower in the social dominance order are forced to eat at odd hours of day, including late at night.

Feeding bouts are defined as distinct eating periods, which are separated by longer intervals. However, the bout includes shorter breaks when animals stop eating. The duration of such pauses are generally less than 5 minutes. The interval between two feeding bout is known as inter-bout duration. A meal includes one or more feeding bouts. Two meals are separated by inter meal interval.

Drinking Behaviour

Water constitute about 70-75% of total body weight of animal; 60-65% of total water remains within the cell (intracellular) and remaining 35 to 40% remains as extracellular fluid like as intravascular fluid (blood), interstitial fluid (space between cells) and transcellur fluid (lumens of organs). Body water deficit can be corrected by three means.

- *Drinking*: domestic animal, unlike wild animals usually satisfy a significant portion of their water requirement by drinking.
- *'Free water' in succulent foods*: water can be physically squeezed out (eg. Water melon) and taken by animals.
- *Oxidation water*: chemically released during oxidative metabolism of foodstuffs.

Drinking Behavior Depends Upon

1. Age and body development: Body water content per unit of weight is higher in young than adult and thus more susceptible to desiccation by heat and low humidity. Drinking behaviour in the form of suckling or actual drinking is present immediately after birth. When the animal approaches weaning age, it learns drinking. Aged animals drink more and lose more due to I) age dependence decrease of ADH hormone and II) decrease renal function i.e. renal absorption of water.

2. Effect of diet: Carnivores generally eat meat or fish, which contain more than 70% free water by weight, so these animals can live for longer period without drinking. Water intake becomes necessity if the same foods are dried to water content of around 60%. The substitution of commercial feed to the domesticated carnivores (like dog) requires that drinking water be made readily available. On the other hand, foods of the herbivorous animals do not provide sufficient free water to meet the normal body water demand. Protein food requires water for excretion of nitrogenous wastes and ingested sodium salts and thus protein food stimulate drinking. Carbohydrate foods provide more water on oxidation than do either fat or protein and thus require lower water intake.
3. Relation between water and feed intake: A constant ratio of water intake to dry matter intake is maintained when all other factors being constant. Restriction of water causes a reduction in dry matter intake. Restriction of water in zebu cattle causes larger decrease in feed intake of zebu breeds than in that of European breeds.

Satiation for drinking: When an animal has voluntarily stopped drinking and has begun to engage in some other behaviour, we can say that water need has been satiated. For many animals, as satiation approaches, the rate of drinking slows and there are pauses which become longer and longer, until finally drinking does not occur.

Rumination

Grazing permits the animal to gather maximum amount of feed with a minimum amount of exposure in an open field situation. Rumination permits the animal to finish eating of the gathered feed at its leisure, perhaps in a less vulnerable environment. Rumination is the act of regurgitating, re-masticating and re-swallowing of previously ingested feed. During rumination a bolus of feed is regurgitated which is re-chewed with lateral grinding jaw movement and then it is re-swallowing. During rumination, animal lies with its foreleg bent under its chest; Its hind limb brought forward so as to lie partly under body. Rumination also does occur while the animal is standing or walking slowly. Rumination bout in cattle varies from 2 min to 1 hour. Normally Rumination (R): Grazing (G) ratio is 0.75:1. It depend upon quality of food, rumination time decreases with high quality fodder.

Social Behaviour

Social behaviour represents the reciprocal interaction of two or more animals and the resulting modification of individual activity. It serves purposes like species cohesion, social facilitation, ecological integration etc. For understanding the influence of social behaviour on food intake, the knowledge of the following terms are needed.

Leader: A leader is an individual who generally initiates a new activity and remains at the head of a moving group. Others respond to the initiative of a lead animal by following (follow reaction). Generally leader is in front during an orderly group progression. Leadership has no relationship with dominance-subordinate relationship. Old animal generally become leader in sheep whereas animal in middle of dominance system become leader in cattle. Sometimes leadership is also shared.

Initiator: The initiator is the animal who first to react in a way that elicits a new group activity. The new activity may or may not be similar type to that of the initiator. For example, feeding by the initiator may elicit feeding in other animals (similar response) whereas alarm call by initiator will elicit freezing behavior by others (non-similar response). When an individual initiate activities different from the remainder, it returns to the activities of the rest of the herd if others do not follow. When a neighbor begins to follow, the behavior policy of the whole group changed slowly.

Controller : Controller is an animal who determine I) whether or not a new group activity occurs and II) which activity will take place and when it will happens. Group members look to the controller before changing activity or moving as a group. Initiator may try or start new activity like group movement but will not be unsuccessful unless the controller decides to move. If required, the controller may exert force or threat and compel others to follow the new activity by exerting force or threat.

Competition: The situation when individuals seek to obtain the same resource (like food, mating partner etc). This need not involve any physical confrontation between rivals. The fastest mover may often succeed in obtaining a food item. The cleverest rather than the strongest or the fastest may be successful.

Social Facilitation

If a behavior that is initiated or increased in rate or frequency by the presence of another animal carrying out that behavior, the individual is known as social facilitator.

Allelomimetic or Allomimetic Behavior

Any behavior where animal perform the same activity with some degree of mutual stimulation and consequent coordination. Such synchronous behaviour is strengthened by activities having diurnal rhythm so that a group may feel hungry or sleepy at the same time. Group synchronization of behaviour is more likely where there is

- Adequate association
- Ability to communicate and react
- A potential for mimicking activities
- Similarity of motivational states
- Suppression of intra-species aggression

Example: When one animal starts Grazing after lying in one cattle, other also start grazing. Even a pig which is already satiated and stop feeding, may again start eating after seeing eating in another pig.

Social Dominance or Rank Order or Aggression Order

The term 'dominance' (in the context of social system) should be restricted to the phenomenon that can occur in every pair of animals in which one member can inhibit the behavior of other. When the dominance is established, then dominant animal need not be aggressive subsequently. Dominance relationships are the result of learning. Once learned, dominance relationship persists for long time. Social dominance is not usually exerted when sheep or cattle are grazing or are resting.

The Avoidance System

Avoidance order is maintained mainly through the behavior of the subordinates. They performed the most social acts, mainly submissive and flight behavior. Both 'retreat' and 'head tilt 'are mainly used in avoidance order. They are usually preceded by nose-to-nose or head-to- head threats and are usually followed by 'no reaction'. Avoidance behavior diminishes aggression in the group.

Social Order

When animals are living together for some time, there is often a clearly established hierarchy. Although it is merely the state of settled out relationship between individuals but it is not an inviolable structure. This order creates social stability and maximal group bonding. It also ensures minimal aggression, which are vital requirement in good animal husbandry. Formerly it was thought that social order was the result of social dominance through aggression. But this view is not tenable. It is now accepted that the avoidance system is mainly responsible for development of social hierarchy and generates social stabilization. If the available area is too small to permit the animals to perform subordination behavior, this will cause settle pair-relations to break down in frustration.

Dominance Hierarchies

Dominance bring both i) advantage in choice of food, sleeping site and receptive females and ii) responsibilities- dominant animal will assume lead in group defense. Within group conflicts are frequently settled by the intervention of dominate male. Dominance does not always correlate with leadership. Many species of domestic cattle are typically led by old females. The dominant cow will not be the first into the milking parlor but it will the first to a feeding area and spend more time eating and less time moving from place to place than low ranking cows. In group feeding the weak and submissive animals can not compete with aggressive and dominant animals while eating the food.

In some studies, social dominance was found to be correlated with food intake, growth, milk production or egg production. The presence of competition for feeding space exerts a major influence on feeding behaviour; rate of eating is increased when there are more animals per feeder. Many farm species shows social synchrony of feeding and it is possible that few submissive animals unable to reach the feeding manger due to the fear of dominant animals. In many cases the deprived animals compensate by increasing their feeding rate.

References

Albright, J.L., Arave C.W. and Albright A. 1997. Book: *The behaviour of Cattle*. CABI Publishing, CAB International.

Andrew, F. F. Farm animal Behaviour: An introduction to behaviour in the common farm species. Baillaiaere Tindall.

Appleby M. C. and Hughes B. O. 1997. Book: *Animal Welfare*. CABI Publishing, CAB International.

Broom D. M. and Fraser A.F. 1997. Book: *Farm Animal Behaviour and Welfare*. CABI Publishing, CAB International, 3rd Edition.

Hafez, E.S.E. 1969. Book: *The Behaviour of Domestic Animals*. Bailliere-Tindale, 3rd Edition.

Jenscen, P. 2002. The Ethology of Domestic Animals: An Introductory Text. CABI Publishing, CAB International.

❑❑❑

Chapter 23

Post Harvest Handling of Fodder Crops in Development of a Fodder Bank

P. K. PATHAK AND P. N. DWIVEDI
Indian Grassland and Fodder Research Institute, Jhansi-284003 (U.P.), India

Introduction

Drought, floods, earthquake, and cyclone are becoming common phenomena in India. During such natural calamities many efforts are done for survival of human beings and ignoring most of the times animals due to poor management of forage resources and an unorganized sector of feed and fodder resource management. Tremendous amount of different varieties of crop residues and grasses are available in India for different uses like paper making, bio-energy generation, fiber extraction, briquetting etc. and could also be well utilized for animal feeding. These feed and fodder resources are either in the form of small size like wheat straw, thick stem plants like sorghum stock or in the form of whole dried crops like paddy straw or dried grasses. All of these fodder resources are highly voluminous and having lower density varying from 40-70 kg/m^3 due to which there transportation, storage and handling are very cumbersome and expensive and therefore can not be utilized up to a maximum extent. Further, the available fodders resources may also be categorized as low-grade roughage, which could be well enriched through ammoniation (liquid ammonia or urea treatment), mixing molasses or changing their physical shape and blending them with leguminous herbaceous additives for enhancing their nutritive value and digestibility as per requirement of different groups of animals.

Need of the Day

Conserving the fodder resources scientifically, is therefore a need for animal feed sector to mitigate the losses occurred in various steps, reduce the cost of handling, storage and transportation, serve timely the needy group in case of natural disasters, to enrich the value of roughages and strengthen the economic power of the farmers. This is possible only by creating fodder bank in different zones of the country after assessing the balance sheet of fodder production. This will not only provide the timely supply of the feed to the most affected area in the country but it will also serve as community based bank from where a needy farmer can deposit his share, withdraw as per his requirement, sale his share or even he can take loan from the bank with a promise to return with interest in term of fodder only and not in cash. The bank would have to maintain its own products in the form of densified blocks, pellets, silage and hay and if agreed by the members leguminous green fodder would also be produced and procured either for sale or conserving as raw material for adding the value to the roughages. It has been observed that various post harvest operations viz. material handling, drying, storage and transport and marketing system influences the total fodder production and its fruitful utilization both in terms of quantity and quality. Being agricultural commodity they are also affected by the surrounding environment resulting into loss of quantity and nutritional quality. An urgent need to conserve the available forage resources is therefore felt for developing a fodder bank in the different regions of the country.

This paper presents briefly about status of some important unit operations involved in developing fodder bank.

Practices in Post Harvest Handling of Fodder Crops

Various unit operations including size reduction and drying are involved in the development of a fodder bank but material handling and transport and storage are the most important. Drying, size reduction and baling/ densification/pelleting etc. are practiced for specific product formulation or when crop residues are wet or green fodder needs to be dried for adding in various products.

Material Handling and Transport

It includes a number of operations that can be executed either by hand (manual) or by mechanical means or devices to convey materials and to reduce human drudgery. After harvesting the agricultural commodity are moved, transported or conveyed from place to place *e.g.* grasses after harvesting are conveyed for drying, size reduction, storage, feeding etc. Similarly wheat straw is collected after harvesting and threshing wheat crop and is either conveyed for storage, marketing or for feeding. Thus the material handling and transporting should aim to lighten the work of human labour. The important material handling equipments for most of the agricultural commodities are belt conveyer, bucket elevator, screw conveyer and pneumatic conveyer. Selection of these equipments depends upon the characteristics of the products to be handled and design details of these equipments are available mostly for grains/ granular materials (Sahay and Singh, 1994). For horizontal moving of bagged or bulk material the belt conveyer may have flat, V-shaped or some other enclosed shape. Bucket elevators with belts are employed in food industries for vertical conveyance of grains, derivatives and flours. However the screw conveyor is widely used in grain handling as well as in animal feed industries for conveying products generally for distances and could be well applied for mixing of different products. Thus, treatment of straw with urea (solid and liquid), steam, molasses etc. is possible using screw conveyor and could be used for mechanizing the process of conveying as well as for treatment of the straw. The traditional conveying equipments or a method for on farm conveyance with their conveying capacity was however not reported in the literature, which may play greater role in costing of the process.

Similarly for rural transportation carts drawn by various animals viz. bullocks, horses, camels etc. are available. According to agriculture research data book (2004) the growth of population of bullock drawn carts has shown an increasing trend from year 1966-67 to 1991-92, which indicates the importance of rural transportation by animal drawn carts specially for form operations. Bisen (1977) designed, fabricated and tested an improved bullock cart and later Deshpande and Ojha (1984) presented the theory and design of animal drawn vehicles. The design details of animal drawn carts are presented by Sahay and Singh (1994) with the detailed design of bullock cart's wheel, falleo, spoke, hub, yoke, cart frame and cart body. They said that the design criteria should suit the size of bullocks, load to be carried and the kind of the road. They however, did not define the effect of these factors in designing and also did not mention the optimum load to be carried out by the different size of bullocks.

Crop residues being voluminous in nature and have lower bulk density than grains need more space, labour, time in handling and farm operation causes more cost in the transportation. Traditionally the wheat straw in loose form is transported by trucks, tractor trolleys and by bullock carts depending upon the distance traveled. Transporting loose straw violated the traffic rules due to bulging and causes losses. Transport losses of loose straw were not reported in the literature. However, transporting bales of different products (grasses, wheat straw and paddy straw) causes 1.5 to 4.0% losses in weight and 15.5 to 43.9% reduction in volume. The study reveled that about five times of the paddy straw when densified can be transported in comparison to loose paddy straw in a single trip of truck (Pathak et al. 2008). The average weight losses during transportation of bale crop residues for 200km distance is 4.51%. In a simulation test at PAU Ludhiana a loss of 6% in the weight of paddy straw bales and 7-15% for wheat straw bales was observed which could be accounted for the conveyance loss (Annon, 2003). Therefore it could be said that type of crop transported, carrying capacity of transport modes, physical dimensions of transport modes, filling methods, time and labour required in filling and emptying, distance to be transported and physical dimensions of filled wagons/trucks/trolleys are the important factors, which affects the economics and efficiency of material handling and transport.

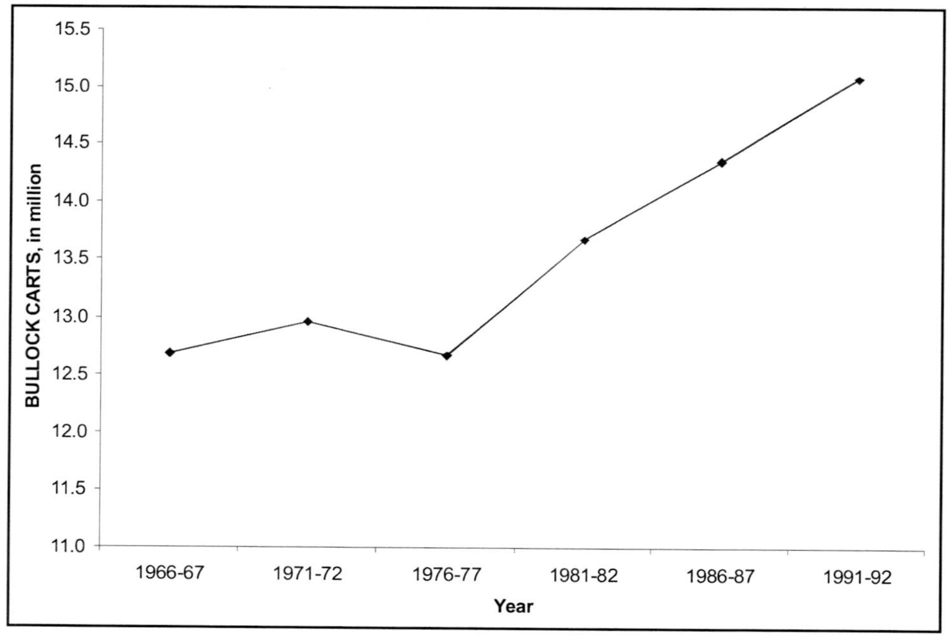

Fig: Trends in Growth of Population of Bullock Drawn Carts
Source: Agriculture research data book (2004)

Storage

Storage is a repeated phase during transit of agricultural produce and the product needs to be stored from one harvest to next thus, demanding additional carry over as safe guard, against speculation in price and market demand or against shortage and famine. The storage structures and or methods for crop residues, grasses and fodders are different to those of grains due to variation in physical characteristics. Several structures have been traditionally used for storing these crop residues, grasses and fodders and may be classified into different groups. They may be permanent and temporary structures depending upon the constructional material used, underground or above the ground depending upon ultimate use of the product, on farm or at other places depending upon their marketing and use. However, most farmers prefer to construct their own storage structures depending on availability of infrastructure and economic considerations. The method of storing crop residues, grasses and fodders also differs depending upon their size and could either be stacked in case of long straw like from rice in the north and finger millet and sorghum in the south or stored in the structures like room, 'bonga' or 'dhar' etc. as in case of wheat straw. Farmers use various others storage structures like earthen pits, wooden or cemented clamps, cemented silos, and sacks. Earthen pits could be lined and plastered with mud and cow dung and covered with jute bags. Storing straw in pits has difficulties for contamination of straw with soil and seepage of water from the side of pit especially during the rainy season and is difficult to fill or to unload and may not be suitable for digging in rocky lands. The traditional method of storage of rice straw - making stacks on the ground in open area by putting layers of straw and making the shape like the dom of mosque was reported by Mamun et.al. (2002) in Bangladesh. While surveying the existing practice of storing rice straw they also quantified the losses of straw. The straw losses were reported in three stages - during harvesting (about 8 and 10% for Boro and T. aus straw, respectively), processing (25 and 23% of Boro and T. aus straw respectively) and storage condition (about 18-20% of straw). Loss of straw in storage was mainly due to earthen evaporated gas, rat, termites, *anjona* (a reptile pest), poultry birds and excessive rainfall and advocated to improve storage system. Suggestion for improving storage condition to some extent is to build the stacks under trees to give some protection or on raised wooden platforms. Polythene, corrugated iron or coconut tree leaves can also be used to improve storage condition.

In temperate zones of India bundles of dry grass from the grassland is carried by both men and women on their back and is stored in pyramid shaped structure called *toil*. Farmers in the hilly region have evolved safe, cheap and protective methods to store hay and fodder for use in dearth period. Hay or fodder is stored in open fields from September to October for meeting needs in winter. These are piled up in a circular shape of pyramid or *telang*. A cloth made of yak hair, called *thobi* is used for covering these structures. Stones of heavy weight are used for pressing it and to keep the hay/fodder in place.

Paddy straw stack (on ground)

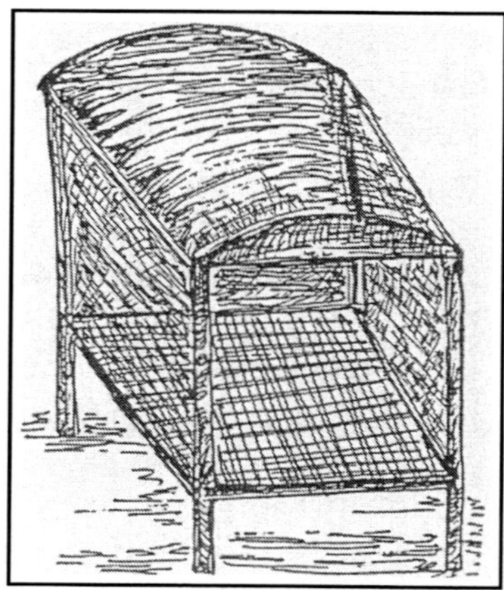

Paddy straw stack (above ground)

Bonga for wheat straw

Dhar for wheat straw and polythene covering

Some improved storage structures for crop residues, straw and grasses were also used. A improved store houses of gable type tin shed with raised slate about 1½ ft height from the ground was built (Mamun et. al. 2002) and compared the quality with that of traditional method in Bangladesh. The size of the storage was length- 22.5 ft., height- 9ft., and width- 13ft. It was found that improved storage method significantly increased nitrogen free extract, in vitro dry matter digestibility and in vitro organic matter digestibility of rice straw. No significant difference was reported in crude protein and organic matter. A cover and plinth storage structure for storing bales/blocks of five different products of grasses, wheat stubbles and paddy straw was made at IGFRI, Jhansi, which includes platforms of size 9×3 m with 0.76 m height from ground and polythene sheet of 1000 gauge for covering the stored bales (Pathak *et al* 2007). Each platform had storage capacity ranging from 1100 to 1600 bales weighing 13-32 t per storage platform depending upon the density of individual block up to a height of 3 m. Ensiling green fodder is done in underground fodder preservation cum storage structure of 900x1250 mm size having capacity 300 kg (Malaviya, 2002) which could be used for making available the product for round the year and in lean period.

Economics of the Process

Any process developed should be economically sound for success of any sector and it is true for fodder bank too. The economics of fodder bank could be evaluated depending upon individual products as well as the whole unit. In view of assessing the economics, the capacity of the bank could be established first, depending upon the size of the targeted herd to be served and category of the herd like milch animal, draft animal, small or large ruminants etc. The purpose of the bank could then be clearly identified according to the product type, time of storage and ingredients added. The process line could then be drawn specifying the different products and material and energy balance for each unit operation be given. Machine for each operation be identified with the matching capacity of previous and next operation. In between two operations suitable material handling equipments should also be identified depending upon the nature and quantity of the material to be handled in view to mechanize the process. A list of all such equipments and machine needs to be prepared with their cost. Housing of these equipments and storing the raw material and final product is also essential which could be identified as per requirements and nature and type of the materials. Such houses are to be well electrified with power backup and water supply. The expenditure on purchasing all raw materials, manpower requirement (labour and supervisor), energy requirement etc. are noted for cost assessment. Expected loans on all these expenditure are also recorded with different taxes and insurances to be paid. The expected output with an assumption of 75-80% plant efficiency is recorded for assessing the profit. All these expenditure are grouped into following headings and are used for calculating process cost, cost benefit ratio, profit, break even point, pay back period etc. The rates of banks, insurances, taxes should be taken as per applicable in the area. Cost of safety should also be taken care.

Conclusions

Many methods for crop residues handling, transport and storage are available in the country at farmers threshold varying with location, type and size of crop residues, type of need (long term or short term), animal type and economic condition of the farmer *etc.* and are modified as per the requirement by researchers and farmers. An inventory for such methods is required to compare there benefits and bottle necks. These methods could finally be selected depending upon cost economics, use and loss in quality and quantity for development of a crop residues based fodder bank.

Example of Cost Analysis: A Work Sheet

Plant Details

Plant capacity	------- T/ day (to be decided depending upon herd size, type product, ingredients etc.)
Working days /year	300 with 80% plant efficiency (assumption)
Working hours	24 in 3 shifts (decided by management depending upon machinery available and target etc.)
Operation type	Continuous/batch (depending upon process)

1. Expenditure on Equipments

a. Baling machine 1 no. cost Rs........

b. urea molasses mixture 1 no. cost Rs........

c. conveyors 1 no. cost Rs........

d. water tank 1 no. cost Rs........

e. molasses tank 1 no. cost Rs........

f. other equipments like harvester, choppers etc 1 no. cost Rs........

g. security machines/ equipment (because fire may attracts during operation)

(the no. of all these will be decided as per process line and matching capacity)

Total cost Rs........

10% of all above for installation Rs........

Total cost including installation Rs........

2. Expenditure on Fixed Capital

a. Building for – –m X – –m space @ Rs. – – –/sqm of plinth area
Rs. ....

b. Electrification @ 12.5% of (a) Rs. ....

c. Internal water supply @5% of (a) Rs. ....

Total Rs. ....

3. Total Fixed Capital Investment Including Building and Machines = 1+2 (Rs)

4. Expenditure on Annual Working Capital

(all material required for continuous working of the plant is added here) like below

a. cost of molasses – – –l/day @Rs. – – –/kg = Rs. ....

b. cost of other materials like leaf meal, concentrate, mineral mixture etc added separately =
Rs. ....

c. cost of crop residues – – –kg/day @Rs. – – –/kg = Rs. ....

d. Miscellaneous charges for handling etc. @ 10% of (a+b+c+ –)
Rs. ....

Total Rs. ....

A. Annual Fixed Cost

a. Depreciation
Machinery @10% Rs. ....
Building @ 5% Rs. ....

b. Interest on loans
60% of total fixed capital @ 18% Rs. ....
75% of working capital Rs. ....

c. Interest on remaining money
40% of total fixed capital @ 10.5% Rs. ....
25% of working capital @ 10.5% Rs. ....

d. Taxes and insurances
@ 2% of total fixed capital Rs. ....

Total Rs. ....

B. Daily Variable Cost

a. cost of raw material like straw

b. cost of binding agent (viz. molasses, bentonite, gwargum powder etc) should be given separately) Rs. ....

c. cost of water Rs. ....

d. cost of value aided ingredients other than molasses like concentrate, leaf meal, cake, mineral mixture, salt etc.

Rs. ....

e. Electric charges Rs. ....

(add the total power consumption/day from all the sources in term of kwh/day)

f. repair and maintenance @ 5% of machine cost Rs. ....

g. man power

supervisor 1 no.@ Rs.— — —- Rs. ....

Labour —no. @Rs.— — —- Rs. ....

h. miscellaneous @10% of all above Rs. ....

Total daily variable cost Rs. ....

C. Annual Variable Cost = Total daily variable cost X no. of days of operation in a year (say 300as given in plant detail) =

Rs. ....

D. Annual Total Revenue

It is calculated in terms of production unit per day (or capacity of plant) X rate of the product per kg or per unit X no. of days of operation (say 300as given in plant detail)

E. Net Profit is Calculated using Expenditure Incurred in the Process and the Return Obtained.

References

Annon. 2003. NATP final report on baling, densification and storage of grasses/ fodders/ crop residues. Submitted to Irrigated Agro-Eco System Directorate (NATP), Directorate of Maize Research, IARI (ICAR), New Delhi by Indian Grassland and Fodder Research Institute, Jhansi-284003 (U.P.).

Annon. 2004. Agricultural Research Data Book 2004. Indian Agricultural Statistics Research Institute (ICAR), New Delhi.

Bisen, H.S. 1977. Design, fabrication and testing of an improved bullock cart. Unpublished M.Tech. Thesis, IIT, Kharagpur.

Deshpande, S.D. and Ojha, T.P. 1984. Theory and design of animal drawn vehicles. Technical bulletin No. CIAE/85/46, CIAE, Bhopal.

Mamun, M.A., Akabar, M.A. and Shahjalal M. 2002. Rice straw, its quality and quantity as affected by storage systems in Bangladesh. *Pakistan Journal of Nutrition* 1, 153-155.

Mamun, M.A., Akabar, M.A., Shahjalal, M. and Pramanik, M.A.H. 2002. Rice straw losses and its impact on livestock rearing in Bangladesh. *Pakistan Journal of Nutrition* 1,179-184.

Pathak, P.K., Dwivedi, P.N. and Gupta, P.D. 2008. Comparative transport cost of loose and baled paddy straw. Paper No. APE-2008-ACP-03. Proceedings of the XLII ISAE Annual Convention and Symposium held at Central Institute of Agricultural Engineering, Bhopal, February 01-03, 2008. pp APE-1.

Pathak, P.K., Dwivedi, P.N. and Gupta, P.D. 2007. Cover and plinth storage system for stacking bales of different crop residues. *Range Management and Agroforestry* 28, 208-209.

Sahay, K.M. and Singh K.K. 1994. Unit operations of agricultural processing. Vikas Publishing House, Pvt. Ltd. New Delhi. Pp 273, 299.

Schiere, J.B. and Sewali, V.J.H. 1988. Differences in field application of straw treatment. In: Singh, Kiran and Schiere, J.B. (eds) Fibrous crop residues as animal feed, aspects of treatment, feeding, nutrient evaluation, research and extension. pp 8-16

Collins, M., Paulson, W.H., Finner, M.F., Jorgensen, N.A. and Keuler, C.R. 1987. Moisture and storage effects on dry matter and quality losses of alfalfa in round bales. Trans. of the American Society of Agricultural Engineering 30, 913-917.

Coblentz, WK, Fritz JO, and Bolsen KK. 1994a. Performance comparisons of conventional and laboratory-scale alfalfa hay bales in small haystacks. *Agronomy Journal* 86, 46–54

Coblentz W.K., Fritz, J.O., and Bolsen, K.K. 1994b. Performance comparisons of conventional and laboratory-scale alfalfa hay bales in isolated environments. *Agronomy Journal* 86, 811–819

Collins, M., Swetnam, L.D., Turner, .M., Hancock, J.N., and Shearer, S.A. 1995. Storage method effects on dry matter and quality losses of tall fescue round bales. *Journal of Production Agriculture* 8, 507-513.

Coblentz, W. K, Fritz, J.O., Bolsen, K. K. and Cochran, R.C. 1996. Quality changes in alfalfa hay during storage in bales. *Journal of Dairy Science* 79, 873–885.

Hlodversson, R. and Kaspersson, A. 1986. Nutrient losses during deterioration of hay in relation to changes in biochemical composition and microbial growth. *Animal Feed Science and Technology* 15, 149–165.

Pathak, P.S. 2003. Prospects of feed crops in India: the role of CGPRT crops. CGPRT centre, UN/ESCAP.

Singh, Punjab, Pathak, P.S., Roy, M.M. 1994. *Agroforstry Systems for Land Use*. Oxford and IBH Publishing Co. PVT. LTD. New Delhi.

Yadav, R.L. and Subba Rao, A.V.M. 2001. *Atlas of Cropping Systems in India*. PDCSR, Modipuram, Meerut-250110, India.

□□□

Chapter **24**

Technologies of Quality Silage and Hay Making from Forages

A. B. MOJUMDAR
Indian Grassland and Fodder Research Institute, Jhansi- 284 003 (U.P.), India

Introduction

India is basically an agrarian country with large livestock population making dairy and livestock industry an important subsidiary occupation of farmers. It contributes to the economy of the country by providing milk, meat, wool etc. India has recently emerged as largest producer of milk in the world but livestock productivity is very low as compared to the developed countries. Low productivity of the animals is ascribed chiefly due to inadequate supply of nutrients. Both quantitatively and qualitatively, there exists a huge gap in availability and supply of feed nutrients which is further compounded during lean and scarcity period (Singh and Mojumdar, 1992). Poor supply of nutrients to livestock during scarcity period is a matter of concern. So, there is urgent need for preservation of nutrients from forages including fodder tree leaves available during the flush period for feeding livestock during lean period (Mahanta and Pachauri, 2005).

India being a tropical monsoon bound country, large quantity of surplus *Kharif* forages are produced which are frequently excess of need. It is therefore essential to preserve the nutrients available from excess forage both during

Kharif as well as *Rabi* season at proper stage of maturity to provide nutrients during lean period. Drought of 2002-2003 affected 62% bovine population in 18 states of the country. Due to inadequate nutritional supply during scarcity, milk production was reduced by 7 % in Tamil Nadu, 8% in Maharastra and 22% in Rajasthan. Poor nutritional support during scarcity also caused adverse effects on dairy animals which included poor fertility and reproductive function, breeding cover, bovine population and draft energy. Successful animals production requires an adequate supply of nutrients throughout the year. The nutrients from forage can be preserved either as silage, hay or high temperature dehydration. There is high energy out put in high temperature dehydration and therefore can not be termed as economically feasible. Forage can be profitably preserved either as silage or hay for providing nutrients during scarcity. Besides hay and silage, attempts have also been made to preserve the nutrients particularly protein in from of leaf meal from the leguminous forages such as *Lucerne, Stylosanthes* etc and leguminous top feed, such as *Leucaena, Sesbania, Gliricidia* etc.

Ensilage-A Potential Mean of Preservation of Nutrients

Ensilage has many advantages over the other methods for preservation of nutrients, particularly from forages. Nutrients from the forages have been preserved using acids for many years since ancient time. Silage is the material produced by controlled fermentation of nutrients under an anaerobic condition. The process is referred to as ensiling. Ensiling of forage requires precautions for proper preservation of nutrients as lack of understanding of the factors associated with ensiling process may produce silage of poor quality leading to the poor animal performances. The fermentation process is governed by microorganism present in fresh herbage or by additives to maintain anaerobic conditions and discourage clostridial growth with minimum loss of nutrients. More recently, this process has been used to preserve carbohydrate rich materials, either alone or through fermentation with other materials, as well as storage of protein rich materials used as animals feeds (Machin, 1990).

Maturity of Forage Crop at Harvest

Time of harvest has a major impact on the nutritive value of silage. Protein content, available energy, daily nutrient intake and digestibility decrease with advancing crop maturity (Mojumdar and Rakib, 1980), while later cutting represents lower carbohydrate and more lignin. Since dry matter yield per unit area are lowered by early harvest, time of harvest is a compromise between nutritive value and yield. High prices for energy and protein tend to favour early harvest despite of lower dry matter yield. Table 1 indicates the optimum stage of harvest of forage crops when nutrients are properly preserved.

Table 1: Recommended Stage of Harvest of Forage Crops

Forage Crop	Stage of Harvest	Reference
Maize	50%flowering to dough stage	Thakur and Sharma, 1998; Mojumdar *et al.*, 1999
Sorghum	50%flowering to dough stage	
Oat	Boot to dough stage	
Grasses	Early flowering	

Ways of Controlling Fermentation for Quality Silage

To have the quality silage, lactic acid type of fermentation need to be encouraged. For achieving this type of desirable fermentation, the forage should be rich in water soluble sugar (more than 5% on dry matter basis). On the other hand, clostridial type of fermentation spoils the quality of silage. There are several ways in which clostridial type of fermentation can be checked. One of the commonly practice is to increase the dry matter of the herbage, prevention of entry of oxygen and increase in sugar content. Promotion of lactic acid production is important path way for controlling clostridial growth. Other ways of controlling clostridial fermentation is by addition of preservatives and additives.

Advantages/ Disadvantages of Ensiling Over Hay

The nutrients in the forages can be preserved as silage or hay. Silage has many advantages over hay and other methods of preservations, chiefly because of less loss of essential nutrients.

- Lower field losses particularly of leafy portion which is relatively rich in protein and minerals
- Lower probability of rain damage and thus leaching of nutrients
- Storage over longer period, if properly packed under optimal ensiling conditions
- Provide more succulent feed to livestock
- Ideal technology for preserving nutrients in temperate conditions
- Less dependence over weather conditions, particularly availability of sun lights

Disadvantages of Silage Making

- Being mechanized technology, requires considerable capital investment
- Limits the preservation of high CP containing forages such as leguminous fodders e.g. cowpea, berseem, lucerne etc
- Losses of nutrients can be high if not properly preserved with exclusion of air and water. Clostridial fermentation spoils the quality of silage and its feeding value. Formation of butyric acid makes silage unpalatable.
- High moisture silage leads to greater seepage losses
- Less marketable
- Voluntary intake by animal is a limiting factor if acid production is high (Demarquilly, 1973)
- Must be fed as soon as possible after removal from silo to avoid secondary fermentation
- Chopping of forage is must otherwise good packing of silo is not possible and allows the air to be trapped which in turn allows mould formation
- Poor technical knowledge of storage

Ways to Reduce Nutrient Losses

Reduction in the nutritive value of silage fermentation with respiratory losses, silage heating and clostridial fermentation is minimized by limiting air and moisture contact with silage (Bolsen et al, 1996). Minimizing oxygen exposure to silage is essential for obtaining good quality silage. Air allows the respiration process to continue using soluble carbohydrates essential for acid production, which generates heat and increases the temperature. Process of respiration results in loss of nutritionally valuable dry matter and energy. Air exposure during preservation tends to progress towards mould formation and leading to rottened silage. The increase in the temperature of silage as a result of heating also reduces its palatability when fed to livestock (Pelz and Hoffman, 1997). Uniformly compacted silage and properly sealing aid in air exclusion.

Dry matter concentration of the forages plays a vital role in minimizing the nutrient losses during ensilage. High moisture silage leads to effluent losses. Another disadvantage of high moisture silage is a clostridial fermentation, which leads to excessive dry matter losses, high butyric acid concentration and lower nutrient intake (Henderson and McDonald, 1971). Harvesting forages at the proper stage of the growth and dry mater content maximizes the nutritive value of silage (Mojumdar and Rekib, 1980). Wilting of high moisture forage to 30% dry matter is a safe way, which inhibits the clostridial fermentation. Clostridia bacteria degrade sugars and also convert lactic acid to butyric acid and elevate ammonia concentration and thus causing pH to rise. They also break down protein to amines. Thus, clostridial fermentation has an undesirable effect on the nutrient leading to their decomposition to undesirable end products, dry matter loss and reduced palatability (Nikolic and Jovanovic, 1986).

The heat caused during fermentation plays vital role in preservation of nutrients. Higher temperature silage (100^{0}F) has been found to be poor in quality. The over heated silage produced at a temperature above 120^{0}F have been found to be resulting into heat damaged protein having brown to dark brown colour with a tobacco type fowl smell. Protein of heat-damaged silage forms a complex with carbohydrates and is not digestible. The part of protein and energy is not available to livestock and resulting in to lower DCP and TDN values (Rodriguez *et al,* 1985). Higher temperature also increases aerobic spoilage and reduces stability of silage.

Water soluble carbohydrate content of forages constitutes the primary nutrient that is fermented to lactic acid and acetic acid by *Lactobacillus* bacteria to produce a low pH (4.5) and stable silage. Maize, sorghum, oat and other cereal fodders usually have higher soluble sugar concentration and a good stable silage having lactic acid as percent of total acid to the tune of 60 is obtained while legume forages having low soluble sugar content are not repeated to produce stable and good quality silage chiefly because of low lactic acid production mostly below 3% of dry matter (Singh and Rekib, 1986a). Carbohydrates in the forages may be naturally occurring or may be added as a separate ingredient such as molasses obtained as sugar industry by-products (Evers and Carrell, 1998), which act as a fermentable substrate. Relatively more lactic acid is produced from glucose present in the ensiling forage than fructose. Hemi-cellulose after acid hydrolysis produces pentoses, which is then fermented to lactic acid and acetic acid (Hendersan *et al,* 1979). An index has been developed on the basis of proportion of lactic acid, acetic acid and butyric acid for the grading of silage quality, which is chiefly governed, by the amount of lactic acid (Zimmer, 1971). Besides carbohydrates, the protein content of the ensiling forage plays an important role in determining the quality and feeding value of silage. High CP content in the leguminous forages leads to ammonia production during fermentation leading to rise in pH (5 and above), buffering action and temperature. The high moisture content (more than 75%) causes more protein loss due to proteolysis by clostridia. Nitrates present in the plant are reduced to nitrites which in turn release ammonia (Singh *et al,* 1983).

Additives for Effective Ensiling of Nutrients

Various types of additives can be used to improve or inhibit the fermentation or supplement nutrients needed by ruminants to be fed as silage. Adding acids such as sulfuric acid, formic acid and other acids decreases the pH of the forage ensiled and helps to preserve it. But corrosiveness of these acids is the limiting factor for their use. Propionic acid reduces aerobic deterioration, heating and mould formation at the top of silage layers. The use of acids has also financial implications for the economic viability of their use. Formaldehyde has been used for effective preservation as it inhibits the fermentation. Addition of formaldehyde @ 5.0 litre per ton of fresh maize fodder has been found to produce good quality silage with higher feeding value when fed to cross-bred calves (Verma and Mojumdar, 1984). Addition of formaldehyde has also been reported to improve the DMI when fed to ruminants (Barry *et al.*, 1973). Forages with marginal concentration of soluble carbohydrate may benefit from enzymes such as cellulase, pectinase and amylase that can break down complex plant structural carbohydrates such as cellulose, pectin and starch present in forage to simple sugar which then can be fermented to lactic acid. McHan (1986) reported increase in water soluble carbohydrate in cellulase and hemi-cellulase treated silage. An increase in soluble sugar content resulted in more lactic acid (10%) and lower ammonia-N (less than 6% of total nitrogen) and pH 4.5 in enzyme treated silage (Van Vauren *et al.*, 1989). Commercial bacterial inoculants have also been used in developed countries which increase the rate of lactic acid fermentation and produce stable silage but such system may not be profitable.

Carbohydrate sources such as molasses, whey, yeast and other energy rich ingredients, have also been used as additives to increase the fermentation and feeding value of silage. Most commonly used carbohydrate sources are molasses which is used to add fermentable sugars to forage low in sugar. It can be added @ 5-10% depending upon the sugar content of ensiling forage. Urea is the most important source of non-protein nitrogen used to elevate CP content of cereal forage silage low in protein. Addition of urea @ 0.5-1.0% has been found to increase CP content and lactic acid content of silage (Verma *et al.*, 1982, Singh and Rekib, 1986b). Nutritive value, particularly CP content of graminaceous forage silage can be improved by mixing legumes forages such as cowpea, berseem and *Leucaena leucocephela* leaves (Verma and Mojumdar, 1985, Mojumdar *et al.*, 1980).

Nutrient Losses During Ensilage

Generally loss of dry matter, carotenes, carbohydrate and proteins occur due to respiration, fermentation and aerobic deterioration. The other losses of nutrients arise from field, harvesting and affluent losses. The field losses may occur due to shattering of leaves and other nutritious portions because of poor harvesting managements. The extent of loss in dry matter depends on the time at which the forage is ensiled. Over the period of 48 hours, losses of DM may occur which may be as high as 6.4 percent after 5 days. Loss of carbohydrates and protein also occur due to respiration and proteolysis by plant enzymes. Studies have been revealed that the loss of nutrients during ensilage was drastically minimized with increasing dry matter content of ensiling material (Honig, 1968). The fermentation losses chiefly depend upon the moisture content. The clostridial type fermentation is deleterious for most of the nutrients. The clostridia are responsible for the loss of protein. Losses thus are dependent upon pH, moisture content of siling material and type of micro-organism growing during course of fermentation. Forages of low dry matter content (less than 22.9%) leads to effluent production with a considerable loss of nutrients (Castle and Watson, 1973). Haigh (1999) observed positive relationship between moisture content of ensiled forage and effluent production. This could be minimized by wilting. The high moisture herbages preserved with the use of additives such as formic acid, sulfuric acid and formaldehyde to inhibit or manipulate the fermentation has been found to increase effluent production. After the silo is opened for feeding to livestock, the silage surface is exposed to air and thus leading to aerobic secondary fermentation. During aerobic degradation, the temperature and pH rises while lactic acid content reduces. Loss of DM and nitrogenous substances occur due to escape of volatile fatty acid, lactic acid and ammonia. Loss of nutrients arising out of secondary fermentation could be 0-15% and could be minimized by management practices such as use of cover, propionic acid etc (Wyss, 2000). The Table 2 below summaries the losses of nutrients during preservation of herbages as silage.

Table 2: Nutritive Losses During Ensilage

Biological process	Judgment	Approx loss (%)
Respiration	Unavoidable	1-2
Fermentation	Unavoidable	1-4
Effluent	Mutual	5-7
Pre-wilting	Unavoidable	2-5
Secondary fermentation	Avoidable	0-5
Aerobic transformation	Avoidable	0-15
Total losses		7-35

Source: Zimmer 1971

Silage Quality

Silage quality is determined by mainly the odour, physical state, pH, ammonia nitrogen, volatile acids and lactic acid.

It should be of pleasant smell and semi dry in nature. It should be of green colour.

pH: 4.5 to 5.0

Ammonnical nitrogen of total N: less than 12% of total N

Lactic acid: Above 3%

Acetic acid: up to 5%

Butyric acid: less than 0.2%

Nutritive Value of Silage

Voluntary intakes of silage has been a limiting factor and lower than that of green forage (Pachauri and Mojumdar, 1994) which is more prevalent with high moisture silage. The main reason of low intake could be ascribed to low pH and high lactic acid content. Wilting has been reported to increase intake of silage considerably (Singh and Rekib, 1986b). Use of formic acid as additive has been reported to increase intake, body weight gain as well as milk production (Waldo and Derbyshire, 1971; Castle and Watson, 1970). Nutritive value of cereal forages can be improved by supplementing them with 0.5% urea or mixing with either leguminous forages such as cowpea or berseem (Upadhayay *et al.*, 1983, Mojumdar *et al.*, 1980) or with top feeds such as *Leucaena leucocephala* (Upadhayay and Mojumdar, 1991). Nutritive values of promising silages are presented in Table 3.

Table 3: Nutritive Value of Silages

Forage material	Ratio (%)	CP (%)	DM (%)	DCP (%)	TDN (%)	DMI (% b.wt.)
Sorghum PC-6	-	4.94	35.0	5.98	56.6	2.25
Hy. Napier NB -21	-	4.30	51.5	0.9	54.1	2.67
Hy. Napier + sesbania	3:1	7.10	53.2	3.9	54.7	2.76
Maize + formaldehyde	0.5%	10.10	56.9	5.9	55.5	2.56
Maize + C. ciliaris	3:1	6.10	46.3	1.8	43.7	1.97
Maize + 1% urea	-	12.5	43.7	7.7	45.4	1.30
Maize + cowpea	1:1	12.6	69.4	8.4	64.1	2.86
Berseem + paddy straw	1:5	7.40	47.6	3.0	46.2	2.42
Berseem + dry grass	1:5	4.00	52.6	1.4	54.4	2.37
Berseem + sorghum straw	1:5	5.70	59.5	3.8	56.4	2.56
Berseem + oat	1:1	14.3	51.8	6.7	56.4	2.88
Lucerne + wheat straw	2:1	8.10	53.1	5.1	50.9	2.95

Source: Rekib and Singh, 1988

Preservation in Form of Hay

Another way of preserving nutrients is practiced in the form of hay. The principle of hay making is to preserve nutritional value of forages through drying it to a level at which the activity of microbial decomposers is inhibited. Forages can be harvested at the stage of proper nutritive value and be preserved as hay for feeding it during lean period. A moisture content of 10-12% is optimum level for halting the microbial activity (Jones and Harris, 1980). The hay stored at 20% moisture level, may favour mould growth. It also increases hay temperature and may lead to spontaneous combustion which leads to loss of nutrients. The growth of microorganisms can be prevented with the help of preservatives such as propionic acid @ 10kg/ton. However, the use of organic acid is uneconomical (Knapp *et al.*, 1976). In the country like India where ample sunlight is available, hay making is popular and economical way of preservation. The thin stemmed forage crops such as oat, lucerne, berseem, cowpea, clovers and grasses are highly suitable for hay making.

Factor Affecting the Quality of Hay

1. Plant species
2. Stage of harvest
3. Leaf stem ratio
4. Chemical composition
5. Deterioration during storage

Nutrient Losses During Hay Making

The field losses during haymaking include respiration, leaching, shattering of leaves and mould growth. About 15-40% of dry matter loss, mostly as leaves was reported when stylo is sun dried for hay production (Amodu, 2004). A precautionary handling of forages particularly legumes is necessary to prevent the shattering of leaves. The leaves are rich in CP, carotenes and minerals. The loss of nutrient is much high during field curing in which the forage is spread on fields. The respiratory enzymes continue to function till moisture in the plant is available and the soluble sugar is degraded to carbon dioxide. Similarly proteases act on the protein of the forage after harvesting and cause the loss of nitrogen. Carotenes are most adversely affected during hay making. Dry matter losses are detailed in the Table 4.

Table 4: Dry Matter Losses in Hay Preparation

Particulars	DM loss (%)
Respiration during wilting	4-15
Leaf shattering (legumes)	5-15
Leaf shattering (grasses)	2-5
Rainfall and damage	10-17
Heating	4.5-5.5

Source: Mc Donald and Clarke, 1987

Thus, the loss of nutrient is variable in various processes of hay making. There is more loss in dry matter if hay is stored in open and humid conditions while very little (around 5%) if stored indoor and dry conditions. Digestibility of protein of hay prepared at high temperature is lowered owing to deleterious effect of heat brought about by reaction of amino acids with carbohydrates (Butler and Baily, 1973). Loss in CP, carbohydrates and vitamins as well as digestibility has been reported during hay making (Madan Mohan *et al.*, 1977).

Preservation in Form of Leaf Meal

The use of top feed as component of ruminant's diet is a widespread practice, particularly in tropical regions. For last two decades, efforts have been initiated to use the leaf meal prepared out of top feeds and leguminous forages as an animal feed stuff because of high concentration of protein of high biological value and other nutrients such as carotene and minerals. There exist a big deficit of concentrate in the country to the tune of 60% (Pathak and Mojumdar, 2002) and this deficit can be partially bridged by replacing the concentrates feeds by such leaf meals. Supplementation of leaf meal can meet the requirement of livestock particularly during dry periods when availability of other green protein rich fodders is scarce. Leaf meal being prepared from legume forages such as *stylosanthes, lucerne* and top feed rich in protein such as *leucaena, sesbania, gliricidia* etc. could act as a replacer of feed concentrates for livestock (Mahanta *et al.*, 1999, Mojumdar *et al.*, 2003). Leaf meal production technology is simple as well as profitable enterprise for the farmers. The outstanding features of these leaf meals are their high CP content (Table 5) and essential amino acid contents (Table 6).

Table 5. CP Content (% DM basis) in Leaf Meals

Leaf meal	CP contents
Leucaene leucocephala	18-21
Sesbania aegyptiaca	20-58
Sesbania sesban	18.00
Prosopis cineraria	12.36
Albizia	15-16
Zizyphus numalaria	13-16
Stylosanthes species	12-18
Cassaava	23.9
Lucerne	20.61

Source: Pathak *et al.*, 2005

Table 6. Eessential Amino Acid Composition (mg/g DM) of Leaf Meals

Amino acid	*Leucaena leucocephala*	*Stylo. scabra*	*S. guianensis*	*S. seabrana*
Lysine	1.76	0.74	0.80	0.71
Leucine	2.17	1.06	1.16	1.04
Iso-leucine	1.37	0.56	0.64	0.58
Theronine	1.21	0.64	0.69	0.61
Methionine	0.46	0.2	0.22	0.20
Cysteine	0.20	0.18	0.16	0.19
Valine	1.44	0.73	0.81	0.72
Histidine	0.54	0.31	0.35	0.31
Arginine	1.51	0.81	0.93	0.80

Source: D' Mello and Fraser, 1981; Changjun, 2004

Leucaena leaf meal (LLM) was safely incorporated in diets of growing lambs by replacing mustard cake to meet up to 45% to the protein requirement without affecting nutrient and energy utilization and feed cost per kg of weight gain was 19% less than the control diet (Mahanta *et al.*, 1999). Dry matter intake of *Leucaena* and other leguminous leaf meal by the ruminants were found comparable with oil cakes. Replacements were also done for groundnut cake with LLM in diet of lactating buffaloes (Garg and Kumar, 1994) and goats (Rai et al, 1994). However, lambs fed iso-nitrogenus complete diet in which LLM replaced 15% of GNC resulted into higher DM intake (Vahidulla and Reddy, 1987). Higher effective degradability of DM and OM as well as UDP (144.62 vs 76.89/ kg DM) in LLM than on cotton seed cake has been observed (Barman and Rai, 2003, Garcia et al, 1996). Use of LLM has been found to contain condensed tannin (1.44%) which is beneficial in term of providing by-pass protein to ruminants (Barman and Rai, 2003). Tangendjaja and Wina (2001) reported higher *in vivo* CP digestibility (77.70%) in sheep fed in LLM. Economical benefits of leaf meal supplementation and body weight gain depends upon intake of nutrients. Rai *et al.* (1994) observed that it was economically advantageous to replace GNC or sunflower cake with LLM. *Sesbania grandiflora* leaf meal was used to completely replace GNC nitrogen in complete diets for Nellore rams without any adverse effect on N-retention and nutritive value (Balapeeraiah *et al*, 1993). Stylo meal prepared from *S. sebrana* species has been found to be rich in amino acids particularly lysine, cysteine and methonine and can be incorporated in the diet of livestock including poultry. Stylo meal was incorporated in the diets of sheep (Mojumdar et al, 2004) replacing concentrate mixture at 50 and 100% level on iso-nitrogenous basis and was found to increase ($P < 0.05$) dry matter digestibility. It could therefore be inferred that leaf meal provides sufficiently enough nutrients for replacement of concentrate mixture in the ration of ruminants with reduced feed cost.

Thus, preservation of nutrients from surplus forages available during flush growth period will pave the way for judicious utilization during lean period and also minimize the feed deficiency. This process will help in sustaining the animal production throughout the year. Furthermore, numerous leguminous forages and top feeds contain high level of important nutrients such as protein, minerals and carotene etc. The leaf meal prepared out of such materials could act as replacer of concentrate feeds for livestock in a profitable way.

References

Amodu, 2004. Stylosanthes; A promising legume in Africa. In "ACIAR Report (S. Chakrabarty, ed.). Canberra, Australia. pp 225: 233

Barry, T.N, Fennesy, P. F. and Duncans, S.J. 1973. Effect of formaldehyde treatment on the nutritive value of silage. *New Zealand Journal of Agriculture Research* 16, 64-68.

Bolsen, K.K., Ashbell, G. and Weinberg, Z.G. 1996. Silage fermentation and silage additives review. *Asian Australasian Journal of Animal Science* 9, 483- 493.

Butler, G.W and Bailey, R.W. 1973. *Chemistry and Biochemistry of Herbage*. Academic Press, New York.

Castle, M.E. and Watson, J.N. 1993. The relationship between the dry matter content of herbage for silage making and effluent production. *Journal of British Grassland Society* 28, 135-138.

Demarquilly, C. 1973. Chemical composition, fermentation characteristic, digestibility and voluntary intake of forage silage. *Annales de Zootechnic*, 22 : 1-10.

Evers, D.J. and Carroll, D.J. 1998. Ensiling salt-preserved waste with grass straw and molasses. *Animal Feed Science and Technology* 61, 241-249.

Haigh, P.M. 1999. Effluent production from grass silage treated with additives and made in large-scale silos. *Grass and Forage Science* 54, 208-218.

Honig, H. 1968. *Die volkenroder bilangaulage. Seigene Futter*, 14 : 304-315.

Hendersan, A.R. and Mc Donald, P. 1971. Effect of formic acid and the fermentation of grass of low dry matter content. *Journal of Science Food and Agriculture* 22, 152-163.

Henderson, A.R., Ewart, J.M and Robertson, G.M. 1978. Studies on the aerobic stability of commercial silage. *Journal of Science Food and Agriculture* 30, 223-228.

Jones, L. and Harris, C.E. 1980. Plant and swath limits to drying. In: Forage Conservation, occasional Symp. No 11, Brit Grass land Soc. (Thomas, C ed.). London, pp 53-60.

Knapp, W.R., Holt, D.A. and Lechtenberg, V.L. 1976. Propionic acid as a hay preservative. *Agronomy Journal* 68, 120-123.

Madan Mohan, Bhatia, D.R. and Patnayak, B.C. 1977. A note on chemical composition and nutritive value of *Pennesetum typhoides* hay. *Indian Journal of Animal Sciences* 47, 504.

Machin, D.H. 1990. Alternative feeds for outdoor pigs. In: Principles and Practices (Stark, B.A., Maduin, D.H and Wilikinson, J.M., eds.). Charcombe Publication, Marlow, U.K. pp 103-114.

Mahanta, S.K. and Pachauri, V.C. 2005. Nutritional evaluation of two promising varieties of forage sorghum fed as silage. *Asian-Australian Journal of Animal Science* 18, 1715-1720.

McHan, F. 1996. Cellulose treated coastal bermuda grass silage and production of soluble carbohydrates, silage acids and digestibility. *Journal of Dairy Science* 69, 1837-1846.

McDonald, A.D. and Clarke, E.A. 1987. Water and quality loss during field drying of hay. *Advance Agronomy* 41, 407-438.

Mojumdar, A.B. and Rekib, A. 1980. Physical and chemical changes in laboratory silage of *Kharif* cereal forages. *Forage Research* 6, 75-82.

Mojumdar, A.B., and Verma, N.C. and Rekib, A. 1980. Ensilage of oat and berseem with dry sehima grass (*Sehima nervosum*). *Forage Research* 6, 221-224.

Nikolic, J.A. and Juvanovic, M. 1986. Some properties of apple pomace ensiled with and without additives. *Animal Feed Science and Technology* 15, 57-67.

Pachauri, V.C. and Mojumdar, A.B. 1994. Nutritional evaluation of two rations of green and ensiled sorghum (PC-6) as basal roughage in milch cows. *Indian Journal of Dairy Science* 47, 610-611.

Pathak, P.S. and Mojumdar, A.B. 2002. Current status of field and forage: Prospects and constraints. Proceed. UPCAR Conference, Lucknow (Oct. 23, 2002). pp 1-35.

Pelz, D. and Hoffman, S. 1997. Dewatering, compacting and ensilaging of spent grains. *Brauwelt*, 15 : 436-439.

Redriguez, A., Riley, J.A and Thorpe, W. 1985. Animal performance and physiological disturbances in sheep fed diets based on ensiled sisal pulp. *Tropical Animal Production* 10, 23-31.

Singh, Panjab and Mojumdar, A.B. 1992. Current status of feeds and forages in management of livestock in India. *Agric. Situation in India*, 28 : 375-382.

Singh, A.P. and Rakib, A. 1986 b. Feeding value of grass silage prepared with different levels of urea. *Indian Journal of Animal Sciences* 56, 572-575.

Singh, A.P. and Rakib, A. 1986a. Effect of wilting and mixing berseem and oat fodder on fermentation pattern of silage. *Indian Journal of Animal Sciences* 56, 85-88.

Singh, A.P., Prasad, J. and Rakib, A. 1983. Note on high nitrate paragrass silage given to rabbits. *Animal Feed Science and Technology* 9, 325-331.

Thakur, S.S. and Sharma, D.D. 1998. Bacterial and volatile acid production rates in crossbred calves fed maize fodder preserved with urea plus salt treatment. *Indian Journal of Animal Sciences* 63, 456-460.

Upadhyaya, V.S. and Mojumdar, A.B. 1991. Studies on quality, intake and digestibility of sorghum silage as influenced by addition of *Leucaena, Sesbania* and urea. *Forage Research* 17, 69-72.

Upadhyaya, V.S., Verma, N.C. and Mojumdar, A.B. 1983. Quality of mixed forage silages and their effect on milk production. *Indian Journal of Animal Sciences* 53, 162-1163.

Van Vauran, A.M., Bergsma, K. and Van Beers, J.A.C. 1989. Effect of addition of cell wall degrading enzymes on chemical composition and the *in sacco* degradation of grass silage. *Grass and Forage Science* 44, 223-230.

Verma, N.C. and Mojumdar, A.B. 1984. Quality and feeding value of maize silage as influenced by formaldehyde treatment. *Indian Journal of Animal Sciences* 54, 50-53.

Verma, N.C. and Mojumdar, A.B. 1985. Studies on quality and feeding value of maize cowpea silage. *Indian Journal of Nutrition Dietetics* 22, 194-198.

Verma, N.C., Mojumdar, A.B. and Rakib, A. 1982. Quality of maize silage on influenced by urea treatment. *Indian Journal of Animal Sciences* 52 , 50-53.

Waldo, D.R. and Derbyshire, J.C. 1971. The feeding value of hay crop silage. Proceed. Forage Conservation Workshop, Washington, pp. 141-148.

Wyss, Y. 2000. Silage additives and aerobic stability. *Revue Swiss di Agric.*, 32, 29-32.

Yu, Y and Thomas, J.W. 1976. Estimate of the extent of heat damage in alfalfa by laboratory measurement. *Journal of Animal Science* 42 , 766-774.

Zimmer, E. 1971. Factors effecting silage fermentation in silo. Paper presented at International silage research conference held at Washington, D.C. (Dec 6-8, 1971).

❑❑❑

Chapter **25**

Leaf Meal Processing Technology and Feeding Strategies to the Livestock

P. N. DWIVEDI AND P. K. PATHAK
Indian Grassland and Fodder Research Institute, Jhansi-284003 (U.P.), India

Introduction

The use of tree leaves as component of animal diet is a widespread practice in many tropical countries and the enormous biological diversity within tropical trees provides enormous opportunity for their use for animal feeding, yet only some 20–30 species have been used and out of these *Leucaena, Gliricidia, Erythrina* and *Acacia* species have predominated. In India, tree leaves of many species viz. *Ficus* sp., Neem, *Grewia* sp., *Celtis* sp., *Sesbania, Prosopis cineraria, Zizyphus* sp., *Flacourtia indica* are used traditionally as fodder. *Sesbania sesban*, a fast growing legume fodder tree of humid warm climate, is rich in crude protein to the tune of 18-25% (Maity *et al.*, 1997).

Majority of livestock population in India subsists on vegetation of natural grasslands where grasses are cheap source of forage. Crop residues, straw and dry grasses form the basal roughage for feeding to livestock which are poor in quality being deficient in protein, available energy and minerals (Singh *et al.*, 1998). Therefore, supplementation of green fodder and concentrate become essential for sustaining livestock productivity. However, there exists a huge gap between demand and supply of both green fodder as well as concentrate in our country (Pathak and Mojumdar, 2002). Moreover, green fodder is not available throughout the year. Therefore, it becomes imperative to search for alternate source of protein rich forage supplement. Supplementation of leaves of leguminous crops is an important and most practical feeding strategy for improving the feeding value of such poor quality roughages. Leaf meal being rich in CP, carotene and minerals could be an appropriate answer for providing nutritious diet to the livestock throughout the year. Supplementation of leaf meal can meet the requirement of livestock particularly during dry periods when availability of other protein rich forages is scarce. Thus it can bridge the gap between the demand and supply of the green fodder and concentrate. In India, traditionally also tree leaves particularly *Zizyphus nummularia* (Pala), Khejri (*Prosopis cineraria*), *Ficus* sp. and Neem leaves were lopped, dried and fed to different categories of livestock.

Leaf Meal Production Technology

The leaf meal production technology is as simple as a farmer can also produce and sell. Thus for production of leaf meal following operations required to be performed.

Harvesting

Harvesting of crop/tree leaves is important in view of time of cutting, which influences the dry matter yield and quality. Cutting too early will give high quality meal but low yields and cutting too late will decrease nutrient content and increase fiber content but give higher dry matter yield. Among the leguminous fodder crops, *Stylosanthes seabrana* is first cut after 55 days whereas *S. scabra* and *S. guianensis* after 60 days and 80 days respectively for leaf meal preparation, depending upon soil fertility and water availability (Guodao *et al.* 2004). The *Leucaena,* a leguminous fodder tree is cut first at 10–15 cm above soil level when it has grown to a height of 1.2 metres. Thereafter 4–6 cuts can be taken per year when the plant reaches a height of 1 metre. The approximate yield under Cuban farming conditions without irrigation or fertilization is 12 tonnes DM/ha with 22–25% crude protein. In some experimental conditions 20 tonnes DM/ha has been achieved (Leng, 1997). Pathak (1988) compared salient characteristics of three basic types of *Leucaena* plants (Hawaiians, Peruvian and Salvador) of age 5 years and reported that leaf production varied from 0.15 to 2.73 kg/tree and forage production was 6.0, 8.72 and 9.86 t/ha/year, respectively. Rajput *et al.* (2004) assessed a *Leucaena* field harvested 15 months back and found the population of the *Leucaena* plants in the field as 1,32,000 per hectare out of which 30,000 plants/

ha having plant base diameter above 2.5cm (mean diameter 3.25cm) was recommended to harvest for production of leaf meal. Yield of fresh *Leucaena* leaves was estimated as 12.105 t/ha/year by them.

The leguminous crops harvested from the field could be directly used for making leaf meal from the whole plant. However, being different in nature, the tree leaves are detached from the stem/twigs and then the leaf meal is prepared. For harvesting leaves from trees, the branches/twigs are lopped manually using axes or saws or mechanically.

Drying

Drying is the way to reduce moisture from the harvested plants and simplest way is sun drying. In sun drying the materials should be spread in a thin layer on a clean surface preferably concrete and dried for two days, then turned over and dried for another day (Guodao *etal.* 2004). Alternatively, drying racks can be built using local materials such as bamboo to dry the materials above the soil surface and avoid spoilage. Indian Grassland and Fodder Research Institute, Jhansi has been developed and evaluated two tier drying structure made up of Subabool *(Leucaena)* logs for shed/ventilated drying of green fodder and tree leaves. Also, extending the drying period diminishes the nutritional quality of the product. Hay production from stylo using sun drying is resulted in loss of 15-40% of the dry matter, mostly in the form of leaves. (Amodu, 2004).

Size Reduction

smaller particle size is required for small animals and birds whereas larger particle size (straw) may be fed to the larger ruminants. Forages and fodder crops, being fibrous material, could be reduced in their sizes mostly by using hammer mill type of grinder or beater type of pulveriser with blower connected to a cyclone dust collector as reported for *Stylosanthes* (Guodao *et al.* 2004, Amodu, 2004 and Chaisang *et al.*, 2004). The lost dry leaves could however be utilized as whole leaf meal if it is possible to collect without much deterioration in the quality. In the process the sun dried stylo crop was threshed by available commercial thresher to reduce its size equivalent to that of wheat straw. The capacity of the thresher was 500 kg/h at average crop moisture content of 7.6%. The process is simple and no special gadgets are required especially for the farmers having the thrashers.

Standardization of Particle Size for Stylo Meal: *Insacco* Digestibility

An *In sacco* experiment was conducted to observe the digestibility of Stylo plant cut in different sizes (Dwivedi and Pathak, 2004). The sizes of cut of

Stylo plant were 1, 2, 5, 10, 15, 20, 25, 30 mm. Nylon bags filled with 5 gm cut material were incubated in the rumen of fistulated cattle for different time intervals (0, 12, 24, 36, 48 and 72 hours). After incubation period nylon bags were dried and analysed the dry matter digestibility as well as CP digestibility. For the different cut size (whole Stylo plant) as 1, 2, 5, 10, 15, 20, 25 , 30 mm, the DM digestibility at 72 hours were 64.40, 64.32, 59.71, 46.88, 43.02, 39.67, 35.60, 33.06 percent, whereas CP digestibility were 67.81, 67.23, 63.36, 50.15, 45.11, 43.20, 39.36, 36.62 percent, respectively. These results indicates that small cut size shown better digestibility and the digestibility percent decreased with increase in cut size (Table 1). The result for the particle size of 1 and 2 mm was almost similar and shown highest digestibility percent. The particle size of 5mm with 59.71 percent digestibility is also considerable. The average CP content of whole Stylo plant samples was 12.87 percent on DM basis. The CP digestibility (range 67.81- 36.62%) was found higher than DM digestibility (range 64.40-33.06%) in all the cut size treatments (Table 2).

Table 1 : *In Sacco* DM Digestibility of Stylo Cut in Different Sizes

Cut size	Digestibility (%) at different hours					
mm	0 hour	12 hour	24 hour	36 hour	48 hour	72 hour
1	14.07	34.23	46.41	57.02	63.24	64.40
2	13.51	32.62	44.96	56.87	63.07	64.32
5	11.97	29.45	41.82	50.77	58.39	59.71
10	11.01	24.73	37.36	41.94	44.63	46.88
15	9.56	23.01	32.28	38.30	41.54	43.02
20	9.11	21.79	30.47	35.82	37.15	39.67
25	8.92	20.58	27.78	32.14	34.58	35.60
30	8.30	20.12	26.35	30.64	33.91	33.06

Table 2: *In Sacco* CP Digestibility of Stylo Cut in Different Sizes

Cut size	Digestibility (%) at different hours					
mm	0 hour	12 hour	24 hour	36 hour	48 hour	72 hour
1	13.81	39.15	51.07	59.22	67.36	67.81
2	13.60	37.06	50.16	57.80	66.75	67.23
5	12.11	33.47	45.02	53.11	62.17	63.36
10	11.23	29.26	39.71	43.87	49.75	50.15
15	09.72	26.60	36.00	40.42	44.06	45.11
20	09.35	24.28	33.58	38.74	41.91	43.20
25	9.07	23.59	30.11	36.02	38.16	39.36
30	8.66	22.37	29.05	35.28	36.00	36.62

Production Process of Stylo Meal

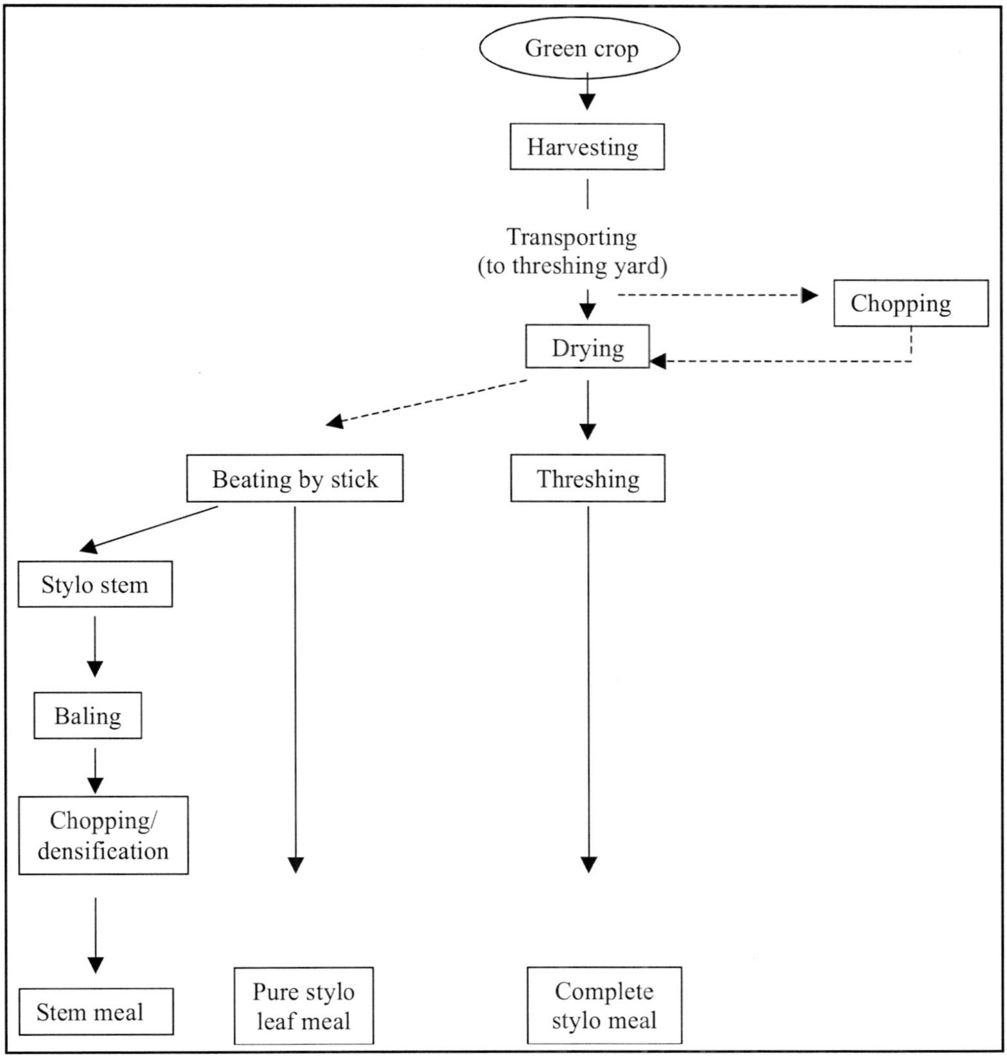

Green crop
↓
Harvesting
|
Transporting (to threshing yard)
↓
Drying (optional: Chopping → Drying)

Drying → Threshing → Complete stylo meal

Drying (optional) → Beating by stick → Pure stylo leaf meal

Beating by stick → Stylo stem → Baling → Chopping/densification → Stem meal

(Dotted line shows optional operations while bold line shows compulsory operations)

Package for Harvesting, Collection and Processing of Stylo Leaf Meal

The stylo plant (Fig: 1) harvested manually by sickle in the month of November at full flowering stage from 1ha field which had moisture content in the range of 32-49% w.b. At this stage 35.56% leaves and 64.46% stem was observed at respective moisture content of 13.5% and 10.44%. Harvesting operation accomplished before the leaves attain the moisture of 13.5% as at this moisture the leaf shatters automatically or by slight jerk. At the shattering moisture about 61.9% of leaves fall and remaining 38.1% leaves intact with the stem. The harvested crop is then transported to threshing yard and was dried up to a moisture range of 8-12% in open sun. Drying of crop accomplishes in 3-4 days when spread on floor at an average surrounding temperature of 13°C in the November. The threshed material, which has similar size as that of wheat straw, could directly be utilized for feeding. Stylo crop grown for seed purpose was also threshed (8.40 q/ha) after harvesting the seeds and meal prepared. This meal had 8.83% CP which is suitable for maintenance ration for ruminants.

Fig 1: *Stylosanthes hamata* Crop in the Field

Processing of Stylo Crop for Making Leaf Meal

The harvested stylo were dried from an average moisture of 37.13%(w.b.) to a final moisture of 12% in open sun in view to extract the leaf meal. A thresher was modified for extracting the leaf meal from harvested, dried stylo crop. The machine consists of a rotating drum on which spikes were mounted in staggered manner, a concave through which the required size of leaf meal passed a feed chute and on outlet for collecting threshed material (Fig: 2). The machine was operated at the capacity for threshing the stylo crop as 6 q/h. The size of some particles of the threshed material was however bigger (6-10 cm) than required for feeding (2-3cm).

Storage of Stylo Meal

The prepared meal were stored for one year in locally made structures (Fig 3&4) of about 25 m^3 capacity and in gunny bags and in polythene packets for storage study. The meal had 12.91% CP initially. It was slightly decreased in all the three storage conditions up to one year of storage. CP content of stored stylo meal reduced from 12.90 to 11.30% in local structure, 11.41% in gunny bags and was minimum in polythene bags as 11.70 per cent after one year storage.

Fig 2: Processing of Stylo Meal with Tresher

Fig. 3: Traditional Outdoor Storage Structure for Storing Stylo Meal.

Fig 4: Storage of Stylo Meal in Gunny Bags

Evaluation of Stylo Species

The chemical composition of stylosanthes hamata for OM, CP, NDF, ADF, Hemicellulose, ADL and ash were measured and found as 91.65, 9.81, 67.32, 44.57, 22.75, 9.08 and 8.35 respectively. About 200lines of different germplasm of *Stylosanthes* were also evaluated for its CP content. CP contents of these lines were ranged from 15-22% in leaf and 11.50 to 14% in stem part.

Economics of Leaf Meal Preparation

Stylosanthes hamata weighing 4.35 t was harvested and meal prepared. The dry matter yield was1.98 t giving CP yield as 0.27 t. Cost economics of leaf meal preparation for consecutive four years calculated, which reveals that average expenditure in cultivation and processing of stylo leaf meal preparation was Rs. 9,139 and 1,935 per hectare. The average gross return was Rs. 20,016 per hectare with net return of Rs. 10,014 per hectare.

Leucaena Leaf Meal: Change in CP and Mimosine Content During Storage

During the year 2004-05,Leuceana plants (53q) were harvested with wood (41.5q) in it for energy and leaves (11.5q) for nutrition on fresh basis. Out of which 4.52 q leaf meal prepared. The prepared leuceana leaf meal (LLM) was stored in polythene lined bags, gunny bags and open under shed.

The leaf meal had 22.08% CP and 2.91% mimosine at the time of storage. After one year storage of *leuceana* leaf meal (LLM) in different storage condition CP content of LLM stored in polythene lined bags, gunny bags and open under shed was reduced from 22.08 to 22.11, 20.01 and 19.86% during storage (Dwivedi and Pathak, 2006). Minimum CP loss was observed in LLM stored in polythene lined bags, then in gunny bags followed by open under shed (Fig 5). Mimosine content did not vary much during storage (2.50 to 2.89%) in all the treatments and it was least varied in polythene lined bags (Fig 6).

The study concludes that LLM can be safely stored in all the three stated conditions with best keeping quality in polythene lined bags. Out of the total harvested plants about 28q air dried wood was collected for making charcoal. From this wood, 24% charcoal was recovered using ICD 200.

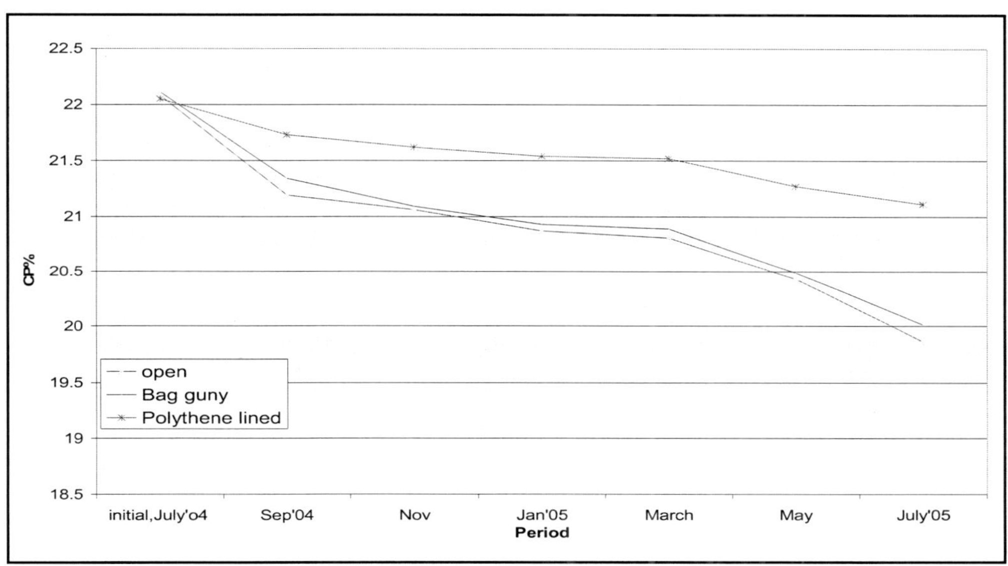

Fig 5: Change in CP Content of Stored LLM in Different Storage Conditions

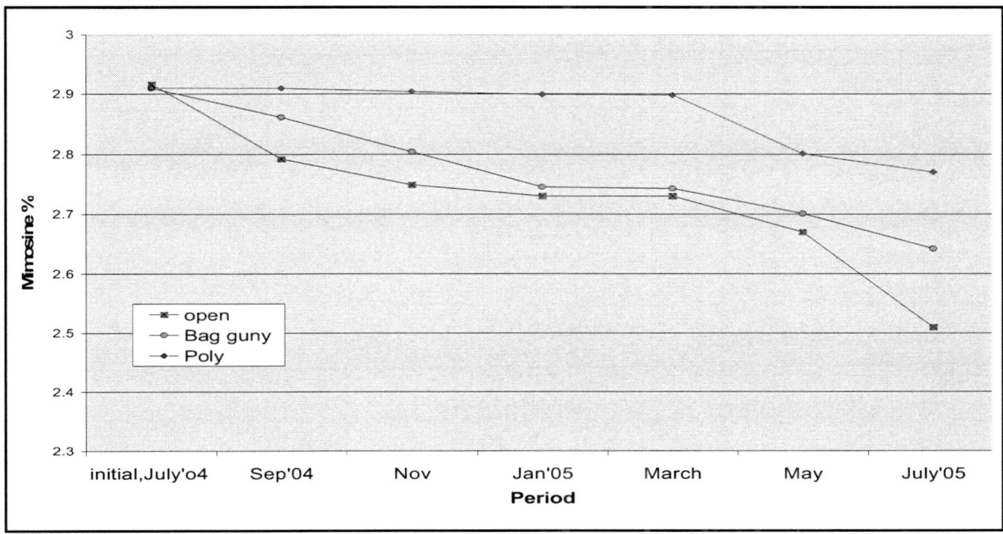

Fig 6: Change in Mimosine Content of Stored LLM in Different Storage Conditions

Leaf Meal as a Ration Component for Livestock

For the last two decades, attempts have been made to use the leguminous leaf meal as an animal feed stuff because of high concentration of protein, high biological value and other nutrients such as carotene and various other minerals.

Nutrient Composition and Level of Feeding Leaf Meal

Several studies have been conducted to demonstrate the usefulness of leaf meal as a component of feed rations of many farm animals (Changjun *et al.* 2004). Chaisang *et al* (2004) however summarized the optimum level of feeding of stylo leaf meal to different animals in various feed formulations and are presented in the table 3. Any assessment of the potential of leaf meals in non-ruminant nutrition necessitates a review of nutrient content. A vast number of plants/shrub/trees are available which have great potential in making leaf meal (Pathak *et al,* 2005). The outstanding feature of these leguminous leaf meals is their high CP contents and essential amino acid contents in the leaf protein especially in the diet of the non-ruminants. High yielding milch animals need green fodder and concentrate mixtures containing considerable amount of cereal grains (about 40%) and oil cake (30-35%). There exists a deficiency of both green fodder as well as ingredients of concentrate mixture. Tree leaves and leaves of leguminous forages can be blended with roughage as complete feed or transformed into leaf meal for feeding it to livestock.

Table 3: Level of Feeding of Leaf Meal to Different Animals

Animals	SLM (%) in the ration
Pig	5-10
Broiler	2-5
Duck	8-12
Goose	15-20
Rabbit	30-40
Cattle, goat and deer	40-60

Nutrient Utilization of Stylo Leaf Meal (SLM)

The leaf meals have been found a good replacer of the concentrate mixture including oilseed cakes. Stylo meal prepared from *S. hamata* harvested at 100% flowering stage was found to replace concentrate mixture when low-grade dry grass based ration was fed to sheep. There was no adverse impact upon dry matter intake (2.73-2.97%) and digestibility co-efficient of other nutrients. However dry matter digestibility (69.29%) was found significantly higher ($p < 0.05$) when 50% of concentrate ration was replaced with stylo meal (Majumdar *et al.*, 2004).

Nutrient Utilization of *Leucaena* Leaf Meal (LLM)

Leucaena leaves incorporation (20%) by replacing 30% crude protein of mustard cake in concentrate mixture and its subsequent usage during block making with natural grass did not interfere with nutrient utilization in Barbari goats and can be incorporated by replacing mustard cake safely without affecting nutrient utilization in Barbari goats. (Samanta *et al.*, 2003). The LLM can be safely incorporated in diets of growing lambs by replacing mustard cake to meet up to 45% to the protein requirement without affecting nutrient and energy utilization and feed cost per kg of weight gain was 19% less than the control diet (Mahanta et al, 1999).

In a comparative evaluation of cotton seed cake and LLM, Barman and Rai (2003) observed higher effective degradability of both DM and OM as well as un-degradable protein (144.62 vs 76.89 g/kg DM) in LLM than that on cotton seed cake. Thus, leaf meal provides scope for replacement of concentrate mixture in the ration of ruminants, without any adverse impact upon nutrient utilization.

Conclusions

Successful animal production requires an adequate supply of feed throughout the year. Increasing demand for cereals, legumes and other food grains for direct human consumption restricts their use in animal production sector. Leaf meal being enriched in protein and having essential amino acid has a wide scope for its commercialization and acceptance. The animal industries must therefore adjust themselves by making changes in their feeding management and production programmes. There exists a big deficit of concentrate in the country to the tune of 60% (Pathak and Mojumdar, 2002). This deficit can be partially bridged by replacing the concentrates in feeds by protein-enriched forages, tree leaves, forage legumes etc. Utilization of such supplements would have a grain-saving effect and would notably increase the currently low feed supply for animals.

References

Amodu, 2004. Stylosanthes : a promising legume in Africa Chapter 22. In: High yielding anthracnose-resistant stylosanthes for agricultural system. Edited by: Sukumar Chakraborty. Australian Centre for International Agricultural Research, Canberra. Pp 225-233.

Barman, K. and Rai, S.N. 2003. Comparative evaluation of cotton seed cake and Leucaena leaf meal on profiles of amino acids, tannins and their influence on digestion kinetics. *Indian Journal Animal Nutrition* 20, 378-388.

Chaisang Phailkaew, Ramesh CR, Kexian Yi and Stur, W. 2004. Utilization of stylosanthes as a forage crop in Asia. In: High yielding anthracnose-resistant stylosanthes for agricultural system. Edited by: Sukumar Chakraborty. Australian Centre for International Agricultural Research, Canberra. Pp 65-76.

Changjun Bai, Guodao Liu, Wang Dongjun, Krishna Daida, Qudratullah S., Prasad VLK, Rama Rao SV, Rao P. Parthasarthy, Ramesh CR, Balagopal R and Gopalan A. 2004. Stylosanthes leaf meal for animal industries in China and India. In: High yielding anthracnose-resistant stylosanthes for agricultural system. Edited by: Sukumar Chakraborty. Australian Centre for International Agricultural Research, Canberra. Pp 243-252.

Dwivedi, P.N. and Pathak, P.K. 2004. Insacco evaluation of stylosanthes hamata meal threshed in different sizes. V[th] Biennial Conference of animal Nutrition Association. Banglore. 24-26 November, 2004. pp 67-68

Dwivedi, P.N. and Pathak, P..K. 2006. Change in CP and Mimosine content of Leucaena leaf meal during different storage conditions. XIIth Animal Nutrition Conference. Anand. 7-9 January, 2006. pp 137

Guodao, Liu Changjun Bai, Dongjun Wang, Ramesh CR and Rao P. Parthasarthy. 2004 . Leaf meal production from stylosanthes in China and India. In: High yielding anthracnose-resistant stylosanthes for agricultural system. Edited by: Sukumar Chakraborty. Australian Centre for International Agricultural Research, Canberra. Pp 253-256.

Leng, R.A. 1997. Bypass proteins from tree foliages (Chapter 5). In : Tree foliage in ruminant nutrition FAO Animal Production and health Paper 139). http://www.fao.org/docrep/003/w7448e/w7448E00.htm#TOC.

Mahanta S.K., Singh, S., Kumar, A. and Pachauri, V.C. 1999. Subabul leaf meal as a replacement of mustard cake in lamb diets. *Small Ruminant Research* 32, 37-42.

Maity S.B., Rekib, A. and Shukla, N.P. 1997. Sesbania sesban as replacement for conventional concentrate mixture in silage based ration for crossbred lactating cows. *Indian Journal Animal Science* 67, 158-159.

Mojumdar, A.B., Pathak P.S., Das, M..M., Mahanta, S.K. and Bhatt, R.K. 2004. Effect of replacing concentrate mixture with stylo leaf meal in ruminent utilization and blood metabolites in ewes. *Proceedings. ANA conference on New Dimensions of Animal Feeding (Editors Sharma et al).* pp 50.

Pathak, P.S. 1988. Ecology and potentials of Su-babool in agroforestry systems. In : Pasture and Forage Crop Research (Ed. Panjab Singh). RMSI, Jhansi. Pp 332-341.

Pathak, P.S. and Mojumdar, A.B. 2002. Current status of feed and forages: prospects and constraints. Base Paper presented at UPCAR, Lucknow. 23rd October, 2002.

Pathak, P.S., Pathak P.K., Mojumdar A.B. and Dwivedi P.N. (2005) Leaf meal production and utilization. National workshop on Animal feed science and technology in Indian perspective(Lead paper). IGFRI, Jhansi. 11-12th Feb2005.

Rajput, D.S., Dwivedi, P.N., Tamhankar, M.B. and Pathak, P.K. 2004 Prediction of production potential of leucaena for energy and animal nutrition. Paper presented in 38th ISAE Annual Convention and Symposium held at DR. B.S. Sawant Konkan Krishi Vidyapeeth, Dapoli from 16-18 January, 2004.

Samanta, A.K, Singh, K.K, Das, M..M., Maity, S.B. and Kundu, S.S. 2003. Effect of complete feed block on nutrient utilization and ruman fermentation in Barbari goats. *Small Ruminant Research* 48, 95-102.

Singh, A.K., Upadhyay, V.S., Singh, K.K. and Mishra, A.K. 1998. Effect of dietary levels of *Abezia lebbek* leaves on the utilization of matured dry grass in goats. *Indian Journal of Animal Nutrition* 15, 290-292.

❑❑❑

Chapter **26**

Practical Aspect of Mineral Nutrition in Livestock

K. K. SINGH, M. M. DAS, S. K. MAHANTA, ANIL KUMAR AND A. K. MISRA
Indian Grassland and Fodder Research Institute, Jhansi-284003 (U.P.), India

Introduction

Minerals, inorganic matters that are not synthesized in the body are essential for the proper functioning of the animal system. A problem arises when the feed does not supply enough quantity of mineral to meet the animal's requirements. This may occur when the feed is low in mineral content, the availability of the mineral is low or when other minerals or nutrients are interfere with the ability of the animal to absorb or utilize the mineral. Essential minerals have been divided into two groups depending upon their requirements, the macro and the microelements. Macro-elements are those that are required in relatively large amounts i.e., greater than 100 ppm (mg/kg diet). This group consists of calcium, phosphorus, magnesium, sulphur, potassium, sodium and chlorine. However, minerals that are required at less than 100 ppm are called trace or microelements. This group consists of cobalt, copper, iron, manganese, iodine, molybdenum, selenium and zinc. Although these minerals are required in very small amounts, they still play important roles in normal body function of an animal.

Mineral in Feedstuffs and their Importance

Calcium and Phosphorus

Animals' calcium (Ca) requirements cannot be considered independently of phosphorus (P) since these two minerals work hand in hand. Calcium and phosphorus are the major mineral constituents of bone. Calcium also plays an important role in muscle function, whereas phosphorus is key element to major metabolic functions throughout the body.

For grazing ruminants, Ca is generally adequate in leguminous forages (Table 1). Phosphorus, however, can be deficient in these forages, and since legumes contain higher levels of Ca, supplemental phosphorus generally is needed in forage-based diets (Singh *et al.* 1997). When animals are fed or allowed to graze legumes, provide a mineral supplement that is higher in phosphorus than calcium to keep the Ca:P ratio in a desired range. If the Ca:P ratio exceeds 10:1, or phosphorus remains deficient, then it will result in reduced growth, feed efficiency and reproduction.

For cattle on a high-concentrate diet, the opposite problem exists. Most cereal grains are rich in P than Ca; therefore, Ca must be added to the ration (Table 2). When Ca:P approaches 1:1, or P intake exceeds Ca, it may result in reduced feed intake and urinary calculi (water belly) in steers and bulls.

On the contrary, if Ca is not supplemented in adequate amounts, Ca deficiencies may occur. However, these deficiencies do not become apparent, until there are occurrence of broken bones, convulsions and death. The optimum performance occurs when the Ca:P ratio in dairy cattle diets ranges from 1.5:1 to 2:1. However, performance typically is not adversely affected unless the dietary Ca:P ratio exceeds 6:1. Disturbed Ca: P ratio has a blocking action on the pituitary gland and consequently on ovarian function (Herrick, 1977), thus prolonging the interval of first ovulation.

Magnesium

Grass tetany, a metabolic disease in cattle grazing on lush, cool-season grasses, has a primary association with magnesium (Mg). Magnesium is a key component in the initiation of many metabolic enzymes activities. It is also important in neuromuscular function. Magnesium deficiencies reduce calving rates, calf vigor and rate of gain in calves. Low Mg in the grass doesn't cause poor calf performance directly, but it reduces the cow's milk production. If the cow's Mg requirements are not met, reduced conception rates would be the primary deficiency symptom.

Table 1 : Mineral Status of Dry and Green Forages (on DM basis)

Feed stuffs	Ca	P	Mg	Na	K	S	Co	Cu	Fe	Mn	Mo	Zn
			(Percent)						(ppm)			
Local grass	0.44	0.09	0.24	0.08	1.16	0.13	0.15	5.91	265	49	0.43	27.50
Maize stover	0.37	0.16	0.25	0.02	1.68	0.08	0.10	6.39	239	39	0.40	35.15
Sorghum stover	0.38	0.12	0.29	0.02	1.62	0.10	0.08	6.04	186	37	0.10	35.74
Rice straw	0.28	0.09	0.20	0.20	2.21	0.10	0.44	5.98	356	168	0.10	28.62
Wheat straw	0.33	0.06	0.15	0.18	1.57	0.13	0.22	4.35	309	48	0.03	11.57
Soyabean straw	0.91	0.09	0.39	0.03	1.35	0.11	0.30	6.68	553	39	0.22	18.76
Urd straw	1.12	0.18	0.20	0.04	1.00	0.24	0.22	10.80	922	80	1.50	32.40
Groundnut straw	2.22	0.17	0.97	0.21	1.12	0.29	0.23	9.56	1510	68	0.74	21.80
Gram straw	0.68	0.14	0.65	0.09	0.63	0.24	0.23	33.39	198	48	0.35	15.53
Arhar straw	0.64	0.09	0.68	0.06	0.03	0.16	0.21	40.46	191	55	0.65	15.78
Cowpea green	1.87	0.25	0.52	0.30	1.94	0.42	0.16	10.90	254	54	0.48	31.40
Lucerne green	2.11	0.31	0.41	0.39	2.70	0.34	0.26	11.88	503	55	1.27	24.81
Guar green	1.89	0.24	0.72	0.14	2.57	0.40	0.24	7.64	282	24	0.23	16.54
Oat green	0.42	0.29	0.26	0.67	2.91	0.30	0.15	11.42	287	80	2.09	30.92
Bajra green	0.38	0.29	0.30	0.04	2.04	0.27	0.14	4.00	531	27	0.83	17.15
Hybrid napier	0.36	0.12	0.23	0.05	2.16	0.15	0.11	4.04	237	32	0.64	14.52

Source: Singh et al. (2005), Garg et al. (2005)

Table 2 : Mineral Status of Common Concentrate Ingredients (on DM basis)

Concentrates	Ca	P	Mg	Na	K	S	Co	Cu	Fe	Mn	Mo	Zn
			(Percent)						(ppm)			
Barely grain	0.07	0.37	0.16	0.04	0.62	0.15	0.50	9.12	120	23.55	0.75	29.55
Maize grain	0.02	0.31	0.12	0.01	0.40	0.14	0.30	5.36	93	9.21	0.22	26.82
Wheat bran	0.21	0.91	0.29	0.01	0.44	0.16	0.31	8.11	113	38.65	0.01	34.20
Oat grain	0.10	0.36	0.16	0.01	0.42	0.15	0.39	13.15	163	91.15	0.98	61.58
Bajra grain	0.04	0.34	0.15	0.03	0.41	0.17	0.10	7.10	119	15.50	0.01	50.35
Sorghum grain	0.15	0.28	0.40	0.03	0.45	0.16	0.15	8.50	276	35.8	3.85	28.00
Guar grain	0.22	0.46	0.24	0.01	1.27	0.27	1.14	15.37	481	15.74	2.76	57.45
Guar chuni	0.21	0.58	0.28	0.02	1.40	0.32	0.63	16.10	464	24.31	2.12	57.36
Mustard cake	0.88	1.18	0.46	0.02	1.45	1.78	0.70	13.80	560	126	5.00	41.00
Sesame cake	2.62	1.09	0.58	0.02	1.13	0.66	0.30	16.50	451	48.50	3.20	28.00
Cotton seed cake	0.31	0.49	0.34	0.02	1.45	0.32	0.30	32.20	350	31.50	1.45	101
Linseed cake	0.61	0.52	0.15	0.02	1.11	0.52	0.70	13.80	560	126	5.00	41.00

Source: Singh et al. (2005), Garg et al. (2005)

Grass tetany, also termed hypomagnesaemia, simply means a deficiency of Mg. Prolonged Mg deficiency results in excessive urination, erratic and nervous behavior (also called grass staggers), twitching of facial muscles, convulsion and death. While all ruminants are susceptible to this disorder, older lactating cows are at the greatest risk. Grazing on lush, rapidly growing, cool-season grasses usually are associated with grass tetany (Smith *et al.* 1999).

Many grasses and cereal hays, such as oat hay are good source of Mg and high in K (Table 1). When Mg levels in hay fall below 0.15%, cattle may become vulnerable to Mg deficiency. In addition, if calcium levels are low and potassium levels are high in these feeds, winter tetany may occur in temperate region. Legume-based hays, however, generally have adequate (0.41-0.72%) magnesium (Garg *et al.* 2005). Drought-stressed annual forages typically are higher in K, which also contributes to Mg deficiency. Feeding a mineral supplement high in Mg should prevent problems associated with grass or winter tetany. Magnesium oxide (MgO) typically is used as a source of Mg in most supplements. It is not palatable and it requires blending with other more palatable ingredients to achieve sufficient intake.

Sulfur

Sulfur (S) is unique in the sense that it is the only trace mineral incorporated into amino acids (specifically methionine and cystiene) and vitamins (thiamin and biotin). Amino acids are the building blocks for protein. Rumen microorganisms use inorganic sulfur to form their own sulfur-containing amino acids. Sulfur has many dietary sources. Soybean products, alfalfa hay and corn byproducts have relatively high levels of sulfur.

Cereal grains such as corn and oats generally contains 0.14-0.17% sulphur, protein sources such as soybean meal, can contain as much as 0.5% sulfur. Forages tend to be more variable in S contents (Singh *et al.* 2002). Legumes, for example, typically have higher content of sulphur (0.27-1.78%), whereas grass hay and crop residues contain very little sulfur. Cattle on pasture require 0.15% sulphur in their diets. Deficiencies in S are not common. However, when a deficiency occurs, signs include poor appetite, emaciation and dullness. In addition, low levels of dietary S can result in poor utilization of non-protein nitrogen (NPN), which in turn reduces microbial growth and fermentation in the rumen.

The sulfur level is critical in growing and finishing rations. These diets, which typically are high in S and low in fiber, can induce sulphur toxicity. Sulfur toxicity can interfere with selenium, copper, molybdenum and thiamin metabolism of ruminants.

Sodium and Chloride

Sodium (Na) and chloride (Cl) work together to maintain cellular volume, pH and osmolarity of body fluids, such as blood. Sodium chloride (NaCl) promotes water intake, which maintain or improve milk production and overall animal health. Individually, Cl is involved primarily in hydrochloric acid production in the abomasum to aid in digestion. Sodium works in conjunction with potassium (K) for nutrient transport into and out of cells. On an average, cattle should consume 11 to 15 grams of salt per day to meet nutritional requirements.

Both the elements can be consumed, and usually in relatively high amounts without any negative effects, but dietary levels of NaCl should not exceed 8%. Excess dietary salt simply is excreted in the urine. Most of forages in India are low in sodium and chloride, therefore, there is no question of their toxicity however, if toxicitiy occur, symptoms include reduced milk production, weight loss etc. Adequate water availability and intake can reduce the risk of salt toxicity. To reduce salt intake and chances of salt toxicity use a salt block rather than a loose granular supplement.

Potassium

Like sodium (Na), cattle need potassium (K) in large amounts to maintain normal body and organ function. Potassium works in conjunction with Na in the body to transport nutrients in and out of cells. It also helps to maintain health by maintaining the cellular water balance (osmotic pressure). Increased levels of K in the animal's system will demand an equal increase in Na to maintain homeostasis in the body's cells.

Most forages typically supply adequate amounts of potassium. In fact, K can be so high in some cool-season grasses and legumes that it can accelerate the onset of grass tetany by inhibiting magnesium (Mg) absorption from the gastro intestinal tract.

In high-concentrate diets, K may become deficient since corn grain and most corn byproducts are low in K. Corn silage, however, is an excellent source of K. Cattle require 0.6-0.7% K in the ration (on DM basis). Deficiencies in K can result in reduced intake, weight loss and stiff joints. Cattle stressed due to long shipping distances may require increased levels of K to replenish lost body reserves.

Cobalt

Cobalt's (Co) primary function in ruminant animals is to provide a substrate for the rumen microbes to produce vitamin B_{12}. Vitamin B_{12} is used in the rumen microbes' metabolic processes to produce propionate (a volatile fatty acid and an important energy source for the animal). Although the animal does not absorb Co readily, the involvement of Co in vitamin B_{12} production makes it an important mineral. Deficiencies of Co are far more common than toxicities, as cattle can tolerate upto 300 times their requirement (0.10 ppm). When deficiencies occur, signs include depressed appetite, listlessness, decreased growth, reduced milk production and rough hair coat. Fortunately, Co typically is adequate in most of the forages and silages. However, inclusion of this mineral in feeding programs of ruminants, particularly when fed on dry forages, provides added insurance with little risk.

Copper and Molybdenum

Red-blood cell health, collagen development, maintenance of reproduction and immunity are just a few of the important roles of copper (Cu). Copper also works in conjunction with molybdenum (Mo) and inorganic sulfur for various important enzymes involved in nucleotide and vitamin metabolism. Consequently, a balance between Cu and Mo is important. A ratio of 2:1 to 4:1 (Cu:Mo) allows for optimal performance in cattle.

In addition, if cattle are exposed to high levels of zinc (Zn), iron (Fe) or phosphorus (P), Cu utilization is reduced. This, in turn, can result in reduced performance in animals. Typically, primary Cu deficiencies (deficiencies that result from low dietary copper) are common in India. Deficiency signs include lighter or faded hair coats, reduced conception rates, severe diarrhea, brittle bones and reduced immune response.

Since Cu interacts with many other minerals (Fe, Mo and S), secondary deficiencies may occur. Secondary Cu deficiencies occur when dietary Cu is made unavailable by interactions with antagonists such as Fe, Mo or S present in the feed or water. The dietary levels of 8 -10 ppm copper may be adequate, but deficiency signs still are evident if elevated levels of sulphur (>0.4 %), molybdenum (>150 ppm) and iron (>500 ppm) are present in the diet. When these minerals are in excess, they compete with Cu for the same absorption sites in the intestinal mucosa. They also form thiomolybdate complexes, which renders Cu unavailable to the animal.

Copper toxicities can occur in cattle when dietary levels exceed 100 ppm, although toxicities are much more common in sheep (toxicity occurs at 25 ppm). Chronic exposure to high levels of Cu results in rapid breathing, elevated temperature, anorexia, dehydration and jaundice. Soluble forms of Cu such as copper sulfate ($CuSO_4$) or copper chloride (CuCl), are absorbed more readily than copper oxide (CuO) and can be used as better supplementation source.

Iodine

Although required at relatively low levels (0.5 ppm), iodine (I) is key in maintaining a normal metabolic rate through its role in producing the hormone thyroxin (T4) from the thyroid gland. Deficiencies of this mineral are more prevalent in the *Tarai* region of India. When iodine is deficient, thyroxin production is reduced because the body attempts to conserve iodine. Lower levels of thyroxin result in lower metabolic rates, which in turn affects milk production, weaning weights and overall animal health. Cows deficient in iodine may have calves that are born blind, weak, hairless or stillborn. In addition, goiter development is common. A goiter is an enlargement of the thyroid gland. It was a significant problem in humans and livestock for many years, and eventually was linked to high levels of fluoride inhibiting the uptake of iodine.

Manganese

Manganese (Mn) plays an important role in growth and reproduction. As dietary levels of Mn increase, concentrations of the mineral increase in reproductive tissues, offering a direct link between Mn and fertility. Manganese is linked to growth through its involvement in specific enzyme functions related to skeletal cartilage formation. Ironically, Ca and P (which are linked to bone growth) can inhibit Mn absorption when fed in excess.

Requirements of Mn may vary depending on the stage of production. For example, growing and finishing cattle require 20 ppm in the diet (or approximately 200 mg/d), while pregnant, lactating cows require 40 ppm (or 400 mg/d). As with Co, Mn has a fairly large "safe" feeding range. The maximum tolerance for the mineral can be as high as 1000 ppm before it causes noticeable negative effects. However, Mn interacts with many different minerals, which can alter the animal's tolerance level for Mn and the minerals with which it interacts. If Mn requirements are not met, reduced conception rates may occur. Other adverse effects include poor growth rates, low birth weights and increased abortion in animals.

Selenium

Selenium (Se), in conjunction with Vitamin E, boosts immune function and the development of antioxidant enzyme systems. In the past, farmers of

Punjab snd Haryana have had trouble managing problems associated with Se because its tolerance range is quite narrow (< 0.2 ppm is considered deficient and 5 ppm is toxic). The source of Se appears to influence tolerance. When the source of Se is organic (associated with either methionine or cysteine amino acids), Se has a greater impact on the liver and small intestine, compared with Se from an inorganic source such as selenate ($SeNO_3$). Although the maximum dietary level of Se tolerance in beef cattle has been estimated to be 2 ppm (NRC, 1980), recent evidence suggests that dietary levels as high as 3 ppm have no ill effect on performance.

Toxicities and deficiencies of Se are very dependent on soil composition. In the Northern States, soil Se content is relatively high (> 2 ppm). In some of these regions, toxicitiy of Se (blind staggers, sloughing of hooves and hair, anorexia and a wide range of birth defects) are common.

Zinc

Zinc (Zn) plays important roles in immune response, enzyme systems and hoof health, DNA, RNA and protein synthesis. Because of its wide scope and the body's demand, it is required at levels of 30 ppm. Green forages, grains and protein sources all are good sources of Zn and on an average contain 35 ppm and 65 ppm of Zn. When cattle are fed forage-based diets or under stressed condition they require supplemental zinc since most of the common feeds and forages are deficient in zinc (Table 1).

Signs of deficiency include reduced feed intake and weight gain, excessive salivation, rough hair coat and eventually swelling of the feet and legs. Critical Zn deficiencies result in hair loss, thickening of skin, and lesions around the nose and mouth. Zinc levels should not exceed 1000 ppm, otherwisw it will affect performance and toxicity may develop.

Mineral Requirement

Mineral requirement is dependent on factors like the level of productivity, physiological stage and species of animal. Marginal mineral deficiency at low levels of production becomes more severe with increased levels of production (Mc Dowell, 1997). To meet out the requirement of P, Na, Cl, Cu, I and Zn appears to be major issue in India (Table 3).

Table 3 : Requirements of Mineral (on DM basis) in Animals

Minerals	Young stock (Cattle/ buffalo)	Lactating stock (Cattle/ buffalo)	Sheep and goats
Ca (%)	0.50	0.70	0.50
P (%)	0.22	0.32	0.22
Mg (%)	0.18	0.20	0.18
K (%)	0.60	1.00	0.60
Na (%)	0.10	0.18	0.10
Cl (%)	0.15	0.20	0.15
S (%)	0.16	0.32	0.20
Co (ppm)	0.10	0.20	0.20
Cu (ppm)	8.00	10.00	8.00
I (ppm)	0.25	0.60	0.25
Fe (ppm)	50	50	50
Mn (ppm)	20	40	20
Mo (ppm)	<2	<2	<2
Zn (ppm)	30	40	30
Se (ppm)	0.10	0.10	0.10

Source: Mc Dowell, 1997

Diagnosis of Mineral Deficiencies

Mineral nutrition disorders range from acute deficiency or toxicities characterized by well-marked clinical sign and pathological changes to mild and transient conditions difficult to diagnose. This latter condition assumes greater importance because they occur over large areas and affect a large number of livestock population.

To diagnose mineral deficiency or excess, clinical sign, pathological and biochemical examination along with mineral analyses of soil, water, plant and animal tissues and fluids are used. Among these methods, animal tissue and fluid levels of mineral along with concentration of particular enzymes, metabolites with which the mineral in question is functionally associated are important indicators of mineral status (Table 4). For example, liver tissue examination is particularly good for evaluating the animal's status of Co, Cu, Mn and Se and blood plasma is good indicator of Ca, P and Mg status.

Table 4 : Specific Diagnostic Organ or Fluids and Critical Levels of Mineral

Element	Tissue or fluid	Critical level
Calcium	Plasma	8 mg/100 ml
Phosphorus	Plasma	4.4 mg/100 ml
Magnesium	Plasma	2 mg/100 ml
Potassium	Saliva	150 mg/100 ml
Sodium	Saliva	150 mg/100 ml
Cobalt	Liver	0.06 ppm
Copper	Liver/ serum	0.65 μg/ml
Iron	Transferrin	14% saturation
Manganese	Liver	0.25 ppm
Zinc	Serum	0.60 μg/ml

Source: Mc Dowell (1997)

Mineral Supplementation Strategies

Mineral deficiency, imbalances and toxicity can be avoided by adopting various mineral supplementation strategies. The strategic choice varies greatly with different elements, climatic condition, animal species and economic status (Underwood and Suttle, 1999). A complete mineral mixture usually contains all the essential minerals such as calcium, phosphorus, magnesium, sulphur, potassium, Iron, copper, cobalt, manganese, and zinc. But the supplementation of all the above mineral elements may not needed always, because many of them such as magnesium, sulphur, potassium, iodine, cobalt, iron and manganese are already present in required quantities in feeds and forages to satisfy the need of dairy animals producing 8-10 kg of milk daily. Reports suggest that some elements such as potassium, and iron are available in excess to that of required by livestock. Therefore, inclusion of these elements in mineral supplements will create an imbalance. It has been shown that K reduces the absorption of Mg and Fe reduces the utilization of copper.

There are various ways of mineral supplementation. The indirect method of providing mineral to grazing livestock include the use of mineral containing fertilizers, altering soil pH and growing specific pasture species. But these methods are not always economical and practically feasible. Therefore, other methods such as direct administration of minerals to animal's drinking water, feed, rumen preparation (such as pellets and bolus) and injection are commonly used for mineral supplementation. Among these methods, mineral supplementation in feed mixtures is the most practical way.

Formulation of mineral mixture

In order to formulate a mineral supplement for ruminant, it is necessary to have an idea of the requirements of the target animals for various elements (Table 4). This includes the age of the animal, stage of current production, relative bioavailability of the nutrients from the source (Table 5), approximate daily dry matter intake and concentration of the element in the mineral mixture. Selection of source compound depends upon the cost and availability of compound in that region.

Table 5 : Some Common Mineral Sources and Relative Bio-availability of Element

Element	Source compound	% of element in compound	Relative bio-availability
Calcium	Calcium carbonate	40.0	Intermediate
	Lime stone	38.5	Intermediate
	Di-calcium phosphate	23.3	High
Phosphorus	Di-calcium phosphate	18.5	Intermediate
	Sodium phosphate	25.0	High
	Steamed bone meal	12.6	High
Magnesium	Magnesium sulphate	14.0	High
	Magnesium oxide	55.0	High
Cobalt	Cobalt chloride	24.7	High
	Cobalt sulphate	21.0	High
Copper	Cupric sulphate	25.0	High
	Cupric chloride	53.0	Intermediate
Iodine	Potassium iodide	63.5	High
	Calcium iodate	69.0	Available
Zinc	Zinc oxide	55.0	High
	Zinc chloride	48.0	Intermidiate
	Zinc sulphate	30.0	High

Source: McDowell (1997)

Characteristics of a Mineral Mixture

An acceptable good quality mineral mixture supplement should have following characteristics:

1. Final mixture should contain a minimum of 6-8% total phosphorus. However, in the areas where forages are consistently lower than 0.2 %, P level may be increase up to 8-10%.
2. Ca: P ratio should not substantially over 2:1.
3. Provide a significant portion i.e., around 50-100% of the trace elements and avoid the excess supplementation of already available minerals from feeds and forages.
4. Formulation should be sufficiently palatable to allow adequate intake in relation to requirements. Being palatability enhancer, it is recommended that an ideal mineral mixture should contain 30-40% common salt.
5. The particle size must allow adequate mixing without segregation.
6. High quality mineral salts/ compounds should be used to ensure optimum bioavailability. Salts containing high fluorine concentration should be avoided.

Types of Mineral Mixture

Keeping in view the important role of mineral on livestock production the Bureau of Indian Standard (BIS, 1992) has given certain guidelines for the formulation of mineral mixtures.

Table 6 : BIS Requirements (%) of Mineral Mixture

Characteristics	Type I (with salt)	Type II (without salt)
Moisture, maximum	5	5
Calcium, minimum	18	23
Phosphorus, minimum	9	12
Magnesium, minimum	5	6.5
Salt as sodium chloride, minimum	22	0
Iron, minimum	0.4	0.5
Iodine as KI, minimum	0.02	0.03
Manganese, minimum	0.10	0.12
Cobalt, minimum	0.09	0.012
Zinc, minimum	0.3	0.38
Sulphur, maximum	0.4	0.5
Fluorine, maximum	0.05	0.07
Acid insoluble ash, maximum	3	2.5

Source: Mudgal *et al.* (2003)

Area Specific Mineral Mixture

As a low cost insurance to provide adequate mineral nutrition, "area specific" mineral supplements should be designed and be available free choice to livestock. In this way one can avoid the imbalances occurring due to the excess supplementation of critical minerals in the animals diet. An area specific mineral mixture usually includes salts of calcium, phosphorus, zinc, copper, iodine and other elements as per their deficiency in feed/ forages (Table 7). It should be given @ of 30-50 g/head/animal depending upon the type of animal and production status or it can be mixed with concentrate ingredients @ of 1-2%. Calcium, copper, iron and selenium, when in excess, can be more detrimental to ruminant production than any benefit derived by providing a mineral supplement. In regions where high forage molybdenum predominates, 3-5 times the copper content in mineral mixture is needed to counteract molybdenum toxicity.

Table 7 : Area Specific Mineral Mixtures Developed at IGFRI, Jhansi

Compound	Quantity (g)
Area specific mineral mixture-I	
Di-calcium phosphate	1960
Copper sulphate	32
Zinc sulphate	8
Common salt	1000
Total	3000
Area specific mineral mixture-II	
Di-calcium phosphate	1950
Copper sulphate	41
Zinc sulphate	8
Potassium iodide	1
Common salt	1000
Total	3000

Thus, it can be concluded that mineral mixture supplementation in the animals diet is vital to optimize animal production and reproduction. Mineral deficiency, imbalances or toxicity not only causes reduced growth and production but also hampers the animal health and reproduction. Therefore, precautions should be taken in the correction of mineral deficiencies and imbalances. An Area specific mineral supplementation could be a cost effective and practical way of correcting mineral deficiencies and imbalances. Further, mineral supplementation programme are not to be considered in isolation but as a part of whole nutrient management system which includes protein, energy and vitamins.

References

Garg, M.R., Bhanderi, B.M. and Sherasia, P.L. 2005. Assessment of adequacy of macro and micro mineral content of feed stuffs for dairy animals in Semi arid zone of Rajsthan. *Anumal Nutrition and Feed Technology* 5, 9-20.

Herrick, J.B. 1977. A paper contributed to 1st All India Symposia on Animal Reproduction, PAU, Ludhiana, India.

Mc Dowell L.R. 1985. Nutrition of Grazing Ruminant in Warm Climate, Academic Press, New York.

Mc Dowell, L.R. 1997. *Mineral of Grazing Ruminants in Tropical Regions.* 3rd edition, Institute of Food and Agriculture Science, University of Florida, Gainesville, USA.

Mudgal, V.D., Singhal, K.K. and Sharma, D.D. 2003. *Advances in Dairy Animal Production.* 2nd edition, IDBC, Lucknow, India.

N.R.C. 2001. *Nutrient Requirement of Dairy Cattle.* 7th edition, National Research Council, National Academy of Science, Washingto, D.C.

BIS, 1992. Mineral mixture for supplementing cattle feeds- Specification (3rd revision). Bureau of Indian Standard, New Delhi.

Singh, K.K., Nag, S.K., Pailan, G.H., Kundu, S.S., Garg, M.R. and Bhanderi, B.M. 2002. Sodium and potassium contents of some common forages. *Indian Journal of Animal Nutrition* 19, 378-380.

Singh, K.K., Srinivas, B. and Ramchandra, K.S. 1997. Chemical changes with maturity and its impact on *in sacco* dry matter degradation of some range grasses and legumes. *Indian Journal of Animal Nutrition* 14, 254-257.

Smith, K.F., Rebetzke, G.J., Eagles, H.A., Anderson, M.W. and Easton, H.S. 1999. Genetic control of mineral concentration and yield in potential ryegrasses wit special emphasis on mineral related to grass tetany. *Australian Journal of Agricultural Research* 50, 79-86.

Suttle, N.F. 1987. The nutritional requirements for Cu in animals and man. Copper in animal and man. Vol. 1 (Eds) Howell, Mc C. and Gawthorne, J..M. CRR Press, Boca Raton, Florida.

❏❏❏

Chapter 27

Forage Based Feeding of Livestock

S. K. MAHANTA, K. K. SINGH, M. M. DAS AND N. DAS
Indian Grassland and Fodder Research Institute, Jhansi-284003 (U.P.), India

Introduction

Forages and roughages are coarse bulky feed with more than 18% crude fibre and low in digestible nutrients (nitrogen free extract around 40%) such as crude protein and energy. They have high content of cell wall materials. The green forages that are cut and fed in fresh condition to the animals are also termed as soiling crops.

Historically, cattle lived on forages and dry roughages during most of the year with some concentrates during limited periods. Thus they are specially well adopted to utilization of forage nutrients to meet their nutritional needs. This is possible because of the anatomy and function of complex four compartment ruminant stomach, where the microbial and other vital process take place. Cattle even require a substantial amount of fibre for good health and optimum performance. Although there are other sources, most of the fibre consumed by cattle comes from forages. In addition to fibre, cattle obtain large amounts energy, protein, minerals, vitamins, lipid and water from forages. Usually forages are the most economical source of nutrients, especially energy but the cost relationship varies enormously in different areas, times and individual situations.

Few Principles of Forage Feeding in Ruminants

- Ruminants are adapted to use forage because of microbes in their rumen
- To maintain ruminant health and productivity, feed the rumen microbes, which in turn will feed the ruminant
- Ruminant nutritional needs change depending on age, stage of production, and weather
- Adequate quantities of green forage can supply most —if not all— the energy and protein, a ruminant needs
- Forage nutritional composition changes depending on plant maturity, species, season, moisture, and grazing system
- Supplementation may be necessary when available forage is mature or if animal needs require it (i.e., high-producing dairy animals)
- Excessive supplementation may reduce the ability of the rumen microbes to use forage

The common practice of feeding milch cows and buffaloes is to feed them on crop residues with little supplementation of green forages (if available) and concentrates. But concentrates feeding results in increased cost of feeding. Feeding green forages, however, reduces the cost and has the potential of higher level of milk production. Nearly 65% of the total expenditure of milk production in crossbred cattle is attributed to the feeding of animals when both concentrates and green fodders were fed as mixed ration. But when the milk production depend upon concentrate based feeding, the cost of feeding towards milk production was 83% while on forage based feeding, it could be reduced to only 40% of the total expenditure. Although, milch animals specially require good quality forages (which have many additional benefits) for expression of full genetic potential of milk production.

Benefits of Green Forages

- Require to maintain normal health and reproduction of all herbivores
- Over-all cooling effect on the body, being-
  - Easily digestible
  - More palatable
  - Slightly laxative
  - Provide fresh nutrients in a most natural form resulting in efficient utilization of the feed without any strain on the body organs

- Longevity and production are adversely affected when cattle are reared either without green or poor quality forages, even though provided with best quality of concentrates
- In absence of greens, Cows reported to give birth to weak, deformed and blind calves
- Feeding concentrates at higher level without sufficient green forages, adversely affect the digestive system of the animals in the long run
- Animals yielding as high as 10 litres of milk can easily be maintained solely on green fodders without any complaint and reducing feed cost by 20% over normal dry roughages and concentrates ration

Benefits of Being a Ruminant

Feeding of forage based rations to animals requires a sound understanding of ruminant nutrition. First we should understand how a ruminant animal (cattle, buffalo, sheep, goats) digests plant matter. Ruminant comes from the word "rumen," which is the first major compartment in the four-compartment stomach of the cow, buffalo, sheep, and goat. This structure is the "furnace" where microbial fermentation takes place. Millions of bacteria, protozoa, and fungi live in the rumen and break down energy-rich plant parts, making them digestible for the host animal. After the forage has been digested in the rumen and is broken down into small pieces, it can pass through the *reticulum* and *omasum*, which function as strainers that keep large pieces of material from passing into the *abomasum*, or "true stomach," where digestion continues. From the abomasums onward, the ruminant digestive system closely resembles non-ruminant animal digestive systems with a small and large intestine, colon, and anus.

As stated earlier, ruminant animals have the ability to harvest forages and convert plant energy, especially cellulose, from grasslands/rangelands and cultivable lands. Cellulose is the portion of the plant structure that comprises the walls of the plant's cells, and is very fibrous and indigestible. Monogastric (single-stomach, nonruminant) animals do not have the ability to digest cellulose. Rumen microbes, however, produce *cellulase,* the enzyme that breaks down the chemical bonds in cellulose, making it digestible to the microbe and, subsequently, to the ruminant animal. Another advantage of rumen fermentation is microbial synthesis of important vitamins and amino acids. All the vitamins the animal needs are synthesized by microorganisms, except vitamins A, D, and E. However, animals fed high quality hay or green forages get their requirement of vitamins A and E. Vitamin D is supplied through exposure to sunlight, which is another advantage of forage production. Amino acids are the building blocks of protein—a crucial nutrient for growth and reproduction in animals. Rumen microbes synthesize these building blocks from ammonia, a by-product of fermentation in the rumen. Given this fact, even poor quality forage can supply some protein for the animal.

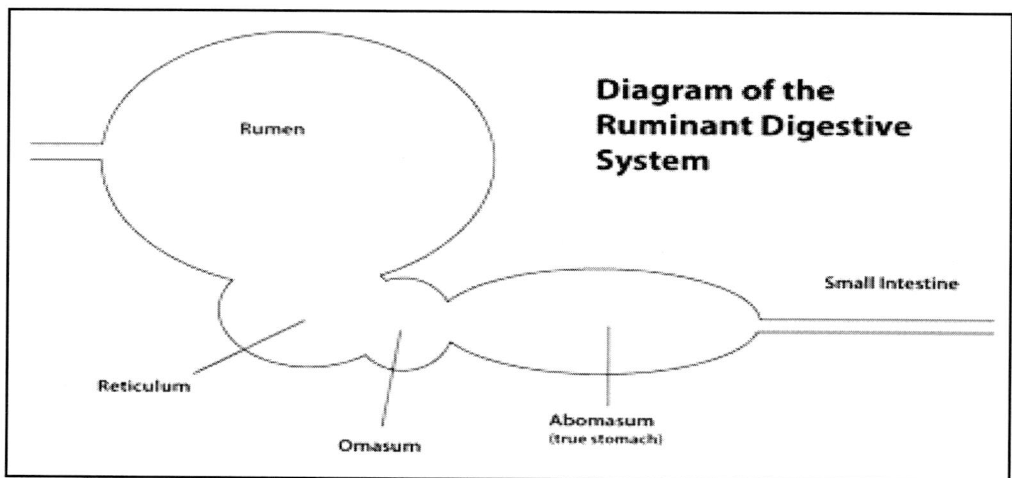

Nutrients absorbed from the digestive tract include volatile fatty acids, amino acids, fatty acids, glucose, minerals, and vitamins. These are used in the synthesis of the many different compounds found in meat, milk and wool, and to replace nutrients used for maintaining life processes including reproduction. Digestion begins when an animal takes a bite from the forages. As the animal chews the feed is formed into a bolus—a packet of feed capable

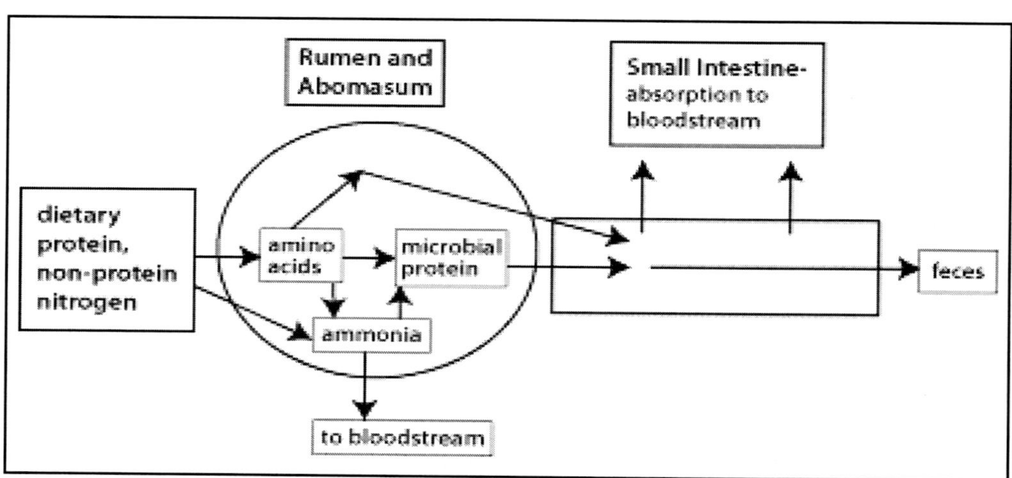

of being swallowed. Saliva is excreted, which further aids in swallowing and serves as a pH buffer in the stomach. Once in the rumen, the feed begins to undergo fermentation. Millions of microorganisms ingest the feed, turning out end products which serve as a major source of nutrients for the animal. Some of the principle products formed are ammonia, methane, carbon dioxide, and *volatile fatty acids* (VFAs). VFAs are absorbed and used as energy by the animal. Ammonia can be absorbed into the animal's system through the rumen wall, or can be consumed by bacteria to become microbial protein. This microbial protein is then passed through the digestive system to be absorbed in the small intestines.

Feeding Value of Forages

Typically it is desirable to make the maximum use of the forages, which can be effectively grown or obtained locally. Generally supplemental feed is needed for milch animals and calves. When formulating the additional feed, a reasonable knowledge of the nutrients provided by the forages is essential.

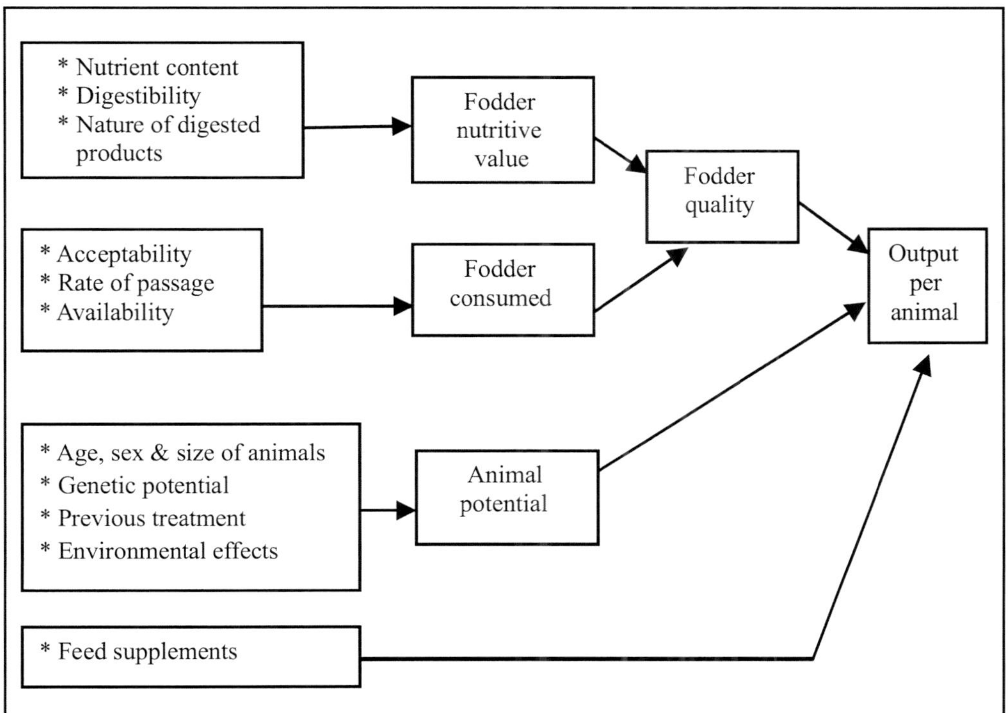

Fig. 1. Fodder Quality and Output per Animal- an Interaction

a) Variation in Nutrients Content and Feeding Value of Forages

A forage plant is a complex form of organic and inorganic compounds which may be essential or nonessential and toxic or nontoxic chemicals. Forage plants essentially supply, energy, protein, minerals and vitamins to the animals at varying degrees depending upon several factors. Major factors affecting the nutrient content and feeding value of forages-

- The species, variety and strain of the plant
- The parts of the plant viz., leaves or stem
- The stage of maturity at harvest
- The fertilization and soil characteristics
- Climates, weather and seasons
- Changes during preservation, processing and storage

b) Voluntary Intake (VI)

The amount of forage dry matter, cattle will voluntarily consume often has more influence on the ability of a specific forage to supply the needed energy of cattle than the usable energy contents. A major reason for the difference in VI among the forages is the varying rates of digestion, specially in rumen fermentation. With faster digestion, the materials leave the rumen more quickly making space for additional forages. Usually a more rapidly digested forage has a grater feeding value because it can be used in large amounts with less concentrates needed.

c) Digestibility of Nutrients

The various plant nutrients ingested by the animal are absorbed from the digestive tract in different proportion. The digestibility may therefore, be defined as the amount of nutrient which has been absorbed from the digestive tract per unit of nutrient ingested by the animals. Many factors influence the digestibility of forage nutrients in the animal system and are associated with the plants or animals. Some of the important factors are as follows:

- Chemical composition of forage
- Stage of maturity of forage
- Physical form of the forage
- Method of processing and preservation
- Species of animal
- Physiological condition of the animal

Besides digestibility, the efficiency of utilization of digested nutrients is also important in determining the feeding value of a forage. The animal utilizes various nutrients for many functions like maintenance, production and reproduction. The health and production functions are the reflections of the quantity of nutrients available to the animal per unit of time (Fig 1).

Forage Feeding Systems

The forge feeding systems of livestock in vogue in different parts of India and other developing countries are as follows-

1. Exclusive Grazing

Nondescript small size cattle and their progenies are mostly let out for grazing for 6 to 8 hours daily. The milk yield from such cows, is very low and farmers get about 0.5 to 1.0 litre milk daily. They are kept mainly for male calves for draft/ traction purposes and dung for manure and fuel. Rearing of such cows under stall-feeding is uneconomical.

2. Grazing Supplemented with Roughage

This system is found in low rainfall areas, where grazing is limited on poor pastures, roadsides and bunds. Animals on return from grazing are offered agricultural crop residues like wheat straw, paddy straw, ragi straw and chaffed stovers of sorghum/ maize. However, milch animals get first priority over the others.

3. Grazing Supplemented with Concentrates

Limited livestock farmers capable to spare grains or purchase brans/ oil cakes from the market practice this system. Although, oil cakes like mustard cake, groundnut cake, linseed cake etc. are commonly and preferably fed to milch animals and working bullocks when they return home from grasslands.

4. Stall Feeding on Exclusive Forages

Animals may be fed on single forage or a mixture of two or more forages. The combinations mostly are as follows:

(a) Dry Forage Feeding

Poor quality forages viz. cereal straws and stovers, mixed dry mature grasses and jungle hay etc are fed to animals either long hay or chaffed ones. However, in certain areas these dry forages are first soak into water and then fed to animals. The soaking period ranges from just dipping of chaffed forages in water to several hours (1 to 4 h). Again the ratio of chaffed dry forage and water is narrow for cattle and wide (more water) for buffaloes. This system is commonly known as 'Sani' feeding which also supplemented with kitchen wastes.

(b) Green Forage Feeding

During kharif season when there is abundant availability of green forages like chari, bajra, maize including various monsoon grasses, the animals are fed green forages for a limited period. While during rabi season very few farmers cultivate forages like oats, berseem, lucerne and barely for sole feeding. No efforts are made to balance the nutrients of forages and feeding completely depends upon the availability of green forage in the fields.

(c) Mixed Forage Feeding

Depending upon the availability of green forages with the farmers, the animals are fed mixture of forages. However, no consideration is given for balancing the ration, the palatability and feeding value of forages.

5. Stall Feeding on Forage and Concentrate

Medium and large livestock farmers in the areas where arable lands are under cultivation for grain and cash crops and there is no scope for grazing practice this system of feeding. The concentrates are available from the cleaning of food grains, barns of wheat and rice, home grown coarse grains and oil cakes of groundnut, mustard and linseeds. While forages are available as agricultural by-products including green weeds.

6. Stall Feeding of Compound Feeds

Compound feeds consisting of basal roughage of wheat straw or chaffed paddy straw and concentrates are fed to animals largely in organized or unorganized dairies of town and cities. This system is also popular in the milk-shed areas.

Limitations of Existing Systems

Due to lack of knowledge about the feeding value of different feeds and nutrient requirement of different categories of farm animals by the livestock farmers, the above-mentioned feeding systems suffer from many limitations.

- Animals are either underfed or overfed
- Supply of nutrients is unbalanced
- Inefficient utilization of feed nutrients
- Animals suffer from deficiency disease syndromes

- Adverse effect on milk production and growth of animals
- Most uneconomical when considered in terms of health, growth, milk production and reproductive efficiency of animals

Feeding of Dairy Animals

Feeding of Calves

Adequate feeding of young calves during the first three months is important on a dairy farm for their optimum growth and to raise them as herd replacement. Although, the growth of calf actually starts at fertilization and pre-natal growth affects post-natal growth very significantly and the latter is correlated with the birth weight. As the newborn calf does not have the functional rumen, its nutrient needs are more similar to those of simple stomach animals than the matured ones. Hence the newborn calf is a baby which requires special nutritional and management cares. Guidance for feeding schedule of calves from birth to 3 months has been presented in Table 1. The milk should be heated to luke warm (body temperature) and given to calves twice daily i.e. morning and evening.

Table 1: Feeding of Calves from Birth to 3 Months of Age

Age (day)	Whole milk (kg)	Calf starter* (g)	Legume forage (g)
Day 1-3	1/10th of body weight	nil	nil
Day 4-7	1/10th of body weight	nil	nil
Day 8-35	1/10th of body weight	*ad. lib.*	*ad. lib*
Day 36-45	1/15th of body weight	*ad. lib.*	*ad. lib*
Day 46-56	1/20th of body weight	*ad. lib.*	*ad. lib*
Day 57-90	nil	*ad. lib.*	*ad. lib*

Legumes should include berseem, lucerne, cowpea, etc. either in green or hay form. Green fodder should be free from soil and other external contamination. It should be mixed with good quality wheat straw @ 500g wheat straw or chaffed rice straw per kg of green fodder. This is important to avoid rapid consumption of green berseem/ lucerne/ cowpea as it may cause tympani/ bloat in growing calves.

* Calf starter constituent	Parts
Oat /Barley/Maize grain (crushed)	50
Groundnut cake/Linseed cake/Til cake	30
Wheat bran/Rice bran	08
Good quality fish meal/dried skim milk	10
Mineral mixture	02

To 100 kg of above mixture ingredients like Rovimix- 10 g, Common salt- 0.5 kg and Aurofac/ TM-5 (antibiotics)- 20 g may also be added to fortify the quality of calf starter:

Feeding of Growing Calves/ Heifers

Compared with baby calves, growing calves/ heifers are relatively easy to feed and this is one reason that these categories of dairy animals are often neglected. Initially the growing calves should receive substantial share of its nutrients from concentrates. As they become older the proportion of forage can be increased gradually in the ration. But the amount of concentrate depends upon the quality of forage. With better quality forage, less concentrate is needed. However, with more typical quality forage, some concentrate may be needed for their good growth and early maturity. A dairy heifer can even obtain all her digestible energy from good quality forage, until just before calving. Around 2 to 4 weeks prior to calving, it is desirable to feed substantial amount of concentrates to heifers so that they can get accustomed to the type of rations needed for lactation. When substantial concentrates are given prior to calving, heifers with high potential often show to have higher peak as well as total lactation milk yield.

The calculation of nutrients requirement of growing calves requires specific knowledge of chemical composition of each feed ingredients and nutrients requirement for a particular body weight. Some examples (Table 2) are given for the guidance in formulating suitable forage based feeding regimes for growing calves.

A livestock farmer desirous of producing concentrate mixture of his own should know the various types of feed ingredients required for making concentrate mixture ideal for animal feeding. If the farmer is not well conversant with the quality of raw feed ingredients and unconventional feedstuffs available in the region, he may not be able to formulate an ideal concentrate mixture for his animals economically. The various feed ingredients and their proportion of mixing in brief are as follows-

(a) Primary source of energy (30-40%): Cereal grains like maize, barley, wheat, sorghum, oats etc.

(b) Primary source of protein (25-30%): Oil seed cakes like groundnut cake, mustard cake, cotton seed cake, linseed cake, sesame cake and animal protein supplements like fish meal, meat meal, skim milk powder etc.

(c) Diluents (5-25%): Cereal by-products like wheat bran, rice bran etc. and pulse chunies like mung chunni, arhar chunni, massoor chunni, gram chunni etc.

(d) Common salt (1-2%)

(e) Mineral supplements (1-2%)

Table 2: Forage based rationing of growing cattle/buffalo calves (after 3 months to maturity)

Category of animal (age)	Concentrate mixture** (kg)	Forage mixture
1. 3-4 months (b.wt. 70-90 kg)	a) 1.0-1.5	Green oat/maize/silage 5 kg + good quality hay 0.5 kg
	b) 0.5-1.0	Green berseem 5 kg + good quality hay 1.0 kg
2. 4-5 months (b.wt. 100 kg)	1.5-2.0	Green fodder 5-10 kg + good quality hay 1.0 kg
3. 5-6 months (b.wt. 130 kg)	2.0	Green fodder 5 kg + good quality hay 1.0 kg + straw 0.5 kg
4. 6-9 months (b.wt. 130-160 kg)	2.0	Green fodder 10 kg + good quality hay 0.5 kg + straw 1.0 kg
5. 9-12 months (b.wt. 190 kg)	2.0	Green fodder 15kg + hay/straw/dry mixed grass 1.5 kg
6. 12 months to age of conception (b.wt. 200-300 kg)		
Cattle*	a) 1.5-2.5	Green maize/oat/jowar 20 kg + straw/dry mixed grass 1 kg
	b) 1.0-2.0	Green berseem 20 kg + hay/dry mixed grass 4 kg
	c) 2.0-3.0	Straw/dry mixed grass 4 kg + green fodder 5 kg
Buffalo	2.0-2.5	Green maize/oat/jowar 20 kg + straw/dry mixed grass 2 kg

*Crossbred cattle require more concentrate to sustain higher growth rate than indigenous ones
**Concentrate mixture

Characteristics of a Good Concentrate Mixture

- A good concentrate mixture should contain 17-22% crude protein (CP) and 65-75% total digestible nutrients (TDN)
- Final concentrate mixture should contain 1 % common salt and 2% mineral and vitamin mixture
- Moisture content should not be more than 12-15%
- Crude fibre content should not be more than 18%
- Concentrate ingredients should be properly ground before mixing
- It should not be too dense or too bulky

Feeding of Milch Cows/ Buffaloes

For the development of practical feeding guides, the values on nutrient requirement (Table 3) are compiled on the assumption that fat content of cow milk varied between 4-5% and in buffalo milk between 6-8%. While the body weights ranged from 150 to 400 and 200 to 500 kg in cows and buffaloes, respectively.

Table 3: Nutrient Requirement of Dairy Animals

Body weight (kg)	Daily milk yield (kg)	Requirement			
		Cows		Buffaloes	
		CP (g)	TDN (kg)	CP (g)	TDN (kg)
150	2	311	1.95		
	3	388	2.29		
	4	464	2.63		
200	2	382	2.34	422	2.56
	3	458	2.68	518	3.01
	4	535	3.02	615	3.46
250	3	489	3.04	549	3.37
	4	566	3.38	646	3.82
	5	631	3.72	743	4.27
	6	708	4.06	843	4.72
300	4	610	3.72	691	4.16
	5	688	4.06	788	4.61
	6	765	4.40	891	5.06
	7	842	4.74	982	5.51
400	5	775	4.74	877	5.28
	6	852	5.08	974	5.73
	7	929	5.42	1077	6.18
	8	1006	5.76	1174	6.63
500	5			938	5.94
	6			1037	6.39
	7			1134	6.84
	8			1231	7.29
	9			13.28	8.74

Feeding of leguminous fodder like green berseem or lucerne and cowpea during winter and summer/ rainy seasons, respectively along with 1-2 kg of dry roughage to lactating cows and buffaloes sustained a milk yield upto 10 kg/ day (Table 4). But animals yielding more than 10 kg/ day required supplementation of energy rich concentrates @ 1 kg for every 2.5 - 3.0 kg of extra milk. Although, feeding of leguminous fodder exclusively resulted in wastage of protein, the most deficient nutrient in cereal fodder based feeding regime. But a mixture of leguminous and cereal fodder at the ratio of 2:1 or 1:1 were found to be more promising. A balanced fodder mixture can be produced by growing legume fodder viz., cowpea, berseem, lucerne, stylo, dolichos, siratro etc. and cereal fodder viz., maize, oat, sorghum, hybrid napier, guinea, paragrass etc. Cereals like maize/ sorghum can be cultivated along with legumes like cowpea during kharif, and berseem & oat during rabi season to supply a balanced fodder mixture. Cows consuming fodder mixture containing one part of berseem and oat each (fresh weight basis) with 1 kg concentrate (for letting down of milk) yielded 7-8 kg of milk/ head/ day. Similarly, hybrid napier and lucerne (3:1) or maize and cowpea (1:1) mixture sustained upto 7 kg milk/ head/ day. Thus, the intake and utilization of nutrients from comparatively low quality fodder were improved by judicious inclusion of legumes.

Table 4: Performance of Milch Cows/ Buffaloes Fed all Forage/ Forage based Rations

Kind of forage*	Level of supplementation (head/ d)	Milk yield sustained (kg/ d)
1. Berseem	1-2 kg dry roughage	8-10
2. Berseem	Concentrate @ 1 kg/ 2.5-3.0 kg extra milk	> 10
3. Cowpea	1-2 kg dry roughage	6-8
4. Cowpea	Concentrate	> 8
5. Hybrid napier + Lucerne (3:1)	1-2 kg dry roughage	6-7
6. Maize + Cowpea (1:1)	1-2 kg dry roughage	6-7
7. Berseem + oat (1:1)	1-2 kg dry roughage	7-8
8. Oat (10% flowering)	1-2 kg dry roughage	7-8
9. Guinea grass	1-2 kg dry roughage	> 5
10. Maize/ sorghum/ bajra/ hybrid napier	1 kg concentrate	4-6
11. Sorghum + Cowpea (2:1) silage	Nil	> 4
12. Oat + berseem (1:1) silage	Nil	5-6
13. Berseem/ Lucerne hay	1 kg concentrate (barley)	> 10
14. Cowpea hay (at 50% flowering)	1 kg concentrate (barley)	8-10

*Offered *ad libitum*

Cereal fodders including grasses are poor in quality, having low CP and high fibre contents coupled with low voluntary intake. Therefore, performance of lactating animals fed on such fodders like maize, sorghum, guinea and hybrid napier grass etc. alone, were adversely affected which could meet only the maintenance with little production needs of animals. But supplementation of concentrates improved the level of milk production. However, oat at early stage of growth (upto 10% flowering) sustained 6-8 kg milk yield daily.

Again the surplus green forages available during the flush period of growth can be conserved as hay or silage to provide some quality forage to milch animals during the scarcity period of October-November or May-June. Berseem / Lucerne hay with I kg of concentrates sustained upto 10 kg daily milk yield during summer months of May-June, when there is complete scarcity of cultivated green forage. Although, milk yield was low (5 kg/ head/ day) in cows fed silages of jowar & cowpea (2:1) or oat & berseem (1:1) without any concentrate supplements. Hence, there is a need of supplementing quality concentrate also, in silage based diets of milch animals to sustain their optimum milk yield under hot dry condition of summer months.

The complete ration of milch cows and buffaloes producing less than 10 kg milk daily should contain 12-14% CP and 60-62% TDN provided the minimum intake is 2% of the body weight. Important leguminous forages contain 18-20% CP and 58-60% TDN, and cereals contain 6-8% CP and 60-65% TDN in dry matter except the pre-flowering oat which contain 12-15% CP (Table 5). Thus a mixture of equal amount of cereals and leguminous forages contains the desired level of protein and energy for milch animals. However, all forage rations of milch animals should be supplemented with phosphorous rich complex mineral mixture. In a condition in which only one type of green forage (legume or non-legume) is available for the feeding of milch animals, the leguminous forage can be balanced though the supplementation of small quantity of high energy grains like maize, barley, jowar etc., and protein deficiency on cereal forage feeding may be balanced through high protein oil cakes.

Feeding of High Producing Milch Animals on Mixed Rations

The genetically superior livestock needs balanced diets for superior feed efficiency to fully express their production traits. Cows with average daily milk yield up to 10 kg, may be fed a mixed ration of crop residues, green fodder and a concentrate mixture (of average quality containing 18% CP and 70% TDN), so as to meet the nutritional requirements. Otherwise, another approach could be to increase the quality of concentrate mixture if green fodder is not available only straw or stovers are the forage sources.

Table 5 : Nutritional value (% dry matter basis) of some common forage for dairy animals

Forage	Species	DMI (kg/ 100 kg b.wt.)	CP (%)	TDN (%)
A. Greens				
1. Berseem	Buffalo	2.41	24.97	70.24
	Cattle	2.10	21.63	57.81
2. Cowpea	Cattle	2.90	20.46	61.05
3. Guar	Cattle	3.25	17.00	47.49
4. Lucerne	Cattle	2.25	19.58	53.14
5. Stylosanthes sp.	Cattle	2.42	15.10	55.75
6. Bajra/ pearl millet	Buffalo	-	8.84	63.81
7. Barley (pre-bloom)	-	-	12.56	60.47
8. Sorghum	Cattle	3.04	6.52	56.08
	Buffalo	-	4.57	63.33
9. Hybrid Napier (NB-21)	Cattle	2.20	4.69	57.99
	Buffalo	-	4.58	63.56
10. Maize	Cattle	2.43	9.32	62.99
Green stalk	Cattle	2.02	4.33	58.57
11. Oats	Cattle	2.57	9.47	62.42
	Buffalo	-	7.81	59.58
12. Sudan grass	Buffalo	3.22	6.35	52.74
13. Teosinate	Cattle	-	9.27	52.46
	Buffalo	1.90	9.23	57.90
14. Dinanath grass	Cattle	2.00	5.13	56.26
15. Dub grass	Cattle	2.50	8.46	55.00
16. Sain grass	Cattle	1.95	2.55	46.68
Hay				
1. Berseem	Cattle	-	15.84	65.80
2. Cowpea	Cattle	2.80	12.30	60.00
3. Lucerne	Buffalo	1.88	14.38	54.40
	Cattle	2.58	14.86	50.04
4. Oats	Buffalo	-	5.61	64.38
	Cattle	-	3.69	54.10
5. Sain grass hay	Cattle	1.95	1.96	48.85
Silages				
1. Hybrid napier (NB-21	Cattle	1.91	4.18	54.11
2. Sorghum	Cattle	1.60	3.75	62.16
3. Maize	Cattle	-	5.12	61.33
4. Oats	Buffalo	-	5.43	63.40

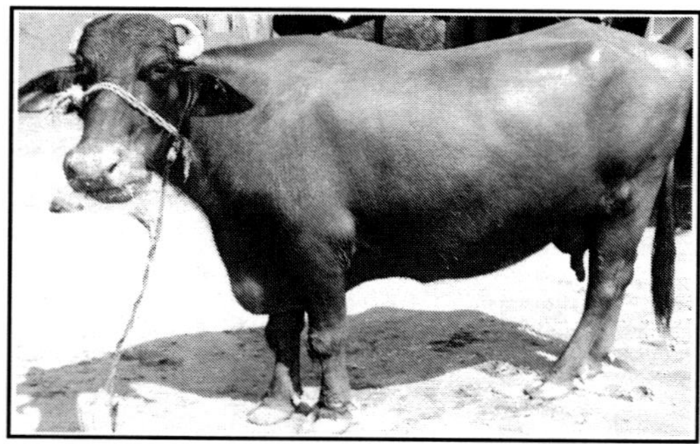

Fig. 2. A Murrah Buffalo Yielding 8-10 kg of Milk on Forage based Rations

But cows with 11 to 20 kg daily milk yield, are the improved ones and the farmers can expect a good economical return from them. They require a well balanced diet for effective persistency, reproduction and feed efficiency. They may be given a ration (Table 6) comprising of little quantity of straws/ stovers, mixture of legume & non-legume (cereal) green forage and good quality concentrate mixture (20% CP and 72% TDN). While cows yielding >21kg milk per day, needs special attention to meet their nutritional requirement. Such high yielding cows will always be deficit in energy in first two trimesters of lactation on conventional diets. They should be given a ration (Table 6) comprising of legume and nonlegume fodders, and a very good quality concentrate mixture (22% CP and 74% TDN). But they should not be fed poor quality crop residues like straws and stovers.

Table 6: Rationing (as such basis) of High Yielding Milch Cows

Feedstuffs/ items	Quantity (kg/ head/ day)
A. Category-I (350-400 kg b.wt, 15 kg daily milk yield & 4-5% milk fat)	
1. Maize/ Sorghum/ Oat	20
2. Berseem/ Lucerne/ Cowpea	25
3. Wheat/ Rice straw	1
4. Concentrate mixture	5
B. Category-II (350-400 kg b.wt, 25 kg daily milk yield & 4-5% milk fat)	
1. Maize/ Sorghum/ Oat	20-25
2. Berseem/ Lucerne/ Cowpea	20-25
3. Concentrate mixture	8-9

Adequate Energy and High Protein Diet Concept

Beyond a particular level of milk yield, say 15 kg/ day in case of cows and 12 kg/ day in case of buffalo, it is very difficult to meet their energy requirements through ordinary nutrient balancing techniques. Usually all such cows and buffaloes will remain under negative energy balance during first 5 months of lactation. Incase the energy requirement is met fully during lactation through addition of high proportion of grains or fat, there may be further increase in milk production and persistency from such animals will be better.

Similarly, protein content and its quality in the ration are also receiving attention now a day. For cattle protein content as high as 24% is being recommended, fulfilling the requirements of both rumen degradable and undegradable proteins. During the period of first trimester when protein and energy balances are negative, two-third increase in milk yield can be achieved due to adequate protein and one-third increase owing to optimum energy level in the rations.

Limitation of High Forage Rations

The major limitation on production in dairy animals fed high forage rations is voluntary intake. After the first meal, a milch cow can eat more forage only to the extent she can digest the forage already eaten. Therefore, the best indication of forage quality is the amount of forage she can eat in a 24 hours period per unit of body size. This takes into account per cent digestibility, rate of digestion, size of the animal and any influence palatability may have on voluntary intake. The factor which influences voluntary intake, and thus forage quality, to the greatest extent is stage of maturity when harvested either as green fodder, pasture, hay or silage. As forage crops mature, the plant becomes more and more lignified making the cellulose (energy) inaccessible to the rumen bacteria, this slows down rate of digestion and reduces voluntary intake.

It has been observed that on average good milch animals are capable of producing 8 to 12 kg of milk per day every month of the twelve months of the year, if they are fed year round at a level of nutrient intake equal to that provided during February-March (when forage quality was at its peak). That high temperature is not the main limiting factor on production is indicated by the fact that the months lowest in average yield of milk do mot coincide with the hottest weather (May-June).

In order to encourage harvesting of all forage at the proper stage of maturity, which should permit the same levels of production year around that is now obtained only in February-March. It is suggested that the producers/ farmers be motivated to accomplish this by changing his measure of accomplishment from tons of forage per year to a forage of such quality as could maintain the same or nearly so level of milk production as could be obtained in February-March with a minimum of concentrate feeding.

In an all forage diet, that is on a ration composed of succulent forages and small quantities of dry feeds, the total intake of feed may not be sufficient to supply all the energy required. The intake of dry matter can be increased with forages like berseem or lucerne by wilting to a certain extent which will bring down the moisture content from 80 to 90 per cent to about 55 to 60 per cent. It nevertheless has to be borne in mind that the voluntary intake from an all-forage ration may not be sufficient to meet the total requirement of digestible energy which has to be met through supplementation of a high energy concentrate mixtures. As the feed intake increases, the total digestibility of the nutrients in the ration diminishes. So the intake of digestible nutrients, as calculated from tables of nutritive value of feedstuffs, will be over valued. In other words, increasing feed intake (especially in high producing animals) a much larger supply of digestible nutrients will have to be provided in the form of extra concentrate mixtures.

Another problem which may crop up in excessive forage feeding with limited grain supplementation; especially at early lactation, is ketosis. This is a very complex disorder and cannot be predicted just how serious this problem might become, nor can firmly be ascertained at varying levels of grain intake, limited or liberal. The high producing cows are usually in an acute need for glucose. Some increase in ketosis in them seems to be a reasonable assumption. What, however, is required is a glucose product protected in the small intestine for direct absorption. A lowering in the conception rate may be a problem if on excessive forage feeding, negative energy balance becomes prolonged. The results of an investigation carried out in Lousiana (USA) showed that cows losing weight had considerably lower conception rate (of 43 per cent as compared to 65 per cent in cows gaining weight). However, it is difficult to interpret this and other similar studies because to often what appears to be a simple energy deficiency is complicated by deficiencies in several other major nutrients. This subject needs much further work but it does provide a caution signal for those who might go too far with the high forage diets.

References

Arora, S.P. 1978. *Feeding of Dairy Cattle and Buffaloes*. ICAR, New Delhi.

Banerjee, G.C. 1988. *Feeds and Principles of Animal Nutrition*. Oxford & IBH Publishing Co. Pvt. Ltd., New Delhi.

Kundu, S.S., Singh, S., Mahanta, S.K. and Pailan, G.H. 2005. *Feeding Farm Animals.* Satish Serial Publishing House, Delhi.

Mahanta, S.K. and Pachauri, V.C. 1999. Importance of quality forage to milch animals for economical milk production. Milcow 21: 26-28.

Miller, W.J. 1979. *Dairy Cattle Feeding and Nutrition.* Academic Press, London.

Minson, Dennis J. 1990. *Forage in Ruminant Nutrition.* Academic Press, Inc., NY.

Mott, G.O. and Moore, J.E. 1969. Forage evaluation techniques in perspectives. In: Barness, R.F., Clauton, D.C., Gordon, C.H., Kloptenstein, T.J. and Waldo, D.R. (eds.) Proceed. Natl. Conf. on Forage Quality Evaluation and Utilization. Nebraska Cent. Educ., Lincoln, NE.

Mudgal, V.D. 1992. Formulation of economic rations for dairy cattle under field conditions. Proceed. All India dairy husbandry officers' workshop, Oct. 22-23, 1992 at NDRI, Karnal.

Pathak, N.N. and Jakhmola, R.C. 1983. *Forages and Livestock Production.* Vikas Publishing House Pvt. Ltd., New Delhi.

Ranjhan, S.K. 1993. *Animal Nutrition in the Tropics.* 3rd rev. edn. Vikas Publishing Pvt. Ltd., New Delhi.

Thakur, S.S. 2001. Scientific feeding for higher milk production. In Short Course on 'Biotechniques to augment lactation in bovines, May 22-31, 2001 at NDRI, Karnal.

❑❑❑

Chapter **28**

Organic Forage Production Management and Production Issues

S. B. MAITY
Indian Grassland and Fodder Research Institute, Jhansi-284003 (U.P.), India

Introduction

Organic farming in its broader sense is a system of farm design and management to create an ecosystem, which can achieve sustainable productivity without the use of synthetic inputs such as chemical fertilizers and pesticides (Speedy *et al.*, 2001). Organic forage production system is holistic approach for sustainable forage production utilizing available organic inputs with soil, water, fertilizers, seed and pesticides management. It promotes and enhance the agro-ecosystem health, including biodiversity, biological cycles, and soil health. Organic forage production system should adopt locally available inputs and adapted systems for their efficient utilization. This is accomplished by using, wherever possible, agronomic, biological and mechanical methods, as opposed to using synthetic materials, and harnessing the synergies and beneficial interactions among various components to fulfill any specific function within system. The primary goal of organic farming is to optimize the health and productivity of interdependent communities of soil life, plants, animals and people. Any organic system is designed to achieve few goals like:

- Enhance biological diversity within the whole system;
- Increase soil biological activity;
- Maintain long term soil fertility;
- Recycle waste product within the system to return nutrient to soil;
- Rely on renewable resources in locally organized agricultural system;
- Promote the healthy use of soil, water as well as to minimize all forms of pollutions;
- Maintain organic integrity in all steps of production system.

Organic farming is not new to Indian farming community. Several forms of organic farming are being successfully practiced in diverse climate, particularly in rain fed, tribal, mountains and hilly areas of the country. But much of the organic produce were of economic importance like herbs, medicinal plants etc. Presently organic farming is gaining attention among farmers, entrepreneurs, policy makers and agricultural scientists for the reason that it minimizes the dependence on chemical inputs (fertilizers, pesticides, herbicides and other agro-chemicals) to improve the quality of resources. The Green revolution technologies involved greater use of synthetic agrochemicals with the adoption of nutrient responsive, high-yielding varieties of crops that boosted the production per hectare in most of the cases. In pace of time, demand for production increased but actual production either slowed down or in a fixed plate. Priorities in agriculture are gradually focused on individual crop performance followed by total system productivity. Now, organic forage production is very limited due to limited scope of organic dairying in the country. With more demand of organic milk, meat and dairy products from elite consumers, organic forage production process gaining momentum in the country.

Large numbers of farmers opted organic farming either individual or in group but organized way of farming is limited. In Rajasthan, Gujarat, Tamilnadu, Chattisgarh and North Eastern states have started sporadic organic agriculture. Organic way of livestock rearing also gaining popularity in Gujarat and Rajasthan. But system certification yet to be started in India.

Organic systems of forage production for feeding ruminants are based on a grass plus legume based sward with regular reseeding, placing a high demand on seed of appropriate varieties. At present, organic forage production is limited due to different reasons. The seed of forage species used by organic farmers is produced conventionally using inorganic fertilizers, herbicides, pesticides and fungicides to produce economic seed yields and maintain the seed quality and purity required by the seed certification agencies. As per IFOAM, seed for use in organic production must be produced organically. Conventional grass seed crops require mineral nitrogen at precise stage of crop development to regulate the balance between vegetative and reproductive growth, stimulates flowering, ensure good seed filling. In organic systems, nitrogen can be supplied by application of animal manure or by using forage legume to fix atmospheric nitrogen, either by relying on the residual N in the soil or by using forage legumes as companion crops.

The livestock sector plays an essential role in agriculture and economic development as well as food security. Global livestock output grew at a rate of 2% in 1999 to surpass 233 million tones during 2000. The global demand for livestock product expected to be doubled by 2020 (Speedy, 2001). As a global average, animal products provide about 16% of the calories in the diet. Livestock industry therefore can be seen to have great economic and nutritional significance in world as a whole. To support the livestock sector, development of fodder sector is inevitable and in new era organic forage production will support organic milk production to satisfy consumers demand of healthy dairy milk and its product.

Fertilization Application Policy Under Organic Standard

The general principle of fertilization is that sufficient quantity of biodegradable material of microbial, plant and animal origin should be returned soil fertility having biological activity within it. As per the National Standard, the followings (Table 1) are some of the materials allowed either fully or in restricted way under organic production.

State of Standards and Regulations

The Codex Alimentarius Commission(FAO, 2001), a joint FAO/WHO Food Standards Programme, began in 1991 elaborating guidelines for the production, processing, labeling and marketing of organically produced food. The requirements in these Codex Guidelines are in line with IFOAM Basic Standards and the EU Regulation for organic food. In the view of IFOAM, which was actively involved in the elaboration of guidelines, this Codex document is an important step in the harmonization of international rules in order to build up consumer trust.

The IFOAM develops the IFOAM Basic Standards (IBS). These are presented as general principles, recommendations, basic standards and derogations. The IFOAM Basic Standards define how organic products are grown, produced, processed and handled. They provide a framework for certification bodies and standard setting organization world wide to develop their own certification standards, but are not used for certification on their own. However, the country specific certification standards take into account specific local conditions and provide more specific requirements than IFOAM Basic Standards. In India, Codex Standards followed for quality and food safety of food for human consumption. Codex has considered several pesticide residues in milk and recommended MRLs for 83 pesticide residues and several veterinary drugs (Table 2) in milk.

Government of India, in its Gazette of India has (March 24, 2006) fixed standards for 7 bio-fertilizers and given the guidelines for certification in respect of manufacturing organic bio-fertilizers.

Table 1. Organic Nutritional Inputs as Permitted/Restricted by National Standard.

S.No.	Origin	Materials	Production level	Status of acceptance
1	Plant	-Crop residue	On farm unit	Permitted
		-Green manure	On farm unit	Permitted
		-Plant residue compost	On farm unit	Permitted
		-Sea weed extracts	On farm unit	Restricted
		-Saw dust	On farm unit	Restricted
2	Animal	-Dung and urine of livestock	On farm unit	Permitted
		-Poultry manure	On farm unit	Permitted
		-FYM and slurry	On & out farm unit	Out farm restricted
		-Blood, meat, bone and leather meal	---	Restricted
3	Microbial	- Bio-fertilizer	-	-
4	Minerals	- Rock Phosphate	-	Restricted
		- Trace Elements		Restricted
		- Basic slag		Restricted
5	Others	-Industrial by-products		Restricted
		-Saw dust		Restricted
		-Organic house compost		Restricted

*Source: Bhattacharya and Kumar, 2005

Control of Pesticide Residues in Forages

Through green revolution agricultural production has been increased manifold but use of insecticide and pesticide has also been increased in crops as well as in forage crop production. Synthetic organic insecticides including organochloro, organophosphate and organocarbamate exhibits a high degree of persistence in environment as compared to other classes of pesticides

(Sharma et al., 2002). The organochloro is not only persistence but also lipophilic, therefore, accumulates in the plant and animal tissue (Stan, 1990). Consumers are thus constantly exposed to pesticides and concerns for organic forage production to safe guard milk, meat and their products as human food. With the implementation of WTO, clean milk production are now an obligation to the international market. Following steps are now recommended to control pesticides effects:

- *Use of Bio-pesticides:* Which are living organisms or a product derived from an animal or plant sources and which kill the pesticides. Bio-pesticides are safer to human being and also do not leave any toxic residues in crops and hence have importance.
- Use of pest resistant varieties of seed through genetic engineering.
- *Integrated pest management (IPM) techniques:* The FAO, through its regional projects extending technical support in the production of IPM.
- *Integrated Nutrient Management (INM) techniques.*
- Organic forage production is a route to meet the INM. Organic farming is a system in which the maintenance of soil fertility and the control of pest and diseases are achieved through the enhancement of biological process and ecological interactions.

Table 2: Codex Maximum Residue Limits (MRLs) of Veterinary Drugs in Milk

Veterinary Drugs	Recommended MRL, ppb
2,4,5-Trichlorophenoxy acetic acid	0.05
Benzylopenecillin	4.0
Ceftiofur	100
Dihydrostreptomycin	200
Gentamycin	100
Neomycin	500
Oxytetracyclin	100
Spectinomycin	200
Sulfadimine	25
Thiabendazole	0.1

* Sources: Indian Dairyman, 2000

National Standards of Organic Farming

Ministry of Commerce under the "National Programme for Organic Production," has prescribed National Standards for Organic Production. These standards are grouped under following six categories:

- Conversion
- Crop production
- Animal husbandry
- Food processing and handling
- Labeling
- Storage and transport

Conversion Requirements

The time between the start of organic management and cultivation of crops or animal husbandry is known as the conversion period. The whole farm including the livestock should be converted to the standards over a period of time. All standard requirements should be met during conversion period. Regular inspections during conversion period should be carried out. Full conversion period is not required where organic farming practices are already in use. But this has to be verified by the inspection agency.

Maintenance of Organic Management

Organic certification is based on continuance. The certification programme should certify the production, which is likely to be maintained on a long term basis. The converted land and animals shall not get switched back and forth between organic and conventional management.

Landscape

Organic farming should contribute beneficially to the ecosystem. Areas which should be managed properly and linked to facilitate biodiversity are:

- Extensive grassland
- All areas which are not under rotation and are not heavily manured
- Pastures, meadows, orchards, hedges, hedgerows etc.
- Ecologically rich fallows or arable land.
- Ecologically diversified field margins
- Waterways, pools, springs, ditches, wetland and swamps.

The certification programme should be set standards for a minimum percentage of the farm area to facilitate biodiversity and nature conservation.

Crop Production

- *Choice of crops and varieties* : All seeds and planting materials should be certified organic, well adapted to local climatic conditions and resistant to pests and diseases. If certified organic seed or planting material is not available then chemically untreated conventional materials can be used. Uses of genetically engineered seeds, pollens, transgenic plants are not allowed.
- *Duration of conversion period* : The minimum conversion period for plant product produce annually is 12 months prior to the start of the production cycle. For perennial plants, the conversion period is 18 months from the starting of organic management. Depending on ecological situation, the certification agency can extend or reduce the minimum conversion period.
- *Diversity in crop* production is achieved by a combination of versatile crop rotation with legumes and by appropriate coverage of the soil with diverse plant species during the year of production.
- *Fertilization Policy* : Biodegradable material of plant or animal origin produced on organic farms should form the basis of fertilization policy. Emphasis should be given to generate and use own on-farm organic fertilizers. Brought in fertilizers of biological origin should be supplementary but not replacement. In case of deficiency, mineral fertilizers can be used as supplementary source and should be applied in their natural composition. Biofertilizers can be used safely under all ecosystems and in all the crops.
- *Pest disease and weed management* : Weeds, pest and diseases should be controlled by a number of preventive cultural techniques, such as suitable rotations, green manuring, balanced fertilization programme, mulching, mechanical control and the disturbances of pest development cycles. Botanical pesticides prepared at farm from local plants, animals or microorganisms are allowed. Use of synthetic chemicals such as fungicides, herbicides, synthetic growth regulators and dyes are prohibited. Use of genetically engineered organisms or products is prohibited.
- *Contamination control* : All attempts should be made to minimize contamination and from outside and within farm.
- *Soil and water conservation* : Soil and water resources should be handled in a sustainable manner to avoid erosion, salination, excessive and improper use of water and the pollution of surface and ground water.

Organic Forage Production and Management

Selection of Area for Organic Forage Production

1. Total area- as per the requirement of feed and fodder
2. Land should have deep well or residue less irrigation facilities.
3. Should contain minimum synthetic residues.
4. There should be natural grassland as part.

Requirement of Feed and Fodder

Organic forage production should be determined as the number of animal in the system. Forage requirement should be calculated on thumb rule considering DM requirement of the animal and their followers on long term basis as follows. For a 50 numbers cattle herd, farming could be initiated with 35-36 young heifers of 2-3 years of age and after 3rd year 50 nos. herd will be maintained following culling and replacement basis (Table 3).

Table 3: Requirement of Feed and Forages for 50 Cattle Unit.

Animals	1st year	2nd year	3rd year	4th year	5th year
Large animal	35	35	35	35+F	35+F
Total DM requirement	800 qt	1100 qt	1420 qt	1420 qt	1420 qt
Requirement of green fodder	1820 qt	2520	3250 qt	3250 qt	3250 qt
Requirement of Bhoosa /straws	200 qt	250	300 qt	300 qt	300 qt
Requirement of conc.	300 qt	350	470qt	470 qt	470 qt

Concentrates will be reduced to 40-50% strategically with the replacement of surplus leguminous hay.

Availability of clean water for human, animal and for fodder production is decreasing day by day. Irrigated water with heavy metal (Table 4) may pollute the soil, feed and forages and thereby animal system and its products.

Table 4: Maximum Residue Limits (ppm) of Heavy Metals in Water

Category	MRL	Heavy Metals
Highly Toxic	10	Cadmium, Mercury, Selenium
Toxic	40	Barium, Cobalt, Cupper, Lead, Molybdenum
Moderately Toxic	100	Arsenic. Iodine
Slightly Toxic	1000	Magnesium, Aluminum

Nutrient Management in Organic Farming

Organic manure will be the main source of nutrients in the system.

1. Use of FYM on the basis of nitrogen requirements for fodder produced from organically adopted animals.
2. Liquid manure through fermentation of green leafy materials, cattle urine from system etc.
3. Use of composting dung, vermi-composting, vermin-wash. Locally adopted earth worm will be utilized for vermin-compost production.
4. Biologically derived nutrients- mulching, composting, N-enriched Phospho-composting, biological pesticide, weedicide and herbicides.
5. Green manuring

Vermi-compost production

Earth worm (*Icenea foetida*) for vermi-compost

Crop Rotation

Maize/Sorghum/Sesame+ Cowpea- Berseem/ Oat/ Barley/ Mustard-Sorghum/ Maize+ Cowpea

Inter-Culture

Guinea + Cowpea/ Guinea + Berseem

Berseem-Guinea interculture

Round the Year Fodder Production Plan under Organic System

Under organic farming fodder production should be planned in advance considering following recommendations:

- Combinations of leguminous and non-leguminous forage crop in the system.
- Scheduled crop rotations (R) considering local weather conditions.
- Soil conditions, preferably red laterite, minimum pesticide residues.
- Under assured irrigations, selection of forages suitable under irrigation, however, land should be at least 50% irrigated for more production.
- Availability of sufficient bio-fertilizers like Farm Yard Manures, vermin-compost, green manure etc.
- Use of local varieties of forages under rain fed condition and improved varieties under irrigated conditions.
- Minimum 2 years conversion period for land from conventional soil to organic soil.
- Use of integrated forage production system.
- Preferably forage seed should be produced organically.

Fodder Production Plan under Assured Irrigations

Crop rotation (R)	Sowing time	Availability	Green Yield (qt/ha)
R1: NB21+ Cowpea- Oat+ Mustard- Multicut Sorghum	June- Nov-March	Round the year	2000.00
R2: Maize+ Cowpea-Lucern+Mustard-Sorghum +Cowpea	June-Nov-March	Round the year	1800.00
R3: Sorghum + Cowpea-Berseem+Mustard-Maize +Cowpea	June- Nov-March	Round the year	1800.00
R4: Guinea + intercrop(Cowpea-berseem-cowpea)	June+ Intercrop(June- Nov-March)	Round the year	1300.00

Fodder Production Plan for Rain-fed Areas

Crop rotation (R)	Sowing time	Availability	Green Yield (qt/ha)
R1: Sorghum + cowpea-Oat+ Mustard	June end- Nov	July- Feb next year	1000.00- 1200.00
R2: Maize+ Cowpea-Barley+ Mustard	June end- Nov	July- Feb next year	1000.00
R3: Sorghum + Cowpea-Barley	June end- Nov	July- Feb next year	1000.00

Conclusions

As we entered into the third millennium, the problem for high agricultural production still remains with the increase of population thereby demands. To meet the food requirements farmers' aim is to increase food grain production followed by dairy milk and its products as well as meat. But the pest problems will become more serious with the intensification of agriculture. Therefore strategic use of pesticide and agricultural process are necessary. Organic forage production will be next alternatives for organic milk and meat production to meet the challenges of market demand and to satisfy the health conscious consumers.

References

FAO, 2001. Codex Alimentarius- Organically Produced Foods. FAO, Rome. Govt. of India. 2006. *The Gazette of India*. New Delhi, Friday, March 24.

Speedy, A.W. 2001. FAO and pre harvest food safety in the livestock and animal food industry. *WHO Consultation Report* at Barlin, 2001.

Stan, H.J. 1990. Pesticide in, Principles and applications of Gas Chromatography in food analysis. *Ellis Horwood Ltd Publication*.

Sharma, V; Sharma, R and Wadha, B.K. 2002. Status and control of pesticide residues in milk and milk products. *Indian Dairyman*. 54, 59-63.

Wadhwa, B.K. 2000. Pesticide residue in milk and milk products; safety aspects of quality assurance. *Indian Dairyman* 52, 17-19.

□□□

Chapter **29**

Nutritional Improvement of Crop Residues

M. M. DAS, K. K. SINGH, A. K. MISRA, ANIL KUMAR AND S. K. MAHANTA
Indian Grassland and Fodder Research Institute, Jhansi-284003 (U.P.), India

Introduction

Fibrous crop residues constitute the major component of ruminant feed in most of the developing countries like India. Dairy production is mainly based on the use of agricultural by-products and crop residues as feed resource with cereal straws contributing 45 - 66 % of the feed consumed by the animals. There is regional variation in the type of cereal straw commonly used for the feeding of animals. In Northern India, wheat straw is commonly used, while feeding paddy straw is common in Eastern, Southern and most parts of Western region. The value of cereal straws to bovines is limited by low voluntary intake, slow rate of digestion and low content of available energy, protein, minerals and vitamins and do not provide adequate amounts of nutrients even to maintain the animal body because of their low intake and digestibility. Extensive research has been carried out, for several decades, on improving nutritive value of cereal straws for livestock using physical, chemical and biological treatments. Since the straws and other crop residues are deficient in both nitrogen and available energy, their improvement as animal feed can be achieved through treatments or supplementation of limiting nutrients. The need for treatment and or supplements depends on the economic feasibility and which in turn depends on feed costs and price of produce.

Physical Method

Water Soaking

Water soaking is assumed to affect the internal physico-chemical characteristics of the straw through swelling of the fibre, softening of particles, loosening the linkages within structural carbohydrates and ultimately the palatability. Small increases in straw consumption are observed in cattle and buffaloes due to soaking of wheat straw with water. Water soaked wheat straw based diet had significantly higher ADF and NFE digestibility than unsoaked wheat straw. Soaking of straw overnight before feeding improved intake as well as digestibility.

Chopping

Physical treatment generally affects the ultra structural make up of fibrous materials like straw, stovers etc. Intake is influenced by the physical form and chemical composition of the feed. Chopping and grinding reduce particle size and may increase digestibility. However, DMI and rate of passage of liquid digesta may not be affected by chopping. Increased *in vitro* digestibilities by physical separation of cell wall components is possible by extreme reduction in particle size through ball milling. The effect of grinding on intake is inversely proportional to the quality of roughage, the intake being more in case of crop residues as compared to dried grass. These effects may result due to increased rate of passage as well as better fermentation in the rumen (Sundstol and Owen, 1984). A major purpose of chopping in farmers conditions is to reduce the wastage and also to reduce possibilities for selective consumption.

Chemical Methods

Most widely applied chemical treatments that have been used in the country include use of sodium hydroxide, anhydrous ammonia treatment, hydrogen peroxide and urea ammoniation techniques.

One of the oldest chemical methods is soaking straw in a 1.5% NaOH solution overnight and then rinsing in fresh water. This technique is further modified later in which straw is placed in chamber/ container and NaOH solution is sprayed over straw for 7 to 8 hr. In the subsequent modification, the concentration of NaOH in the straw was increased to have 4 to 4.5% NaOH in the straw DM. In ruminants NaOH treatment improves dry matter digestibility to a greater extent than feed intake (Wanapat *et al.*, 1985). Sodium hydroxide like other alkali dissolves hemi cellulose and swells cellulose there by increasing the extent and rate of cellulose and hemi cellulose digestion by ruminants. However, the degree of increase in digestion may be dependant on the concentration of the alkali used, treatment procedure and time of exposure. The high price of sodium hydroxide, its limited availability, handling difficulty, risk of soil pollution and possible physiological disturbances due to sodium ions in animals body make it difficult to use under practical conditions.

The technique of treating straws with the anhydrous ammonia at 30 gm/kg of straw followed by 8 weeks of incubation. In this method aqueous ammonia treatment was also extensively tried using 120g of 25% NH_3 solution per kg straw with a storage period of 8 weeks. Anhydrous ammonia and aqueous have the same effect in improving the nutritive value of cereals straws. Another most popular chemical treatment that has been tried earlier was calcium hydroxide [Ca(($OH)_2$] as spray or in the form of soaking at the rate of 10 to 15g per 100g straw DM for 24hr.

Treating straws with hydrogen peroxide is relatively of recent wherein 3-5g of H_2O_2 per 100g straw was used for one h followed by addition of 0.001 normal H_2SO_4 solution prior to sealing. Most of these treatment procedures are not cost effective and required infrastructure facilities and some chemicals are also hazardous. The use of sodium hydroxide on small farms at village level is uneconomical and dangerous. Therefore, these methods are practically not feasible and hence not in use at present.

Urea-Ammoniation

Urea is generally the most practical and often the cheapest chemical available for treating straws in many areas of the tropics. This treatment is most suitable in tropics because of its ready availability, familiarity of farmers with its transport, storage and application in addition to its good effect on intake, digestibility, growth and milk production. The urea treatment involves release of ammonia by the hydrolysis of urea and its utilization by the microbes for their proliferation thereby enriching the straw with microbial protein and rise in stack temperature to 55°C which facilitates the penetration of ammonia in to cell wall and breaks the alkali labile lingo-cellulose bond. The lignin content of the straw however not affected. The recommended method for urea treatment of paddy straw is 4 kg urea dissolved in 65 liters of water is sprayed with garden sprayer on 100 kg paddy straw (ICAR, 1985) and then stored for 21 days in anaerobic conditions either by ensiling or by stacking.

This is one of the technology that has been strongly recommended for field application and tried extensively. Urea treatment is reported to improve the nutritive value of cereal straws by increasing digestibility, palatability and crude protein content. Large numbers of farmer trials on straw treatment with urea have been conducted throughout the world including India. Urea treatment is not used on a wide scale by the farmers because of inadequate extension efforts to popularize the technology, non-availability of sufficient straw and the limited availability of liquid cash with farmers for purchase of urea.

Effect on Nutrient Utilization and Performance

The effect of urea ammoniation of straw may be due to either added nitrogen of solubilising cell wall structures or increasing the digestibility of both. Straw treatment with caustic soda improves digestibility and intake by 24 and 40%, while straw treatment with urea enhances digestibility (table1) between 20 and 27%, respectively. Thus urea treated straw would be able to support maintenance requirement and alkali treated wheat straw, supports an additional gain of about 100-200 g/d (table 2) and feeding of ammoniated straw improved milk production in cows (table 3). Therefore, ammoniation alone will lead to some increase in production, but the full potential will only be realized when the correct supplements are added. A supplement of bypass protein is the most important. Very economical responses are obtained to 1 kg/d of cottonseed cake or similar, more conventional sources of protein. A one hundred to one hundred and fifty percent increase in daily live weight gain has been quite common, with this level and type of supplement but with much smaller responses to further increments. Supplementation is expected to correct any imbalances in the nutrients presented for metabolism. It is well established that ruminants fed low-N roughages respond to a source of protein. However, it is often assumed that such protein should remain relatively undegraded in the rumen.

Long term feeding of cows and calves on treated straw has no adverse effects on milk yield, gain of calves or on number of services per conception, birth weight of calves, and interval from calving to first estrus, conception rate and semen quality or fertility of sheep and cattle.

Biological Treatment of Lignocellulosics Materials

Lignocellulose consists of three major components: cellulose, hemicellulose and lignin. Lignin is particularly difficult to biodegrade, and reduces the bio-availability of the other cell wall constituents. Cellulose is a long chain of glucose molecules, linked to one another primarily with glycosidic bonds. The simplicity of the cellulosic structure, using repeated identical bonds, means that only a small number of enzymes are required to degrade this material.

Hemicelluloses are branched polymers of xylose, arabinose, galactose, mannose, and glucose. Hemicelluloses bind bundles of cellulose fibrils to form microfibrils, which enhance the stability of the cell wall. They also cross-link with lignin, creating a complex web of bonds, which provide structural strength, but also challenge microbial degradation. Lignin is a complex polymer of phenyl propane units, which are cross-linked to each other with a variety of different chemical bonds. This complexity has thus far proven as resistant to detailed biochemical characterization as it is to microbial degradation, which greatly impedes our understanding of its effects. Nonetheless, some organisms, particularly fungi, have developed the necessary enzymes to break lignin apart. The initial reactions are mediated by extra cellular lignin degrading enzyme and manganese peroxidases, produced by white-rot fungi. Actinomycetes also decompose lignin, but to a lesser extent. Lignin degradation is primarily an aerobic process, and in an anaerobic environment lignin can persist for very long periods. The effect of lignin on the bioavailability of other cell wall components is thought to be largely a physical restriction, with lignin molecules reducing the surface area available to enzymatic penetration and activity (Haug, 1993).

Table 1. Digestibility, TDN/ME and DMI of Untreated Wheat Straw and Treated/ Supplemented, as Reported by Various Workers

Treatment/feed	Species	Straw DMI/100 kg BW	StrawDMI g /w $^{0.75}$ kg BW	Straw DM Digestibility %	Straw TDN Kg
Wheat straw	*Sahiwal x HF*	1.27	48.77	49.54	44.92
Wheat straw	*Karan Fries*	1.12	44.0		40.12
Anhydrous ammonia (3%w/w) treated straw	*Karan Fries*	1.56	62.5		56.04
Wheat straw	*Bos Indicus*	1.10	49.4		46.85
Urea treated wheat straw 1.60 (4% urea)	*Bos Indicus*	72.20		54.77	
wheat straw (4% urea treated)	Haryana calves	65.3	48.15	47.98	
wheat straw (+0.4 kg DM/100 kg Body weight)	Buffalo Heifers		53		46
wheat straw (2% urea)	-		65		48
wheat straw (NaOH 5%)	-		51		55
wheat straw (NaOH 5%+2% urea)	-		81		57

Table 2. DMI, Digestibility and Average Daily Gain (ADG) in Animals Fed Urea Treated Straw Rations

	Digestibility			DMI(g/ kg W$^{0.75}$)			ADG (g)
	UT	T	Increase %	UT	T	Increase %	Average daily gain (g)
(1)	(2)	(3)	(4)	(5)	(6)	(7)	(8)
1) Rice straw Buffalo calves. 4 week treated, 4% urea (40% concentrate in ration)	31	62	100	69	95	38	
2)Rice straw 5% urea	41	52	27			42	
3) Rice straw (3%) urea treated	41	51	24			33	
4) Rice straw (5%) urea treated	45[2]	56	24	46	61	33	
5) Rice straw (3%) urea treated	45	54	20	46	52	13	
6) Rice straw (5%) urea treated	48[2]	55	15				
7) Rice straw	42[1]		26	65		46	
8) Rice straw	53[1]		11			34	
9) Rice straw 5% urea (Buffaloes)	50[1]	52	4	2.11[3]	2.34	11	182-79
10) Rice straw 5% urea (cattle)	44[1]	52	18	2.68[3]	2.70		34-7
11) Rice straw 3% urea (Buffaloes)	43[1]	53	23	2.03[3]	2.17	7	130-50
12) Rice straw 6% urea treated (Buffaloes)	43[1]	55	28	2.03[3]	2.52	24	130-210
13) Rice straw 4% urea treatment (Cross-bred cattle)	41[1]	48	17			44	
14) Rice straw treated with 6% urea+1.5 kg concentrate						21	79-431
15) Rice straw 5% urea treated (6 Buffalo bull)				79.6	86.7	9	90-97
16) Wheat straw (Treated ammonia in commercial oven)	45[2]	56	24				
17) Wheat straw Treated 3.2% urea (Buffaloes)	41.2[1]	49.2	19.4	1.31	1.69	29	
18) Rice straw (Urea treated in calves)	40	51	27.5	1.7	1.9	11.76	110

1-DM digestibility, 2- OM digestibility, 3- DMI/100 kg body weight
UT= Untreated, T= Treated

Table 3. Effect of Feeding Untreated and Ammoniated Rice Straw to Cows of Different Level of Milk Production (liter per day).

Level of production (l/d)	Untreated		Ammoniated		% increase in percent
	No. of cows	Milk/day	No. of cows	Milk/day	
0.50 to 1.00	6	0.58	6	1.42	145
1.01 to 1.49	13	1.15	13	2.15	87
1.50 to 1.99	9	1.50	9	2.51	66
2.00 and above	15	2.05	15	2.97	45

Saadullah *et al.* (1988)

Microbiological Processes that Upgrade Straw Quality

Single Cell Protein Production from Lingo-cellulosic Substances

Since straws are normally resistance to microbial growth, pre-treatment is done before yeast cultivation. There are two processes: In the first process the straw is hydrolyzed with three parts of 0.5 N H_2SO_4 and 1 part of the straw which is steam treated in a pressure cooker under 10-15 *psi* of steam pressure for 30 minutes. The acid hydrolyzed straw is then treated with NH_3 to raise the pH to 4.5 and it has been estimated that at this stage it contains 20% fermentable sugars and 2.3% N. This is then incubated with suitable organisms like *Candida ntilis* or *Aspergllus pullulans.* In the second process, straw is treated with 4% (dry weight) NaOH and after neutrilisation of excess alkali, ammonium sulphate is added. This pretreated straw is then inoculated with the cellulolytic organisms like mixed culture of *Cellulomonas* sp. or *Alcalegenes faecalis.*

Utilization of Spent Straws as Animal Feed after Mushroom Production

Mushrooms are generally produced using straw as partial or complete growth medium. The use of large used fertilizer bag was suggested as they could be stuffed with the straw and then steamed as such. After cooling, the bags could also be inoculated with the white rot fungus (*Pleurotus* sp.) and incubated for several weeks. When the bags had been filled with mycelium they were slit open at the top which allowed fruiting. The residual straw after the harvest of mushroom could be fed to ruminant animals. At lower temperatures *Ganoderma applanatum* preferentially degrades lignin in wheat straw and helps to increase the digestibility of the fermented product.

How White Rot Fungi Improve the Nutritional Value of Crop Residues

In primary phase active mycelia growth takes place and the available nutrients are utilized. The secondary phase starts when scarcity of essential nutrients like carbon, nitrogen, and sulphur occurs. During this phase, distinct changes in the behavior of fungus occurs which leads to sporulation or a decline in the growth rate of fungus. Lignin degradation triggered only in the secondary phase of growth and that to in response to nitrogen limitation. When the filamentous white rot fungi are grown on lignocellulosic materials, the tubular filamentous grow along side of solid particle utilizing liquid film covering the solid particle as moisture and nutrients. The growing tip of filaments (hypha) secretes powerful extra cellular lignin degrading enzymes, which degrade the lignin into in its metabolic products. The presence of lignin in the culture medium is not essential for the synthesis of lignolytic enzyme system, but increasing the partial pressure of oxygen in the gas phase significantly increasing its synthesis. The white rot fungi attack and cause oxidative changes in the phenylpropane subunits and cause ring cleavage while the lignin polymer is intact i.e. no de-polymerization has occurred.

Specific Properties of Lignolytic Fungi

Fungi can be classified into different groups: Sweet-rot, brown-rot, and white-rot. Sweet-rot fungi only metabolize soluble sugars or other easily digestible cell constituents of crop residues. Brownrot fungi primarily decompose cellulose and hemicellulose at an extremely low pH (2.0-3.0) and decomposed material is characterized by lignin accumulation. White-rot fungi are able to decompose and mineralize all plant cell constituent-cellulose, hemicellulose and lignin. The *in vitro* digestibility of fungal substrates decreases at the initial stage of colonization of white rot fungi and increases afterwards. During the incubation period, the content of soluble substances are increased. The increase in digestibility depends on the fungal species, cultivation time, temperature and composition of substrate.

Fungi may also be divided in to four groups according to the different physiological behavior.

a) Fungi that degrade little or no lignin and decompose cellulose and hemicellulose selectively. The fungi of this group are *Agrocybe aegerita,Flamulina velutipes, and Volvariella volvacea.*

b) Fungi that decompose lignin selectively and other substrate components (cellulose, hemicellulose) only partially IVDMD of these substrates is increased. This groups includes *Dichomitus squalens, Abortiporus biennis, Pleurotus spp.,and Stropharia rugosaannulata.*

c) Fungi, that decompose lignin, cellulose and hemicellulose. IVDMD is decreased or remains unchanged. Strains of this group are *Polyporus Ganoderma and pleurotus spp.*

d) Fungi that decompose lignin and other substrates (cellulose and hemicellulose.) rapidly and change digestibility of the substrate only partly. The main fungus of this group is *phanerochaete chrysosporium*.

White-rot Basidiomycetes fungi degrade the polymers cellulose, hemicellulose and lignin at different rates and extents. In general, strains of white-rot fungi are the most effective lignin-modifying enzyme, of which lignin-Peroxidases (liP), manganese dependent Peroxidase (mnP), laccase and H_2O_2-produsing oxidase (aryl alcohol oxidase and glyoxal oxidase) are the most important. The oxidative enzyme ligninase, manganese peroxidase and laccase may be defined as phenol oxidase belong to the class of peroxidases.

Depending on the pattern of extracellular lignolytic enzymes, white rot fungi may be divided in to four groups: (1) theLiP-MnP group, (2) the MnP-Laccas group, (3) the laccase-AAO group. *P.chrysosporium* is a very efficient degrader of lignin. Typical fungi producing (1) LiP-MnP are *Phlebia radiata, Coriolus versicolor and Trametes gibbosa*, (2) fungi producing MnP-Laccase are *Panus tigrinus and Dichomitus squalens*. (3) LiP-Laccase producing fungi are *Junghuhnia separabilima and Phlebia ochraceofulva*, and (4) fungi producing AAO are *Pleurotus ostreatus* and *Pleuotus Sajor-caju*.

Fermentation Process for Bioconversion

1) *Solid-state fermentation:* It can be defined as a process in which solid substrates are decomposed by fungi, which are able to grow on and through the substance (lingo-cellulose crop residue) under controlled environmental conditions with the aim of producing a high quality product. It is the oldest fermentation process for the bioconversions, being used by man since centuries ago.
2) *Surface fermentation :* In this process, the fungi were allowed to grow as a mycelial mat on the surface of liquid medium.
3) In submerged fermentation process, the submerged culture was allowed to grow on high-diluted substrate. The process is more efficient with higher performance and easy operations. But due to high degree of dilution, the process is costly as compared to solid-state fermentation where there is no dilution, also per unit processing of substrate a high reactor volume is required.
4) *Gas/solid fluidized bed fermentation*: This process has number of advantages in term of heat and mass transfer and substrate is in continuous operation. But again these processes are very expensive especially for the delignification of the ligocellulosic substrate because delignification process is very slow.

Among these methods, solid state fermentation is widely accepted method for bioconversions of ligmnocellulosic feed and is very close to nature. Secondly since there is no dilution,therefore small reactor volume is required and it is fed directly to animal. Solid-state fermentation with white-rot fungi is influenced by various factors such as species of the fungus, substrates, temperature, moisture content and composition of the gas phase. The major constrains of microbial methods are difficulties in process control (SSF), organic matter losses(table 4), and isolation of substrate specific microorganism. Feeding of biologically treated straw in different species reported to decrease in organic matter digestibility and digestible organic matter intake (table 5).

Table 4. DM Loss, Chitin and Biomass Content and DM Loss During Fermentation of Wheat Straw by *Coprinus Fimetarius*

Period of fermentation (days)	Chitin (%)	Fungal biomass(%)	DM loss (%)
3	2.93	25.02	6.60
5	3.37	28.78	7.50
7	4.27	36.46	23.35

Walli *et al.* (1988).

Table 5. Animal Performance on Biologically Treated Straw as Reported by Various Workers

Animal species	Treatment	OM (%)	OMD (%)	DMI (g/d)	DOMI (g/d)
Goat	UTS (Rice straw)	79.4	71.5	-	747
	FTS	72.2	60.2	-	582
Goat	UTS (wheat straw)	85.0	48.5	445	183
	FTS	80.0	46.5	550	204
Heifer	UTS (wheat straw)	87.6	61.6	4280	2300
	FTS	78.2	49.5	4320	1700

The main problem of biological upgrading of low quality roughages into feed is to find suitable microorganism with metabolic patterns different from those of the rumen flora and fauna for use in the process of delignification and upgrading. "Ideal microorganisms" for upgrading crop residues in to animal feed should have a strong lignin metabolism with poor degradation of cellulose and hemicellulose so that in the treated material, carbohydrates are preserved for their utilization by rumen microbes as an energy source in ruminants.

Use of Fibrolytic Enzymes for Improving Ruminants Diet

The major limitations in the utilization of crop residues are the crystalline nature of cellulose and recalcitrance of lignin along with low content of nitrogen and minerals. Under ideal feeding conditions cell wall digestibility in the digestive tract is still generally less than 40-50 percent. Further when the condition in rumen are sub-optimal for fiber digestion, as can be the case for animal fed poor quality roughages, low in nitrogen or animal fed high grain diets, cell wall digestion in the total tract may be only 30-40 % (Beauchemin *et al.* 2001). In the past, numerous physical, chemical and biological treatment method have been tried to improve the feeding value of crop residues. Despite encouraging results, particularly from the use of sodium hydroxide and urea/ ammonia, none has yet been proved to satisfy biological, practical and environmental requirements. In spite of these efforts, cell wall digestibility continues to limit the intake and utilization of crop residue by ruminants.

Although ruminant animals posses huge microbial population in the reticulorumen but due to very complex nature of feed, animals are not able to utilize nutrients such as starches, proteins, minerals etc. that are enclosed within fiber rich cell walls. In spite of high enzyme activity in the rumen, the utilization of grass/ is limited due to insufficient endo-beta-1,4- glucanase activity (Wallace *et al.* 2001). The feed is the biggest single cost in most of the animal production system. Therefore, in view of the increasing the price and gap between availability and requirement of cereal grains, the energetic potential of the these lingo-cellulosic material should be exploited using exogenous fibrolytic enzymes.

Exogenous Enzymes for Ruminants

Cellulose and hemicellulose from wood or straw in purified form is almost completely digestible in the rumen and has a high net energy value for ruminant animals. However, due to the complex nature of cell wall it is partially utilized by ruminants. Thus the exogenous fibrolytic enzymes might be the key to improve the utilization of crop residues for ruminants. The use of fibrolytic enzymes is biological as well as environmental friendly. These enzymes mainly contain cellulase and hemicellulase activities and are of mainly fungal origin (Table 6). These enzymes can be in powder, granular or liquid forms. When stored at 5-20 ^{0}C, the stabilized enzyme products typically retain their enzyme activity for at least one year.

Table 6. Sources of Exogenous Enzyme from Fungal Species

Fungal species	Enzymatic actvity
Aspergillus niger	*Cellulase, hemicellulase, xylanase*
Trichoderma viride	Cellulase, hemicellulase, xylanase
Trichoderma reesei	Cellulase, hemicellulase, xylanase
Penicillum funiculosum	Beta-glucanase, xylanase
Trichoderma konigii	Cellulase, xylanase

The focus of most enzyme related research for ruminants has been on polysaccharidases that degrade plant cell wall. Cellulose and hemicellulose, the major structural polysaccharides in plants are converted to soluble sugars by enzymes collectively reffered to as cellulases and hemicellulases.

Mode of Action

The mode of action of enzymes in ruminants is not fully understood. The enzyme addition increases feed intake due to pre and post ingestive enzyme effects such as increased palatability due to increased released of sugars, in creased rate and extent of digestion or increased passage rate. These factors reflect on the hydrolytic capacity of the rumen which indirectly reduces the gut fill and enhances feed intake (Beauchmin *et al.,* 2000). In some studies, enzyme supplementation has increased animal performance without increasing feed intake. This may be due to reduced digesta viscosity (Hristov *et al.,* 2000) and alterations in ruminal fermentation. Xylanase addition increase carbohydrate utilization and VFA production and addition of combination of enzymes increased carbohydrate digestion and increased the acetate: propionate ratio. Further enzyme additives have been shown to enhance ruminal cellulytic bacteria and increase the rate of colonization of feed by ruminal microorganisms and increase the rate of degradation in the rumen (Yang *et al.,* 1999). This may be due to chemo tactic responses to soluble sugars released from the plant fibers by the polysaccharide enzymes.

Exogenous enzymes are most effective when applied by spraying enzyme solution directly on to the forages prior to ingestion. The exogenous cellulase was more effective to release soluble sugar from wheat straw followed by black gram straw, grass and groundnut haulm (table 7). Improvement in animal performance due to the use of feed enzymes has been attributed to increase in feed digestion of dry matter and fiber *in situ* or *in vitro*. Exogenous enzymes in the rumen are generally more stable, particularly when applied to feed prior to ingestion. Application of enzymes to feed enhances the binding of the enzyme with the substrate, which increases the resistance of the enzymes to proteolysis and prolongs their residence time within the rumen. In the rumen, the close association between bacteria and feed particles concentrates digestive enzymes close to their specific substrates.

Applying enzymes to feed also provides a slow-release mechanism for enzymes in the rumen (Beauchemin *et al.*,1999). Thus, the greater the proportion of the diet treated with enzymes, the greater the chances that enzymes endure in the rumen. Without this stable feed-enzyme complex, the enzymes are solubilized in ruminal fluid and flow rapidly from the rumen.

There is evidence for pre-consumptive effects of exogenous enzymes causing the release of soluble carbohydrates and in some cases, partial solubilization of NDF and ADF. Application of enzymes to feed causes structural changes, thereby making feed more amenable to degradation. Cell wall hydrolysis in the rumen proceeds in an erosive manner and it is well recognized that a major constraint to digestion is the limited colonization and penetration of cellulolytic microbes and their hydrolytic enzymes onto the exposed surfaces of feed particles.

Adding exogenous enzymes to the diet increases the hydrolytic capacity of the rumen mainly due to increased bacterial attachment (Yang *et al.*, 1999), stimulation of rumen microbial populations (Wang *et al.*, 2001), and synergistic effects with hydrolases of ruminal microorganisms. The net effect is increased enzymatic activity within the rumen, which enhanances digestibility of the total diet fed. Increased hydrolytic capacity of the rumen can also lead to an increased digestibility of the non-fiber carbohydrate fraction, in addition to increasing digestibility of the fiber components of a diet, which explains why fibrolytic enzyme can be effective in high-concentrate diets.

Exogenous enzymes appear to survive in the small intestine for a time sufficient to have an effect on target substrates. However, postruminal effects of exogenous enzymes on digestion are likely only a significant factor when enzymes are in fused into the rumen (Hristov *et al.*, 2000) or added to feed in a manner that allows for easy solubilization from feed and rapid passage from the rumen (Beauchemin *et al.*, 1999).

Table 7. Effect of Fibrolytic Enzyme Application on the Release of Soluble Sugar from Different Roughages

Roughages	Soluble sugars released (µ mol/g substrate)
Wheat straw (untreated)	171.66
Wheat straw (enzyme treated)	283.88
Grass (untreated)	175.53
Grass (enzyme treated)	279.43
Groundnut haulm (untreated)	203.33
Groundnut haulm (enzyme treated)	292.76
Black gram straw (untreated)	217.76
Black gram straw (enzyme treated)	325.53

Dose of Enzyme

An enzyme product that is ideally formulated can be ineffective if provided in insufficient amounts or in excess (table8).

Table 8. Effect of Dose of Enzyme on Nutrient Utilization and Animal Performance

Forages/ concentrates	Dose of enzyme	Effects
Alfalfa hay	0.25-1.0 ml/kg	Increased digestibility
Timothy hay	4.0 ml/kg	Increased daily gain
Ryegrass silage	3.5 ml/kg	Increased daily gain
Total mixed ration	1ml/kg	Increased milk production
Alfalfa hay	2-4.0 ml/kg	No effect
Supplementation	4-16 g/kg	No effect

Method of Enzyme Application

Applying fibrolytic exogenous enzymes in a liquid form to feeds prior to consumption can have a positive effect on animal performance (Yang *et al.* 2000). In contrast, infusion of enzymes into the rumen has not been effective. The close association of enzymes with feed may enable some form of pre-ingestive attack of the enzymes upon the plant fiber and/or enhance binding of the enzymes to the feed, thereby increasing the resistance of the enzymes to proteolysis in the rumen (Beauchemin *et al.* 2004)

There is apparently little or no requirement for a reaction phase or incubation time between treatment and feeding of forages. Enzyme application to dry hay showed no difference in NDF digestibility when enzyme applied immediately before feeding and a 24-h incubation period (Lewis *et a1.* 1996).

Beauchemin *et al.* (1999) suggested that enzymes should be applied to a large portion of the diet to increase the chances that enzymes endure in the rumen. Adding enzyme to a small portion of the diet may allow rapid passage of enzyme from the rumen, lessening the enzyme effect in the rumen. Rapid passage of the enzyme is not an issue in vitro. Thus, in vitro batch culture assays used as a bioassay to predict the effects of exogenous enzymes on digestibility may not accurately predict variations in the in vivo response caused by method of providing the enzyme to the animal.

Effect of Enzyme Application on Chemical Composition

Enzyme treatment of various roughages viz., oat hay, wheat straw and berseem straw did not affect the protein content(table 9). The NDF and ADF contents however, reduced with enzyme application, though not statistically significant. The reduction in fiber contents results from the partial solubilization of fiber portion and is confirmed by lower contents of cellulose in treated roughages. The effect was more pronounced with oat hay as compared to wheat straw or berseem straw.

Table 9. Changes in Chemical Composition Due to Enzyme Application

Attributes	Oat hay		Wheat straw		Berseem straw	
	Untreated	Enzyme Treated	Untreated	Enzyme Treated	Untreated	Enzyme Treated
CP	6.61	6.80	3.36	3.52	9.60	9.70
NDF	75.52	73.59	74.00	73.12	66.08	65.47
ADF	52.10	49.95	50.84	50.24	51.65	51.80
Cellulose	41.42	40.15	39.56	38.96	34.44	33.64
Hemicellulose	23.42	23.64	23.16	22.88	14.43	13.67
OM	91.06	91.00	89.24	90.54	90.40	90.30

Ruminal Effects

Feedstuffs consumed by the ruminants are degraded by the rumen microorganisms through the action of various enzymes in to different products like volatile fatty acids (acetic, propionic, butyric acids etc), ammonia and gases (carbondioxide and methane). The concentration of these metabolites indicate the quality of the roughage and their further utilization in the reticulo-rumen, which in turn govern the production potential of the animal. Supplementing the diet with exogenous enzymes increased the total enzyme activity in the rumen (table 10). Further, when the condition in rumen are sub-optimal for fiber digestion, as can be the case for animal fed poor quality roughages, low in nitrogen or animal fed high grain diets, cell wall digestion in the total tract may be only 35-50 % (Beauchemin *et al.*,2001). Under such circumstances exogenous enzymes application would improve the fiber digestion. Applying enzymes to feed causes the release of soluble sugars from feed, thereby increasing the chemo tactic attraction and eventual attachment of fibrolytic rumen bacteria to the plant surface. Lactating dairy cows showed no difference in ruminal pH receiving a diet containing feed enzyme (Beauchamin, 2000). TVFA concentration however, improved significantly ($p<0.05$) by enzyme treatment both in goats and buffaloes (table10). Cows fed diets containing enzymes exhibited higher VFA concentration with no effect on molar proportion than control diets. Higher TVFA concentration with enzyme treated diets supports the theory that applying enzyme to roughages resulted in higher ruminal digestion which is further conformed by higher DM digestibility. Ammonia N concentration was not affected by enzyme application in goats and buffaloes. In lactating dairy cows no effect of enzyme on ammonia N or free amino acids concentration was noticed (Broderick *et al.* 1997). In heifers, exogenous polysaccharides degrading enzyme supplementation improved the fibrolytic activity in the rumen. The highest ruminal xylanase and reducing concentration occurred at 2 hr post feeding/ administration of enzyme (Hirstov *et al.* 2000).

Table 10. Effect of Fibrolytic Enzyme on Rumen Metabolites

Attributes	Adult goat		Buffalo	
	Untreated	Enzyme Treated	Untreated	Enzyme Treated
PH	6.41	6.48	6.30	6.41
TVFA (m mol/l)	147.90	159.40	74.50	84.83
Ammonia N (mg/dl)	27.72	28.84	21.09	22.96
Total N (mg/dl)	95.76	96.88	76.531	75.60
CMCase (μg glucose/h/ml)	702.10	820.40	973.00	1116.90
Xylanase (μg glucose/h/ml)	1965.60	2482.67	1697.42	2208.89

References

Beauchemin K A, Colombatto D, Mogavi D P, Yang W Z and Rode L M. 2004. Mode of action of exogenous cell-wall degrading enzymes for ruminants. *Canadian Journal of Animal Science* 84: 13-22.

Beauchemin K A, Rode L M, Maekawa M, Morgavi D P and Kampen R. 2000. Evaluation of non-starch polysaccharide feed enzyme in dairy cow diets. *Journal of Dairy Science* 83: 543 - 553.

Beauchemin K A, Yang W Z and Rode L M. 1999. Effect of grain source and enzyme addition on site and extent of nutrient digestion in dairy cows. *Journal of Dairy Science* 82: 378-390.

Beauchemin K A, Yang W Z and Rode L M. 2001. Effect of Barley grain processing on the site and extent of digestion on beef feedlot finishing diets. *Journal of Animal Science* 79: 1925-1936.

Broderick, G.A., Derosa, R. and Reynold, S.1997. Value of treating alfalfa silage with fibrolytic enzymes prior to feeding the silage to lactating dairy cows. US Dairy Forage Research center,1997 *Research Summaries* pp 71-73

Haug, R.T. 1993. The Practical Handbook of Compost Engineering. Lewis Publishers, Boca Raton, Fl. 717 pages.

Hristov A N, McAllister T A, and Cheng K J. 2000. Intraruminal supplementation with increasing level of exogenous polysaccharides-degrading enzymes: Effect on nutrient digestion in cattle fed a barley grain diet . *Journal of Animal Science* 78:477-487.

ICAR. 1985. Nutrient Requirement of Livestock and Poultry. ICAR, New Delhi.

Lewis, G. E., Hunt, C. W. ,Sanchez, W. K, Treacher, R., Pritchard, G. T. and Feng, P.1996. Effect of direct fed fibrolytic enzymes on the digestive characteristics of a forage based diet fed to beef steers. *Journal of Animal Science* 74,3020-3028.

Saadullah, M. 1984. Supplementing Ammoniated Rice Straw for Native Cattle in Bangladesh. PhD Thesis. The Royal Veterinary and Agricultural University, Institute of Animal Science, Denmark

Saadullah, M., Hossain, B., D'silva, D., Dolberg, F.1988. Utilization of fibrous crop resdues as feed for ruminants in small holder farming in Bangladesh. Pp294(2.17). In: VI World conference on Animal Production, Helsinki, Finland.

Sundstol, F. and Owen, E. 1984. Straw and Other Fibrous By-Products as Feed. Elsevier

Walli, T.k., Gupta, B.N., Rai, S.N. and Singh, K.1988. Biological treatment of straws, pp. 75-81. N transaction during fermentation and subsequent utilization by ruminants. In: Kiran Singh and J.B. Schiere(eds.): Fibrous crop residues as animal feed. Proc. International Work shop 27-28 October,1988, SRS campus , Bangalore, NDRI, India.

Wanapat,M.,Sundstol,F. and Garmo,T.H.1985. A comparison of alkali treatment methods to improve the nutritive value of straw. I.Digestibility and metabolizability. *Animal Feed Science and Technology* 12,295-309.

Wallace,R.J. and Hartnell,G.F.2001. Technical note:Methods for detecting liquid enzyme additives added to animal feeds. *Journal of Animal Science* 79,2731-35.

Wang,Y., McAllister,T.A., Rode,L.M., Beauchemin,K.A., Morgavi,D.P.,Nsereko,V.L., Iwaasa,A.D. and Yang,W.2001. Effects of an exogeneous enzyme preparation on microbial protein synthesis , enzyme activity and attachment to feed in the rumen simulation Technique. *British Journal of Nutrition* 85,325-332.

Wongsrikeao, M., Ibrahim,M.N.M. and Pearce,G.R.1984. *Agricultural Wastes* 9,87.

Yang W Z, Beauchemin K A and Rode L M. 1999. Effects of enzyme feed additives on the extent of digestion and milk production of lactating Dairy Cows. *Journal of Dairy Science* 82:391-403.

Yang, W. Z., Beauchemin, K.A. and Rode,L.M.2000. A comparison of methods of adding fibrolytic enzymes to lactating cow diets. *Journal of Dairy Science* 83,2512-20.

Zadrazil, F., Puniya, A. K. and Singh, K.1995. Biological upgrading of feed and feed components. In: Biotechnology in animal feeds and animal feeding of feed and feed components (Ed: Wallace, R.J. and Chesson, A.), VCH,Weinheim. Pp.55-69.

❑❑❑

Chapter **30**

Challenges and Priorities for Improvement of Forage Crops Biotechnological Approach

D. R. MALAVIYA, A. K. ROY AND P. KAUSHAL
Indian Grassland and Fodder Research Institute, Jhansi -284003 (U.P.), India

Introduction

A few decades back Mendel's law and their direct utilization were basic for plant breeding programmes and the first half of the century dominated the exploitation of quantitative genetics allowing improvement of crops for polygenic traits as well as the success of cytogenetics for interspecific hybridization contributed to constant increase of performances in new cultivars (Spragues *et al.*, 1959; Demarley, 1977). But, the tremendous technological development in recent past has equipped the plant scientists with enormous options to tailor the plants according to need.

The working group of IInd World Crop Science Congress held at New Delhi India (November,1996) has also stressed on genome mapping and marker assisted selection in plant breeding recognizing the importance of synteny. It recognized that information obtained on genome maps of different crops should be widely shared among all bonafide users. It also mentioned that there are large number of genes available e.g. Osmotin gene for moisture stress, Ama-1 for quality traits, bar gene for herbicide resistance, bxn gene for broomoxynil resistance, 'crrl' gene for chlorosulfuran resistance, 'ifda' gene for 2,4-D resistance, cowpea trypsin inhibitor gene 'CpTI' for resistance against caterpillars and Barnase Baster for male sterility which can be incorporated in to the target species.

There is always a debate on basic vs applied biotechnology. It is an important issue particularly in view of the huge amount of money involved in biotechnological research. As regards research in agricultural biotechnology, it is well understood that the focus should be mainly on the applied aspects of the biotechnology. No doubt, to understand various systems it would be necessary to get the basic information but that cannot be sole objective.

As regards improvement of forage crops through non-conventional breeding methods or biotechnological approach at IGFRI, India efforts in this direction started in late eighties but biotechnological approach for forage crop improvement worldwide has made remarkable headway. IGFRI is also engaged in addressing many such problems of great importance which otherwise were coming as bottlenecks in improvement programme.

An overview of various modern biotechnological methods has been described hereunder with an objective that understanding of these new techniques will not only help in opening up the thinking of the breeders but also the scientists of other disciplines in properly addressing the problems. Various techniques, which are frequently being used, can be grouped in following three parts:

- Molecular techniques for better understanding of the genetic structure of the plants
- Inserting foreign genes directly into the plant genome.
- *In-vitro* regeneration of plants from any plant part.

For the convenience of presentation biotechnological approach of forage crop improvement here has been presented in following heads:

A. Biotechnology as an aid to conventional breeding.

B. Molecular techniques.

C. *In virto* screening.

A. Biotechnology as an Aid to Conventional Breeding

A number of techniques have been developed which just only help at one or more steps involved in conventional breeding methods. These techniques save time and energy required for conventional methods.

Embryo rescue: Interspecific hybridization has been one of the most effective tool in crop improvement but in many cases the interspecific crosses are not successful due to incompatibility barriers at different levels. In case of post fertilization incompatibility barriers embryo rescue is most effective and successful techniques. In this case the cross is made in natural condition but at an appropriate stage the embryo is taken out and placed on culture media and supplemented with nutrition in absence of which the embryo would have aborted after few days. The technique has been well exploited in *Lolium/ Festuca* complex for production of hybrids (Zwierzykowski, 1987*), Lolium/ Dactylis* hybrids have also been developed by this method (Oertal et al 1996) and many interspecific hybrids in *Trifolium.*

Micropropogation: This a means of multiplying the plants or elite genotypes from any plant part. The process helps in developing true to type plants at one hand and exploiting the somaclonal variation at other. The process has been extensively used in *Cynodon dactylon* and *Dactylis glomerata.* The production of somatic seeds in *Dactylis glomerata* has been recommended by Conger (1991). Although somatic seeds production has been achieved in *Medicago sativa* but this is not cost effective (Conger, 1991).

Andogenic haploid plant production: Regeneration of the plantlets from reproductive part such as anther has resulted in haploid plant production. The process is very effective in developing plants with double set of genome in otherwise tetraploid tropical grasses. In diploid outbreeding grasses the further utilization of double haploids for the development of inbred lines may be hampered by the presence of incompatibility system, however even this can be overcome by use of self compatibility alleles (Hayward 1990). At a higher ploidy level the andogenic haploid plant production is more successful. From a *Lolium multiflorum x Festuca arundinacea* hybrid (2n=5x=35) anther culture has been used to stabilize polyploid hybrid combination ranging from 14 to 21 chromosomes (Humphreys et al 1996). This has produced novel genetic material exhibiting enhanced tolerance to environmental stress. do Valle and Savidan, (1996) has reported that apomixis in *Cenchrus* and *Bracharia* seems to be single gene control mechanism and if so there are ample chance of developing sexual types by producing androgenic haploids.

Creation of novel variation: Creation of novel variations in forages has ample opportunity which is as such not possible through conventional methods.

The various means of creating variation in forage grasses and achievement are described hereunder:

Somaclonal variation: These variations arise during culture of the tissues *in vitro* because of unpredictable random process and are often genotype and ploidy level dependent. In majority cases the variations created are not of much economic importance but in a few instances useful variations have been created for traits such as productivity and disease resistance (Conger, 1991). At IGFRI too the somaclonal variants created in *D. annulatum* has been found to be disease resistant and better in quality.

Somatic hybridization: The species which show incompatibility even at pre-fertilization stage, if required can be hybridized using protoplast fusion technique. In the process the protoplast of the two target species are separated and fused in artificial conditions and then regenerated into a plant. To date this process has shown very little success in perennial grasses although Spannenberg et al (1995) have recently transferred male sterility in this manner in some *Lolium x Festuca* hybrids.

Genetic transformation: Artificial introduction of some foreign gene in the plant genome is genetic transformation. Th insertions of genes may be by chemical, electrical, physical or micro-projectile transfer. In grasses it has been only limited success so far (Hayward 1996). To date successful transformation has been reported in perennial grass including *Lolium sp., Festuca sp. Agrostis sp., Dactylis, Paspalum and Dichanthium.*

B. Molecular Techniques

Various molecular techniques such as RFLP, AFLP, RAPD, SNP and isozymes have been used from time to time for characterization of germplasm, cultivar identification, detection of hybrids and genetic mapping and gene tagging. The molecular characterization of agriculturally important plants is equally important to that of many grasses and weeds because the through transformation techniques, genes idetntified from any plant or living organism can be transferred to the target species. There exists a lot of similarity in DNA sequence across various species and genera. DNA sequencing has revealed that around 30% sequence of human genome is common with plant genome of *Arabidopsis thaliana* which grows from arctic to equator.

Germplasm Characterization

Characterization of germplasm is one of the most important aspect particularly in view of the changing scenario with regard to Plant Biodiversity act. Forage germplasm needs to be classified with two objectives; first for developing DNA fingerprints and secondly for identifying the duplicates in the germplasm. So far the forage germplasm characterization has been mainly on isozyme basis. Information is available on isozyme characterization of *Lolium* (Hayward et al 1995a) *Bracharia* and *Cenchrus*. These surveys have identified regions of geographical distribution of the species where maximum genetic diversity can be obtained. Recently IGFRI has also completed isozyme characterization of various *Trifolium* species and also RAPD characterization of *Dichanthium annnulatum*.

Identification of Varieties

The varieties can also be subjected to molecular characterization in order to avoid any dispute regarding in the use of germplasm in coming years. The major problem encountered with these molecular markers of the forage species is that most of the cultivars are synthetic populations and variability exists within the population. Still, the efforts have been made for characterizing the varities based on RAPD, RFLP markers. Hayward and McAdam (1977) have suggested discrimination between the varieties based on gene frequencies.

Reproductive Behaviour

The molecular techniques also help in identifying the mode of reproduction in any species and is more important in the species with complex breeding mechanism and poor genetic markers. Marker methods such as RFLP or isozyme are most appropriate for this situation. Based on this technique *Cenchrus* germplasm was evaluated and the study concluded that the grass which is supposed to be obligate apomict possesses a number of accessions with intra-accession variability indicating the presence of sexuality.

Genome Analysis

The major application of genetic markers in crop breeding today is for the development of genetic maps, gene tagging, marker assisted selection and map-based cloning. The genetic maps offer the means to identify regions of the genome concerned with the determination of specific agronomic traits (QTLs) and the ability to manipulate them in breeding programs (Hayward, 1990).

Genetic Mapping

Genetic mapping in forage species has not been attempted much. At IGER genetic mapping of the *Lolium* based on back cross type of segregating family from the cross of an hybrid *L. perenne x L. multiflorum* to double haploid *L. perenne* has been attempted (Hayward 1996). This map construct using RFLPs, RAPDs and isozymes markers revealed seven linkage groups which have been allocated to specific chromosomes of the complement of seven by the use of trisomics. GISH has also been successfully utilized to *Loliu/Fesuca* hybrids which enables the specific chromosomes to be identified (Hayward et al., 1995b). Genetic maps are now being developed of ryegrass, fescues, bracharia and *Paspalum* species.

Gene Tagging

For single gene control traits gene tagging is important but in case of forages most of the desirable agronomic traits ate multigenic and hence are difficult to tag. The *S* locus of two locus incompatibility system in ryegrass has been linked to an isozyme locus (Cornish et al 1983). This region of the genome was mapped in more detail by Hayward (1996) but as yet family could not be classified for incompatibility genotypes. A possible probe for S locus has been identifies by Langridge (See Hayward 1996) from *Phalaris coeruulescens* which should enable more precise mapping of this gene. Recently genes controlling apomixis have been tagged. Some other major genes in grasses which are the subject of tagging programs include the pairing control genes in Lolium (Hayward et al 1995b) and the crown rust resistance gene(s) also in *Lolium* (Adams et al 1996).

Comparative Genetic Mapping

The development of genetic maps of major graminaceous crops using common sets of genetic probes has lead to identification of conserved segments of the genomes across a wide range of species (Moore et al 1995). It means, similar traits across the range of species are likely to be controlled by the similar sets of genes (or QTLs) and thus the marker identified for one crop can well be used to tag homeoloci in another species. For example the flowering time QTLs on linkage map in *Lolium* is located in the same region of genome which is syntenic to that controlling vernalization/ photoperiod loci in wheat and barley. Hayward 's effort to search for markers linked to apomixixs gene in *Bracharia* was based on the set of probes regularly distributed across the maize genome. This segments were homed in on that segment with other probes from the same syntenic region of maize.

Quality Breeding

The plants can be bred for increasing animal production by the improvement of aminoacid balance of the plants used in intensive feeding production system. For example Lupin meal is deficient in sulphur aminoacid content and Methionine is added to the feed. At CSIRO gene normally expresssed in sunflower seed was taken out and construct made so that it is expressed in lupin grain. This has increased the sulphur aminoacid content of the transgenic lupin. In the first animal test on rats fed on transgenic lupin has showed increased body weight gain.

Controlling Fungal Diseases

Antifungal compounds are either proteins or secondary metabolites. Antifungal proteins not only include polypeptides affecting the integrity of the fungal cell wall, but also ribosome-inhibiting proteins and (plant produced) antibiotics raised against essential fungal products. Antifungal metabolites include phytoalexins.

As regards biotechnological approach for developing plant types having fungal resistance efforts have been made and success have been achieved For example, although it is not known whether PR8 proteins (class III chitinases) have antifungal activity in vitro, transgenic plants expressing PR8 genes show enhanced fungal resistance in many cases such as in from sugarbeet to tobacco (Nielsen et al 1993), cucumber to tobacco (Lawton et al 1993) and from tobacco to tobacco (Lawton et al 1993). Similarly PR11, a tobacco endochitinase homologous to bacterial exochitinasses, displays antifungal activity in vitro (Melchers et al, 1994).

In 1991 the first report came regarding development of transgenic with PR protein producing ability but the these tobacco plant with PR3 (class I chitinase) did not show much difference for fungal resistance (Neuhaus et al 1991). However in the same year Broglie and coworkers achieved resistance against Rhizoctonia solani in tobacco and canola expressing a class I chitinase gene. Since then a number of reports have been published dealing with fungal resistance of transgenic plants constitutively expressing single PR-protein genes e.g. tobacco, rice, carrot, soybean and potato(Alexander et al 1993, Lawton et al 1993, Yoshikawa et al 1993, Jongedijk et al 1995, Liu et al 1995, Stuiver et al 1996 and Lin et al 1995). PR2 (class II glucanase) protein producing isolated from Alfalfa has been transferred to tobacco plant for Cercospora nicotiane resistance (Zhu et al 1994).

Transgenics

In Australia, the first transgenic crop has been planted commercially (in October 1996) cotton in 75,000 acres. This transgenic crop is based around Bt gene, which gives protection against *Heliothis* damage 68% reduction in the usage of insecticides is expected because of introduction transgenics.

The major threat being posed regarding transgenics being the transfer of resistance genes to some other plant. Progress has been made in developing a system of sterilizing the trees based on the new knowledge of flower development gained largely from *Arabidopsis* research in last few years. *Arabidopsis* has a major role in today's molecular biology. It is a self pollinating and has a rapid life cycle and can easily be grown in laboratories in large number. It possesses only 20,000 to 25,000 genes and a lot of them have been mapped so far and 75% DNA have been sequenced. Peacock (1996) reported that the group has been able to isolate counterpart genes to the *Arabidopsis* flower development from Eucalyptus. And have tested the control of flower formation using antisense Eucalyptus sequence in *Arabidopsis* and thus have developed transgenic Eucalyptus which is expected to be flower sterile.

As regards forage improvement programme, it is at a stage when formulating any programme on transgenic improvement should be in the light of increasing awareness regarding genetically modified organisms(GMO) because at some places people have shown resistance in accepting the GMO as such for large scale cultivation.

Genome mapping and MAS

In the last few years lot of progress has been made in mapping and tagging of many ariculturally importtant genes with mmolecular markers which forms the foundation for marker assisted selection (MAS). Molecular linkage maps - a prerequisite for MAS , have been developed for many crop plants using different kinds of molecular markers.

These markers have proved their worth and are considered to be better than other markers because unlike morphological physiological, colour, protein or isozyme markers which are few in number these are large in number hence have better resolution. Secondly, these do not require to be expressed i.e. the particular environment for example biotic or abiotic stress is not required and can be detected in lab conditions. Screening is non-destructive for presence or absence.

Molecular markers are used to identify and tag the desired genes. Linkage analysis is one of the basic and indispensable methods in genetics. The linkage can define the distance between polymorphic traits which are recognized as different in appearance, such as enzyme activity or nucleotide sequence on the allelic locus. Previously for example, the appearance of culm length, leaf length, panicle, seeds and isozymes of esterase or peroxidase were chosen as markers for genetic system. The linkage maps based on these markers have been constructed for rice, maize, wheat etc. In rice, 185 markers were positioned on 12 chromosomes (Kinoshita, 1995).

Polymorphism in the nucleotide sequence satisfies a requirement for it to function as a molecular marker in mapping. These polymorphism are revealed by molecular techniques such as RFLP, AFLP, microsatellite or simple sequence repeat (SSR), RAPD and single strand conformation polymorphism (SSCP). Among these the RFLPs were the first to be used in human genome mapping (Botstein et al 1980) and later on they were used for plant genome mapping (Weber and Helentzaris, 1989).

QTL Analysis

Most of the agronomic traits like yield and yield components (grain number, grain weight), plant height, days tto flowering etc are controlled by several genes. The number of genes and their interactive effects controlling the expression of quantitative traits are poorly understood. The work of Osborn et al (1987), Tanksley and Hewitt (1988) and Paterson et al (1991) in tomato has established the utility of RFLP markers in improvement of qualitative traits. Molecular markers linked to quantitative traits such as solid content in tomato (Osborn et al 1987), fertility restoration and wide compatibility (Zhu et al 1995), drought resistance (Nguyen et al 1995(, root morphology (Yadav et al 1995), heading date (Yano et al 1995a), shattering habit (Fakuta et al 1995) in rice and to apomictic mode of reproduction in *Pennisetum squamulatum* (Ozias-Akins et al 1993) have been reported. In forages comparatively less work has been reported. Some work on identification and transfer of QTL for drought resistance in *Pennisetm* has been done at IGER UK by Yadav at el (Personal discussion).

Molecular markers identified together with breeding behavior can be utilized for MAS if (i) markers cosegregate with desired tait, (ii) markers are closely linked with no or very low cross overs with the genes govering the trait, (iii) an easy means of screening the large population is available e.g. PCR based technologies are better and easy, (iv) Screenign techniques is having high reproducibility. MAS in plant breeding is extensively being used psarticularly for disease and pest resistance. Many genes, QTLs have been identified in rice, wheat, maize Sorghum, Barley, Soybean, Pea, Alfalfa, common bean, Mungbean, tomato, beet and potato.

C. *In vitro* Screening

Selection of germplasm for salinity tolerance is very important for the production of large scale material for cropping. Germplasm can be screened *in vitro* at two levels. One at the seedling level and one at the tissue or cell level. Asraf (1989) has reported that screening 1000 seedlings of each cultivar of *Medicago sativa* for shoot growth at three concentrations of NaCl resulted in selection intensities of <1% being achieved in each case. He advocated the recurrent selection for seedling NaCl tolerance.

Studies have shown significant interspecific and intraspecific variation among legume and clover species for salt tolerance. Intracultivar variation has been identified in Lucerne and *T. repens,* allowing selection programmes to be undertaken. Response to selection on the basis of agronomic or physiologic criteria has varied between species and criteria (Rogers,1992).

Mezentev *et al.* (1987) have reported that regenerants with high salt tolerance were obtained *in vitro* using B5 medium supplemented with a 1% NaCl solution. It has been found that control of Cl^- uptake in Lucerne (*Medicago sativa)* and white clover (*T. repens*) was found to be an effective criterion for *in vitro* selection for salt tolerance.

Improving some species for salt tolerance have met with limited success. Cell culture techniques, as supplementary procedures, may enhance the potential for achieving that success (Rains *et al.,* 1986). Cell culture systems offer several advantages to research programs studying salt tolerance at the cellular level. The application of these techniques to breeding can provide alternative methods for screening germplasm and identifying useful material for incorporation into crop improvement programs. A unique property also exists to study cellular level mechanisms and functions of salt tolerance. Complications due to difference in morphology and stages of development are reduced when using cell culture because of relatively homogenous and undifferentiated nature of the cultured cells. A large number of cells can be screened rapidly in a relatively small area. Genetically similar material can be used so that observations will be related to the tolerance of the cells. Traits can be selected at the cell level (i.e. salt tolerance) or somaclonal variability for that trait can be evaluated in regenerant plants and their progeny (Rains *et al.,* 1986). Maheshchandani (1991) has suggested that cell/tissue culture may be of use in the identification of putative salt tolerant genotypes. In order to exploit such biotechnological techniques, it is necessary to develop criteria which would help in identifying salt tolerant lines. Rice callus culture grown under increasing salt (NaCl) stress, were found on analysis to accumulate considerable amounts of free proline compared with unadapted cells. Variant cell lines selected for increased tolerance to saline environments, could possibly regenerate plantlets that could sustain their unhindered growth on such soils (Narayanaswamy, 1994). It has reported that salt resistant plants of *D. innoxia* can be recovered from a cell line selected from haploid derived callus cultures. Sathish *et al.*(1997) have reported that rice lines resistant to salt stress were obtained by *in vitro* selection of anther derived cultures when exposed to NaCl. Winicow (1991) has reported that plants (Lucerne) regenerated from the salt tolerant cell cultures showed improved salt tolerance compared to parent plants. The regenerated plants were vigorous, flowered and were self fertile. He further advocated that the cellular salt tolerance characteristic passaged through the regenerated plants since callus cultures initiated from immature ovaries of salt tolerant regenerant plants were salt tolerant with out additional selection of 1% NaCl. The tolerance phenotype appeared dominant in F_2 seeds obtained from self fertilization of the tolerant plants.

Most of the transgenic improvements in plant salt tolerance reported to date have been achieved through the detoxification strategy. This is obvious in the case of transgenic plants overexpressing enzymes involved in oxidaative protection, such as glutathione peroxidase, super oxide dismutase, ascorbateperoxidase and glutathione reductase (Roxas *et al.*, 1997; Allen, *et al.*, 1997). More recent engineering with the regulatory protein NPK 1, a mitogen activated protein (MAP) kinase, appears to mediate oxidative stress responses. Engineering with osmolytes such as mannitol, fructans, trehalose, ononitol, proline, glycinebetaine and ecotone also probably work through oxidative detoxification. These osmolytes are active in scavenging ROS (Shen *et al.*,1997). In addition targeted production of the osmolytes in the chloroplast by placing a signal sequence in front of engineered enzymes, results in better protection. This is consistent with the notion that the chloroplasts are the primary sites of ROS production. Finally, the osmolyte producing transgenic plants are improved in tolerance not only to salts but also to various other stresses such as chilling, freezing, heat and drought, which also generate ROS (Kalir and Poljakoff-mayber, 1981). *Arabidopsis* plants transformed with a modified bacterial gene for choline oxidase accumulated glycinebetaine in chloroplast and were more tolerant to salt, cold and heat stresses than wild type plants (Sakamoto *et al.*, 2000)

Apart from isozyme being used as a marker for the population and evolutionary studies, it can also be used as markers for plant adaptations to extreme environmental conditions like salinity. For example, the NaCl tolerant cell line of cotton (cv Coker 312) that was grown on 150 mM NaCl exhibited significantly greater catalase (341%), peroxidase (319%), glutathine reductase (287%), ascorbate peroxidase (450%), gamma glutamyl cysteine synthetase (224%), and glutathione-s-transferase (500%) activities than in the control (Dalton, 1996). Gangopadhyay *et al.*(1997), have reported that a significant increase of an acid phosphatase isoymic band in adapted calli, irrespective of NaCl or mannitol stress, indicated that it may be used as a marker for osmotic stress tolerance.

Initiatives and Achievements at IGFRI in Forage Crop Improvement

Berseem (T. alexandrinum)

The most important winter season forage crops suffers from susceptibility to root rot and stem rot and most of the improved cultivars are susceptible. Improvement programme is hampered due to narrow genetic base, lack of intraspecies donors for different traits including disease resistance, interspecific incompatibility and non availability of molecular markers for important traits and molecular maps. At IGFRI, Jhansi interspecific hybrids between *T. alexandrinum x T. apertum , T. alexandrinum x T. constantinopolitanum, T. alexandrinum x T. resupinatum* through embryo rescue. The progenies show variability for the traits such as days to flower, inflorescence shape, flower colour, seed colour, disease resistance.

Oat (Avena sativa L.)

An importanta winter season forage crop needs develoment with regard to development of lines tolerant to salinity/ alkalinity tolerant, diseases like crown rust, dual-purpose oat varieties, exploiting tertiary gene pool for introducing desirable traits,

Lucerne (Medicago sativa L.)

The major problem of the crop is Lucerne weevil and non-availability of intraspecies donor for resistance. However, screening of the germplasm (at IGFRI, Jhansi) identified *Medicago* species showing resistance to Lucerne weevil. Interspecific incompatibility in these crosses using conventional breeding methods have not been successful. At IGFRI tissue culture and transformation protocols for Lucerne have been developed successfully. Work in direction of developing transgeneic plant with resistance against Lucerne weevil is in progress.

Sorghum

Due to limited response of phenotypic selection for fodder-character improvement, there is need of precise selection for disease resistance utilizing marker technology. However, limited information on important QTLs is available on important traits such as *in vitro* digestibility of stover, protein content, resistance to foliar diseases and stem sugar content in dual purpose sorghum. At IGFRI work on identification of low HCN possessing lines, moisture stress tolerant lines, and identification of QTLs for the genetic improvement of stover quality, foliar diseases and drought tolerance in dual purpose sorghum is in progress.

Stylosanthes

The species is suited to range conditions. At IGFRI work on development of lines tolerant to drought and anthracnose is being attempted. Additionally six QTLs linked with resistance have been identified at different levels of LOD score located on six chromosomes. Twelve SSR markers have been developed using microsatellites sequences through enriched library prepared with degenerate primers. Utilizing molecular markers (SSR, STS, RAPD and ISSR) to estimate genetic relatedness between accessions of Stylo species.

Cenchrus Ciliaris

Breeding in predominantly apomictic species is limited. This species has been well studied to understand the apomixis. At IGFRI focus has been on identification and characterization of a sexual plant from the global collection, development of intraspecific hybrids utilizng sexual plants of *C. ciliaris*, identification of two RAPDs markers for apomixis, one converted to SCAR "OPF-8-600" and A 97 Kd protein fraction associated with sexuality in *C. ciliaris*.

Guinea Grass (*Panicum maximum*)

The apomictic mode of reproduction limits utilization of germplasm. At IGFRI major achievements have been rapid discrimination technique for apomictic and sexual plants on the basis of modified ES analysis, development of population segregating for apomixis generated, recombinants expressing high level of partitioned apomixis components identified, 2 RAPD markers linked to apomixis identified, Species specific markers for guinea grass identified and ploidy series generated from a single progenitor.

Pennisetum

Hybrid sterility in potential interspecific crosses is major linmitation in developing forage types. Ploidy level differences for obtaining interspecific hybrids in interploidal crosses is another constraint. Hence, utilization of techniques, such as embryo rescue, to overcome post-fertilization incompatibility barriers in interspecific crosses can be attempted. At IGFRI major achievements are production of interspecific hybrids of *P. squamulatum* and *P. orientale* with pearl millet, production of trispecific GOS hybrid, Establishment of protocol for Embryo cloning, production of perennial, thin stemmed, grassy, high tillering, apomictic fodder pearl millet.

Future Thrust

In India, biotechnological approach for forage crop improvement has started only recently and gradually gaining momentum. There are lot more areas which can be duly addressed with the help of biotechnology. Some of them are described hereunder:

Characterization

Molecular characterization of forage germplasm is one of the most important area which needs immediate attention because India with wide range of climatic variation hold a sizeable number of germplasm of various grasses and legumes which grow in varied conditions. The germplasm growing in very harsh climates i.e. extremely high and low temperatures, plants to alpines, very long days to very shot days, hot and cold arid areas are likely to posses genes for resistance to these abiotic stresses. Hence, characterizing these germplasm will ensure in future that the country will hold claim on use of these germplasm lines in breeding elsewhere in the world. Secondly, if we are able to identify these genes and get them patented, this can fetch tremendous amount of money as royalty in future. Chances of presence of these genes in these grass and legume species including weeds is more as these are growing in non-arable lands as compared to various cultivated crops which are grown on arable lands.

Apomixis

Most of the tropical grasses are apomictic in nature, hence, restrict the breeding efforts. Identification of gene (s) controlling apomixis is one area which can pay larger dividends in grass breeding. Identification and cloning of these genes can well be patented and can also be used in transferring in other cross pollinated crops for fixing heterosis and thus saving tremendous amount on producing hybrid seed every year. This problem is being paid attention globally but with little success. The mechanism of control of apomixis is not yet very clear. Another aspect associated with study on apomixis is identification of sexual lines in grasses. If some reliable molecular technique is available for screening of the grass species for existence of sexuality, it would accelerate the breeding process in these grasses because many of these plants can be used in crossing and further the plants with better agronomic traits and apomixis can be selected and advanced for developing varieties.

Creating Variability

Certain forage crops like Berseem lack genetic variability. Hence, progress in breeding has not been much in last few decades. Creation of variability through somaclonal variation and through developing interspecific hybrids using embryo rescue can be a faster and effective means for creating variation. Some other crops too like Lucerne could also be attended by this method as plants with desirable traits such as weevil resistance and persistence are not readily available.

Quality Breeding

Quality breeding involved two aspects. One is improving the quality of forage in terms of protein, IVDMD and aminoacid balance and secondly reducing/removing the antiquality factors in them. The first aspect is in general but with major emphasis for improving quality of forage grasses. Most of the grasses at 50% flowering stage are poor in quality which deteriorates further with advancing maturity. If some check can be imposed on its deterioration by some molecular technique by altering the biochemical pathways, it would be major breakthrough in grass improvement.

Some other general problems which can be dealt with biotechnology are:

- Limitation of lignin synthesis would significantly improve digestibility. Indeed several gram negative bacteria possesses the enzymes which can operate lignin biodegradation (Demarley, 1989). The genes coding for their enzymes could be made available. Specific esterases and antiperoxidases could be used to prevent lignin polymerization or to inhibit covalent links with the wall carbohydrates. But the importance of adjusted promoters in essential for such an objective.
- The two important families of nitrate-reductase genes could be made more active by selection in the most efficient organism and reinsertion in forage grasses.
- It would be stimulating goal to try to induce soybean seed proteins in the leaves of forage grass species as synthetic gene <<HEAAE>> has shown improvement in potato protein content.

- Bovine growth hormone gene has been isolated and is available. Provided that carefully selected promoters induce the expression only under special signals, it would be very stimulating to produce BGH in the forage plant itself.
- Several genes which encode auxins are available. Linked to an appropriate promoter they can be included in the genome to be enhanced for rapid recover of the canopy after a cut.
- Bloat problems might be partially solved when biochemical pathways are known.

References

Adams, E., Van Bockstaele, E., Depicker, A.and De Loose, M. 1996. Molecular analysis of the pathogen response of rye grasses upon infection with crown rust. Proc. 2nd International Conference on Harmful and Beneficial Microorganisms in Grassland Pasture and Turf. *IOBC wprs Bulletin* 19: 211-213.

Alexander, D.; R.M. Goodman; M. Gut-Rella; C. Glascock,; K.Weymann, L. Friedrich, D. Maddox, p. Ahl-Goy; T.Luntz, E.Ward and J. Ryals. 1993. Increased tolerance to two oomycete pathogens in transgenic tobacco expressing pathogenesis-related protein la. *Proc. Natl. Acad*. USA 90: 7327-7331.

Allen, R.. *et al*. 1997. Use of transgenic plants to study antioxidant defenses. *Free Radic. Biol. Med.* 23, 473-479. Quoted by Zhu, 2001

Ashraf, M. 1989. Salinity effects on some cultivars of Berseem (*T. alexandrinum*). *Tropennal and wirt*. 90: 93-104.

Botstein, B., White, R.L., Skolnick, M. and Davis, R.W. 1980. Construction of a genetic linkage map in man using restriction fragment length polymorphism. *Am. J. Hum. Genet.* 32: 314-331.

Conger, B.V. 1991. Biotechnology in forage grass breeding. In: *Fodder Crops Breeding: Achievements, novel strategies and biotechnology.* Proceedings of the 16th Meeting of Eucarpia Fodder Crops Section, Wageningen, the Netherlands, pp. 147-153.

Cornish, M.A., Hayward, M.D. and Lawrence, M.J. 1983. Self- incompatibility in ryegrass III. The joint segregation of S and Pgi-2 in *Lolium perenne*. *Heredity*. 44: 55-62.

Den Nijs, A.P.M. and Elgersma, A.(eds). 1991. Fodder crops breeding: Achievements, novel strategies and biotechnology. In: *Fodder Crops Breeding*. Op. Cit.

do Valle, C.B. and Savidan, Y. 1996. Genetics, cytogenetics and reproductive biology of Brachiaria. In: Brachiaria: Biology, Agronomy and Improvement. Eds. Miles, J.W., Maas, B.L. and do Valle, C.B. CIAT Publicationno.259, pp. 147-163.

Fukuta, Y., Harushima, Y. and Yano, M. 1995. Genetic analysis of shattering habit using molecular marker in rice (*Oryza sativa* L.). In: *Proceeding of Inter Rice Genetics Symp.*, IRRI, Manila, the Philippines (in press).

Gangopadhyay; Basu, S. and Gupta, S.1997. In vitro selection and physiological charecterization of NaCl and Mannitol adapted callus lines in *Brassica juncea*. *Plant cell, Tissue and Organ culture* 50: 161-169.

Hayward, M.D., G.H. Degenaars, F. Balfourier and F. Eickmeyer. 1995a. Isozyme procedures for characterization of germplasm, exemplified by the collection of *Lolium perenne* L. *Genetic Resources and Crop Evolution* 42: 327-337.

Hayward, M.D., G.H. Degenaars, F. Balfourier and F. Eickmeyer. 1995b. Exploitation of visible and promiscuous homoeologous recombinations as a key to genome analysis in grasses. BBSRC Plant and Animal Genome Analysis Programme, First annual conference, Warwick 1995, p. 15.

Hayward, M.D., 1990. Genetic strategy and future prospects of breeding cross pollinated crops. *Norwegian Journal of Agricultural Sciences. Supplement* no. 9: 77-86.

Humphreys, M.W., A. Ghamari-Zara, I. Pasakinskiene and M.C. Kerlan 1996. The potential of anther culture derived plants from a *L. multiflorum x F. arundinacea* (2n=5x=35) hybrid as a novel forage grass crop. *Proc. Of the Eucarpia Fodder Crops Section Meeting*, Radzikowa, Poland.

Jongedijk, E.H., Tigelaar, J.S.C. van Rockel, S.A. Bres-Vloemans, I. Dekker, P.J.M. van den Elzen, B.J.C. Cornelissen and L.S. Melchers. 1995. Synergistic activity of chitinases and b -1-3 glucanases enhances fungal resistance in transgenic tomato plants. *Euphytica* 85: 173-180.

Kalir, A and Poljakoff-Mayber, A., 1981. Changes in activity of malate dehydrogenase, catalase, peroxidase and super oxide dismutase in leaves of *Halimione portulacoides* L. Allen exposed to high sodium chloride concentration. *Ann. Bot.* 47, 75-85.

Kinoshita, T. 1995. Report of committee of gene symbolization, nomenclature and linkage group. *Rice Genet. Newslett.* 12: 9-153.

Lawton, K.,S. Uknes, L. Friedrich, T. Gaffney, D. Alexander, R. Goodman, J.P. Metraux, H. Kessmann, P.Ahl Goy, M. Gut Rella, E. Ward and J. Ryals. 1993. The molecular biology of systemic aquired resistance. In: B. Fritig and M. Legrand (Eds.) *Mechanism of Plant Defense Responses*. Kluwer Academic Publishers, Dordrecht, the Netherlands. Pp. 422-432.

Lin, W., C.S. Anuratha; K. Dutta; Potrykus; S. Muthukrishnan and S.K. Datt. 1995 Genetic engineering of rice for resistance to sheath blight. *Biol/Technol*. 13: 686-691.

Liu D., K.G. Raghothama, P.M. Hasegawa and R.A. Bressan. 1994. Osmotin overexpression in potato delays development of disease symptoms. *Proc. Natl. Acad, Sci. USA* 91: 1888-1892.

Maheshchandani, N 1991, callus culture and adult plant response tpo salt stress. *Annals of biology(Ludhiana)*, Haryana Agricultural University, Hisar.

Melchers, L.S., M. Apotheker-de Groot, J. A. van der Knaap, A.P. Ponstein, M.B. Sela-Buurlage, J.F.Bol. B.J.C. Cornelissen, P.J.M. van den Elzen and H.J.M. Linthorst

1994. A new class of tobacco chitinases homologous to bacterial exo-chitinases displays antifungal activity. *Plant J.* 5: 469-480.

Mezenstev, A.V., Golyskina, N.A., Solodkaya, A.A. 1987. Producing initial breeding material of a clover and Lucerne using cell and tissue culture methods. *Plant Breeding Abstracts*, 1987, July, pp 648

Moore, G., Devos, K.M., Wang, Z. and Gale, M.D. 1995. Grass line up and form a circle. *Current Biology* 5: 737-739.

Narayanaswamy, S. 1994., *Plant cell and tissue culture.*, Tata Mc Graw Hill publishing company Ltd., New Delhi.

Neuhaus, J-M, P. Ahl-Goy, U. Hinz, S. Flores and S. Flores and F. Meins Jr. 1991. High level expressions of a tobacco chitinase gene in Nicotiana sylvestris. Susceptibility of transgenic plants *Cercospora nicotanae* infection. *Plant Mol. Biol.* 16: 141-151.

Nguyen, H.T. , C.R. Babu, M.S. Pathan, E.H. Shashidhar, H.G. Zheng and A. Blum. 1995. Molecular mapping of draught resistance traits in rice. *Proc. of Third Intern. Rice Genetics Symp.* op. Cit.

Nielsen, K.K.; J.D. Mikkelsen, K.M.Kragh and K. Bojsen. 1993. An acidic class III chitinase in sugar beet: induction by *Cercospora beticola*, characterization, and expression in transgenic plants. *Mol. Plant-Microbe Interact.* 6(4): 495-506.

Oertal, C., J. Fuchs and F. Matzk. 1996 Successful hybidization between *Lolium* and Dactylis. *Plant Breeding* 115 (2): 101-105.

Osborn, T.C., Alexander, D.C. and Fobes, J.F. 1987. Identification of restriction fragment length polymorphisms linked to genes controlling soluble solids content in tomato fruit. *Theor. Appl. Genet.* 73: 350-356.

Ozias- Akins, P., Lubbers, E.L., Hanna, W.W. and McNay, J.W. 1993. Transmission of the apomictic mode of reproduction in *Pennisetum*: Co-inheritance of the trait and molecular markers. *Theor. Appl. Genet.* 85: 632-638.

Paterson, A.H., Tanksley, S.D. and Sorrells, M.E. 1991. DNA markers in plant improvement. *Adv. Agronomy* 46: 39-90.

Rains, D. W.; Cruogham, S. S. and Crougham, T.P., 1986. Isolation of mutant cell lines and plant: salt tolerance.. In: *Cell culture and somatic cell genetics of plants* vol 3, 527-547 (Ed. Vasil, I. K.)., Academic Press, Inc., New York.

Roxas, V.P. *et al.*, 1997 Overexpression of glutathione S-transferase/glutathione peroxidase enhances the growth of transgenic tobacco seedlings during stress. *Nat. Biotechnol.* 15,988-991. Quoted by Zhu, 2001.

Sakamoto, A. *et al.* 2000. Genetic engineering of biosynthesis of glycinebetaine enhances tolerance to various stress. In: *Plant tolerance to abiotic stresses in agriculture: Role of genetic engineering* (vol. 83) (Cherry, J.H. *et al.*, eds.). Quoted by Zhu, 2001.

Sathish, P., Gamborg, O.L., Nabors, M.W. 1997. Establishment of stable NaCl resistant rice plant lines from anther culture pattern of K^+/Na^+ in callus and plantcells., *Theoretical and applied genetics*. 1997, 95(8).

Shen, B. *et al*. 1997 Increased resistance to oxidative stress in transgenic plants by targetting mannitol biosynthesis to chloroplasts. *Plant Physiol*. 113, 1177-1183. Quoted by Zhu, 2001.

Spangenberg, G., Wang, Z.Y., Wu, X.L., Nagel, J. Iglesias, V.A. and Potrykus, I. 1995. Intergeneric symmetric and asymmetric somatic hybridization in *Festuca* and *Lolium*. *Euphytica* 85: 235-245.

Stuiver, M.H., H. Tigelaar; L. Molendijk; E. Troost-van Deventer, M.B. Sela-Buurlage, J.Storms; L.Plooster, F.Sijbolts, J. Custers, M. Apotheker-de Groot and L.S. Melchers. 1996. Broad spectrum resistance in transgenic carrot plant. In. G. Stacey, B. Mullin and P.M. gresshoff (Eds.) 8th International Congress Molecular Plant Microbe Interactions. Op. Cit. B-93.

Tanksley, S.D. and J. Hewitt. 1996. Use of molecular markers in breeding for soluble solids content in tomato - a re-examination. *Theor. Appl, Genet*. 75: 811-823.

Weber, D. and Helentzaris, T. 1989. Mapping RFP loci in maize using B-A translocations. Genetic 121: 583-590.

Winicow, I. 1991. Charecterization of salt tolerant Alfalfa (Medicago sativa) plats regenerated from salt tolerant cell line. *Plant cell reports* 10: 561-564.

Yadav, R.S., Courtois, B. and Huang, N.1995. Maping genes controlling root morphology in a double haploid population of indica x japonica cross of rice. In: *Proc. of Third Intern. Rice Genetics Symp. op. Cit*.

Yano, M., Y. Harushima, Y. Kuboki, S.Y. Lin, Y. Nagamura, N.Kurata and T. Sasaki. 1995 QTL mapping as an aid to tagging genes conferring heading date in rice. In: *Proc. of Third Intern. Rice Genetics Symp. op. Cit*.

Yoshikawa, M.M. Tsuda and Y. Takeuchi. 1993. Resistance to fungal diseases in transgenic tobacco plants expressing the phytoalexin elicitor-releasing factor, b -1-3 endoglucanases, from soybean. *Naturwissenschaften* 80: 417-420.

Zhu, L., C. Lu, P.Li, L.Shen, P.He, Y.Chen. 1995. Using DH populations of rice for QTL mapping and geographical genotyping. In: *Proc. of Third Intern. Rice Genetics Symp. op. Cit*.

Zhu, Q.; E.A. Maher, S. Masoud, R.A. Dixon and C.J. Lamb. 1994. Enhanced protection against fungal attack by constitutive co-expression of chitinase and glucanase genes in transgenic tobacco. Biol/Technol. 12: 807-812.

Zwierzykowski, Z. 1987. Interspecific and intergeneric hybridization in the *Lolium- Festuca* complex. *Proc. Intern. Conf. Hladke Zivotice*, Czechoslovakia, pp. 29-38.

Crop Wise Selected Publications of IGFRI

Berseem

Malaviya D. R., A. K.Roy, P. Kaushal, B. Kumar and A. Tiwari. 2008. Phylogenetic studies based on isozyme similarity among *Trifolium* species. *Plant Systematics and Evolution* (in press) DOI 10.1007/s00606-008-0070-7

Kaushal, P., A. Tiwari, A. K. Roy, D. R. Malaviya and B. Kumar. 2006. *In vitro* regeneration of *Trifolium glomeratum*. *Biologia Plantarum* 50: 693-696.

Malaviya, D. R., B. Kumar, A. K. Roy, P. Kaushal and A. Tiwari. 2005. Estimation of variability of five enzyme systems among wild and cultivated species of *Trifolium*. *Genetic Resources and Crop Evolution* 52: 967-976.

Malaviya, D. R., A. K. Roy, A. Tiwari, P. Kaushal and Bijendra Kumar. 2006. *In vitro* callusing and regeneration in *Trifolium resupinatum*- a fodder legume. *Cytologia* 71: 229-235.

Kaushal, P., D. R. Malaviya, A. K. Roy, B. Kumar and A. Tiwari. 2005.*Trifolium alexandrinum* x *T. resupinatum* - interspecific hybrids developed through embryo rescue. *Plant Cell, Tissue and Organ Culture* 83:137-144.

Roy, A. K., A. Tiwari, D. R. Malaviya, P. Kaushal and B. Kumar. 2005. Regeneration efficiency in *Trifolium apertum* as influenced by explant media interaction. *Cytologia* 70: 153-159

Roy, A. K., D. R. Malaviya and P. Kaushal. 2005. Pollination behaviour among different breeding populations of Egyptian clover. *Plant Breeding* 124: 171-175.

Malaviya, D. R., A. K. Roy, P. Kaushal, B. Kumar and A. Tiwari. 2004. Development and characterization of *Trifolium alexandrinum* x *T. apertum* using embryo rescue. *Plant Breeding* 123: 536-542.

Roy, A. K., D. R. Malaviya, P. Kaushal, B. Kumar and A. Tiwari. 2004. Interspecific hybridization of *T. alexandrinum* with *T. constantinopolitanum* using embryo rescue. *Plant Cell Reports* 22: 705-710

Malaviya, D. R., A. K. Roy, P. Kaushal and B. Kumar. 2004. Affinity between *Trifolium alexandrinum* and *T. apertum* – cytological investigations in embryo rescued hybrid. *Cytologia* 69: 425-429.

Lucerne (*Medicago*)

Tiwari, K.K., and A. Chandra. 2008. Bulk genomic DNA PCR analysis - a rapid approach to estimate genetic relatedness among heterogeneous lucerne *(Medicago sativa*. L.) populations. *Ind. J. Agric. Biochem*. (in press).

Chandra, A. 2008. Transferability of SSR markers across twelve species of forage legumes for germplasm characterization and evaluation. *Indian J Genet*. (in press).

Chandra, A. and K. C. Pandey. 2008. Genetic relatedness among weevil resistant *Medicago* species and Indian susceptible cultivars of lucerne (*Medicago sativa* L.) using SSR and RAPD markers. *Indian J. Genet.* (in press).

Chandra, A. and K. C. Pandey. 2008. Effect of proteinase inhibitors on Indian alfalfa weevil (*Hypera postica Gyll*) growth and development. *Acta Physiol. Plant.* 30: 501-505.

Kumar S., A. Chandra and M. G. Gupta. 2008. Plantlet regeneration *via* multiple shoot induction in Indian cultivars of Lucerne (*Medicago sativa* L). *J. Plant Biochem. Biotech.* 17: 181-184.

Chandra, A. 2007. Bulk genomic DNA PCR analysis - A rapid method to estimate genetic relatedness among heterogeneous lucerne (*Medicago sativa* L.) cultivars. *Cytologia* 72: 363-368.

Suresh Kumar, A. Chandra and K. C. Pandey. 2007. Screening of *Bt* genes against lucerne weevil (*Hypera postica* Gyll.). *Range Mgmt. & Agroforestry*. 28: 291-292.

Suresh Kumar and M. G. Gupta. 2007. Direct plantlet regeneration in lucerne (*Medicago sativa* L.). *Range Mgmt. & Agroforestry*. 28: 283-284.

Chandra, A., K. C. Pandey and U. P. Singh. 2006. *Medicago scutellata* – A possible source of weevil resistance for lucerne improvement. Ind J. *Plant Genet. Resour.* 19: 291-293.

Baronia, C. and A. Chandra. 2006. Isozyme analysis of five Indian cultivars of Lucerne (*Medicago sativa* L.). *Indian J. Agric. Biochem.* 19:75-77.

Suresh Kumar, V. Bhat, B. V. Bhat and M. G. Gupta. 2002. *Agrobacterium*-mediated transformation of Lucerne (*Medicago sativa* Linn.): optimizing biological and physical parameters. *Ind. J. Biotechnol.* 1: 298 - 300.

Stylosanthes

Chandra, A., 2008. Genetic relationship among accessions of *Stylosanthes hamata* based on seed proteins profile. *Indian J Genet.* (in press).

Chandra, A. 2008. Developing biotic and abiotic stress tolerant *Stylosanthes* cultivar: molecular aspects. *Ind. J. Agric. Biochem* (in press).

Chandra, A. and R. K. Bhatt. 2008. Carbon isotope discrimination function analysis and drought tolerance of stylo species grown under rain-fed environment. *Acta Physiol. Plant.* 30:63-69.

Chandra, A. 2008. Diversity among *Stylosanthes* species: Habitat adaphic and agroclimatic affinities leading to cultivar development. *J. Env. Biol.* (in press).

Tewari, S. and A. Chandra. 2008. Genetical assessment of diploid progenitors of *S. scabra* by Isozyme, RAPD and STS markers: a possible strategy for improvement of drought tolerant allo-tetraploid *S. scabra* species. *Euphytica* 162: 39-50.

Chandra, A. 2007. Molecular assessment of variation in *Stylosanthes fruticosa* accessions employing STS and RAPD markers. *Cytologia* 72: 251-258.

Chandra, A. and S. Tewari. 2007. Isolation of genomic DNA from stylo species without liquid nitrogen suitable for RAPD and STS analyses. *Cytologia* 72: 287-293.

Chandra, A., P. S. Pathak and R. K. Bhatt. 2006. Stylo research in India: Prospects and challenges ahead. *Curr. Sci.* 90: 915-921.

Chandra, A. 2006. A partial genetic map of *Stylosanthes* developed using inter-specific F2 cross. *Cytologia* 71: 353-358

Ma, Z. Y., A. Chandra, J. M. Musial and C. J. Liu. 2004. Molecular evidence that *Stylosanthes angustifolia* is the third putative diploid progenitor of the hexaploid *S. erecta* (Fabaceae), *Plant Syst. Evol.* 248: 171-176.

Chandra, A., P. S. Pathak, R. K. Bhatt and A. Dubey. 2004. Variation in drought tolerance behavior of different *Stylosanthes* accessions. *Biol. Plant.* 48: 457-460.

Thumma, B. R., B. P. Naidu, A. Chandra, D. F. Cameron, L. M. Bahnisch and C. Liu. 2001. Identification of causal relationship among traits related to drought resistance in *Stylosanthes Scabra* using QTL analysis. *J. Exp. Bot.* 52:203-214.

Oats

Roy, A. K. and R. N. Choubey. 2005. Utilization of genepool in genetic improvement in forage oats – Achievements and future prospects. Ind *J. Plant Genet. Resour.* 18: 108-109.

Choubey, R. N., S. N. Zadoo and A. K. Roy. 2003. Performance of back cross derived progenies of *Avena sativa* x *Avena sterilis* for forage yield and related traits. *Indian J. Genet.* 63: 251-252.

Sai Prasad, S. V., R. N. Choubey, S. N. Zadoo and A. K. Roy. 2003. Nature of genetic variability and diversity in a large exotic collection of forage oat (*Avena sativa* L). *Indian J. Agric. Res.* 37: 23-28.

Choubey, R. N., S. V. Sai Prasad, S. N. Zadoo and A. K. Roy. 2001. Assessment of genetic diversity and inter relationships among yield contributing traits in forage oat germplasm. *Forage Research* 27: 149-154.

Choubey, R. N., S. N. Zadoo and A. K. Roy. 1995. Analysis of forage yield and related traits in backcross derived progenies of interspecific matings (*Avena sativa* x *A. sterilis*) of oats. *Crop Improvement* 23: 155-157.

Pearl Millet

Kaushal, P., Aarti Khare, S. N. Zadoo, A. K. Roy, D. R. Malaviya, A. Agrawal, S. A. Siddiqui and R. N. Choubey. 2008. Sequential reduction of *Pennisetum squamulatum* genome complement in *P. glaucum* (2*n*=28) x *P. squamulatum* (2*n*=56) hybrids and their progenies revealed its octoploid status. *Cytologia* (in press).

Kaushal, P., A. K. Roy, A. Khare, D. R. Malaviya, S. N. Zadoo and R. N. Choubey. 2007. Crossability and characterization of interspecific hybrids between sexual *Pennisetum glaucum* (pearl millet) and a new cytotype (2*n*=56) of apomictic *P. squamulatum*. *Cytologia* 72: 111-118.

Kaushal, P., A. K. Roy, S. N. Zadoo and R. N. Choubey. 2004. Cytogenetic analysis of thermosensitive genic male sterility (TGMS) recovered from a *Pennisetum glaucum* (L.) R. Br. x *P. violaceum* (Lam.) L. Rich cross. *Cytologia* 69: 409-418.

Kaushal P, A. K. Roy, S. N. Zadoo, R. N. Choubey. 2003. Embryo cloning in *Pennisetum glaucum*: production of multiple plantlets from immature zygotic embryo. *Range Management and Agroforestry* 24: 156-158.

Roy, A. K., P. Kaushal, S. N. Zadoo and R. N. Choubey. 2003. Identification of a new cytotype of *Pennisetum squamulatum* Fresen. with 2*n*=56 chromosomes. *Range Management and Agroforestry* 24: 71-73.

Cenchrus

Bhat, B. V., V. Bhat, m. G. Gupta and S. Gupta. 2007. Isozyme based genetic similarity in *Cenchrus* (Poaceae). Indian J. *Range Mgmt. Agroforestry* 28: 285-286.

Dwivedi, K. K., S. R. Bhat, V. Bhat, B. V. Bhat and M. G. Gupta. 2007. Identification of a SCAR marker linked to apomixis in Buffel grass (*Cenchrus ciliaris* L.). *Plant Sci.* 172: 788-795.

Bhat, V., S. J. Dalton, S. Kumar, B. V. Bhat, M. G. Gupta and P. Morris. 2001. Particle in-flow gun mediated genetic transformation of Buffel grass (*Cenchrus ciliaris* L.), optimizing biological and physical parameters. *J. Appl. Genet.* 42: 405-412.

Gupta, S., M. G. Gupta, V. Bhat and B. V. Bhat. 2001. Status of apomixis and sexuality in four species of *Cenchrus. J. Plant Biol.* 28: 153-159.

Dicanthium

Kumar, J., S. M. Shukla, V. Bhat, S. Gupta and M. G. Gupta. 2005. *In vitro* regeneration and genetic transformation of *Dicanthium annulatum. DNA and Cell Biol.* 24: 670-679.

Gupta M. G., V. Bhat, B. V. Bhat, C. N. Neerja and S. Gupta. 2003. Phylogenetic relationships in tetraploid agamospecies of *Dicanthium* complex based on isozyme phenotypes. *J. Plant Biol.* 30: 61-64.

Dalton, S. J., A. J. E. Bettany, V. Bhat, M. G. Gupta, K. Baily, E. Timms and P. Morris. 2003. Genetic transformation of *Dicanthium annulatum* (Forssk.) – an apomictic tropical forage grass. *Plant Cell Reports* 21: 974-980.

Gupta, M. G., S. Gupta, B. V. Bhat, V. Bhat and S. T. Ahmed. 1998. Estimation of frequency of apomixis by auxin induced parthenocarpy technique in *Dicanthium annulatum* (Forssk.) Stapf. *Indian J. Range Mgmt. Agroforestry* 19: 149-153.

Gupta, M. G., S. Gupta, B. V. Bhat and V. Bhat. 1998. Plant regeneration from callus cultures initiated from nodes of a grass-*Dicanthium annulatum* (Forssk.) Stapf. *Acta Botanica Indica* 26: 31-36.

Gupta, M. G., S. Gupta, B. V. Bhat, V. Bhat and Bhagmal. 1998. Somaclonal variation for facultative apomixis in marvel grass (*Dicanthium annulatum* Forssk. Stapf.). *Forage Research* 24: 111-114.

Gupta, M. G., S. Gupta, B. V. Bhat and V. Bhat. 1997. *In vitro* regeneration and somaclonal variation in a tropical pasture grass *Dicanthium annulatum*. *Ind. J. Range Mgmt. Agroforestry* 18: 25-30.

Apomixis

Kaushal, P., A. Agrawal, D. R. Malaviya, S. A. Siddiqui and A. K. Roy. 2008. Ploidy manipulation in guinea grass (*Panicum maximum* Jacq., Poaceae) utilizing a Hybridization-supplemented Apomixis-components Partitioning Approach (HAPA). *Plant Breeding* (Accepted).

Kaushal, P., D. R. Malaviya, A. K. Roy, Shalini Pathak, A. Agrawal, Ambica Khare, S. A. Siddiqui. 2008. Reproductive pathways of seed development in apomictic guinea grass (*Panicum maximum* Jacq.) reveal uncoupling of apomixis components. *Euphytica* 164:81-92.

Jain, A., A. K. Roy, P. Kaushal, D. R. Malaviya and S. N. Zadoo. 2006. Isozyme banding pattern and estimation of genetic diversity among guinea grass germplasm. *Genetic Resources and Crop Evolution* 53: 339-347.

Kaushal, P., S. N. Zadoo, D. R. Malaviya and A. K. Roy. 2005. Apomixis research in India: past efforts and future strategies. *Current Science* 89: 1092-1096.

Kaushal, P., D. R. Malaviya and A. K. Roy. 2004. Prospects for breeding apomictic rice: a reassessment. *Current Science* 87: 292-296.

Jain A., S. N. Zadoo, A. K. Roy, P. Kaushal and D. R. Malaviya. 2003. Meiotic system and probable basic chromosome number of *Panicum maximum* Jacq. accessions. *Cytologia* 68: 7-13.

□□□

Chapter **31**

Agroforestry Interventions for Reclamation of Degraded Lands

A. K. HANDA
National Research Center on Agro Forestry, Jhansi-284003 (U.P.), India

Introduction

The traditional land use systems have become very fragile and unsustainable due to unbalanced utilization of natural resources, resulting in large-scale degradation and consequent occurrence of wastelands. Soil degradation is one of the greatest challenges facing the mankind. Most ancient civilizations flourished on fertile soils. Archaeological evidence has shown that soil degradation was responsible for the extinction of Harappan civilization in Western India, Mesopotamia in Western Asia and Mayan culture in Central America.

The term "Wasteland" although being used by the department of Economics and Statistics since the very inception of land use classes adopted by it appears rather a misnomer. The word "Waste" give impression of land being use less or non-usable. This is actually not so . Every bit of land created by nature has its own use. But if mankind is unable to diagnose undertaken and use it appropriate to its capability it is not Nature's fault to call this as wastelands. The technical task group set up by Planning Commission in December, 1985 for formulating a definition of wastelands and their classification suggested that wastelands is defined as the land which is presently degraded and is lying unutilized except current fallows, due to various constraints. This has been divided into two broad classes (1) the culturable land (2) un-culturable lands. Lands which in spite of having the potential for supporting the vegetation are not being properly utilized due to different constraints could be put under culturable wasteland whereas lands which has no potential to develop vegetation cover could be formed non culturable wastelands. Soils are the most precious natural resource of any nation as it takes almost 1000 years to produce an inch of soil. It is, therefore, imperative that we manage and conserve them to meet the growing demands of food, fodder, fiber and fuel.

The primary cause of degradation is the demographic pressure on land, leading to loss of vegetative cover through deforestation. Harvesting of timber, collecting fuel wood, overgrazing, shifting cultivation and encroachments on forests are some of the reasons for deforestation. Besides that, ignorance of proper soil conservation measures, non-judicious use of fertilizer and pesticides, faulty irrigation and water management, discharge of industrial effluents, sewage/sludge are also responsible for soil degradation to a considerable extent.

According to the estimates given by NBSSLUP (Sehgal and Abrol, 1994), 57% of land (187.7 million ha) in the country is potentially exposed to various degradation forces; of which a dominant fraction is under water erosion 45% (149.9 million ha), wind erosion 4.1% (13.5 million ha), chemical deterioration 4.2% (13.8 million ha) and physical deterioration 3.6% (11 million ha). As per record of Ministry of Agriculture (1994), 37% of total geographical area of the country (107.43 million ha) is degraded due to various causes of land degradation.

Steps for Development of Degraded Lands

To Know the Species Suited to the Area

Knowledge of suitable species of grasses, legumes and trees is a pre-requisite before initiating any development programme. Ours is a vast country with a great diversity of climate and variation in topography and soils. Therefore, any blanket recommendation can not be made. One has to choose the right species that can adapt to a given environment.

Measures for Protecting the Area from Animals

Overgrazing has been responsible for the degradation of land. Simple closure of the area would help rapid regeneration of grasses, legumes and trees already available in that area. Protection may be done either by barbed/ woven wire with angle iron/stone/wooden poles or cut thorny shrub materials, loose stone wall, cattle proof trench, live hedge etc. For live fencing species like *Carrisa carandas, Lantana camera, Agave Sisalana, Agave Americana, Lowsonia alba, Lowsonia inermis, Euphorbia tirucalli, Opuntia species, Prosopis juliflora, Ipomea fistula, Parkinsonia aculeate etc.* may be effectively utilized which is economical also. By simple protection, production from natural vegetation, recorded at the tune of 3.5 t/ha as compared with only 0.08 t/ha from unprotected area (Shankarnarayan *et. al.*, 1974).

Eradication of Unwanted Bushes

Under Indian climatic condition, closure of an area leads to bush formation, which occupies over 30% area, and this causes a reduction in the productivity of grassland. Therefore, unwanted woody perennials such as *Lantana karonda,* etc. should be eradicated to get higher production. Eradication may be done either manually by uprooting the bush/trees or chemically by application of 0.8% 2,4,5-T solution on freshly cut stumps or broadcast weedicide picloram at 10-15 kg/ha around bushes. Application of weedicide showed good kill of bushes and prevented resprouting (Shankarnarayan, *et. al.*, 1974).

Suitable Soil and Water Conservation Measures

Soil and water conservation measures are one of the most important6 factor for improvement of degraded lands especially on sloppy lands. On such type of lands the first step is to check the run off improve the water percolation and increase soil moisture status of the soil. Staggered contour bunds/trenches (3x0.5x0.5) are made to arrest the water flow. The soil is deposited along with contour and the trenches are dug facing the slope. 500-600 trenches/ha are recommended for semiarid region with rainfall up to 1000 mm. They retain 95% of the run off even from the rainstorms of high intensities. On the soil mounds, seeds of legumes such *Stylosanthes hamata, Macroptillium atropurpureum* etc. are sown to enrich soil and optimizing the soil moisture availability.

Studies conducted at Jhansi revealed that in *Sehima-Heteropogon* grassland contour furrow at 6 m interval were most effective as compared to 0, 3 and 12 m intervals as regards to highest percentage of cover (66.0%), of most desirable species *Sehima nervosum* as well as optimum moisture storage while *Heteropogon contortus* showed dominance at 3 m intervals (Rai, 1989). Ahuja and Man (1975) also reported that pitting and contour furrow at 8 to 10 m intervals increased forage yield 6 times. Further, Das and Yadava (1979) obtained significantly higher yield in the area pitted with staggered pitting discar at Jodhpur. The increase in yield was 80-115% in natural grassland and 50-70% in reseeded pasture.

Establishment of Grasses, Legumes and Trees

Just by protection of the degraded lands, the regeneration of natural vegetation starts. This is often time consuming process. Therefore, establishment of suitable grasses, legumes and trees of that area is desirable. After selection of suitable species planting and sowing may be done at the onset of monsoon. Pits of 45 cm^3 size at 5x5 m or 6x4 m spacing should be dug during May-June for weathering of soil so as to minimize the attack on seedling by insects and pests. At the time of plantation of sapling, in each pit a mixture of urea, single super-phosphate and potash (125 g in the ratio of 1:4:1) along with 40 to 50 per cent farm yard manure of the total volume of the pits may be given. 10 to 15 g BHC powder per pit may be given besides 15 to 20 kg/ha at the time of soil working to save the saplings from termite attack. 5-11 months old tree seedlings may be planted during monsoon in the pits depending the growth of saplings of tree species.

Immediately after plantation of trees, the interspaces are sown with grasses at the rate of 4-5 kg/ha at 100 cm spacing. In between two rows of grass, small seeded legumes such as Stylo or Siratro should be sown at the rate of 5 to 6 kg/ha. The seeds of grasses and legumes should be in the form of pellet made with clay soil and farm yard manure for better establishment. For quick growth and production of grasses and legumes 40 kg P2O5 and 20 kg N/ha at the time of land preparation followed by split application of 20-60 kg N/ ha annually depending on the rainfall and its distribution has been recommended (Rai and Kanodia, 1981). In arid zones, complete soil working after removal of unwanted bushes etc. followed by sowing of perennial grasses resulted better establishment and higher productivity (Chakravarty and Verma, 1970).

Rehabilitation Strategies

The terms like restoration, reclamation and rehabilitation have been frequently used (Singh and Jha 1994). National Academy of sciences USA (1974) defined restoration as the replication of site conditions prior to disturbance; whereas reclamation refers to rendering a site habitable to indigenous species; and rehabilitation implies that degraded land will be returned to a form and productivity in conformity with a prior land use plan including stable ecological state that does not contribute substantially to environmental deterioration and is consistent with surrounding aesthetic values.

There are following three alternatives in restoration:

1. *Full restoration*: restoration of a site to its pre-damaged condition
2. *Partial restoration*: restoration of selected ecological attributes of the site
3. Creation of an alternative ecosystem type.

According to Jordan et al. (1988), restoration is the recreation of entire communities of organisms, closely modeled on those occurring naturally; while reclamation refers to any deliberate attempt to return a damaged ecosystem to some land of productive use or socially acceptable condition short of restoration. Some environmentalists use the term restoration in a broader sense to refer to many lands of environmental management, including reclamation.

Brinck *et al* (1988) described in detail the following levels of restoration: Active restoration: accurate re-creation of site conditions existing prior to the disturbance activity involving a consideration of all environmental components, whether valued or not;

Passive restoration: disturbed areas to be habitable abandoned in conditions designed to be habitable to organisms that were originally present or others that approximate the original habitat and opportunities are created for natural regeneration.

Rehabilitation: disturbed environment is restored to a form and productivity confirming with a prior use plan defining ecosystem components of value. Also includes a stable ecological state that does not contribute substantially to environmental deterioration and is consistent with surrounding values;

Protected natural regeneration: following disturbance, environment is abandoned and biophysical components are left in stable conditions. Area is left to regenerate with no assistance or valued direction. It is protected from further disturbance activities;

Unstable natural regeneration: major disturbance activities cease and environment is abandoned with no development plans. Area is left unstable but protected from further human activities; and

Unstable unprotected regeneration: major disturbance activities cease but environment may not be abandoned for other activities. Area is left unstable with no development plans.

Land Degradation Through Water Erosion and their Rehabilitation

Water erosion is the most serious degradation problem in the Indian context, resulting in loss of top soil in 130.5 million ha and terrain deformation in 16.4 million ha. Based on the analysis of existing soil loss data, the average rate of soil erosion estimated a first approximation is 16.35 t/ha/yr, yielding a total soil loss of 5.3 billion tones annually in the country (Dhruvanarayana and Ram Babu, 1983). Nearly 29% of total eroded soil is permanently lost to the sea, 61% simply transferred from one place to another and remaining 10% is deposited in the reservoirs resulting in the reduction of their storage capacity by 1 to 2 % annually.

Ravine Lands

In our country there are about 3.67 million hectare of ravine lands of which about 2.63 million hectare are confined to Uttar Pradesh, Madhya Pradesh, Rajasthan and Gujarat. In addition every year due to faulty agricultural practices more than 8.000 hectares of land is converted into ravine. It is not an uncommon sight to see farmers cultivating the land up to the edge of the ravine thus enhancing the process of ravine extension. During 1970's centrally sponsored ravine reclamation programme reclaimed about 2,000 hectare of ravine land in Uttar Pradesh at an enormous cost, varying from Rs. 3,000 - 4,000 per hectare (Anonymous, 1972) and leased it to farmers. It is revealed that farmers find it extremely difficult to maintain some of the gullies leveled for arable farming. Experience have revealed that best use of ravine land is to put it under suitable permanent vegetation cover. Only very small and shallow gullies and undulating land adjoining the village may be reclaimed for arable farming or agroforestry. Medium and shallow gullies may be utilized under silvipasture and hortipasture and deep gullies under tree plantation.

Some ravine area near Agra (Uttar Pradesh) and Kota (Rajasthan) has been successfully been reclaimed by the Central Soil and Water conservation Research and Training Institute, Dehradun (Prajapati, 1979). The most successful trees and shrubs in the salt affected ravines identified are *Acacia tortilis, Albizia amara* and *Dichrostachys cinerea.* In a study of an experimental area in ravines the assessment of production of different trees species showed that *Acacia nilotica* is the most promising species under ravines are for producing the fuel wood. In ravines areas, the natural grasslands are over-grazed and degraded with poor productivity. The main grasses found in gullies of Rajasthan, Uttar Pradesh, Madhya Pradesh and Gujarat are *Dichanthium, Cenchrus and Sehima.* By protection sowing and planting of these grasses the green fodder yield can be achieved in 2-5 years. This practice also reduces run-off and soil loss considerably to the tune of 6-10 tones/ha/yr.

Agroforestry barriers are reported quite effective in controlling run-off and soil loss on 4% slope (Narain *et al.,* 1998). The total sediment deposited along hedgerows (3- year period) and tree rows (9-year period) ranged from 184 to 256 t/ha, equivalent to 15 to 20 mm soil depth. Vegetative measures were found relatively cheaper sustainable and as effective as earthen bunds up to 1.5 per cent slope at Indore and up to 4% slope at Dehra Dun (Bharadwaj, 1988). Performance of vegetative barriers is situation specific. For example vetiver grass barrier has neither been universally effective nor accepted by the people. Therefore, barrier of native grass species should be emphasized.

Productivity of some multipurpose trees in silvipastoral systems on highly degraded lands in semiarid regions was worked out and found that *Albizia amara* based silvipastoral system produced 7.95 tonnes/ha/yr fodder followed by *Acacia tortilis* 7.60 tonnes/ha/yr. The production of *Hardwickia binata* based silvipastoral was quite low (2.32 t/ha/yr) on account of slow growth of the tree (Roy, *et al.*, 1998).

The silvipastoral system is the most appropriate land use system for degraded lands. On a shallow (15 cm) red chalka soil having about 2 % slope, a silvipastoral system comprising *Leucaena leucocephala* (K8) and *Cenchrus ciliaris* and *Stylosanthes hamata* was set up in August, 1981 at the CRIDA, Hyderabad. Contour furrows were thrown open and *Leucaena leucocephala* seedlings planted at 7.5x2m on a drizzling day in contour furrow. Four strips at the upper reaches of the plot were put under *C. ciliaris* and the lower 4 strips were seeded with *Stylosanthes hamata*. The system has come up very well and has shown the way to rehabilitate some of the degraded or marginal lands. On an average productivity (dry matter) of 6 tonnes/ha/year was obtained from *S. hamata* and 3 tonnes/ha/yr from *C. ciliaris*, besides, 6 tonnes/ha/yr (biomass) from *Leucaena leucocephala.* After 3 years an interchanges in strips was affected, *Cenchrus* strips were put to *S. hamata* and vice-versa, the idea being that the soil structure builds up. After 6 years, it is proposed to grow a cereal or a pulse or an oilseed crop on the strip thus rejuvenated by *Cenchrus-stylo* partnership by carrying out pollarding of *Leucaena leucocephala* trees.

In addition to the production of pasture grasses and legumes, wood and leaf fodder from the silvipastoral system under degraded land, the system also improves soil fertility. Deb Roy (1986) studied *Albizia lebbeck-Cenchrus-Stylo* and *A. procera-Cenchrus-Stylo* based systems and observed significant improvement in soil organic matter, available nitrogen, phosphorus and potash. Organic matter filled-up was 38% in the former compared with 25% in the latter system. Available phosphorus and potash were 33 and 16% and 9.5 and 11.0% higher respectively. Singh *et. al.* (1998) reported higher annual turnover of organic matter, total nitrogen and organic sulphur under silvipastoral management in semi-arid region. Research efforts made at

NRCAF, Jhansi for over a decade has resulted in development of appropriate technologies for the rehabilitation of degraded lands of Bundelkhand region (Solanki, et al., 1998). Aonla based agroforestry system has been found highly successful and acceptable to the farmers.

While comparing some agroforestry models for eroded lands in the Shivaliks, Grewal (1993) found that under Eucalyptus-Bhabar grass (*Eulatiopsis binata*) system the soil loss was negligible (0.07 t/ha) followed by *Acacia catechu – Napier grass (0.28 t/ha) and Teak-Leucaena-Bhabar* (0.43 t/ha). Soil loss, run-off and nutrient losses were less under agroforestry systems as compared to arable crops and fallow.

Technically it is feasible to grow trees in wastelands of different categories. Some of the trees have wider adaptability and are thus emerging as commercially viable ventures for the rehabilitation of wastelands. They supply fuel, wood, poles, pulp and other materials that are in great demand. Some of the trees like *Acacia tortilis. Albizia lebbeck, Prosopis cineraria, Prosopis juliflora, Zizyphus* species are financially viable for arid zones of Rajasthan and *Casuarina equisetifolia* and *Anacardium accidental,* are having higher economic value in coastal areas of Andhra Pradesh.

Land Degradation Through wind Erosion and its Rehabilitation

Wind erosion is a serious problem in the arid and semi-arid regions including the states of Rajasthan, Haryana, Gujarat and Punjab. Wind erosion is also prevalent in coastal areas where sandy soils predominate and in the cold desert regions of Leh in extreme North-Western India. In India wind erosion is moderate to severe in arid and semi-arid regions of North-West, covering an area of 28, 600 km^2 of which 68% is covered by sand dunes and sandy plains. However, active wind erosion is observed in the extreme Western sectors of the country. The loss of top soil, terrain deformation and over blowing sand represents 1.9, 1.4 and 0.8% of the total area, respectively. For example, at a wind velocity of 20-km/h soil erosion from bare sandy soil at Bikaner was 273.7 kg/ha/day, whereas from loamy sand soil at Jodhpur the soil lost was 15.6 kg/ha/day (Gupta, 1993). Keeping soil surface cover can effectively control soil loss. From degraded pasture land considerable soil may be lost. Similarly, use of tractors has also accelerated the soil erosion . Measures to control soil loss due to wind erosion have been developed. These include biological stabilization of sand dunes and development of live shelterbelts.

Sand Dunes Stabilization

Sand dunes are the barren stretches of land, formed due to sand movement. The dunes in the desert ecosystem are grouped under new and old dunes.

Old dunes are usually high, relatively stable and support vegetation, and are often under cultivation. These old dunes were formed at least 10,000 to 15,000 years ago (Singhvi and Kar, 1992). In the successful sand dunes stabilization with vegetation there are three most important processes viz. (i) protection against biotic interference by fencing the area with barbed wire or thorny bushes (live fence) (ii) sand dune fixation to check the movement of sand and (iii) afforestation by grasses and trees.

After fixation sand dunes can be permanently established with vegetation. Stabilization with vegetation is important particularly from point of view of build-up and conservation of soil resources and also meeting the fuel and fodder needs of the human beings and animals. Studies on the sand movement show that the stabilization of sand dunes with grasses and tree checked the movement of sand. This could be due to binding of sand particles, soil crust formation, addition of organic matter, and presence of surface cover.

Shelterbelt Plantation

Shelterbelt on field boundaries may effectively control injuries to the tender seedling from sand blasting and hot wind. Similarly, development of windbreaks across the wind direction has been shown to reduce wind velocity by 20-46% and soil loss by about 76%. Usually three row wind breaks, with central row of tall tree like *Albizia lebbeck* with single row of branched trees like *Acacia tortilis, Cassia siamea or Prosopis juliflora* on other sides of *Albizia lebbeck* are very affective. On leeward side of the wind breaks soil retained 14% higher moisture and pearl millet crop recorded 70% higher grain yield (Gupta *et al.*, 1997). The shelterbelts should be continuous without any gap, as the gap has tunnel effect.

Chemical Deterioration of Lands and their Rehabilitation

Chemical deterioration of Indian soils occurs mainly doe to the loss of nutrients or organic matter through accumulation of salt or pollutants.

Saline and Alkalization

Soils suffering from salinity and alkalinity problems (10.1 million ha) are widespread in the country. Worst affected areas lie in parts of Indo-gangetic alluvial plains in Punjab, Haryana, Uttar Pradesh and Bihar, the arid desertic region of Rajasthan and the black soil region (vertisol) of Gujarat, Madhya Pradesh, Rajasthan, Maharastra, Andhra Pradesh and Karnataka.

Soil sodicity has resulted in the degradation of about 2.5 million hectares of agricultural lands due to faulty irrigation method under canal irrigation projects (Singh, 1991). The affected soils are characterized by high pH, low organic matter content, excessive exchangeable sodium, high clay dispersion and hardpan formation (Szablocs, 1989). These factors have contributed to significant loss of vegetation and reduced agricultural productivity (Abrol, 1986). Successful crop production on these soils requires high inputs of gypsum, chemical fertilizers and irrigation to control sodicity (Singh, 1991). Consequently, considerable efforts in research and resources are required to restore and rehabilitate these lands.

Mesquite *Prosopis juliflora* is an extremely hardy species, growing on a wide range of soil types including heavy day and has a good tolerance to drought, water logging, alkalinity and salinity. The tree coppices vigorously and it seldom browsed although mature pods are fed to livestock. *Prosopis juliflora* occurs naturally throughout the 'wastelands' of Tamil Nadu state. It caters for the fuel needs of a great majority of the rural poor (Aiyasmy, 1984) and its availability has greatly reduced illicit felling for fuel in natural forests. In Ramanathapuram districts were rainfall is confined to the northeast monsoon and substantial saline patches occur, farmers use *Prosopis juliflora* as a fallow species for four years. This ameliorates the soil sufficiently to enable the farmers to grow annual food crops for at least two years, after which the trees are allowed to reinvade the site for another four years. In addition to soil amelioration, the other major use of the trees is for fuel wood or charcoal. The return from charcoal, however, is higher than that from fuel wood and thus many families are involved in the manufacture of charcoal. It is estimated that 15,000 tonnes of charcoal, mostly from mesquite, is transported annually to Madras from Ramanathapuram. The system involving the use of *Prosopis juliflora* for soil amelioration offers a cheap and ecologically acceptable way to reclaim not only saline sites but also large areas of wasteland in Tamil Nadu. Direct seeding is cheaper method of its establishment and subsequent survival is also good, as the species is not browsed. Economic return are encouraging and many 'panchayats' (village governing councils) report incomes in excess of Rs. 2000 per ha after six year of mesquite growth (for fuel or charcoal) on wastelands.

A silvipastoral model comprising the growing of pahari kikar (*Prosopis juliflora*) and Karnal grass (*Leptochloa fusca*) together developed and standardized at Karnal, has proved highly successful for the development of community lands (Singh, 1995). In this model, pahari kikar and Karnal grass are grown together for nearly 5 years. Later, when the surface soil is reclaimed, the grass is replaced by better forage such as shaftal (*Trifolium resupinatum*), berseem (*Trifoliun alexandrinum*) and senji (*Melilotus parviflora*). After 8-10 years of growing trees and forage, the soil is reclaimed to the extent that the normal arable crop may be raised either after removing the trees or in association with trees. Singh et al. (1993) identified promising species of trees and herbaceous plants for rehabilitation of salt-affected soils.

In addition to biomass production trees growth on salt affected soils helps in their amelioration by improving physical, chemical and biological properties of the soils. Ameliorating effects of 20 years old tree plantations showed that organic carbon increased by 3-5 times pH on an average decreased from 10.3 to 8.03. A green house pot trial was conducted to assess the impact of *Prosopis* afforestation on the productivity and fertility of degraded sodic soils in Haryana, India. Wheat (*Triticum aestivum, L.* cultivar HD 2329) plants were grown from seed on top soils collected from a chronosequence of 0, 5, 7 and 30 years *Prosopis juliflora* plantation established on highly sodic soils and a non sodic reference soil collected from a local farm. The afforestation improved the physical and chemical properties of surface soil by decreasing pH, electrical conductivity and exchangeable Na levels and increasing infiltration capacity, organic carbon total nitrogen, available P and exchangeable Ca, Mg and K levels (Bhojvaid, *et al.,* 1996).

Physical Degradation

Water Logging

The physical deterioration of soil due to waterlogging or submergence/ flooding has affected around 11.6 million ha land. Area occupying lower topographic positions, area along the rivers (Diara lands) and those lying beyond levees (Tal lands) are few typical examples. It has been estimated that nearly 40 million ha area of the country is flood prone. These lands are generally not remunerative for crop production but offer an alternative to increase the production through silvipastoral and hortipastoral systems. Growing of grasses under waterlogging may yield substantial green forage to meet the shortage of fodder to the animals.

Swampy and Wet Lands

The extent of wet lands is more than 6 m ha apart from the permanent water bodies passing different kind of problems in our country. Marsh lands and swamps are usually found in southern and eastern India. Such type of affected areas can be improved by growing of suitable MPTs like *Barringtonia acutangula, Salix tetrasperma, Bischiffia javanica, Lagerstroemia flosreginae, Eucalyptus robusta, E. rudis, Dalbergia latifolia, Glyricida maculata, Populus euphratica, Syzygium cumini.* To get higher production from these areas in addition to MPTs, suitable grass (*Brachiaria mutica, B. decumbens, Iseilema laxum, Dichanthium caricosum, Paspalum nolatum*) and legumes (*Sesbania mmicrocarpa,*

Latononis bainesii, Glycine wightii) to be established and 20 to 40 t/ha green forage yield can be obtained (Patil and Pathak, 1981).

Rehabilitation of Degraded Lands through Watershed Management

Watershed is a topographically delineated natural unit that is drained by stream system. The watershed represent a hydrological unit of area, but can also be described as a biophysical, socio-economic and sometimes a political unit for planning and management of natural resources. Excessive soil erosion in the catchments, siltation of reservoir, negative environmental impacts of mega river valley projects and reluctance of international agencies for funding of large dams is an other current scenario through resource conservation strategies are major concerns of Agenda-21 of the United Nations Conference on Environment and Development (UNCED) or Rio-Conference (1992). These challenges are expected to be fulfilled through integrated watershed management programme at micro level with the help of community empowerment. A review on the status of watershed research in watershed management (Samra, 1997) revealed that integrated development of watershed reduced the run-off or moderate flooding of down-stream area and improved in situ moisture conservation for increased biomass production. Besides improvement in crop yield, grasses and trees in the catchment, ground water recharge and rise in water table due to integrated watershed management have been spectacular in different regions of the country.

Ram Babu *et al.*,(1997)evaluated watershed-management programme for production, protection, impact and economic viability of 32 watershed ,representing 10 major agro-ecological regions in India. Due to these programmes, the productivity of arable land increased by 4.2 to 15.4 q/ha (average 7.8 q/ha), and that of non-arable land by 1.0 to 8.5 q/ha (average 5.5 q/ha), and milk production increased by three times on sustained basis. Soil loss also reduced by 10 to 80 per cent, and run-off from 2 to 42 per cent, leading to better in-situ moisture conservation. Casual employment opportunities @ 215 mandays/ha/year during implementation phase and 20 man days/ha/year in post intervention period were generated. These programmes were found economically sound (Benefit: cost ratio 1.10 to 2.94 and internal rate of return 12.3 to 37%).

Mine Spoils

It is estimated that nearly 3000 billion tones of mine overburden is dumped annually the world over. At present about 3,86,000 ha land per annum is disturbed by mining which is expected to go up to 9,24,000 ha per year by 2000 A.D (Dadhwal and Katiyar, 1996). Mine spoils result in shifting mass of sand and rubble, denudation of forests, reduction in water holding capacity of reservoirs, rivers, streams, etc. Such areas can be reclaimed by addition of soil amendments such as liming materials for acidic mine-spoils, sewage sludge, paper mill sludge, power plant fly ash, normal soil mulches, manures, compost, etc. so that plant growth can be established. Agroforestry is another viable option for rehabilitation of such sites. Establishment of permanent vegetative cover of suitable trees and grass mixtures will improve their deteriorated conditions and stop further degradation.

For the revegetation of mine spoils, plant species that are native of the place, hardy, have N_2 fixing ability, have aesthetics, provide quick coverage during establishment and are ecologically and economically acceptable should be preferred. Though tree species have been identified for planting in mine spoil areas, but little efforts have been made to develop an agroforestry package that can be recommended at the field scale.

Cold Desert

About 10 million ha in the Westerm Himalayan states of Jammu and Kashmir and Himachal Pradesh, lying between the Greater of Himalayas and the mountains on the edge of the Tibetan plateau, are a cold desert (Rai, 1996). This is an unstable and ecologically fragile region, situated in the rain shadow of the towering mountains ranges and it experiences extremes of cold and dryness. The ever-increasing biotic pressures on the desert vegetation for fuel, fodder and grazing have led to serious ecological degradation. *Hippophae rhamnoides*, an important indigenous multipurpose shrub of the region, is used as fuel and fodder by the local inhabitants. A study conducted in Spiti Valley has revealed that one ha plantation of this shrub can meet the fuel wood needs of about 20 families (ICFRE, 1993). Recently some *Eucalyptus* and *Poplar* clones have been introduced as agroforestry species for revegetation of cold desert in Kashmir Valley.

This is one area where agroforestry will be the most sustainable land use system. Concerted research efforts are required to identify promising forest, fruit and herbaceous species for this purpose. Preference should god for the collection, identification and improvement and multiplication of the local species. Other species of the same genus having more desirable attributes may be introduced from other areas having similar adaphic and climatic conditions.

Bioenergy Sector

The main biomass energy sources in rural areas that are being used in the households, include wood from forest, croplands and homesteads, cow dung and crop biomass. Bioenergy from tree biomass is one significant contribution from agroforestry. Species like Prosopis juliflora are being used for power generation in states like Karnataka and Andhra Pradesh. P. juliflora due to high calorific value of over 5000 Kcal is preferred in these power plants. An estimated 0.51 million ha area is under P. juliflora. Even if 25 percent of the area is meant for power generation leaving the rest 75 percent for fuel and charcoal, the Bioenergy potential works out to 1000 MW. Similarly other tree species can be used for running the biomass based power plants. Thus a total of 5000 MW power could be produced from the biomass sourced to trees under moderate conditions, which meets almost one third of the ultimate potential of 16000 MW from biomass. Establishing power plants of such a capacity attract an investment of Rs. 90830 crores. R& D can help in enhancing productivity and assisting the power plants in captive plantation management on degraded lands. Promoting bioenergy through P. juliflora also encourages tremendous employment generation.

Agroforestry for Carbon Trading

Agroforestry has importance as a carbon sequestration strategy because of carbon storage potential in its multiple plant species and soil as well as its applicability in agricultural lands and in reforestation. Agroforestry can also have an indirect benefit on carbon sequestration when it helps to decrease pressure on natural forests, which are the largest sinks of terrestrial carbon. Another indirect avenue of carbon sequestration is through the use of agroforestry technologies for soil conservation, which could enhance carbon storage in trees and soils. Carbon compounds are sequestered or accumulated by plants to build their structure and maintain their physiological process. The energy captured in the molecular bonds of carbon compounds generally present between 2-4 % of the radiation absorbed by the tree canopy. In the recent past CDM projects have been attracting carbon trading. India has also caught with the trend. At the global level, so far 299 such projects have been registered with UNFCCC that have the capacity to save 577 million t of CO2 annually. Out of these firms 32 % are from India. Some of them are in the agroforestry sector like the two biomass based power projects of LANCO group in Andhra Pradesh. With the full potential of 25.36 million ha the carbon trading potential is very huge.

References

Abrol, I.P. 1986. Fuel and forage production from salt affected wastelands in India. Reclaim Reveget Res. 5, 65-74.

Ahuja, L.D. and Mann, H.S.1975. Rangeland development and management in western Rajasthan. *Annals of Arid Zone* 14, 29-44.

Aiyasamy, S. 1984. Summary Report of Social Forestry Monitoring Advisory Board: III Round II Phase (Mimeo). Department of Agricultural Economics, Centre for Agriculture and Rural Development Studies, Tamil Nadu Agricultural University, Coimbatore, India.

Anonymous. 1972. Centrally sponsored ravine reclamation programme for U.P. Ann. Rep. Government of India Bhattacharya, S.S. 1083. Agroforestry in Arunachal Pradesh. National Workshop, 'Agroforestry', Karnal (eds.) N.K. Mathur and M.G. Gogate FRI, DehraDun

Bharadwaj, S.P. 1988. Effect of zero tillage and living mulch on rainfall disposition and erosion losses at 4% slope Annual Report, CSWCRTI, Dehra Dun pp. 41-42.

Bhojvaid, P.P.; Timmer, V.R. and Singh, Gurbachan. 1996. Relcaiming sodic soils for wheat production by *Prosopis juliflora* afforestation in India,. Agroforestry Systems. 34, 139-150.

Brinck, P. L.M. Nilsson, and U. Svedin, 1988. Ecosystem redevelopment. Ambio 17, 84-89.

Chakravarty, A.K.and C.N. Verma. 1970. Study on the pasture establishment technique. I Effect of seed rate, method of sowing and seed treatments on seeding emergence in *Cenchrus ciliaris* and *Lasiurus sindicus* Annals of Arid Zone 9, 236-244.

Dadhwal, K.S. and Katiyar, V.S. 1996. Land degradation and environmental pollution by mining and land slides. In: Soil Management in Relation to Land Degradation and Environment (Biswas, T.D. et. al. eds.), ISSS, New Delhi, pp 114-126.

Das R.B. and Yadav R.C. 1979. Development of CAZRI "Staggered pitting discar " for water harvesting and increase pasture productivity in arid zone. *Annals of Arid zone,* 18, 96-100.

Deb Roy, R. 1986. Studies on some aspects of renewable energy (fuel) and pasture production on eco system of *Albizia procera* and *Albizia lebbeck* with grass and legume, Ph.D. Thesis Jiwaji University, Gwalior, M.P.

Dhuravanarayana, V.V. and Ram Babu. 1983. *Journal of Irrigation and Drainage Eng., ASCE,* 109, 419.

Grewal, S.S. 1993. Agroforestry systems for soil and water conservation in Shivaliks. In: Agroforestry in 2000 AD for the semi-arid and arid tropics. *NRC for Agroforestry,* Jhansi; 82-85.

Gupta, J.P. 1993. Wind erosion of soil in drought prone area. In desertification and its control in the Thar, Sahara and Sahel Region (Eds A.K. Sen and A. Kar), pp. 91-105. Scientific Publisher, Jodhpur.

Gupta, J.P.; Kar, A; and Faroda, A.S. 1997. Desertification in India: Problems and possible solution. *Yojna* 41, 55-59.

ICFRE, 1993. Annual research report for 1991-92, Indian Council of Forestry Research and Education, Dehra Dun, India, p 81.

Jordan, W.R., R.L. Peters and E.B. Allen, 1988. Ecological restoration as a strategy for conserving biological diversity. *Environ. Manage. 12, 55-72.*

Narain, P.; Singh, R.K.; Sindhwal, N.S. and Joshi, P. 1998. Agroforestry for soil and water conservation in the western Himalayan valley region of India. *Agroforestry Systems* 39, 191-203.

National Academy of Sciences, Study Committee, 1974. Rehabilitation potential of western coal lands. Ballinger, Cambridge, M.A.

Patil, B.D. and P.S. Pathak. 1981. Transformation of tropical wet lands habitat for forage and fuel production in India. *International Journal of Ecology and Environmental Sciences.* 7, 151-156.

Prajapati, M.C. 1979. Silvipastoral system of production in ravines. *Proc. Summer Inst. 'Silvipasture Production for marginal and sub-marginal land,* June 4-July 3, IGFRI, Jhansi.

Rai, P. 1989. Improvement and management of rangelands in India. Seminar on "Problems of rangeland and pastureland development in Himalayan Region" Nov. 20-26, Peshawar, Pakistan

Rai, P. and K.C. Kanodia. 1981. Effect of split application of nitrogen on the quantity and quality of herbage in *Sehima-Heteropogon* grassland. *Forage Research* 7: 83-85.Rai, S.N. 1996. Prominent findings of forestry research in India. FORSPA Publication 11, Bangkok, Thailand, p79.

Ram Babu; Dhyani, B.L.; Agrawal, M.C. and Samra, J.S. 1997. Economic evaluation of watershed management projects, Bulletin No. T-33/D-23, CSWCRTI, Dehra Dun.

Roy, M.M.; Nigam, G. and Kumar, V. 1998. Productivity of semi multipurpose trees in silvipastoral system on highly degraded lands in semi arid region. *Indian Journal of Forestry* 2, 4-8.

Samra, J.S. 1997. Status of Research on watershed management. Presented at 173rd meeting of Governing body of ICAR, March, 1997.

Sehgal, J.L. and Abrol, I.P. 1994. Soil Degradation in India Status and Impact. Oxford and IBH Publishing Co., Pvt. Ltd., New Delhi, India.

Shankarnarayan, K.A., R. Kumar and P. Rai. 1974. Panchayat grassland management poses eight problems. *Indian Farming,* 24, 31.32.

Singh, N.T. 1991. Dryland salinity in the Indo-Pakistan subcontinent, Paper presented at the International Conference on Degradation and Restoration of Arid Lands. Lubbock. TX, Sept. 24-25, 1991.

Singh, Gurbachan; N.T. Singh and Tomar, O.S. 1993. Agroforestry in salt-affected soils. Technical Bulletin No. 17, CSSRI, Karnal, 65p.

Singh, Gurbachan. 1995. An agroforestry practice for the development of salt affected using *Prosopis juliflora* and *Leptochloa fusca. Agroforestry Systems* 27: 61-75.

Singh, Gurbachan; Harinder Singh and Bhojvaid, P.P. 1998. Amelioration of sodic soils by trees for wheat and oats production. *Land Degradation and Development*. 9, 453-462.

Singh, J.S. and A.K. Jha, 1992. Restoration of degraded land – An overview. In : Restoration of degraded land – Concept and strategies (Ed. : J.S. Singh). Rastogi Pub. Meerut, India. Pp 1-10.

Singhvi, A.K. and Kar, A. (Eds) 1992. Thar desert in Rajasthan. Geological of India. Solanki, K.R.; Bisaria, A.K. and Handa, A.K. 1998. A decade of research (1988-98)- National Research Centre for Agroforestry, Jhansi, 44p.

Szablocs, I. 1989. Salt affected soils CRC Press Inc. Boca Raton, FL, 62 pp Timmer VR (1991). Interpretation of seedling analysis and visual symptoms, In: Van den Driessche R (ed) Mineral Nutrition of Conifer Seedlings, PP. 113-134 CRC Press Inc. Boca Raton, FL.

❑❑❑

Chapter 32

Tannin Rich Feeds and Their Strategic Utilization

B. K. BHADORIA
Indian Grassland and Fodder Research Institute, Jhansi-284003 (U.P.), India

Introduction

Domestic ruminants play role of corner stone in the livelihood of farmers providing substance as milk and meat, cart puller, manure for crop production, cash income from sales of their products and safety net of capital assets to face the risks and misfortune in harsh environment. Despite that there animals are undernourished or in malnutrition because of deficit in fodder supply and expensive high density feed like oilseed cakes or concentrates of animal origin which is beyond the reach of farmers. The efficiency of forage based livestock production system is limited by forage quality and quantity during the lean period (NAS 1979, Schultz Krafts 1986). Although not yet fully exploited, the strategy of incorporating proteineous tree leaves in to feeding schemes such as cut and carrier system or protein bank has enormous potential towards solving the severe nutritional limitations (Ranjhan 1986,Davendra 1990, Kass *et al.,* 1992, Osuji *et al.,*1995). Thus, the use of native tree leaves or shrubs may compensate, because of their multiple uses. There are number of advantages to use tree leaves in feeding practices in amalgamation of traditional experiences and modern knowledge such as availability on farm, accessibility, provision of variety in diet, reduction in purchase of concentrate, and reduce losses from soil erosion, greater efficiency of tree species for photosynthesis, improved soil structure and fertility with increasing effects on crop yield, and more closed cycling of organic matter and nutrients and creating better micro climatic conditions for the growth of agricultural crops. Further, trees and forests were always considered as an integral part of the Indian culture. The ancient scriptures and historical records amply support this. The best of Indian culture was born in the forests. The Aryan civilization was started in our forests and our Rishis who evolved the Hindu religion, lived in forests in complete harmony with nature. The ashrams were the centers which harmonized agriculture and pasture with trees, animals and birds. It was widely believed that destruction of forests and cutting of trees created famine conditions, where as planting and maintaining trees were regarded as noble acts. The use of protein rich tree leaves and shrubs as protein supplement for low protein low quality grasses, agriculture bye products may help to overcome this limitation.

The foliage of tree leaves can be used in conjunction with low quality roughage. Besides these there is limited use of tree leaves due to presence of probable anti-quality / anti- nutritional factors which are product of metabolism activities in plant life. Therefore these are often referred as secondary metabolites. Plant secondary metabolites are a large and diverse group of chemicals that have key role in interactions between ruminants and plants both in natural and ecosystem. There are more than 250000 known molecules and many more are waiting for their recognition. Many of these have characterized by structure elucidation. In fact these compounds are produced to defend the plant from predators. These includes group of compounds like alkaloids, terpenoids, non protein amino-acids, cyanogenic glycosides, saponins and phenolics that are known to have anti nutritional and toxic effects on the animals. However there structurally more complex compounds that has physiological effects on ingestion by the animals, both adverse and beneficial; these include peptides carbohydrates, polyphenols terpenoids and others. Carbohydrates such as arabinoxylans, b-glucans are frequently present in plants often in concentration that have adverse effect on animals. The PSMs interact with different molecular and cellular targets including enzymes, hormone receptors neurotransmitter receptors and trans-membrane transporters and can thus mimic a response at the corresponding targets. It has been found that the effect of PSMs on their ingestion is due to structural similarities between them and molecule occurring in naturally with in the animal or reaction with other molecule already present.

In particular, the presence of ployphenolics or proanthocyanidins or condensed tannins in tree leaves is point of concern because of their ability to bind protein, carbohydrate and other molecules (Muller-Harveavy1992) effecting on digestion and intake as consequent effect on production, hence effective nutrition of ruminant is not what it eats and excrete but rather the combination of rumen microbial fermentation and unfermented rumen escape with due absorption of vital nutrients for growth and production. Polyphenolics, are therefore likely to be consumed in all agricultural systems where trees and shrubs are used as livestock feed. However,tannins are not confined only to just trees and shrubs. The term "*tannin*" came into existence from word "*tanning*" for the phenol rich plant extract used to convert animal hides into leather as a process of conversion and chemical structure of such substances were not known. With the development of knowledge, the utilization of such products in leather making declined (Hegatery 1989). Henceforth, the confusion prevailed over the use of term "tannin", so, it was replaced with "*plant polyphenolic*" a more chemically accurate designations (Haslam 1989). Hence the term polyphenolics was accepted widely, however it includes many materials which do not exhibit unique property of tannins. Further, the ployphenolics to be categorized as tannin; it must exhibit certain chemical and functional characteristics. Chemically it must be oligomeric with multiple structural unit containing free phenolic groups. The presence of multiple polycyclic unit decides the lower limit of molecular weight about 500 but there is no upper limit, sometimes values may exceed MW 20, 000 (Williams et al 1983) and with increase of MW, there may be limited solubility in any solvent. The functional property defines their capacity to bind proteins. Although alkaloids and some other secondary metabolites also interact with protein, but tannin protein interaction, is the main basis of biological activity of tannin, resulting in the herbivore food selection and physiology, and other curable properties. The tannin may be regrouped into condensed and hydrolysable tannin.

Condensed tannin: Condensed tannins are most widely distributed in plants and are oligomers of flavanoid linked by C-C bonds, that are not susceptible to acid or enzymatic hydrolysis. On heating with strong acid in presence of trace metal ion (Porter et al 1986), the interflavan bond is cleaved oxidatively yielding degradative product as monomeric anthocyanidin, such as cyanidin or else, therefore condensed tannins can easily be designated as *proanthocyanidin* replacing a term leucoanthocyanidin.

The monomeric unit of proanthocyanidin is catechin or epicatechin with a flavan-3,4-diol or flavan-4-ol molecule as its precursor (Santos-Buelga and Scalbert, 2000). Size of proanthocyanidins is typically described by its degree of polymerization (DP) with DP=1 as monomers and those with DP = 2 to 10 and DP > 10 as oligomers and polymers, respectively (Santos-Buelga and Scalbert, 2000). Proanthocyanidins with an average degree of polymerization up to 85 in grape skins (*Vitis vinifera* L.) (Monagas et al., 2003), and 190 in cider apples (*Malus domestica*) have been reported. Proanthocyanidins typically consists of upper unit (extender unit), middle unit and terminal unit, and based anthocyanidin moiety, these are designated likewise procyanidin or prodelphinidin etc. The stereochemistry at C-4 is highly influenced by hydroxylation on the A-ring. Therefore on the basis of inter flavanoid linkage along with stereochemistry these could be classified often as B type and A type. More than 300 B type proanthocyanidin have been already been reported these are typically singly linked flavanyl must between C-4 of chain extension and C-8 and C-6 of terminal unit. B type proanthocyanidins are subdivided B-1, B-2, B-3, B-4, B-5 and B-6 on the basis of hydroxylation and stereochemistry.

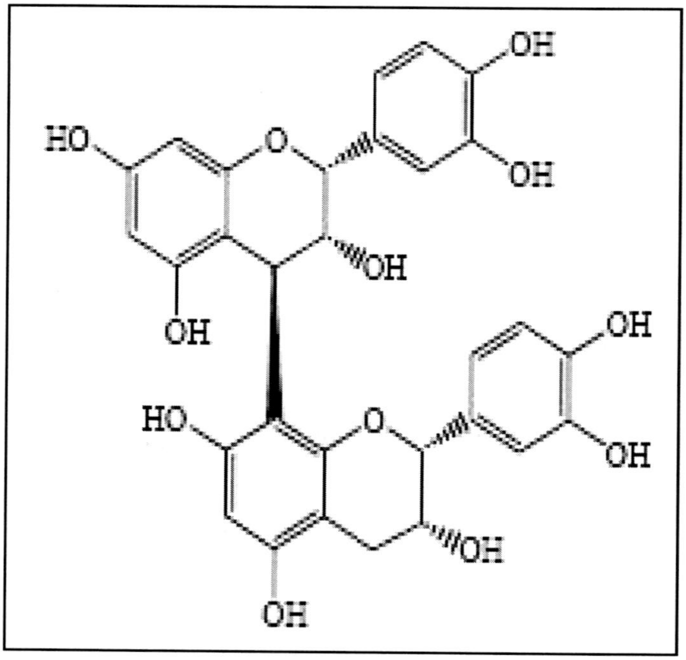

B-type Dimer

A type proanthocyanidin : On the contrary in A type proanthocyanidin besides C-C interflavanyl linkage, there is second ethereal linkage between C-2 of terminal unit and extension chain, producing conformational stability.

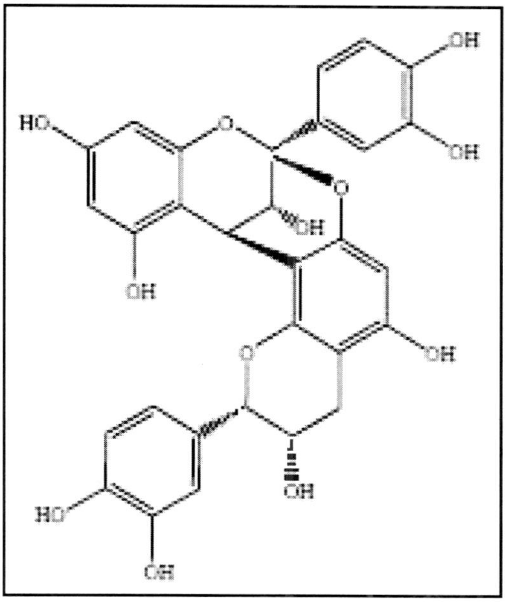

A-Type Dimer

Hydrolysable tannins are easily hydrolyzed with weak aqueous acid or alkali and by enzymatic action yielding the central core polyol, and the constituent phenolic acid (Bravo, 1998; Mueller-Harvey, 2001). Gallotannins hydrolyze to the polyol and gallic acid whereas the phenolic acid liberated from hydrolysis of ellagitannins (B) is hexahydroxydiphenic acid (A)that, in the aqueous acidic or basic environment, converts to the lactone form, ellagic acid.

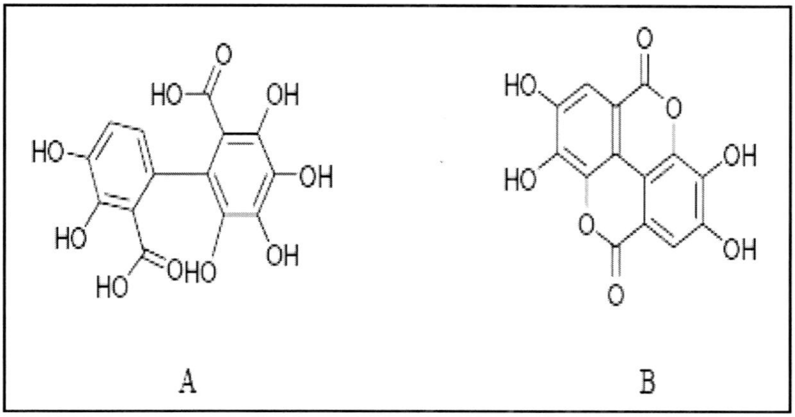

A B

Toxicity of Tannins

Condensed tannins readily combine with dietary protein, salivary protein, digestive enzymes and these protein bound with tannins most unlikely undergo normal metabolism. However continued consumption, of tannin can cause gastritis with irritation, and also odema of intestine. Once the tannin crossed the intestinal membrane thousands of possible harmful interaction could result.

Hydrolysable tannin may also exert similar toxic effect. In addition to this, it can be hydrolysed in rumen to yield phenolic acid which can be absorbed several other toxic manifestation would be reflected. Very rich tannin in feed produce flaky sedimentation in urine with odema in submendibular region.

Quantification of tannin by of the method or assay is ever point of discussion in relational to biological activity. Types of feeds which contain tannins are found widely in plant material, in both leguminous and non-leguminous species. Robbins et al. (1987) correlated crude protein content to the proportion of digestible protein for feeds such as grasses and agriculturally-produced legumes and grains with very low tannin contents. Using the correlation equation obtained to predict protein digestibility and comparing this to the measured protein digestibility of tanniniferous feeds consumed by mule deer, an estimate of the reduction in digestible protein due to tannins was obtained. This was found to be highly correlated with the activity of the tannins as assayed by a protein precipitation assay. Hanley et al. (1992) used predictive equations developed by Robbins et al. (1987) to predict successfully the protein and dry matter digestibilities of tanniniferous forage leaves from seven species and one sample of twigs when fed to black-tailed deer, confirming that a protein precipitation assay gave a useful estimate of the effect of tannins *in vivo*. McKey et al. (1978) have also found that total phenols (by the Folin assay) and extractable condensed tannins (by acid butanol) correlated negatively with the dry matter digested by rumen inoculums during 96 hours incubation. Data were quoted for 30 species of trees from which samples of mature leaves had been taken. For 72 West African fodder trees and shrubs, Rittner and Reed (1992) found that in vitro protein degradability was negatively correlated with total phenols (by ytterbium precipitation assay) and extractable condensed tannins (by acid butanol). However, the behavior of some species deviated greatly from that indicated by the correlation studies. Wood and Plumb (1995) found strong correlations between total phenols (Prussian blue assay) and the inhibition by tannins from Bolivian fodder tree leaves of in vitro fermentation (gas production) by rumen microbes. Similar correlations were found using protein precipitation by the radial diffusion assay, but there was no significant correlation with the acid butanol assay.

Other workers have had less success in finding correlations. Makkar et al. (1989) were unable to find significant correlations between total phenols (Folin assay), condensed tannins (by the vanillin assay) and protein precipitation with *in sacco* dry matter loss for leaves of 10 species of trees from India. Makkar (1993) lists 28 agro-industrial by-products which might be useful as feeds. Ninteen of these by-products, including various seeds and seed cakes, have been shown to contain polyphenols. Some species contain both types of tannin. Ritter and Reed (1992) studied 72 West African fodder tree and shrub species (from Benin, Niger and Nigeria) and found polyphenols in all the species studied. Average polyphenolic contents did not appear to be different in the climatic zones studied. Wood *et al.* (1995) studied 20 tree leaf fodders from Bolivia, 26 from West Africa and 24 from Colombia. The occurrence of tannins is not restricted to particular limited classes of plants or climatic zones. Devendra (1995) has noted some broad estimates of the importance of tree browses (shoots, twigs, leaves, fruits and pods of trees and shrubs), which are generally high in tannins and important in arid and semi-arid regions. In northern Africa browse forms 60 - 70% of rangeland production and 40% of the total available feed, other estimates put such feeds as making up 40 - 50% of the total available feed. In India 60 - 70% of the forage requirements for goats are tree browses, tree legumes being particularly important (Devendra, 1995).

Beneficial Effect of Tannin

The tannins in moderate quantity (less than ~4 % on DM basis) have beneficial effect to the animals. The multiple phenolic hydroxyl groups, with variable size, produces great variation in their interactions primarily with proteins, to lesser extent with metal ions, amino-acids, and carbohydrates. Thus there is considerable variation in flavan3-ol composition at molecular level in position of functional groups and their stereochemistry and proanthocyanidins can react with enzymes and protein in several ways. Importance and allurement of green fodder is well known for ruminants and this lushness is due to presence of the chloroplast in the cells of green fodder and chloroplast contain 69% of their dry weight as protein (Krik & Tilney-Bassett 1967) which account for 75% of total leaf protein (Mangan 1982) as Ribulose-1, 5 biphosphate carboxylase- oxygenase (EC4.1.1.39) and it major soluble protein. It catalyzes the initial step in Calvin's reductive pentose phosphate cycle. Rubisco, the principal dietary protein typically consist of eight large subunits (LSU MW 54000) and eight small subunits (SSU MW16000) these subunits are detectable on stained gel they represent The LSU degrades relatively faster than that of SSU. The ultimate effect on protein metabolism in ruminants appears to be effected by type of proanthocyanidins in diet. Presence of proanthocyanidins/condensed tannins increases duodenal NAN, flow per unit total N eaten, and reduces ammonia concentration in the rumen (Waghorn et al 1987, 1994). The degradation of principal dietary protein (Rubisco in to Large Sub Unit (LSU) and Small Sub Unit (SSU)) is effected by the presence of procyanidins. The digestion of forage in the rumen is a combination of solubilization and degradation. Solubilization is an important action and can be characterized as the release of protein from cell wall in to ruminal fluid with chewing in mouth. Degradation is a catabolic process by microbial proteases producing peptides, amino acids and ammonia. Protein solubilization and degradation in rumen are independent process and solubilization could be defined as loss of principal protein (Rubisco) in rumen mainly. In vitro studies conducted at this institute in our laboratory revealed that maximum solubilization of Rubisco was observed in *F religiosa*, (22.2%) of total followed by *F.racemosa* (17.24%), *A.pendula* (16.36%), and *E. jambolana* (15.01%), however it is noteworthy that the solubilization was comparatively quite higher than that of roughage wheat straw as 8.95% and legume hay 3.02%. On addition of these tree leaves to wheat straw and legume hay there was considerable boost in Rubisco solubilization. Addition of Peepal (*F. religiosa*) and Jamun (*E.jambolana*) to wheat straw, the dietary protein solubilization increased to 27.97% and 10.77% at ratio of 1:1 respectively. The solubilization declined with increase of Peepal leaves as 22.38% and 5.61% at ratio of 2:1 and 3:1 and to Jamun it increased as 23.44 and 28.22% in ratio of 2:1and 3:1 respectively. This indicated synergetic effect in solubilization of dietary protein particular level and more than that there was effect to lower the solubilization. Further addition of peepal leaves to the with legume (*Trifolium alexendrium*) hay the solubilization of dietary protein increased gradually to 26.06% (3:1)from 11.82%(1:1), in Jamun it decreased from 11.89% to 9.91%. Supplementation of Gular (*F.racemosa*) and Kardhii (*A.pendula*) leaves in similar ratio to wheat straw the solublization percentage reached to its highest state as 28.54% (2:1,tree leaves: wheat straw),but in Kardhii the there was increase in solubilization at leaves from 15.74% (1:1) to 23.03%(3:1) The LSU was also degraded more than SSU. The LSU degradation in *Ficus religiosa* was up to 80% whereas in *A.pendula* it was up to 40%. In *F racemosa* and *E.jambolana* the LSU degradation was moderate as 50-60% was in comparison to pure from Baresem. The addition of wheat straw to these top feed the degradation of LSU there was no much change in quantum of LSU degradation. The addition of Barseem to top feed the LSU degradation there was increase in LSU by 5-12%. The SSU degradation was up to 60% in *Ficus religiosa* where as in *F.racemosa* the SSU degradation was bit higher than that of *Ficus religiosa* by 3%. In *E.jambolana* and *A.pendula* the degradation was quite low and remained 30 and 18% respectively. The addition of wheat straw increased the SSU degradation. Rumen microorganism also utilizes a large portion of dietary protein but these could be degraded 0.7% to ammonia rumen (Ulyatt et al.1975). These effects are presumably due to due to interaction between tannin and protein which occur in rumen. From this study it was evident that that tannin rich diet may add rumen escape calibration to enhance AA absorption.

Domestic ruminants grazing forages out fields are exposed to various diseases. Many of them are nutrition oriented. Animal feeding on excessive green legumes are often subjected to frothy bloat and internal parasites in goats, sheep and cattle. However both the problems are controlled by use of chemicals- detergents to disperse the foam, and proprietary synthetic drugs as anti helminthes. These remedies have short term effect causing severe problems in long term use by triggering free radicals and initiating the etiology of many other diseases. In case of helmenthic problem the regular use synthetic drugs lead to resistance towards the organic compound. Therefore, there is imperative need to develop nutrition based control without residual effect and with better prophylaxis to be good performer.

In livestock, three classes of modern synthetic anthelmintics are described as 'broad-spectrum', each class having a different mode of action: (1) benzimidazoles and pro-benzimidazoles; 2)imidazothiazoles and/or tetrahydropyrimidines; (3) macrocyclic lactones.However, as for other pests, resistances to anthelmintics are now widespread, and there is an urgent need for alternative solutions to control gastrointestinal nematodes

Presence of parasites in abomasum, and small intestine cause severe loss of protein and alter the path of protein synthesis from skeletal muscles to repair of gut tissues, leading to reduction in nitrogen retention (Kimambo et al.1988, McRae 1993). The condition could be improved by supplementing protein rich diet (Brown et al. 1991,Coop and Homes 1996) however the condition could also be improved in better way with rapid weight gain by feed natural tannin rich diet (Niezen et al 1993, 1994, 1996). Subsequent studies revealed that lambs fed on tannin rich diet had higher weight gain with low gastrointestinal parasites and lung worm (Niezen et al 1995,1998, Molan et al. 2000,). The burden of *Trchstronylus colubriformis* and *Haemoncochus* could be lowered by feeding incorporating feed containing tannin rich extracts. Gastrointestinal parasites (GIP) cause marked production losses to livestock throughout the world. Control of ruminant GIP over the past decades has been achieved by the use of anthelmintic drugs, but control of GIP is becoming more difficult due to the increased resistance of parasites to common anthelmintics, which has been reported in goats, sheep, and cattle.Alternative parasite management strategies using forages containing condensed tannins (CT) have recently been suggested (Niezen et al., 1995;). Trichostrongyle nematodes, the parasites of the gastrointestinal tractand lungs because major losses in production are parasites of one of the main constraints associated with livestock production. Anthelmintics: Drugs of varying biochemical structure, used to kill helminthes and eliminate the worms from the hosts. The possible antiparasitic role of condensed tannin supported by most *in vivo* studies has usually been confirmed by results of in vitro bioassays that measuredthe effects on different nematode stages of crude plant extracts or specifically extracted condensed tannins fromvarious origins.

It seems possible that consumption of forage CT may reduce GIP numbers and improve animal performance through direct and indirect mechanisms. Alternative, non drug GIP control strategies that are practical and realistic for introduction into farm production systems are required. The recent in vivo and in vitro studies have revealed the importance of tannin rich feed in control of parasitic loads and can prove in eco-friendly nutrition.

The excess consumption of fresh legume forages by animals there occur problem of bloat, because of the soapy foam produced by saponins and protein in them. The foam in rumen of animal triggers automatic closure of sphincter which periodically opens to release the gases formed by fermentation. An animal can produce up to 2 liter of gas perminute. Due to un-relieved gas rumen volume expends beyond the limit and crushes heart and lungs leading to rupturing of capillaries leading to bleeding from eye sockets and rear end. Sub lethal bloat also impacts upon animal; that animal may survive mild bloat episode but reduce their efficiency. The process is so fast some times animal dies in 30 minutes. If we look at our some old traditional way of animal rearing, in such event to protect animal from bloat accidents they were given Jamun or Peepal leaves along with fresh legume forages. The chemistry behind was that plant, on mastication by animals the cell content are mixed with that of legumes and allowing the tannins to bind the protein via hydrogen bonding in reversible manner. If adequate concentration is there to precipitate the soluble dietary protein thus forming the stable protein foam. It has been shown by *in vitro* studies pure tannin dramatically reduce the compressive strength of protein foam produced from leaf protein in dose dependant manner (Tanner et al. 1995) For complete bloat safety 0.2% tannin of dry weight is required.

Browses are particularly important in extensive livestock systems, and become more important when the supply of alternative feeds is restricted such as during dry seasons and periods of drought. Thus tanniniferous feeds can be an important part of the diets of livestock in higher potential areas.

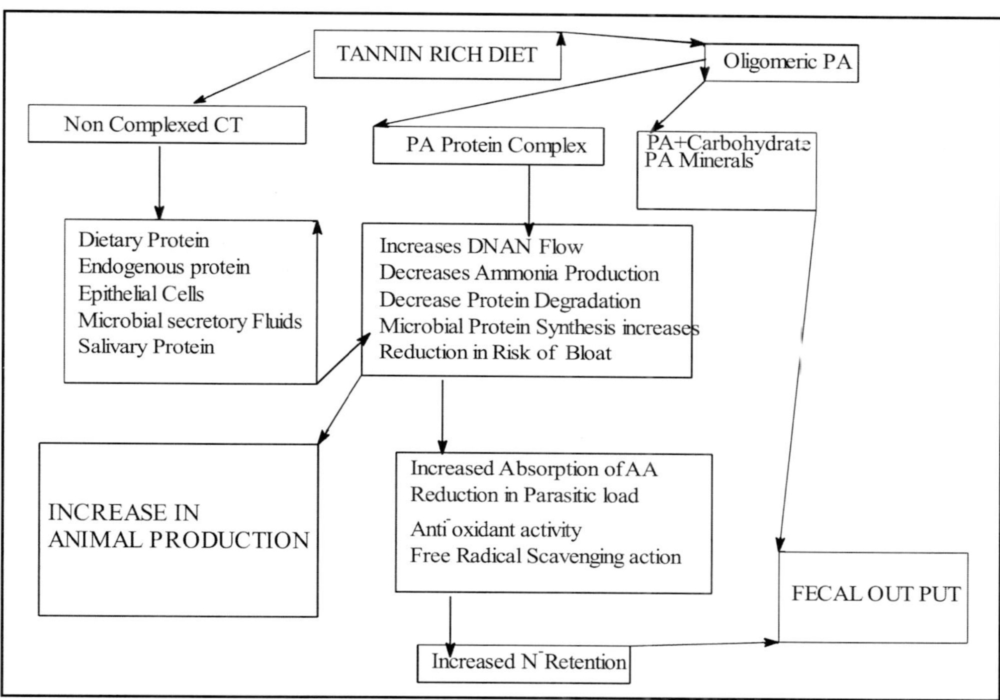

Moderate concentration of Tannin with Change Perspective

References

Devendra C 1990The use of shrubs and tree fodder for ruminants.In: C Devendra (ed) Shrubs and trees fodder for farm animals 42; Proceeding of workshop in Denapsar Indonesia 24-29 July 1990

Hanley T A, Robbins C T, Hagerman A E and McArthur C (1992). Predicting digestible protein and digestible dry matter in tannin-containing forages consumed by ruminants. Ecology 73 537-541

Kass M., Benavides J., Romero F, Pezo D., 1992 Lessons from main feeding experiment conducted in CATIE using fodder trees

Kimambo A.E. MacRae J.C. Walker A. C.F. and Coo R.L. 1988 The effect of prolonged sub-clinical infection with *Trichostrnoglus* columriformis on the performance and nitrogen metabolism of growing lambs Vet Parasit. 28, 191-203

Krik,J.T.O.,& Tilney- Bassett R.A.E, 1967 In The Plastids :Their chemistry,structure growth and Inheritance London W.H. Freeman and Company.

Mangan J.L. 1982 The nitrogenous constituents of fresh forages, In: Forage Protein in ruminant Animal production British society of Animal production occasional publication no 6 [D.J. Thompson, D.E. Beevar and R.G. Gunn editors] Thames Ditton, British society of Animal Production

Makkar H P S, Singh B and Negi S S (1989) Relationship of rumen degradability with microbial colonization, cell wall constituents and tannin levels in some tree leaves. Anim Prod 49 299-303.

Mc Rae J.C. 1993 Metabolic consequences of intestinal parasitism *Proceeding of nutrition Society* 520: 121-130

Molan, A. L., G. C. Waghorn, and W. C. McNabb. 2002. Effect ofcondensed tannins on egg hatching and larval development of *Trichostrongylus colubrifomis* in vitro. Vet. Rec. 19:65–69.

Molan, A. L., G. C. Waghorn, B. R. Min, and W. C. McNabb. 2000.The effect of condensed tannins from seven herbages on on *Trichostrongyluscolubriformis* larval migration in vitro. *Folia Parasitol.*47:39–44.

Mongas M., Gomez- Cordoves C., Bartolome B, Laureaeo O., and Ricardo da Silva 2003, Monomeric, oligomeric and flavan 3-ol composition of wines and grapes fro Vitis vinifera L.cv graciano *J. Sci. food and Agric.* 51: 6471-6481

Mullar – Harvey I, and McAllan AB 1992 tannins and their biochemistry and nutritional properties in advances in Plant cell Biochemistry snd biotechnology JAI press pp151-217

NAS (National Acedemy of Science) 1979 Tropical Legumes Resources for Future NAS Washington DC USA

Niezen, J. H., G. C. Robertson, G. C. Waghorn, and W. A. G. Charleston.1998. Production, fecal egg counts and worm burdens ofewe lambs which grazed six contrasting forages. *Vet. Parasitol.*80:15–27.

Niezen, J. H., G. C. Waghorn, and W. A. G. Charleston. 1998. stablishmentand fecundity of *Ostertargia circumcincta* and *Tricostrongyluscolubriformis* in lambs fed lotus (*Lotus pedunculatus*)or perennial ryegrass (*Lolium perenne*). *Vet. Parasitol.* 78:13–21.

Niezen, J. H., T. S. Waghorn, W. A. Charleston and G. C. Waghorn.1995. Growth and gastrointestinal parasitism in lamb grazingone of seven herbages and dosed with larvae for six weeks. J.Agric. Sci. 125:281

Osuji P.O, Frenadez- Riveras and Odenyo A.1995 Improving fibre utilization and protein supply in animals fed poor quality roughagesILRI nutrition research and plants In Rumen Ecology Research Planning. Proceeding of workshop at ILRI Etheopia

Ranjhan S.K. 1986, Sorces of feed for ruminant production in South East Asia In: B.J. Blair. D.A.,Ivory, T.N Evans (eds) Forages in South east asia and soth pacific Agriculture p24, proceedings of an international Workshop held at Cisarua Indonesia ACIARas apart of N-ration in legume trees as protein source for *livestock* 102,161-176

Rittner U and Reed J S (1992) Phenolics and *in vitro* degradability of protein and fibre in West African browse. *J Sci Food Agric* 58 21-28.

Robbins C T, Hanley T A, Hagerman A E, Hjeljord O, Baker D L, Schwartz C C and Mautz W W (1987) Role of tannins in defending plants against ruminants: reduction in protein availability. *Ecology* 68 98-107.

Santos-Buellga C., and Scalbert A., 2000,Proanthocyanidins and tannin like compounds –nature , occurrence dietary intake and effects on nutrition and health. *J. Sci. food and Agric.* 80,1094-1117

Schultze –Krafts R.,1986 Exotic and native legumes for forage production in Southeast Asia In : GJ. Blair, D.A. Ivory and T.R. Evans (Eds) Forages in Southeast Asia and south pacific Agriculture P.36 *Proceeding of an International workshop* held at Cisarua Indonesia ACIAR

Tanner GJ, Moore A E ,and Larkin PJ 1994 Proanthocyanidins inhibits hydrolysis of leaf proteins by rumen microflora *in vitro British J Nutrition 7*1947-958

Ulyatt MJ, Mac Rae JC, Clarke RT, and Pearce PD, 1975. Quantitative digestion of fresh herbage of by sheep IV *J.Agric.Sci.*(Cam) 84, 453-458

Waghorn GC, Jones WT, Shelton ID, and McNabb WC, 1990,Condensed tannin and nutritive value of herbage Proc.NZ, grassland Association 51: 171-176

Williams V.M., Porter L.J., and Hemingway R.W. 1983 Molecular weights profile of PA polymers. *Phytochemistry*. 22:, 569-572

Wood C D and Plumb V E (1995) Evaluation of assays for phenolic compounds on the basis of in vitro gas production by rumen micro-organisms. *Animal Feed Science and Technology* 56 195 - 206.

❑❑❑

Chapter 33

Secondary Metabolites of Forage in Relation to Animal Production

S. K. MAHANTA, K. K. SINGH, M. M. DAS, S. K. NAG AND N. DAS
Indian Grassland and Fodder Research Institute, Jhansi-284003 (U.P.), India

Introduction

Plants collectively accumulate a wide range of different chemicals in their cells. These can conveniently be divided into primary plant metabolites such as proteins, fats and carbohydrates and those of secondary metabolites such as phenolics, terpenoids and alkaloids (Fig 1). These secondary metabolites (SM) do not appear to play an essential role in the basic life activities of the plant but are important constituents of animal feed.

There are many types of forage/ plants which contain chemicals or which accumulate chemicals as SM that are poisonous to livestock. The results of poisoning can range from minor irritations and slightly lowered animal performance to severe cases where the animal is in a great deal of distress and may die. There are many factors that contribute to the toxic principles in plants. Individual plant species and varieties may differ in their poison content from early growth to maturity. With some plants, there is an increase in their ability to poison with advanced stages of growth, whereas with others the danger lessens. The state of the plant when eaten may also be important. In some cases, damage to the plant or wilting may produce poisonous chemicals in the plant, which were not present in the fresh material. In other cases, such as with buttercups, the poison is contained in the fresh plants but not in dried plants. Certain parts of a plant may be poisonous while other parts may not be poisonous such as the rhubarb stalk/ stem is eatable while the leaves are very poisonous.

Animal factors also influence the ability of plants to poison. Different animal species are susceptible to different plants and poisons. There is evidence that livestock can detect and avoid most plant toxins. Most poisonous plants are bitter and not generally consumed when other forage is available in adequate quantity. The forage species that does contain toxins is less preferred, while livestock show a preference for less toxic plants or species (Table 1). The age of the animal is also important. Young animals are often more susceptible than older ones, but it is not always the case. Animals may build up resistance to certain poisons by being exposed to small quantities at first. Then, if a large quantity is consumed, they are resistant because their metabolism has already adjusted to handle the poison. An animal that is hungry or has certain dietary deficiencies is more likely to eat toxic quantities of a poisonous plant than a well-fed animal.

Table 1: Species Variation for Nontoxic Effects of Plant Toxicants

Plant species	Toxic compound	Least affected animal species
Blue lupine	Alkaloid	Sheep
Reed canarygrass	Alkaloid	Sheep
Crotalaria	Alkaloid	Cattle
Braken fern	Cyanogenic glycoside	Sheep and deer
Sorghum and sudangrass	Cyanogenic glycoside	Cattle and sheep
Alfalfa	Saponin	Swine
Rapeseed meal	Glucosinolate	Ruminant, swine, poultry

Modes of Action of Secondary Metabolites and Counteraction by Animals

The most successful SM would render the plant so unpalatable as to be never eaten. However, it is commonly observed that most plants are eaten to some extent despite containing some form of putative chemical or physical defence. So, from the plant's viewpoint SM may be effective by reducing the probability of herbivory or the amount eaten by a herbivore. Similarly, from the herbivore's viewpoint the main question does not concern avoidance, rather how do herbivores regulate their intake of SM within limits that are tolerable. The effects of, and counter-adaptations to, SM are initially considered as physiological and behavioural, although the final analysis suggests that these two interact and should not be separated (Iason, 2005).

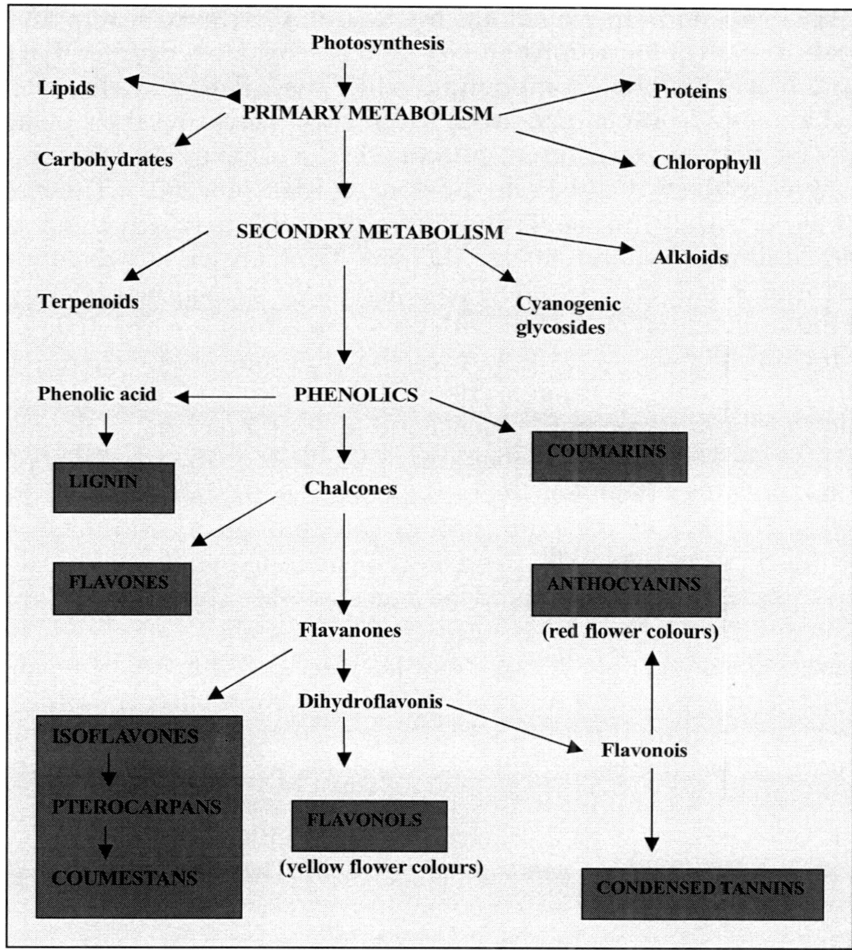

Fig. 1: Interrelationship between primary and secondary metabolites in plants

Toxic Secondary Metabolites (negative)

Toxic SM generally enter the animal via absorption from the gastrointestinal tract and exert their effects via their lipophilicity, causing them to reside in membranes, thus disrupting membrane function, or they enter the cells. Toxic SM can as a group exert a plethora of negative effects on animals, such as induction of cysts and tissue damage, and ultimately lead to death. For example, some phenolics can be carcinogenic and damage to the intestinal wall has been shown in small mammals following ingestion of tannic acid. Such damage is bound to interact with the metabolic functions of the affected tissue. Pyrrolizidine alkaloids can also be carcinogenic and hepatotoxic (Cheeke, 1994), and other alkaloids are neurotoxic. Detoxification in animals is largely a result of the action of mixed function oxidase enzymes that catalyse the oxidation of lipophilic compounds. In the 'phase I' reactions the cytochrome P450 enzymes add functional groups to toxins, increasing their polarity and thereby facilitating excretion in solution via the urine or, for large molecules, their return to the gut in the bile (Rozman & Klaassen, 2001). These enzymes are now classified into several subfamilies within which there is broad functional similarity, although specific catalytic characteristics depend on their precise amino acid sequences. The so-called phase II reactions of detoxification effect the conjugation of the SM with other molecules to form conjugates that have increased polarity. It is known that the main detoxifying chemical reactions can occur in a range of tissues, including the intestinal epithelium, but most of the work on both phase I and phase II reactions in mammals has focused on the endoplasmic reticulum of the liver as the main site of action. In many cases this focus on liver metabolism is likely to be well founded. However, the lack of consideration of the detoxification functions in the gastrointestinal epithelium is surprising considering that this tissue represents the first tissue to be encountered by ingested SM. This omission may be a result of the inadequacy of in vitro cellular models in which to study the processes (Carrie`re *et al.*, 2001). It has been recently hypothesised that P450 acts in the gut epithelium in concert with P-glycoproteins, an efflux system that actively pumps lipophilic substances from cells, thereby together preventing absorption and further effects of toxins. This process implies that faecal excretion of plant secondary metabolites, which is often attributable to gut microbial effects, or biliary excretion of detoxification products, may also be a result of active processes in the gut wall. This challenging hypothesis remains to be tested. Although conjugation with a range of moieties is possible, including with glucuronic acid, sulfate, glycine and glutathione, the conjugate considered most often in studies of mammalian herbivory is glucuronic acid, perhaps because it is the easiest to quantify.

Pro-Oxidants and Antioxidants (positive or negative)

Many phenolics can theoretically act against animals that ingest them by being oxidized and generating free radicals such as super oxide and other free radicals that can attack DNA and inactivate enzymes. Conversely, some flavonoids such as quercetin can, depending on the conditions, act as antioxidants; their unsaturated C-ring structure scavenges free radicals produced from normal oxidative metabolic processes, thereby limiting oxidative damage (Alessio *et al,*. 2002). Despite great interest from the medical world because of the possible use of phenolics for treatment and prophylaxis against mutagenesis and carcinogenesis, the possible advantages of phenolics have not been specifically shown to be a determinant of herbivore animals foraging behaviour or food intake.

Digestibility Inhibition (negative)

Digestibility inhibition is a well-studied series of processes, in which PSM act within the gut of animals by either forming a complex with a food substrate and preventing degradation, e.g. condensed tannins binding with proteins on mastication, or inhibiting digestive degradation by the animal's endogenous or microbially-derived digestive enzymes, e.g. proteinase inhibitors. The complex interactions of condensed tannins during ingestion and digestion have been comprehensively reviewed (Foley *et al.*, 1999), and have been shown in particular to reduce apparent N digestibility in all twenty-six studies of foregut and hindgut-fermenting species of placental mammals in which the diets offered contained condensed tannins. As their name implies hydrolysable tannins can be hydrolyzed to aglycones that are readily absorbed and need to be dealt with in the same way as other absorbable toxins and non-tannin phenolics, although they may also possess protein-binding activity. Dietary tannins can be counteracted by some herbivores, particularly those whose diet is tannin-rich, by the production of salivary tannin-binding proteins. These have a high affinity for, and preferentially bind with, dietary tannins on mastication, rendering them ineffective in forming complexes with dietary proteins, and reducing their negative effects on protein digestibility. Further partial counter-defence against dietary tannins may be provided by microbial populations able to degrade tannin–protein complexes present in the hindgut of some species of herbivore.

Protection of Protein from Microbial Degradation (positive)

The processes that inhibit digestive protein degradation can in some circumstances have positive effects. The binding of condensed tannins with proteins depends on the specific chemical characteristics of both reactants and is pH dependent. For tannins of certain forage species, at low concentrations, the formation of complexes with protein in the forestomach of ruminants can limit the degradation of protein by rumen bacteria and deliver more essential amino acids for absorption in the small intestine, where the tannin–protein complexes dissociate as a result of the change in pH (Min *et al.*, 2003). This process can therefore be viewed as the condensed tannin protecting the protein from microbial degradation in the rumen. However, although this means by which protein by-passes the rumen may be an important advantage in agricultural conditions where the ruminants' diets may be of high quality (protein and energy), these conditions are likely to occur only very rarely in wild, free-ranging or extensively-managed ruminants, which have mixed lower-quality diets.

Anthelminthic Effects (positive)

The positive nutritional role of condensed tannins, proposed to be functional in ruminant livestock, may underlie their suggested effectiveness as anthelminthics. This effect is hypothesized to function either via the enhanced amino acid absorption that increases the effectiveness of the immune system, or via the condensed tannins directly inhibiting parasite larvae (Athanasiadou and Kyriazakis, 2004). If the effect is a result of a general enhancement of the immune system, then future work may demonstrate positive effects of condensed tannins against other pathogens.

Depression of the Activity of Symbiotic Microbial Populations (negative)

Many herbivorous animals that lack cellulase enzymes have populations of symbiotic microorganisms in their gastrointestinal tracts, which ferment cellulose to volatile fatty acids that are absorbed and used as an energy source by the host. The activity of these microbial populations can be depressed by ingested SM (Wallace, 2004), although microbial populations can adapt to become tolerant to SM such as hydrolysable tannins. Furthermore, the presence of a foregut fermentation system is hypothesized to represent an important anti-SM strategy, providing a medium for microbial detoxification of SM before they reach the animal's tissues, e.g. the metabolism of pyrrolizidine alkaloids by the rumen microflora of sheep (Cheeke, 1994). However, conversely, in some cases the process of herbivore foraging can render compounds more toxic or harmful; for example, rumen microbial populations can toxify as well as detoxify some SM, and the hydrolysis of phenolic glycosides in the rumen can increase their toxicity. The action of plant or animal enzymes on precursors that are brought together on mastication can yield toxic products; for example, the action of plant cyanogenic glycosides in generating HCN from a-hydroxynitriles, or the formation of isothiocyanates or nitriles from glucosiniolates catalysed by plant myrosinase. The action of the rumen microbiota on S-methyl cysteine sulphoxide in brassicas yields toxic dimethyl disulphide. On the other hand, in some cases the hydrolysis of phenolic glycosides can yield deconjugated sugar moieties that can themselves be fermented (Murray *et al.,* 1996).

Disruption of the Sodium Mineral Balance (negative)

Disruption of the Na mineral balance in response to ingestion of diets rich in SM has been recorded in animals This effect has been attributed to SM-induced Na wasting via the urine (perhaps a result of impaired kidney function), or to greater faecal loss following poor absorption or enhanced salivary or intestinal mucosal cycling of Na to the gut. The Na loss in animals is related to the general phenomenon of acidosis induced by the detoxification of SM in many species (Foley *et al.*, 1999). The phase I and II detoxification reactions generally lead to the formation of organic acids, the excretion of which may be associated with the loss of cations such as Na that accompanies the excretion of phosphate used to buffer the urinary pH.

The Energy Costs of Detoxification (negative)

The energy costs of detoxification are hypothesized to be small and the effects of SM on whole-animal energy metabolism have been rarely measured. This shortcoming is surprising given the importance of energy as a common currency in animal foraging studies. The energetic effects of infusion of low molecular- weight phenolics into the gastrointestinal tract of sheep, after controlling for their effects on reduced food intake and digestibility, are quite small. However, infusion into the peripheral circulation shows increased energy metabolism, which has been explained in terms of a toxic effect in the animal tissues. Gallic acid ingestion increases the BMR in animals, although a depressed energy metabolism following ingestion of some monoterpene SM has been recorded (Boyle and Dearing, 2003).

Diuresis (negative)

Diuresis results from the ingestion of the monoterpene a-pinene and creosote resin in animals (Dearing *et al.*, 2001). This water loss is an integral part of the general processes of detoxification and excretion of SM, but its importance in the determination of the nutritional ecology of animals occupying water scarce environments such as deserts or high-latitude zones in winter is unexplored.

Integration of Physiological Effects of Secondary Metabolites with Foraging Behaviour

Feedback Mechanisms

A hierarchy of routes by which SM affect foraging behaviour has been identified. These routes range from pre-ingestive stimuli (e.g. taste and smell) to post-ingestive mechanisms that involve feedback from reduced digestibility, or to post-absorptive effects of SM (Foley *et al.*, 1999). The pre-ingestive mechanisms are likely to act rapidly (e.g. via stimulation of the trigeminal nerve) to determine whether or not a food is eaten, whereas the post-ingestive mechanisms would involve a lag before being sensed and take longer to affect foraging. The effects of digestibility reducers such as condensed tannins are probably perceived by the herbivore as a forage of lower quality. Food intake and diet composition can be modified by herbivores learning to associate negative effects. A role for previous learning as a determinant of food choices is particularly likely for SM, since negative stimuli produce stronger learning responses than positive ones (Iason, 2005); although it should be remembered that herbivore diet selection is in many cases probably the result of influences of both the nutrient and SM content of the available foods. However, although these learning processes are probably effective where stimuli are strongly and clearly associated with a particular food, animals foraging in heterogeneous environments may receive a complex set of temporally-overlapping signals from SM. This outcome may limit their ability to associate particular stimuli with particular forages and thereby restrict the power of associational learning (Duncan and Young, 2002). It is hypothesized that this effect may be most acute in foregut-fermenting herbivores, in which the direct effects of ingested SM are buffered by the enlarged forestomach, and in larger herbivores that process SM more slowly than smaller animals because of their slower mass-specific metabolic rates.

Intake Regulation

Although some forage are avoided almost completely by herbivores, it has been argued that these situations are rare, and that SM intake is considered to be regulated to within limits that are tolerable by the herbivore. This form of regulation implies an interaction between behavioural and physiological mechanisms, which are only beginning to be understood. Both ecological responses to SM and the physiological and pharmacological mechanisms on which they are based are likely to be threshold phenomena with ranges of sensitivity and tolerance. Such a threshold effect has been demonstrated in the case of cattle responses to increasing doses of alkaloids in larkspur (*Consolida ajacis*). Considerable progress on the key relationships between the concentration of SM and food intake, the consequent dose of the SM ingested and identification of thresholds of plasma concentration has recently been made using arboreal marsupial folivores. Development and application of this pharmaco-kinetic approach in conjunction with behavioural studies of foraging in this and other appropriate model systems is advocated. In general, as the concentration of a SM in the forage increases the intake decreases, and where this response has been measured it has been shown to be either linear or curvilinear. However, relatively few other studies have investigated a sufficient range of concentrations to quantify the precise relationship. There is some evidence of temporal cyclicity in the intake response to ingested PSM, i.e. the animals follow periods of ingestion of SM containing food with periods of low intake that permit plasma toxin levels to decrease below a threshold level (Pfister et al. 1997).

Diet Choice

There are many correlative studies that have identified SM as determinants of diet choice by animals, but the current view is that the choices made by herbivores reflect both the concentration of SM and nutrients in available food plants. Examples of positive selection for SM are rare. However, the preference for natural forages by roe deer (*Capreolus capreolus*) is positively correlated with the presence of protein-binding phenolics (Tixier *et al.*, 1997). Although roe deer exhibit lower rates of ingestion of diets containing high concentrations of hydrolysable tannins than a control diet with no tannins, they eat relatively more of an artificial diet to which moderate levels of hydrolysable tannins have been added than the control diet. These results probably arise from physiological adaptations of this species to ingestion of tanniferous diets, including the large salivary glands that secrete saliva containing a tannin-nullifying protein (Duncan *et al.*, 1998). A basic tenet is that a herbivore's ingestion of a particular food source is limited by its ability to detoxify and excrete specific secondary metabolites. A reduction in intake is likely to lead to a herbivore seeking alternative foods. Several studies support the resultant prediction that herbivores choose mixed diets in order to minimise the effects of particular SM, thereby avoiding nutritional limitation as a result of saturation of specific detoxification pathways. Lambs are able to eat more food when given the choice of foods with different SM added than they can when offered no choice or a diet containing only one of the SM. It remains a major challenge for future research to translate these known and hypothesized physiological relationships from intake regulation into herbivore food choices under field conditions and to generate behaviourally-based predictable management of plant communities by herbivores.

Animal Adoption to Secondary Metabolites Acting as Anti-Nutritional Factors

To live a healthy life herbivores must consume nutritious plants and avoid toxic or low quality forage if possible. To meet this challenge animals possess various adoptive behaviors that limit toxins intake or possess systems to detoxify it. Selective grazing is the herbovore's first line of defense against the negative effect of plants with toxic attributes. When a grazing animal smell and taste a plant, the flavour is either pleasing or unpleasing depending on the animal's previous experiences. This depends up on the digestive feed back, which begins with in 10-15 minutes of consumption. Successful navigating the seasonal and spatial variation of forage quality in rangelands can be accomplished by knowing how much to eat, when to eat and what else to eat. This behavior increases animal's ability to ingest necessary nutrients and helps in reducing potentially toxins. Animals possess several mechanisms to restrict the toxic effects of plant compounds once ingested. Sheep, goat and cattle can and will vomit in response to toxins, but it is rarely observed. Horse experience diarrhea, which aids in rapid elimination of toxins from the gut. The rumen is generally well adopted to bind, degrade or detoxify plant toxins. The neutral pH of the rumen tries to modify the toxin produced. The rumen bacteria try to break the compound into less toxic compounds. In addition to that, if toxins are absorbed from the gut into blood stream, they are often transported to liver. The liver modifies the compound to less toxic compound thus reduce the toxic effect.

Common Toxic Problems Associated with Forage Feeding

Toxic constituents of forages lead to a great economic loss to the livestock farming. Some toxic factors seriously affect the health of animals while others cause reduced production without visible signs of ill health. The harmful effects of these constituents become more alarming especially in the adverse climatic conditions like drought or famine when animals are forced to eat whatever is available. There are large numbers of toxic constituents found in different species of forage plants causing toxicity problems in animals. Some of the commonly occurring toxic problems have been discussed below.

Nitrate Toxicity

Nitrogen is basic building block of proteins and an essential nutrient for animal and plant. However, nitrogen in the form of nitrate (NO_3) and in excess amounts (>5000 ppm on DM basis) causes toxicity.

The excessive nitrate concentrations usually are triggered by some environmental stress when plant growth is restricted but absorption of nitrate from soil continues. The most common stress is drought. Lack of moisture together with excessive soil nitrogen for existing growing conditions, is a frequent cause of toxic level of nitrate in sorghum. Other stress factors that favour buildup are reduced sunlight from cloudiness; frost, certain herbicides including 2,4D and deficiency of essential nutrients like phosphorus, sulphur and molybdenum. Most nitrates accumulate in plant stem rather than in leaves and concentration tend to be highest in immature forages. The accumulation of nitrate differs in different plant parts like petiole> leaf> stem> root> inflorescence> bulb> fruit> seed in decreasing order. Several factors like genetic, environmental (humidity, temperature, photoperiod) soil-N, fertilizer application, availability of other nutrients etc affect uptake and accumulation of N in plants.

The level of nitrate that causes toxicity in ruminants varies depending on rate of intake, diet type, acclimatization to nitrate and nutritional status. As a rule, forage containing <5000 ppm nitrate on dry matter basis is safe. Forage containing 5000-10000 ppm nitrate is considered potentially toxic, while forage containing over 10000 ppm nitrates is considered dangerous.

Nitrate poisoning of animals is actually nitrite (NO_2) poisoning occurs when nitrate is reduced to nitrite in the gastrointestinal tract. The nitrite is absorbed into the bloodstream where it reacts with hemoglobin to form methemoglobin. This compound, which is brown in color, is incapable of releasing oxygen. In acute cases of poisoning in cattle, 60 to 80% of the total hemoglobin is comprised of methemoglobin. Sheep generally do not develop as much methemoglobin and are therefore more resistant to this form of poisoning. The symptoms of acute poisoning are trembling, staggering, rapid breathing, and death. Chronic poisoning may result in poor growth, poor milk production and abortions.

Some plant species are naturally good accumulators of nitrates. Common plant species that are associated with nitrate poisoning are sorghum, maize, sudangrass, oat rye, millets, alfalfa, sweet clover, fescue and johnson grass (Table 2). The legume and grass species that are used for pastures or hay crops are not considered good nitrate accumulators, but given the right conditions can accumulate concentrations of nitrate that are potentially hazardous.

Table 2: Typical Nitrate Content of Common Forages (DM basis)

Forages	Stage of harvest	NO_3-N (ppm)	NO_3 (ppm)
Fresh alfalfa	Early bloom	400	1760
Alfalfa silage	Early bloom	200	880
Fresh sorghum	50% flowering	1200	5280
Fresh maize	50% flowering	1000	4400
Maize silage	50% flowering	500	2200
Pearl millet	50% flowering	1450	6512
Sorghu-sudan	Dough stage	1650	7260

For the treatment of nitrate toxicity, an intravenous injection of 1-4% methylene blue solution at the rate of 20 ml per 100 kg body weight has been effective in saving the animals. Methylene blue converts the methaemoglobin back to oxygen carrying haemoglobin. Proper management of forage and livestock can avoid problems with nitrate toxicity. Splitting the nitrogen doses into 2-3 applications can minimize nitrate accumulation. Use of low nitrate containing varieties is the best way to avoid nitrate toxicity.

Prussic Acid Poisoning

Prussic acid, also called hydrocyanic (HCN), normally is not present in plants. However, several common plants can accumulate large quantities of cyanogenetic glycoside. When plant cells are damaged by wilting, frosting or stunting, the glycoside degrades to form free HCN. Conditions in the rumen also favor degradation of the glycoside to free HCN. Thus plants that contain the glycoside have the potential to cause HCN toxicity when consumed by ruminants. Plants most likely to cause HCN poisoning are sorghums. The potential is greatest for johnson grass and least for true sudans. Other materials with HCN potential include white clover, vetch seed and chokecherry (Table 3).

As with nitrate buildup, some stress usually triggers accumulation of cyanogenetic glycoside in plant tissue. The potential for accumulation and HCN toxicity increases during drought. Occasionally, poisoning occurs when hot, dry winds induce temporary moisture stress in plants. The potential for poisoning is greater with excessive soil nitrogen and young plants. Toxicity also is more likely when periods of rapid growth are followed by cool, cloudy weather. Lush regrowth after cutting for hay, grazing or frost is particularly dangerous. Unfortunately for the livestock owner, often the only indication of prussic acid poisoning is dead animals. HCN is one of the most potent, rapid-acting poisons known. It interferes with oxygen use at the cellular level. Glycosides hydrolysed by enzymes release HCN, which is a volatile gas. Ruminants are highly susceptible. Excess CN^- inhibits cytochrome oxidase, stops ATP synthesis and animals are deprived of energy causing death finally. When a lethal dose (2.0-4.0 mg/ kg body weight) is consumed, animals die from asphyxiation in a few minutes.

When seen, clinical signs occur in rapid succession. Initially there are excitement and muscle tremors. Rapid and difficult breathing follows. The animal goes down, gasps for breath and may convulse. The pupils are dilated and mucous membranes are bright pink. A characteristic sign of HCN toxicity is a bright cherry-red color to the blood, a symptom that persists for several hours after death. Although blood is oxygenated, HCN interferes with the release of oxygen from oxyhaemoglobin to other tissues. This situation contrasts with nitrate toxicity, where oxygenation of blood is restricted. The rumen may be distended with gas, and the odor of "bitter almonds" may be detected when the body cavity is opened.

Table 3: Cyanide Content in Common Forages

Plant	Mg CN/ 100 g tissue weight
Sorghum forage	250
Wild cherry	370
Arrow grass	77
Bitter almond	250
Lima beans	150
Bamboo tips	800

Treatment of HCN poisoning, with a mixture of sodium nitrate and sodium thiosulfate or with methylene blue, can be successful if administered soon after symptoms appear.

As with nitrate, most problems with prussic acid can be avoided with proper management of forage and animals. Test any forage crop thought to contain HCN before animals are grazed or fed. Sorghums fertilized heavily with nitrogen and stunted by drought or cool, cloudy weather should be suspected. Reduce risk of poisoning from sorghums by using split application of nitrogen. Young plants have a higher HCN potential than more mature ones, so do not graze or harvest sorghum until plants are 2-2.5 feet high. This practice also applies to regrowth that occurs after cutting for hay or grazing. If regrowth occurs following frost, delay grazing until a hard freeze kills the entire plant. Do not graze sorghum following a killing frost until plants thaw and wilt for a few days. Spraying of cyanogenetic plants with a herbicide may increase the toxic hazard. Graze pastures to a uniform height, then remove animals to prevent selective consumption of lush regrowth. Rotation grazing and heavy stocking rates help in this regard. To acclimate cattle to new pasture, fill animals on native grass or hay during the day, then graze sorghums in late afternoon and evening. Proper field curing or ensiling results in considerable loss of HCN. If the forage is questionable as pasture, harvesting for hay or silage reduces the potential for HCN toxicity. However, if hay is poorly cured before baling, extremely high in HCN potential at cutting, or contains Johnson grass, it still may cause problems. Plant varieties differ in their potential for prussic acid poisoning. As with nitrate, chances for HCN toxicity are somewhat lower with true sudans and sudan-sudan hybrids than with sorghum-sudan or sorgo-sudan hybrids.

There is a quick qualitative test for HCN potential in plant tissue. It also can be used to confirm the presence of HCN in rumen contents of animals that die from prussic acid poisoning. Leaves are higher in HCN potential than stems. Glycoside levels increase during the morning, then level off and begin declining in the afternoon and evening. Therefore, samples for prussic acid analysis must include leaf tissue and should be collected in late morning or early afternoon. Randomly sample fresh forage from several locations. For hay, take cores from several bales. Seal two or three handfuls per sample in a plastic bag, store in the dark, refrigerate unfrozen, and deliver to the laboratory without delay.

The level of HCN required to cause toxicity varies, depending on rate of intake and individual animal tolerance. Generally speaking, view as dangerous any forage analyzing more than 200 ppm HCN on an as-fed basis.

Oxalate Toxicity

Most of the tropical grasses contain sufficient levels of oxalate to impair mineral utilization include buffel, pangola, setaria, kikuyu etc. Oxalate is found in plants in two major forms. Soluble oxalate is rapidly broken down to carbonate and formate in the rumen through microbial action, or else precipitated as insoluble calcium oxalate in the rumen contents. Both processes prevent or reduce the uptake of oxalate into the ruminal mucosa and into the bloodstream. If substantial amounts of oxalate are absorbed, calcium oxalate will precipitate in submucosal arterioles and damage the rumen mucosa. Similar damage may occur in lung capillaries, causing pulmonary oedema. In the kidneys, oxalate is secreted by the tubular epithelial cells, causing them damage, and calcium oxalate crystals are precipitated in the tubular lumina.

Plasma calcium concentration falls in the presence of high plasma oxalate levels and clinical signs of hypocalcaemia may occur. Animals dying from acute oxalate poisoning often have terminal pulmonary oedema, with copious amounts of stable froth in the trachea and bronchi and marked hyperaemia of the mucosa of the forestomachs and of the lower alimentary tract. Kidneys are usually swollen and the cortex pale.

Acute oxalate poisoning usually occurs when hungry stock are placed on pasture dominated by oxalate-containing plants such as halogetone, soursob, elephant grass, guinea grass, redroot, kochia and greasewood, that are at the time, high in oxalate levels. Fasting, or low calcium diets prior to exposure predispose to intoxication. Fasting reduces the numbers of methanogenic bacteria, which utilize formate, and if formate is not metabolised oxalate breakdown itself is inhibited. Damage to the ruminal mucosa can be followed by fungal or bacterial invasion leading to acute rumenitis and intractable indigestion. Animals acutely or sub-acutely intoxicated with oxalate respond only transiently or not at all to calcium borogluconate therapy and death occurs due to severe metabolic disturbance complicated by renal, pulmonary and gastro-intestinal injury. Clinical signs included dyspnoea, collapse, loss of consciousness and death without struggling. Slender iceplant (*Mesembryanthemum nodiflorum*) has also been associated with acute poisoning.

Chronic oxalate toxicity usually occurs in animals grazing *Oxalis pes-caprae*. Animals grazing soursob develop a high degree of tolerance to ingested oxalate. There is, however, usually a low level of oxalate absorption with progressive renal injury and deposition in the kidneys of calcium oxalate crystals. Over a period of 2 to 12 months of grazing on such pastures, animals gradually develop renal failure and the occasional death occurs. At necropsy, the kidneys are small, white and fibrotic with large amounts of crystals.

Alkaloids

The alkaloids are of various kinds and they constitute the largest class of plant secondary compounds, occurring in 20-30% of perennial herbaceous species. Alkaloids are a diverse group of basic organic compounds containing nitrogen. There are different kinds of alkaloids (Table 4) present in various range species. When grazing animals ingest these species they cause toxicity.

Table 4: Some Common Plants and their Alkaloids

Type of alkaloid	Plant species	Toxic symptoms
Pyrrolizidine	*Senecio jacobaea, S. vulgaris, S. latifolious, Heliotropium europium, Echium plantagineum*	Head pressing, straight walking and death
Piperidine	*Conium maculatum, Cicuta maculata, Daucus carota*	Nervousness, trembling, death
Pyridine	*Nicitiana attenuata, N trigonophylla*	Excitement, shaking and twitching
Indole	*Sporobolus poiretii*	Arthrogryposis in pigs
Quinolizidine	*Lupines spp., L. mutalilis*	Laboured breething, paralysis, convulsion and death

Alkaloids are organic basic substances with a bitter taste, examples of which are morphine, atropine, nicotine, quinine and strychnine. The alkaloids generally are irritating to the gastrointestinal tract producing nausea, colic and diarrhea and also act on the central nervous system to produce blindness, muscular weakness, convulsions and death. Some of the common plants containing alkaloids are given in Table 5.

Table 5: Toxic Symptoms of Alkaloids in Animals

Plant species	Species affected	Toxic symptoms
Locoweed	Horse, cattle, sheep	Depression, incoordination, nervous
Tall larkspur	Cattle	Muscular weakness, rapid breathing
Senecios	Cattle, horse	Photosensitization, nervous
Lupine	Cattle, sheep	Muscular weakness, respiratory failure
Poison hemlock	Horse, ship, pig	Muscular weakness, respiratory failure
False hemlock	Sheep	Birth defects
Death camas	Sheep	Excessive salivation, lung congestion

Glycosides

Glycosides are natural plant secondary metabolites found in many rangeland plants and forages. They are chemically diverse group of compound that contains the sugar glucose. They bear little resemblance to each other and can form toxic compound upon hydrolysis (Table 6).

Table 6: Toxic Glycosides in Plants and their Effect on Livestock

Types of glycosides	Plant species	Signs of toxicity	Treatments
Nitro-containing glycosides	Legume: Milk vetch (Astragalus spp.), Coronilla, Indigofera and Lotus	Acute clinical signs are, laboured breathing, blues skin or tongue, muscular weakness.	Protein supplementation can reduce the toxicity of glycosides.
Cyanogenic glycosides	White clover, choke cherry, service berry, arrow grass, sudan grass, bermuda grass	The signs of cyanide poisoning are dyspnea, excitement, gasping, staggering, paralysis and death.	Administration of nitrite-thiosulphate is the preferred treatment.
Cardiac glycosides	Foxglove, squill, milk weed dogbanes and lilies	Acute signs of poisoning in cattle and sheep include restlessness, laboured breathing, frequent urination and defecation.	Administration of activated charcoal, potassium chloride, atropine and phenytoin

Contd...

Table 6. Contd...

Saponins	Lucerne, yam, Albizia sp., Sesbania sp.	Photosensitization, sunburn around eyes, ears, nose and udder.	Addition of cholesterol in the diet can reduce the incidence of poisoning
Glucosinolates	Mustard family	Enlargement of thyroid gland (goiter), stinkweed etc.	Drugs such as methimazole, amphenone and chlorpromazine may be used.
Calcinogenic glycosides	Nightshade, jasmine, yellow oat	Calcium deposition in soft tissues	Calcium and phosphorus supplementation may be useful.
Phenolic glycosides	Subterranean clover, lucerne	Reproductive problems like abortion in sheep and cows	Vaccination against phytostrogenic disorders

Photosensitization

This is a disease caused by the sensitization of superficial layers of skin to sun light due to ingestion of photodynamic agents in forages or accumulation of normal metabolic products of forages in body due to liver damage. In animals, photosensitization is known to occur on many pasture and forage plants like *Trifolium alexandrinum, Medicago sativa,* rape and kale. Ingestion of green oat and *Lantana camara* has also resulted in photosensitization in animals. Primary photosensitization due to ingestion of exogenous photodynamic agents usually occurs when forages are growing rapidly in lush green stage. Animals are generally affected within 4-5 days of exposure to such forages.

Although hepatogenous photosensitization is common in animals grazing green pasture, it can also occur on hay or other stored feeds. This type of photosensitization is due to phylloerythrin - an end product of plant chlorophyll which is accumulated in animal body owing to hepatitis or bile duct obstruction. Extensive outbreaks of photosensitization have been recorded in cattle fed on water damaged lucerne hay, but the cause in these circumstances was not known.

The distribution of skin lesions is characteristic and restricted to dorsum of the body which is exposed to sun rays. There is development of erythema and oedematous swelling on ear lobes, eye lids, muzzle, face and lateral aspects of teats. There is some times intense rubbing on these sides due to irritation. When teats are involved the animals may kick. Some times, edema is much severe which may result into drooping of ears, dyspnoea and dysphagia. In much severe cases, necrosis and gangrene is the terminal stage. The skin lesions may be severe enough to cause shock. The diagnosis of photosensitization is based on distribution of lesions and their sudden development in animals grazing in sun light.

The affected animals should immediately be removed from sun to shade and should not be fed on toxic forages. Antihistamines and cortisones should be administered immediately.

Mimosine Toxicity

Mimosine, a non protein amino acid structurally similar to tyrosine, occurs in a few species of mimosa and all species of closely allied genus leucaena. Concern has arisen because of the importance of *Leucaena. leucocephala,* in which the level of mimosine in the leaf is about 1.50-6.05% (Table 7) and varies with seasons and maturity (Kumar, 2003).

Table 7: Toxic Constituents of Some Commonly Available Forage at a Glance

Forage	Toxic constituent	Range	Safe limit
M. P. chari (*Sorghum bicolor*)	Cyanogenetic glucoside	0-1054 ppm as HCN	200 ppm HCN
Teosinte (*Fuchlaena maxicane*)	-do-	12-328 ppm as HCN	
Reonjha (*Acacia leucophloea*)	-do-	400-1094 ppm as HCN	
Oat green (*Avena sativa*)	Nitrate	2.492% KNO_3; 0.42-1.54% as NO_3 -N	1.50% as KNO_3
Oat hay (*Avena sativa*)	Nitrate	3.0-7.0% as KNO_3	
Chari/Jowar (*Sorghum* spp.)	Nitrate	3.0% as KNO_3	
Turnip tops (*Brassica rapa*)	Nitrate	8.0 % as KNO_3	
Para grass (*Brachiaria mutica*)	Nitrate	2.54-4.73% as KNO_3 0.10-3.44% as KNO_3 0.06-3.1% as NO_3	
Anjan Grass (*Cenchrus ciliaris*)	Nitrate	2.916-2.957% as KNO_3	
Bajara (*Pennisetum americanum*)	Nitrate	4.0953-4.1294% as KNO_3	
Hybrid Napier (*Pennisetum purpureum*)	Oxalate	3.06% as Oxalic acid 1.70-4.36% Oxalic acid	Large quantity is needed for toxic effects
Nandi grass (*Setaria sphacelata*)	Oxalate	7.0% as Oxalate	
Sugar beet leaves (*Beta vulgaris*)	Oxalate	17.0% as Oxalate	
Subabool (*Leucaena leucocephala*)	Mimosine	3.47-6.058% 2.10-3.02% 1.50-6.0% 2.2-5.9%	

Source: Prasad (1987)

In ruminant animals mimosine causes poor growth, loss of hair and wool, swollen and raw coronets above the hooves, lameness, mouth and esophageal lesions, depressed serum thyroxine level and goiter. Some of these symptoms may be due to mimosine and others to 3, 4 dihydropyridine, a metabolite of mimosine in the rumen. Toxic signs like skin lesions also resemble Zn deficiency. Reduction in calving percentage due to Leucaena leucophala feeding has also been reported. Postmortem examination of lambs died due to Leucaena toxicity at this institute showed erosions on gums, ulcers on tongue, oesophagus and haemorrhages in different organs. The lambs which died after 65 days of Leucaena feeding also showed comparatively larger thyroids, weighing more than double and ulcers on lateral surface of the tongue.

In case of non-ruminants the toxicity symptoms range from complete alopecia to a slight hair loss. Cataract and 100% alopecia has been reported in rats fed on Leucaena leucocephala having more than 0.75% mimosine.

A solution to the momosine problem could be the development of low mimosine cultivars. However, low mimosine types are found to be unproductive and have low vigour. The other approach is to feed Leucaena mixed with other feeds. Hiremath (1981) suggested that use of leucaena fodder may be restricted to 30% of the green forage in the case of cattle and buffalo and 50% for goats. The effect of leucaena and mimosine can be reduced by heat treatment, by supplementation with amino acids or with metal ions such as Fe2+, Al3+ and Zn+2 (Kumar, 2003; Mahanta and Singh, 2006).

Management Tips to Reduce Toxic Effects of Forages

It should be noted that most of the toxic plants are unpalatable and animals will usually do not consume or graze them if given a choice. One of the most important steps in preventing animal suffering or loss is to follow good forage/ pasture management practices viz.:

- Identification and removal of plants containing anti-nutritional factors from grazing areas before seed setting.
- Burning of affected areas for 2-3 consecutive years.
- Reseeding with desirable vegetation.
- By avoiding hungry animals turning on to rangeland containing poisonous plants.
- Forages which are known to accumulate toxic constituents should not be harvested at young stage.
- Forage crops should also not to be when there is water stress.

- Excessive use of nitrogenous and potassic fertilizers should be avoided.
- The problem of toxicity can be minimized by offering SM containing forages slowly to the animals or by mixing with other quality forages.
- Mineral supplementation containing calcium, magnesium, copper, zinc and cobalt helps in toxicity reduction.
- Ensiling or hay making reduce the degree of intoxication.
- Energy supplementation may over come some of the negative effect of toxicosis.

References

Alessio, H.M., Hagerman, A.E., Romanello, M., Carando, S., Threlkeld, M.S., Rogers, J., Dimitrova, Y., Muhammed, S. and Wiley, R.L. 2002. Consumption of green tea protects rats from exercise-induced oxidative stress in kidney and liver. *Nutrition Research* 22 : 1177–1188.

Athanasiadou, S. and Kyriazakis, I. 2004. Plant secondary metabolites: antiparasitic effects and their role in ruminant production systems. *Proceedings of the Nutrition Society* 63 : 631–639.

Boyle, R.R. and Dearing, M.D. 2003. Ingestion of juniper foliage reduces metabolic rates in woodrat (Neotoma) herbivores. *Zoology* 106 : 151–158.

Carriere, V., Chambaz, J. and Rousset, M. 2001. Intestinal responses to xenobiotics. *Toxicology in Vitro* 15 : 373–378.

Cheeke, P.R. 1994. A review of the functional and evolutionary roles of the liver in the detoxification of poisonous plants with special reference to pyrrolizidine alkaloids. *Veterinary and Human Toxicology* 36 : 240–247.

Dearing, M.D., Magnione, A. and Karasov, W.H. 2001. Plant secondary compounds as diuretics: an overlooked consequence. *American Zoologist* 41 : 890–901.

Duncan, A.J. Young, S.A. 2002. Can goats learn about foods through conditioned food aversions and preferences when multiple food options are simultaneously available? *Journal of Animal Science* 80 : 2091–2098.

Duncan, P., Tixier, H., Hofman, R.R. and Lechner-Doll, M. 1998. Feeding strategies and the physiology of digestion in roe deer. In: Anderson R, Duncan P, Linnell, JDC (eds) The European roe deer: the biology of success. Scandinavian University Press, Oslo. pp. 91–116.

Foley, W.J, Iason, G.R. and McArthur, C. 1999. The role of plant secondary metabolites in the nutritional ecology of mammalian herbivores: how far have we come in 25 years? In: Jung, HJG, Fahey, GC (eds) Nutritional ecology of herbivores. American Society of Animal Science, Savoy, IL pp. 130–209.

Harika, A.S. 1992. Occurrence of toxic constituents in forages and their feeding management. Proceedings All India Dairy Husbandry Officers' Workshop, October 22-23, 1992 at NDRI, Karnal. pp. 78-83.

Hiremath, N. B. 1981. Subabool (leucaena leucocephala)-a wonder plant. *Indian Dairyman*, 33: 351-356.

Iason, G. 2005. The role of plant secondary metabolites in mammalian herbivory: ecological perspectives. *Proceedings of the Nutrition Society* 64 : 123-131.

Kumar, R. 2003. Anti-nutritive factors, the potential risks of toxicity and methods to alleviate them. *http:// www.fao.org/DOCREP/003/T0632E/T0632E10.htm*.

Launchbaugh, K. 2001. Anti-quality Factors in Rangeland and Pastureland Forages. Grazing Land Technology Institute, USDA-Natural Resource Conservation Service, Fort Worth, TX.

Mahanta, S.K. and Singh, U.P. 2006. Strategies of feeding subabul (Leucaena leucocephala) forage in livestock- a review. *Range Management & Agroforestry* 26 : 92-100.

Min, B.R., Barry, T.N., Attwood, G.T. and McNabb, W.C. 2003. The effect of condensed tannins on the nutrition and health of ruminants fed fresh temperate forages: a review. *Animal Feed Science and Technology* 106 : 1–4.

Morris, P. and Robbins, M. 1997. Manipulating the chemical composition of plants. IGER Innovations, IGER, UK. pp. 12.

Murray, A.H., Iason, G.R. and Stewart, C. 1996. Effect of simple phenolic compounds of heather (Calluna vulgaris) on rumen microbial activity in vitro. *Journal of Chemical Ecology* 22 : 1493–1504.

Pfister, J.A., Provenza, F.D., Manners, G.D,, Gardner, D.R. and Ralphs, M.H. 1997. Tall larkspur ingestion, can cattle regulate intake below toxic levels? *Journal of Chemical Ecology* 23 : 759–777.

Prasad, J. 1987. Animal Health Problems Associated with Forage Based Diets. IGFRI, Jhansi.

Rozman, K.K. and Klaassen, C.D. 2001. Absorption, distribution, and excretion of toxicants. In: Klaassen CD (eds) The basic science of poisons. McGraw-Hill, New York. pp. 107–132.

Tixier, H., Duncan, P., Scehovic. P., Yani, J., Gleizes, M. and Lila, M. 1997. Food selection by European roe deer (*Capreolus capreolus*): Effects of plant chemistry and consequences for the nutritional value of their diets. *Journal of Zoology* 242 : 229–245.

Wallace, R.J. 2004. Antimicrobial properties of plant secondary metabolites. *Proceedings of the Nutrition Society* 63, 621–629.

❑❑❑

Chapter **34**

Glucosinolate : Its Estimation and Role in Animal Feeding

M. K. TRIPATHI, A. S. MISHRA AND A. K. MISRA*
Central Sheep and Wool Research Institute, Avikanagar (Rajasthan) India
Indian Grassland and Fodder Research Institute, Jhansi-284003 (U.P.), India

Introduction

The glucosinolates are a large group of sulphur-containing secondary plant metabolites, which occur in all the economically important varieties of *Brassica*. A wide variety of glucosinolates exists owing to modification of the side-chain structure. To date more than 120 different glucosinolates have been identified (Chen and Andreasson, 2001), all shares a common structure comprises a bD-thioglucose group, a sulphonated oxime moiety and a variable side-chain derived from methionine, tryptophan or phenylalanine (Fig. 1). These all belongs to three groups, aliphatic, aromatic and indole glucosinolates with respect to their side chain (Bennett *et al*., 1996). Through, genetic manipulation new double low *Brassica* cultivars are evolved. The double low *Brassica* cultivars contain very low concentrations of antinutritional substances glucosinolates and erucic acid, are referred as canola. Standards of double low cultivars differ between countries to countries. This complicates the levels of rapeseed meals inclusion in feeding. According to Swedish specification for 00 RSM, erucic acid should not account for more than 2% of the total fatty acid and each gram fat free dry matter should not contain more than 40 µmoles total glucosinolates. The Canadian standards are similar for erucic acid but the glucosinolate limit is less than 30 µmoles/ g fat free DM, which does not include indolyl-glucosinolates. The European Community countries allow a limit for glucosinolates per gram air dried rapeseed is 20 µmoles, including indolyl-glucosinolates.

$$R-C\begin{matrix} \nearrow S-\beta\text{-D-glucose} \\ \searrow N-OSO_3^- \end{matrix}$$

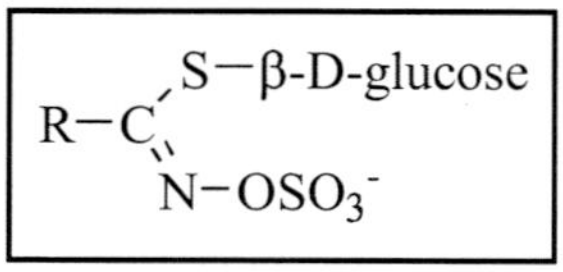

Fig 1. General Structure of Glucosinolate

The ingestion of substantial amount of glucosinolates produces deleterious effects on animal health and production. Upon ingestion, the intact glucosinolates and /or their breakdown products either absorbed from the intestinal lumen and /or converted into other products. The lower content of individual glucosinolates in ileal digesta in comparison to diet of pig (Rowan *et al.*, 1991) supports this hypothesis. The fodder and seed meals of genus *Brassica* are the chief source of glucosinolates in animal feeds and fodders. Inclusions of these have been known for a long to reduce the palatability (Hill, 1991), induce iodine deficiency (Bruel *et al.*, 2000c), hypertrophy of liver, kidney and thyroid (Mandiki *et al.*, 1999b; Tripathi *et al.*, 2001c; Mabon *et al.*, 2000) and at higher levels mortality (CSWRI, 2002). Deleterious effects of glucosinolates are greater in non-ruminant animals compared to ruminants. In general, young animals are more sensitive to glucosinolates than adult and older animals. Higher glucosinolates ingestion in poultry increases mortality and lowered egg production as well as egg weights, whereas glucosinolates are fatal to pigs. Studies with fish also provoked deleterious effects of glucosinolates on growth inhibition and thyroid function. Various processing techniques have been worked out to remove glucosinolate and minimise the deleterious effect of dietary glucosinolates ingestion on animals (Fenwick *et al.*, 1986; Wallig *et al.*, 2002; Minkowski *et al.*, 2002; Smithard and Eyre, 1986; Quinsac *et al.*, 1994; Liu *et al.*, 1994; Huang *et al.*, 1995; Jensen *et al.*, 1995; Tripathi *et al.*, 2001a; Tyagi, 2002; Das and Singhal, 2005). However processing of glucosinolates containing food/feed give rise to a certain degree of glucosinolates breakdown products by enzymatic hydrolysis or other chemical reactions. The glucosinolates degradation takes place at each level of feed processing starting from oil extraction to diet preparations (Wallig *et al.*, 2002). The heating (cooking) reduces glucosinolates level by approximately 30-60 % depending on the type of compound, degree and time of heating, the degradation products are hardly detectable, exception of thiocyanate ions and ascorbigen. Pulping results in complete breakdown of glucosinolates by autolysis, glucosinolates may partly survive depending on the degree of crushing. The degradation products isothiocyanates and nitriles are persistent. The fermentation gradually decreases the glucosinolates level to zero. The degradation products 1-cyno-3-methylsulphinylpropane may remain in the final product (de Vos and Blijleven, 1988). Mandiki *et al* (2002) demonstrated that despite of comparable total glucosinolate content of two rapeseed meal, antinutritional effect was more pronounced in Apex variety than Synergy. Therefore glucosinolate profile, which may vary among variety and strain of rapeseed meal influence adversity of antinutritional effect of glucosinolate. The aim of present review is to bring together the informations available and technologies developed so far for efficient utilization of glucosinolate containing RSM in feed formulation. The presence of glucosinolate metabolites in animal produces and effect on quality traits of animal produce.

Prospects and Problems of Glucosinolate Intake

Rapeseed meal is the chief source of glucosinolates (Gls) in animal nutrition; it is a good source of protein and sulfured amino acids. Content and composition of Gls varies due to plant species, agronomic practices and climatic conditions. The Gls content is generally higher among rapeseed meal (RSM) grown under tropical environment than occur in temperate. The glucosinolate composition is also influenced by growing conditions, the RSM originated from Indian sub-continent contain primarily 3-butenyl, 2-propenyl and 4-pentenyl glucosinolates in proportion of 73, 24 and 1% respectively (Fig. 2), other indoyl and aromatic glucosinolates are in the range of 1.6 to 3.7%, whereas glucosinolates from European and other temperate countries contain basically 2-propenyl glucosinolates, accounting to more than 95% of their total glucosinolates and did not contain 4-pentenyl glucosinolates. Fermentation conditions (pH and cofactors),Gls content and composition of RSM affect the level and type of metabolites (Fig 3), major metabolites are thiocyanates (SCN), isothiocyanates (ITC), nitriles, 5-vinyl-2-oxazolidinethione (VOT) and 5-vinyl-1,3-oxyzolodine-2-thione (5-VOT). Reduced palatability, decreased growth and production are the major deleterious effects of glucosinolate ingestion in animal diets. Apart from total glucosinolate content SCN, nitriles and VOT estimates are the chief attribute of RSM quality as these are produced upon hydrolysis of Gls following the processing of RSM. Progoitrin and epi-progoitrin content is the palatability attributes, palatability impair at level between 2.3-4.65 µmole g^{-1} diet, while at higher a significant feed intake depression occur. Nitriles are known to affect liver and kidney functions.The SCN interferes with iodine availability, whereas 5-VOT is responsible for the morphological and physiological changes of thyroid. Difference in Gls profile among the RSM induces varying levels of glucosinolates metabolites in animal tissues. The indoyl glucosinolates, precursors of thiocyanates will increase the SCN concentration whereas corresponding intake of progoitrin (2-hydroxy-3-butenyl glucosinolate), will increase the concentration of 5-VOT. High intake of glucosinolate progoitrin induces an accumulation of 5-VOT mainly in thyroid and lungs. The 5-VOT has a high affinity for thyroid. The SCN are secreted in milk of RSM fed animals. RSM feeding did not impairs quality traits of carcass, whereas RSM feeding results a higher contents of unsaturated fatty acids ($C_{18:00}$, $_{1\ trans}$, $C_{22:2}$) in carcass and milk fat.

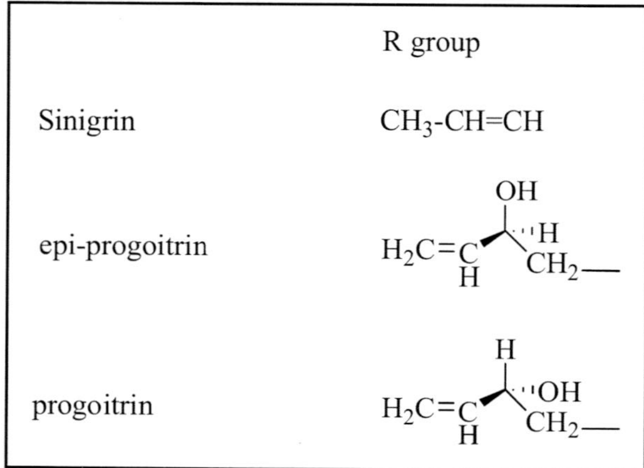

Fig 2. The "R" Group Varies upon the Type of Glucosinolates. The Sinigrin, Epi-progoitrin and Progoitrin are the Major Glucosinolates of Brassica oil Meals

Glucosinolate Tolerance Level and Feed Quality Improvement

Ruminants are less sensitive to dietary glucosinolates than non-ruminants (Table 1). Among non-ruminants, pigs are more severely affected by dietary glucosinolate compared to rabbit, poultry and fish. Young ruminants can tolerate total Gls up to 1.5 to 4.22 µmole g^{-1} diet, whereas pig 0.78 µmole g^{-1}, rabbits 7.0, poultry 5.4 and fish 3.6 µmole g^{-1} diet. Among various treatments to detoxify RSM water extraction, heat and $CuSO_4$ treatments were suitable for RSM quality improvement. Added iodine supplemetation1.0 mg kg^{-1} diet in pigs and 500 mg I each kg RSM included in ruminants diets seems promising because of economic and easiness compared to other treatments. Therefore, a desired amount of RSM can be used for animal feed formulation adopting a suitable technology to minimize or remove Gls related deleterious effects on animals. Additionally, increased contents of unsaturated fatty acids ($C_{18:00}$, $_{1\ trans}$, $C_{22:2}$) in carcass and milk fat and deposition of ally-isothiocynate (AITC) are advantages of RSM feeding because of their human health promoting effects.

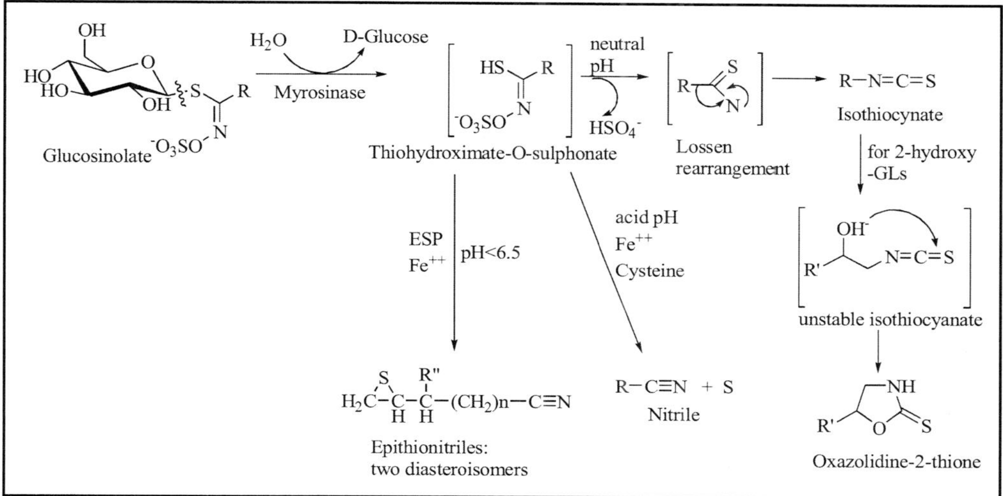

Fig.3. Scheme of Enzymatic Hydrolysis of Glucosinolate

Among techniques worked out to remove glucosinolate and minimise the deleterious effect on animals heating, pulping and solid-state fermentation were found effective (Table 2). Heating reduces glucosinolates depending on the type of compound, degree and time of heating, the degradation products are hardly detectable, exception of thiocyanate ions and ascorbigen. Pulping resulted in complete breakdown of glucosinolates by autolysis, glucosinolates may partly survive depending on the degree of crushing. The degradation products isothiocyanates and nitriles are persistent. The fermentation gradually decreases the glucosinolates level to zero. The degradation products 1-cyno-3-methylsulphinylpropane may remain in the final product. Processing of glucosinolates containing food/feed give rise to a certain degree of glucosinolates breakdown products by enzymatic hydrolysis or other chemical reactions. The glucosinolates degradation takes place at each level of feed processing starting from oil extraction to diet preparations. Isothiocyanates produced during processing or fermentation may either absorb by the intestinal epithelium or under certain circumstances isothiocyanates under go rearrangement reactions to form thiourea or corresponding amines. Allyl-isothiocyanates formed allyl-amine whereas benzyl-isothiocyanates formed benzyl-amine. Free amines formed by rearrangement reactions of isothiocyanates may under go further chemical changes to yield thiourea by reaction with *n*-butyl-amine (Fig 4). Thiourea also known to induce hypothyroidism in small ruminants.

Table 1. Glucosinolate Tolerance Level in Different Animals

Animal species	Glucosinolates level maximum upper limit	Reference
Rat	0.5 µmol per gram diet	Bjerg *et al.*, 1989
Pig	0.78 µmol per gram diet	Opalka *et al.*,2001
	2 µmol per gram diet with added supplementation of at least 1000 µg Iodine/ kg diet	Schone *et al.*, 2001
Poultry	2-4 µmol per gram diet	Mawson *et al.*, 1994
Rabbit meat	7 µmole per gram diet	Tripathi *et al.*, 2003
Growing steers	10 to 15 µmol per gram diet	Bush *et al.*, 1978
Growing calves	7.7 µmol per gram diet	Mawson *et al.*, 1994
Dairy cow	11 µmol per gram diet	Laarveld *et al.*, 1981
Young lamb	2 µmol per gram dry diet	Derycke *et al.*, 1999
	0.88 µmol per gram liquid diet	Corrill *et al.*, 1976
Fish	1.4 µmol per gram diet	Glencross et al., 2004a,b

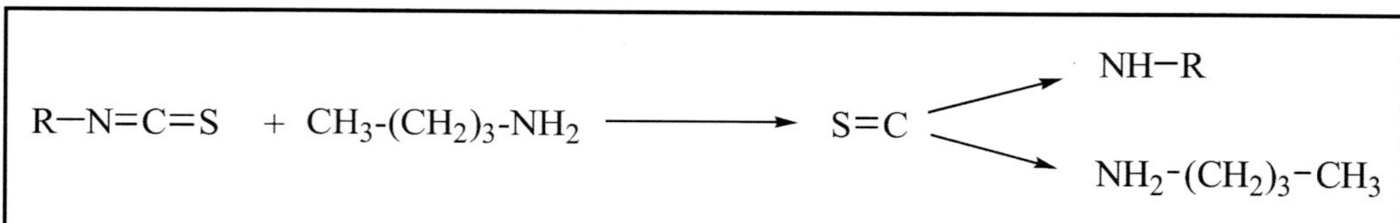

Fig. 4. Conversion of Isothiocyanate to Thiourea

Glucosinolate and Reproduction

High glucosinolates intake is known to induce hypothyroidism due to iodine deficiency in animals. Iodine deficiency in combination with goitrogenic agents has been reported to influence fertility in ruminants. The condition is expressed as increased prevalence of congenital goitre and stillborn calves, retained placenta, ketonaemia and reduced resistance to infection to be more common (Wilson, 1975). Inclusion of RSM containing total glucosinolates 31µmoles/g DM in diets of dairy cows during first calving and amounting a daily intake of 44 mmoles per day per cow caused thyroid disturbances which depressed fertility (Ahlin *et al.*, 1994). Higher levels of low glucosinolate rapeseed meal increased thyroid stimulating hormone response, ovulation days, normal reproduction function, calving to conception period and more numbers of inseminations per pregnancy.

Table 2. Effect of Processing Methods for Glucosinolate Decomposition on Animals Responses

Treatment	Conditions of treatment	Animal Response	Reference
Cu treatment	1 kg RSM soaked with 6.25 g $CuSO_4.5H_2O$ in 0.5 lit hot water	Pigs performance improved terms of intake growth and thyroid hormones status.	Schone *et al.*, 1997
Water soaking	Seed to water ratio; 1:10 for 6 hr and dried at 60ºC	Improved intake and nutrient utilisation in sheep	Tyagi *et al.*, 1997
	Soaking of meal for 24 h, with 2 parts of water and drying at 60ºC	Did not improve any attribute significantly in growing pig.	Schone *et al.*, 1988
Water extraction	Seed to water ratio; 1:8	Detoxified meal improved animal performance	Liu *et al.*, 1994
	Meal was soaked in water, 1:10 ratio, for 10 min filtered with gooch funnel at 25psi	Not determined	Tripathi *et al.*, 2000
Heat treatment	Toasting at 100°C, 2 h	120 min although reduced most of glucosinolates but lowered rat performance. For non ruminant feeding 30 min toasting is optimum for 00 RSM to remove half glucosinolates and maintain protein quality	Jensen *et al.*, 1995
	110°C, 2 h	No significant effect of treatments on milk yield and goitrogenic substance secretion in milk	Subuh *et al.*, 1995
	The acid treatment for 72 hr (16 ml HCl kg^{-1} meal) and heating at 180ºC for 2h	Acid treatment followed by heating improved nutrient utilisation and growth in calves	Tripathi *et al.*, 2001
Heat (toasting)	103-107°C, 30-40 min.	Poultry : Reduced amino acid digestibility	Newkirk *et al.*, 2003
Formaldehyde treatment	0.8g/100g CP	No significant effect of treatments on milk yield and goitrogenic substance secretion in milk	Subuh *et al.*, 1995
Micronization (Heat)	195°C, 90 seconds	Growing Pig : Improved nutrient digestibility	Lawrence, 1978

Reproductive disorders due to higher glucosinolate intake include more numbers of acyclic and cystic ovaries however abortion and retained placentas conditions were not observed in this study. A glucosinolates metabolite thiourea feeding 50 and 100 mg to adult and young female goats reproductive disorders such as small, pale and inactive ovaries, small and infantile uteri along a few inactive mucosal glands were observed (Reddi and Rajan, 1985). These reproductive disorders were due to induced hypothyroidism and were dose dependent, more sever in young females. Negative effects of feeding high glucosinolate RSM on fertility of cows were reported (Ahlstrom, 1978; Papas *et al.*, 1978; Laarveld *et al.*, 1981c; Sanchez and Claypool, 1983) as evident by increased numbers of AI per pregnancy, prolonged calving to conception and calving intervals. Poor reproduction traits were observed by Emanuelson et al. (1993) in cow fed high amounts of very low glucosinolate rapeseed meal. Therefore, not only high glucosinolate RSM but higher inclusion of low glucosinolate RSM in diets of first calvers cows disturbed fertility (Ahlin *et al.*, 1994), indicated that young and still growing, first calving cows are also sensitive to dietary glucosinolates. Ewes fed diets containing 27-30 μmoles low glucosinolate rapeseed meal contributing 1.2 – 1.6 μmoles glucosinolates/ g DM diets reduced plasma reproductive hormone estradiol provoked disturbance in the reproductive function (Mandiki *et al.*, 2002). These ewes consumed 2.5 to 7.6 mMoles glucosinolates per day during late gestation and lactation. The negative influence of a very low intake (9-10 g oil meal/day) of high glucosinolate *Brassica juncea* meal on serum quality and testosterone level of bucks was observed (Pattanaik *et al.*, 2004), which improved when additional I was supplemented.

Effect of Glucosinolates Metabolites on Intestinal Microbial Growth

Studies on the biological effects of glucosinolates metabolites are puzzling. A sizable number of studies advocated the promotive effects of glucosinolates against various human cancers, while other shown negative effects of glucosinolates on animal and human health. There is increasing interest on bacterial bacterial degradation of glucosinolates because dietary glucosinolates escaping stomach digestion are metabolized by colonic microflora. Some of the glucosinolates metabolites are considered to have antibacterial properties. Sinigrin hydrolysis products have been shown highly inhibitor to growth and concentration required for complete inhibition was species dependent (Brabban and Edwards, 1995). The inhibitory effect of glucosinolate degradation products varies with the species of microorganism. The *E. coli* were the most susceptible to the toxic effects of sinigrin hydrolysis products than *Bacillus* strains (Mithen *et al.*, 1986). A higher concentration of isothiocyanates including phenyl and allyl derivatives inhibits O2 uptake in yeast (Kojima and Ogawa, 1971). Allyl-isothiocyanates also inhibited the growth of *E coli* (Rihova, 1982). Possibly glucosinolates metabolites binds to a site with in the organism, thereby inhibiting the growth. This was established using the cell to inhibitor ratio studies (Brabban and Edwards, 1995). The isothiocyanates inhibited the growth of gram-negative organisms by binding the sulphydryl groups (Fenwick *et al.*, 1983). The initial inhibitory effect of sinigrin before its hydrolysis was found to be species dependent *Bacillus subtilis* being most resistant (80 μg / ml) and *Saccharomyces cervisiae* (40 μg / ml) the most sensitive (Brabban and Edwards *et al.*, 1995). Strong inhibitory effect of cabbage juice on *Escherichia* was also reported by Tolonen et al. (2004). The formation of nitriles by human intestinal bacteria *Bifidobacterium* sp. was confirmed (Cheng *et al.*, 2004) using sinigrin and glucotropaeolin in vitro. Nitriles are also considered cyto and neuro-toxicants (Tanii *et al.*, 2004). The antimicrobial activity of different sulfur compounds derived from cabbage was tested, than sinigrin did not inhibited the growth of bacteria up to 1000 ppm, whereas sinigrin metabolite, allylisothiocyante strongly inhibited the growth of oxidative and fermentative organisms (Kyung and Fleming, 1997). The minimum inhibitory concentration was 1 and 4 ppm. Allylisothiocyanates was found to have a minimum inhibitory concentration between 50 and 500 ppm for bacteria, including gram positive and gram negative bacteria.

Glucosinolate Metabolites in Animal Produce and Effect on Quality

The thiocyanate ions are the degradation products from indolyl-3-methylglucosinolates or other glucosinolates which give carbonium ion and the thiocyanate ion in non-enzymatic and enzymatic catalyzed reactions. When these are absorbed into the blood, the thiocyanate ions can be transferred to the milk (Papas *et al.*, 1978, 1979; Laarveld *et al.*, 1981b; Emanuelson *et al.*, 1993, Tripathi *et al.*, 1996) or transformed by liver enzymes. These may also act as competitive inhibitors in the transfer of iodide to the thyroid hormone in the thyroid gland (Laarveld *et al.*, 1981a; Ahlin *et al.*, 1994). The thiocyanate ions and iodide were thus considered to be important in relation to the quality of glucosinolates containing feeds (Danielsen *et al.*, 1994) and studies of antinutritional effects of glucosinolates, more precisely indolyl-3-methylglucosinolates (Loft *et al.*, 1992). Dietary glucosinolate intake is found to increase thiocyanate and nitriles secretion in milk and reduced milk iodine content (Laarveld *et al.*, 1981a; Fig 5). The 5-VOT is responsible for the reduction of animal performance (Maheshwari *et al.*, 1979). It is transferred to serum, milk, muscular tissues and to certain organs such as liver, lung, kidney, and mainly thyroid. The precursor of 5-VOT is known as goitrine (2-hydroxy-3-butenylglucosinolate) or progoitrin, these are found in high proportions in the most of rapeseed varieties, even in 00 varieties and remains in meal after oil extraction. Despite the decrease of the glucosinolate content in meal from 00 rapeseed varieties, their consumption still can decrease yield and meat quality. The level of 5-VOT were in biological fluid and tissues of ruminants ranged from 3-8 μg per kg in milk, 15-200 μg per kg in plasma, 80 – 250μg/ kg in urine (De Brabander and Verbeke, 1982), 70 μg / kg in thyroid and <0.5μg / kg in liver and kidney (De Brabander and Verbeke, 1986).

Cyanide exposure 1, 2 or 3 mg KCN/kg/day, to goat increased both cyanide and isothiocyanate in dam milk. Indirect intoxication of kids through suckling developed microscopic lesions in the brain, thyroid, liver and kidney of kids and dam as well, without alteration on the biochemical panel (Sot-Blanco and Górnaik, 2003). Many reports emanating during 1970-80 shown that high glucosinolates (76.1 μmoles TGLs/ g DM) RSM diets in dairy cows depressed iodide and increased thiocyanates level in milk, however using low glucosinolate (11.5 μmoles TGLs/ g DM) amounting to 14-47 mmol daily glucosinolate intake milk iodine and thiocyanates content did not changed (Laarveld et al., 1981b).Whereas, feeding low glucosinolates ((4.0-15.5 μmoles TGLs/g DM)) RSM at higher levels in dairy cows with 37-63 mmoles daily glucosinolates intake might reduced milk iodine and increased serum thiocyanates as the cows higher plasma thyroid stimulating hormone showing a mild and subclinical hypothyroidism (Ahlin *et al.*, 1994). Increased serum thiocyanates and secreation of thiocyanates in sheep milk was also found (Tripathi *et al.*, 1996) on mustard meal included diets. Thus, instead of type of RSM low or high, total glucosinolates content of the diet should be considered criteria of inclusion RSM in ruminant diets. However, an increased percentage of poly-unsaturated fatty acids in milk (Emanuelson et al., 1991) were observed, when high proportion of rapeseed meal was included in the diet of dairy cow.

In general dietary glucosinolates did not influence carcass quality. Hill (1991) indicated that inclusion of rapeseed meal in animal diets did not have any adverse effect on carcass flavour and quality. However, Australian consumers detected a strong, less acceptable flavour from meat of rapeseed fed lambs (Hopkins et al.,1995) but carcass quality did not impaired. The use of rapeseed meal induces a significant decrease in the proportion of some saturated fatty acids ($C_{12:0}$; $C_{14:0}$) and an increase in unsaturated fatty acids of peripheral fat, mainly $C_{22:2}$ and to a less extent *trans* $C_{18:1}$ (Derycke et al., 1999). Mabon *et al.*, (2000) also reported that inclusion of 25 % rapeseed meal amounting to 2.23 to 4.28 μmoles TGLs/ g DM did not influenced carcass quality apart from a slight increase in the quality of peripheral fat in lambs. Similarly, A linear increase in rabbit carcass fat content was noticed on mustard meal diets (Tripathi et al., 2003). Feeding of low glucosinolate rapeseed meal containing total glucosinolate 9.91 and 21.71 μmol / g DM as sole protein supplement at 25 % in diets of lambs contributed 2.23 or 4.28 μmole glucosinolate/ g diet increased the goitrogens (5- VOT) in muscle, plasma and organs such as thyroid, liver, kidney and lungs (Mabon *et al.*, 2000). The induction level of 5-VOT was corresponding to glucosinolate content of the diets. Glucosinolates intake increased the SCN ions in plasma muscle and target tissues (Mabon *et al.*, 2000; Emanuelson *et al.*, 1993). The SCN concentration did reflected glucosinolate level of the diet however it is correlated to the ingested indolyl glucosinolates (Mabon *et al.*, 2000). Inclusion of 20 % rapeseed meal containing glucosinolates 11 and 20 μmole g^{-1}, deposited 3.8 and 8.3μmole g^{-1} goitrin respectively in loin muscle, the goitrin deposition was affected by the dietary progoitrin concentration. However this goitrin concentration was found to be safe for human. The lung and kidneys accumulates thiocyanates therefore kidney could be considered as the route for SCN excretion. Higher glucosinolate intake induced higher 5-VOT in thyroid, however low glucosinolate intake increase SCN concentration in lungs and kidney (Mabon *et al.*, 2000). Therefore difference in glucosinolate profile among the rapeseed meals induces varying levels of glucosinolates metabolites in animal tissues. The indolyl glucosinolates, precursors of thiocyanates will increase the SCN concentration whereas corresponding intake of progoitrin (2-hydroxy-3-butenyl glucosinolate), will increase the concentration 5-VOT, which is metabolite of progoitrin. It is to be noted that high intake of glucosinolate progoitrin induces an accumulation of 5-VOT mainly in thyroid and lungs. 5-VOT has a high affinity for thyroid (Mabon *et al.*, 2000), which possibly responsible for the morphological and physiological changes of thyroid observed (Mandiki *et al.*, 1999) on glucosinolate ingested diets.

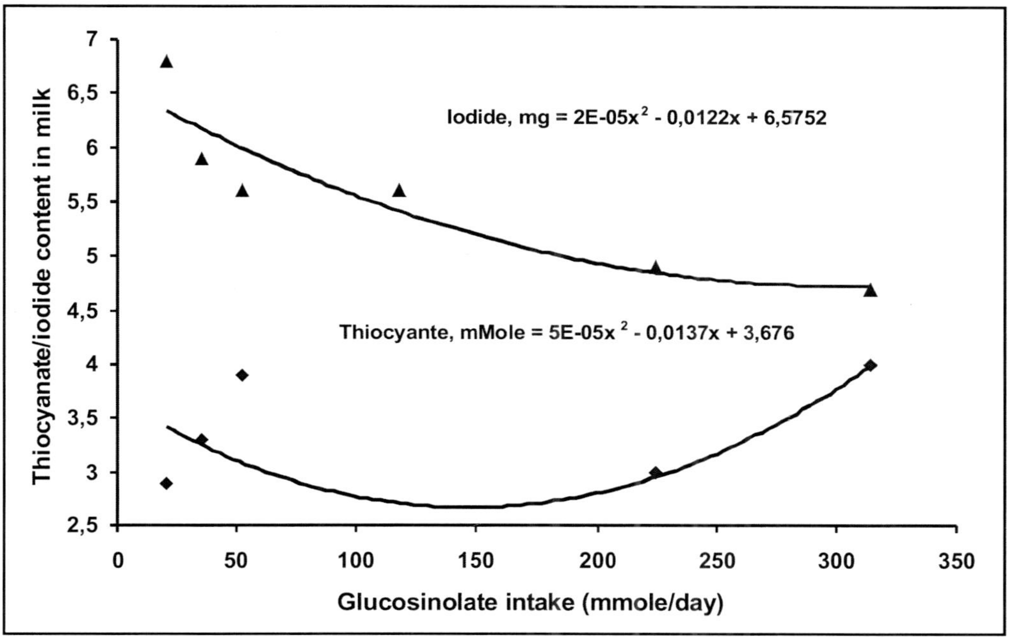

Fig. 5. Effect of Dietary Glucosinolate Intake (mMole day^{-1}) in Dairy Cows on Daily Milk Iodide Content (?, mg) and Thiocyanate (¦, mMole) Secretion

Conclusions and Recommendations

The total glucosinolate content should be taken in to consideration for use of Brassica originated feeds in animal feed formulations. Estimation of Gls derived metabolites in RSM may provide added information with regard its quality and possible influence on animals. A very low glucosinolate (8-10 µmole g^{-1}) rapeseed meal can be included in pig diets up to 12 %, the inclusion of meal should maintain total glucosinolate less than 0.78 µmole g^{-1} diet. Pig diets may contain glucosinolates up to 2 µmole g^{-1} but it requires 1mg added Iodine per kg diet. A10 % RSM can be included in broiler rabbits feeding and the total glucosinolate should not exceed to 7.0 µmole g^{-1} diet. Broiler chicken diets may contain glucosinolate 6 to 10 µmol g^{-1} diet. The rapeseed meal containing glucosinolates 3.6 µmole/g can effectively replace 30 % dietary protein in fish diets. An inclusion of 25 % low glucosinolate rapeseed meal (15 to 33 µmole g^{-1}) contributing a total glucosinolate content 1.5 to 4.22 µmole g^{-1} diet can be possible in growing and fattening lamb diets. Diets with Gls 1.2 to 1.6 µmoles g^{-1} DM shall induce reproductive disturbance in ewes. Among various technologies developed to improve glucosinolate related deleterious effects of RSM, water extraction, the heat and $CuSO_4$ treatments seem effective. Added iodine supplementation 1.0 mg kg^{-1} in pig diets and 500 mg I kg RSM inclusion in ruminant is another effective strategy for effective utilization of RSM.

Glucosinolates Estimation in Biological Samples

Extraction of Glucosinolates in Seed Powder

Place A 0.200 g (+/- 0.1 mg) sample powder was placed in 5 ml polypropylene screw caped tubes. Then place the tubes in an agitating water bath at 75°C for 1 min to inactivate vegetal myrosinase enzyme. ADD 2 ml 70% boiling methanol (V/V) and 200 µl solution of 5 mM glucotropeolin as internal standard, vortexed the tubes to mix methanol and internal standard with sample and placein agitating water bath for 10 min at 75°C, Centrifuge at 10000 × g for 15 min. at 4°C, collected supernatant in 10 ml volumetric flask. Extracted the pellet three more times following the addition of 2 ml 70% boiling methanol (V/V), vortexed to mix methanol, incubation in agitating water bath for 10 min at 75°C, Centrifugation at 10000× g for 15 min. at 4°C, collected supernatant together with the first supernatant and make volume to 10 ml in volumetric flask. Discarded the pellet and store supernatant at -20°C and analysed glucosinolate content was by HPLC with in two weeks.

HPLC Analysis of Glucosinolates

Analyse intact sinigrin as glucosinolate by HPLC following the method described by Helboe et al (1980) and modified by Bjerg and Sorensen (1987).

The separation of sinigrin is carried out using a reverse phase liquid chromatography with a silica graphite octadecyle ion exchange colonne C18 (Colonne Nucleosil N5 C18- 25QS Interchim).The mobile phase consisted of 5mM tetraheptylammonium bromide, 10mM $NaH_2PO_4.H_2O$ and 45 % (v/v) acetonitrile in distilled water, pH adjusted to 7.0 and eluted 60 ml per hr. The sample injection volume is 50µl. Eluted peaks are detected by u. v. absorbance at 229 nm (u.v./ visible detector, Model 168). Under the conditions retention time for sinigrin is 6 minutes. To establish the detector response and column conditions purified sinigrin varing from 0-12 mM concentrations iseluted and analyzed.

Allyl Isothiocyanate Analysis (AITC)

AITC are volatile compound, we used solid-phase microextraction (SPME) in conjugation with gas chromatography (GC). The SPME coupled with GC has been shown to be an effective method for analysis of volatile organic compounds (Steffen and Pawliszyn, 1996; Elisert and Levsen, 1996) including isothiocyanates (Elfoul *et al.*, 2001; Gandini and Riguzzi, 1997). Our analytical method involved exposing an SPME fibre to the headspace of the glass vials containing samples and desorbing the volatile AITC into the fiber coating in the injection port of a gas chromatograph (Elfoul et al., 2001). The vials contained mustard seed powder and whole diluted flora medium. These vials were the stored one after desired incubation periods. The sample vials were thawed at 4°C and 10 or 20µl of 1nM methyl isothiocyanate solution was injected through the butyl-rubber stopper as an internal standard. The vials were thoroughly vortexed to obtain a homogenous solution and warmed at 35°C. After 10 minutes warming, a carboxen/polymethylsiloxane-coated silica fibre (film thickness 75 µm, Supelco, Saint-Quentin-Fallavier, France) fitted into a protective stainless steel needle was introduced through the stopper and exposed to the vial headspace for 10 minutes to allow the absorption of the volatile isothiocyantes. The absorbed compounds were thermally desorbed in the splitless injector port of a gas chromatograph (Perkin Elmer autosystem 1020N) during 20 s at 250°C. The separation of compound was achieved on a non polarCP-sil 8 CB capillary column (30 m × 0.53 mm i.d.; film thickness 1.5 µm, Supelco) with nitrogen as the carrier gas (inlet pressure 70 kPa). The temperature gradient was programmed as follows: basal temperature was stable at 35°C for 2 min, then temperature increased to 50°C at 2°/min, thereafter the rate of temperature increased to 5°C/min until reached to 70°C and finally the temperature was increase rapidly at 20°C/min until 200°C and held for 1 min. Detection was by flam ionization and detector temperature was 250°C. Data were collected and peaks integrated using the logiceil Turbochrom version 4.1 integrator, Perkin Elmer). The peaks identification was based on the identity of retention times with those authentic standards. We used 1 to 30 nM concentrations for MITC and AITC for peaks identifications, under the conditions retention time for MITC and AITC were 6.5 and 13.8 min respectively. Calibration curves were plotted for MITC and AITC using equimolar amounts of 0 to 30 nM, which were injected in the glass vials of identical capacity to sample vials and containing 4 ml sterile semi-synthetic medium. The response coefficient obtained from the ratio of peak area of AITC to peak area of MITC was 1.119. Amount of AITC present in the sample vial was quantified by comparing the peak area of AITC to peak area of MITC in the sample corrected by the response factor, as follows.

Calculations

$RC = (S_{MITC}/\ S_{AITC}) \times$ (concentration of AITC/ Concentration of MITC)

Where, S= Surface area or peak area, RC= Response coefficient

Quantification of AITC

AITC (nM/vial) = ($S_{AITC} \times RC \times$ Quantity of internal standard e.i.10 or 20 n moles) / S_{MITC}.

References

Ahlin, K.A., Emmanuelson, M., Wiktorsson, H., 1994. Rapeseed products from double low cultivars as feed for dairy cow: effects of long term feeding on thyroid function, fertility and animal health. *Acta Veterinaria Scandinavica*. 35, 37-53.

Ahlstrom, B., 1978. By-products from rapeseed protein concentrate processing as feedstuffs.I. Fines to dairy cows. In *Proceedings 5th International Rapeseed Congress*, Malmo, 2, 235-239.

Bennett R. N., Kiddle G., Hick A. J., Dawson G. W., Wallsgrove R. M., 1996. Distribution and activity of microsomal NADPH-dependent monooxygenases and amino acid decarboxylases in cruciferous and non-cruciferous plants, and their relationship to foliar glucosinolate content. *Plant, Cell and Environment*. 19, 801-812.

Bjerg, B. and Sorensen, H., 1987. Quantitative analysis of glucosinolates in oilseed rape based on HPLC of desulphoglucosinolates and HPLC of intact glucosinolates. In Wathelet, J. P. (ed) Glucosinolates in rape seeds; analytical aspects. The Netherlands, 125-130.

Brabban, A. D. and Edwards, C., 1995. The effect of glucosinolates and their hydrolysis products on microbial growth. *Journal of Applied Bacteriology*, 79, 171-177.

Bruel, C., Boujard, T., Escaffre, A. M., Kaushik, S. J., Boeuf, G., Mol, K. E., Geyten, S. Van der., Kühn, E. R., 2000. Dietary low glucosinolate rapeseed meal affect thyroid status and nutrient utilization in rainbow trout (Oncorhynachus mykiss). *British Journal of Nutrition*. 83, 653-664.

Chen, S., Andreasson, E., 2001. Update of glucosinolate metabolism and transport. *Plant Physiology and Biochemistry*. 39, 743-758.

Cheng, D.L., Hosimoto, K. and Uda, Y. 2004. In-vitro digestion of sinigrin and glucotropeolin by single strain of *Bifidobacterium* and identification of digestive products. *Food and Chemical Toxicology*,42, 351-357.

CSWRI., 2002. Long term feeding effect of high glucosinolate mustard meal on rabbits. Annual Report 2002-2003. Central Sheep and Wool Research institute, Avikanagar. India

Danielsen, V., Eggum, B.O., Jensen, S. K., Sorensen, H., 1994. Dehulled protein rich rapeseed meal as a protein source for early weaning piglet. *Animal Feed Science and Technology*. 46, 239-248.

Das, M. M., Singhal, K. K., 2005. Effect of feeding chemically treated mustard cake on growth, thyroid and liver function and carcass characteristics in kids. *Small Rumin. Research* 56, 31-38.

De Brabander, H. F., Verbeke, R., 1982. Determination of oxazolidine-2-thione in biological fluid in the ppb range. *Journal of Chromatography*. 252, 225-239.

De Brabander, H. F., Verbeke, R., 1986. Determination of 5-vinyloxazolidine-2-thione residues in ruminant tissues. Pro. 32th European Meeting of Meat Research Workers. *Gent* 1, 45-48.

De Vos, R. H., Blijleven, W. G. H., 1988. The effect of processing conditions on glucosinolates in cruciferous vegetables. Z. *Lebensm Unters Forsch*. 187, 525-529.

Derycke, G., Mabon, N., Mandiki, S. N. M., Bister, J. L., Wathelet, J. –P., Marlier, M., Paquay, R., 1999. Chemical changes and influence of rapeseed antinutritional factor on lamb physiology and performance 1. Animal performance and thyroid physiology. *Animal Feed Science and Technology*. 81, 81-91.

Elfoul, L., Rabot, S., Khelifa, N., Quinsac, A., Duguay, A., Rimbault, A., 2001. Formation of ally isothiocyanate from sinigrin in the digestive tract of rats monoassociated with a human colonic strain of *Bacteroid thetaiotaomicrone. FEMS Microbiology Letters*, 197, 99-103.

Elisert, R. and Levsen, K. 1996. Solid-phase microextraction coupled to gas chromatography: a new method for analysis of organics in water. *Journal Chromatography* A. 733, 143-157.

Emanuelson, M., Ahlin, K. A., Wiktorsson, H., 1993. Long term feeding of rapeseed meal full fat rapeseed of double low cultivars to dairy cows. *Livestock Production Science*. 33, 199-214.

Emanuelson, M., Murphy, M., Lindberg, J. E., 1991. Effects of heat treated and untreated full fat rapeseed and tallow on rumen metabolism, digestibility, milk composition and milk yield in lactating cow. *Animal Feed Science and Technology*.34, 291-309.

Fenwick, G. R., Spinks, E. A., Wilkinson, A. P., Heaney, R. K., Legoy, M. A., 1986. Effect of processing on the antinutirent content of rapeseed. *Journal of The Science of Food and Agriculture*. 37, 735-741.

Fenwick, R. G., Heaney, R.K. and Mullin, W. J., 1983. Glucosinolates and their breakdown products in food and food plants. *Critical Reviews of Food science and Nutrition*, 18, 123-201.

Gandini, R. and Riguzzi, R. 1997. Headspace solid-phase microextraction of methyl isothiocyanate in wine. *Journal of Agricultural and Food Chemistry*.45, 3092-3094.

Helboe, P., Olsen, O., Sorensen, H., 1980. Separation of glucosinolates by high-performance liquid chromatography. *Journal Chromatography*. 197, 199-205.

Hill, R., 1991. Rapeseed meal in the diet of ruminants. *Nutrition Abstract Review*. 61, 139-155.

Hopkins, D.L., Beattie, A.S., Pirlot, K. L., 1995. Meat quality, carcass fatness and growth of short scrotum lambs grazing either rape or irrigated perennial pasture. *Australian Journal of Experimental Agriculture*. 35, 453-459.

Huang, S., Liang, M., Lardy, G., Huff, H E., Kerley, M. S., Hsieh, F., 1995. Extrusion process of rapeseed meal for reducing glucosinolates. *Animal Feed Science and Technology*. 56, 1-9.

Jensen, S. K., Liu, Y. G., Eggum, B. O., 1995. The effect of heat treatment on glucosinolates and nutritive value of rapeseed meal in rat. *Animal Feed Science and Technology*. 53, 17-28.

Kojima, M. and Ogawa, K., 1971. Studies on the effects of isothiocyanates and their analogues on micro-organism. (1) Effects of isothiocyanates on the oxygen uptake of yeasts. *Journal of Fermentation Technology*, 49, 740-746.

Kyung, K. H. and Fleming, H. P., 1997. Antimicrobial activities of sulfur compounds derived from cabbage. *Journal of Food Protection*, 60, 67-71.

Laarveld, B., Brockman, R. P., Christensen, D. A., 1981a. The effects of Tower and Midas rapeseed meals on milk production and concentration of goitrogens and iodine in milk. *Canadian Journal of Animal Science*. 61, 131-139.

Laarveld, B., Brockman, R.P., Christensen, D.A., 1981c. The goitrogenic potential of Tower and Midas rapeseed meal in dairy cow determined by thyrotropin-releasing hormone test. *Canadian Journal of Animal Science*. 61, 141-149.

Laarveld, B., Brockman, R.P., Christensen, D.A., 1981b. The effect of the level in canola meal concentrate on milk iodine and thiocyanate content and thyroid function in dairy cow. *Canadian Journal of Animal Science*. 61, 625-632.

Liu, Y. G., Zhou, M. Q., Liu, M. L., 1994. A survey of nutrient and toxic factors in commercial rapeseed meal in China and evaluation of detoxification by water extraction. *Animal Feed Science and Technology*. 45, 257-270.

Loft, S., Otte, J., Poulsen, H.E., Sorensen, H., 1992. Influence of intact and treated indoyl glucosinolates on the metabolism in vivo metronidazole and antipyrine in rat. *Food and Chemical Toxicology*. 30, 927-935.

Mabon, N., Mandiki, S. N. M., Derycke, G., Bister, J. -L., Wathelet, J. P., Paquay, R., Marlier, M., 2000. Chemical changes and influence of rapeseed anti-nutritional factor on lamb physiology and performance. 3. Antinutritional factor in plasma and organ. *Animal Feed Science and Technology*. 85, 111-120.

Maheshwari, P.N., Stanley, D.W., Gray, J.I., van de Voort, F.R., 1979. An HPLC method for simultaneous quantitation of individual isothiocyanates and oxazolidinethione in myrosinase digest of rapeseed meal. *Journal of American Oil and Chemical Society*. 56, 837.

Mandiki, S. N. M., Derycke, G., Bister, J. -L., Mabon, N., Wathelet, J. P., Marlier, M., Paquay, R., 2002. Chemical changes and influence of rapeseed antinutritional factor on gestating and lactating ewes. 1. Animal performances and plasma hormones and glucose. *Animal Feed Science and Technology*. 98, 25-35.

Mandiki, S.N.M., Mabon, N., Derycke, G., Bister, J. -L., Wathelet, J. P., Paquy, R., Marlier, M., 1999. Chemical Changes and influences of rapeseed antinutritional factors on lamb physiology and performance. 2. Plasmatic substances and activity of the thyroid. *Animal Feed Science and Technology*. 81, 93-103.

Miñkowski, K., 2002. Influence of dehulling of rapeseeds on chemical composition on meal. *Animal Feed Science and Technology*. 96, 237-244.

Mithen, R. F., Lewis, B. G. and Fenwick, G. R., 1986. In vitro activity of glucosinolates and their products against *Laptoshaeria maculans*. *Transactions of the British Mycological Society*, 87, 433-440.

Papas, A., Ingalls, J. R., Campbell, L. D., 1978. Effect of Tower and 1821 rapeseed meals and tower gum on milk yield, milk composition and blood parameters of lactating dairy cows. *Canadian Journal of Animal Science*. 58, 671-679.

Papas, A., Ingalls, J. R., Campbell, L. D., 1979. Studies on the effect of rapeseed meal on thyroid status of cattle, glucosinolate and iodine content of milk and other parameters. *Journal of Nutrition*. 109, 1129-1139.

Pattanaik, A. K., Khan, S. A., Mohanty, D. N., Varshaney, V. P., 2004. Nutritional performance, cilinical chemistry and semen characteristics of goats fed mustard (*Brassica juncea*) cake based supplement with or without iodine. *Small Ruminant Research*. 54, 51-59.

Quinsac, A., Lessire, M., Krouti, M., Ribaillier, D., Coöc, J. P., Fauduet, H., Rollin, P., 1994. Improvement in the nutritive value of high and low glucosinolate rapeseed meal by aqueous extraction. *Animal Feed Science and Technology*. 48, 265-272.

Reddi, N. M., Rajan, A., 1985. Pathology of the female reproductive organ in experimental hypothyroidism in goats. *Kerala Journal of Veterinary Science*. 16, 119-126.

Rihova, E., 1982. Mutagenic effects of allylisothiocyanates in *Escherichia coli* WP-67. *Folia Microbiologica*, 27, 25-31.

Rowan, T.G., Lawrence, T.L.J., Kershaw, S.J., 1991. Effects of dietary copper and iodine on glucosinolate concentration in ileal digesta and in faeces of growing pigs given diets based on rapeseed meals. *Animal Feed Science and Technology*. 35, 247-258.

Smithard, R. R., Eyre, M. D., 1986. The effect of dry extrusion of rapeseed with other feedstuffs upon its nutritional value and anti-thyroid activity. *Journal of The Science of Food and Agriculture*. 37, 136-140.

Sot-Blanco, B., Górnaik, S.L., 2003. Milk transfer of cyanide and thiocyanate: cyanide exposure by lactation in goat. *Veterinary Research*. 34, 213-220.

Stanford, K., Wallins, G. L., Smart, W. G., McAllister, A. A., 2000. Effect of feeding canola screenings on apparent digestibility, growth performance and carcass characteristics of feedlot lambs. *Canadian Journal of Animal Science*. 80, 355-365.

Steffen, A. and Pawliszyn, J. 1996. analysis of flavor volatiles using headspace solid-phase microextraction. *Journal of Agricultural and Food Chemistry*. 44, 2187-2193.

Tanii, H., Takayasu, T., Higashi, T., Leng, S., Saojoh, K., 2004. Allylnitrile: generation from cruciferous vegetables and behavioral effect on mice of repeated exposure. *Food and Chemical Toxicology*, 42, 453-458.

Tolonen, M., Rajaniemi, S., Pihlava, J. M., Johansson, T., Saris, P. E. J. and Ryhänen, E. L., 2004. Formation of nisin, plant-derived biomolecules and antimicrobial activity in starter culture fermentations of sauerkraut. *Food Microbiology*, 21, 167-179.

Tripathi, M. K., Agrawal, I.S., Sharma, S.D., Mishra, D.P., 2001. Effect of untreated, HCl treated or copper and iodine supplemented high glucosinolate mustard (*Brassica juncea*) meal on nutrient utilisation, liver enzymes, thyroid hormones and growth of calves. *Animal Feed Science and Technology*. 92, 73-85.

Tripathi, M. K., Tyagi, A. K., Karim, S. A., 1996. Effect of mustard cake and its glucosinolates on milk production, milk composition and blood biochemistry of lactating ewes. *World Review of Animal Production*. 31, 32-34.

Tripathi, M.K., Agrawal, I. S., 1998. Effect of HCl treatment with heat on glucosinolates content of rapeseed-mustrad meal. Proc. International Conference on Sustainable Animal Production, Health and Environment. HAU, Hisar, India.

Tripathi, M.K., Agrawal, I.S., Sharma, S.D., 2000. Effect of physio-chemical treatments on glucosinolates content of various rapeseed-mustard meals. *Indian Journal of Animal Nutrition*. 17, 211-216.

Tripathi, M.K., Agrawal, I.S., Sharma, S.D., Mishra, D.P., 2001. Effect of substitution of soybean meal with treated or untreated high glucosinolate mustard (*Brassica juncea*) meal on intake, digestibility, growth performance and body composition of calves. *Animal Feed Science and Technology*. 94, 137-146.

Tripathi, M.K., Mishra, A.S., 2008. Glucosinolate in animal nutrition: A review. *Animal Feed Science and technology*. 132, 1-27.

Tripathi, M.K., Mishra, A.S., Misra, A.K., Mondal, D., Karim, S.A., 2001. Effect of substitution of groundnut with high glucosinolate mustard (*Brassica juncea*) meal on nutrient utilisation, growth, vital organ weight and blood composition of lambs. *Small Ruminant Research*. 39, 261-267.

Tripathi, M.K., Mishra, A.S., Misra, A.K., Prasad, R., Mondal, D., Jakhmola, R. C., 2003. Effect of graded levels of high glucosinolate mustard meal (Brassica juncea) meal inclusion on nutrient utilisation, growth performance, organ weight and carcass composition of growing rabbits. *World Rabbit Science*. 11, 211-226.

Tripathi, M.K., Santra, A., Chaturvedi, O.H ., Karim, S.A., 2004. Effect of sodium bicarbonate supplementation on ruminal fluid pH, feed intake, nutrient utilisation and growth of lambs fed high concentrate diets. *Animal Feed Science and Technology*. 111, 27-39.

Tripathi, M.K., Tyagi, A. K., Karim, S. A., 1999. Performance of weaner lambs maintained on mustard cake and groundnut meal based diets. *Indian Journal of Animal Science*. 69, 91-92.

Tyagi, A.K., 2002. Influence of water soaking of mustard cake on glucosinolate hydrolysis. *Animal Feed Science and Technology*. 99, 215-219.

Wallig, M. A., Belyea, R. L., Tumbleson, M.E., 2002. Effect of pelleting on glucosinolates content of Crambe meal. *Animal Feed Science and Technology*. 99, 205-214.

Wilson, J.G., 1975. Hypothyroidism in ruminants with special reference to foetal goiter. *Veterinary Records*. 97, 161-164.

□□□

Chapter 35

Nutritional Significance and Antimethanogenic Potential of Feed Tannis

M. K. TRIPATHI AND DEEPIKA CHATURVEDI
Central Sheep and Wool Research Institute, Avikanagar-304501 (Rajasthan), India

Introduction

What are the Tannins: Tannins are astringent polyphenols occur naturally in many plants, their products and by-products. Chemical nature and dose of tannins are the deciding factors for their diverse biological activity in safe and efficient use as feed for animals. Tannins can be defined by their physical or biological activities, or by their chemical structure. According to general definition of Horvath (1981): Tannins are 'water-soluble polymeric phenolics of high molecular weight containing phenolic hydroxyls and carboxyls groups to form effectively strong complexes with protein and other macromolecules under the particular environmental conditions. Chemically, tannins are of two classes: (i) condensed tannins (CT) or proanthocyanidins, which are polymers of flavonoid units joined by highly stable covalent carbon–carbon bonds and (ii) hydrolysable tannins (HT), which contain a carbohydrate with hydroxyl groups partially or totally esterified with phenolic acids such as gallic acid (gallotannins) or ellagic acid (ellagitannins).

Tannins Protein Interactions and Nutritional Benefits

Tannins are linked to proteins by hydrogen bonds between the hydroxyl groups of tannins, and the amine and carboxyl groups of proteins. Aromatic cycles of tannins and amino acids can also interact to form van der Waals bonds. When the protein/tannin ratio is low, tannins cover the protein with their hydrophilic groups. The 'protein–tannin' complexes then become more hydrophobic in their outer layer, and precipitate in water solution. When the protein/tannin ratio is high, tannins form chemical bridges between proteins and precipitate them (Haslam, 1988). The same property was explored as a way to slow dietary protein breakdown in the rumen and enhance the amino-acid bioavailability in the small intestine, and reduce ruminal NH_3 production and nitrogen (N) excretion in urine (Aerts *et al.*, 1999; Barry and McNabb, 1999). Tannins during the process of the digestion of protein form reversible complexes with proteins, which is stable in rumen. However, tannins-protein complex can dissociate post-ruminally in response to the extremes of pH. The low pH in the abomasum as well as the high pH in the small intestine can stimulate dissociation. CT have been shown to readily precipitate proteins within the pH range 3-7 but not outside of that range (Diaz-Hemanderz *et al*, 1997). This understanding forms the logic behind the hypothesis that CT can improve the total availability of protein in forages by reducing the wasteful deamination of protein in the rumen in diets where rumen degradable protein is in excess of microbial requirements. This impairment of ruminal protein degradation improves the animal's nutritional status and reduces the amount of N released into the environment. In vitro studies indicated that tannin–protein complexes are formed in the pH range 6 to 7 prevailing in the rumen, and are destroyed at pH, 3.5 in the abomasum and at 7 in the small intestine, thus making proteins available for gastric and pancreatic digestion (Jones and Mangan, 1977). The nutritional effects of tannins vary widely owing to their great chemical diversity. The type and concentration of tannins hinders the 'ideal' evolution of the 'tannin–protein' complex stability according to pH in the digestive tract of animals. Tannins from one type of forage can precipitate more protein than another (on a g/g basis), therefore, tannins of one source are more astringent than others. This raises the possibility that 'tannin-protein complex' containing particularly astringent CT may not dissociate post-ruminally as completely as those with less astringent CT. Consequently, some CT may not improve the supply of digestible protein post-ruminally as much as others. It has been observed that, stability of tannin-protein interactions with *Leucaena leucocephala*, does not occur with *Lotus corniculatus* or *L. pedunculatus*, for which the complexes are too stable, and remain undissociated and undigested in the small intestine (Mueller-Harvey,

2006). When released in the intestines, tannins can have detrimental effects if they bind endogenous animal proteins such as enzymes, mucus or intestinal mucosal cells. For example, quebracho (Schinopsis lorentzii) tannins have no effect on ruminal by-pass of proteins, but can dramatically reduce protein absorption in the small intestine of ruminants (Min *et al.*, 2003; Mueller-Harvey, 2006). The increase in faecal N usually observed with tanniferous plants may originate in undigested complexes formed with dietary and (or) endogenous proteins. The effects of tannins are more pronounced when they are directly administered in the rumen rather than added to the feed. This stresses the capacity of some salivary proteins, especially proline-rich glycoproteins, to bind tannins (Shimada, 2006) and prevent their further activity in the rumen. This action on salivary proteins explains the sensation of astringency recognised during the ingestion of tannin-rich feeds. It also indicates that proteins with high proline content, rich in non-polar amino acids, with high molecular weight and with an open tertiary structure, are more susceptible to interaction with tannins (Mehansho *et al.*, 1987). Owing to their astringent taste, feeds rich in tannins have poor palatability. The lower feed intake, associated with a lower rate of digestion observed in some tannin-rich plants (Stienezen *et al.*, 1996) can offset the positive effects of tannins on N digestion, resulting in a net negative effect on animal production.

Antimicrobial Effects and Microbial Interaction of Tannins

The possible antimicrobial effect of tannins, in both the rumen and the intestine, may modify feed digestion in ruminants. For example, *L. corniculatus* tannins lower the concentrations of bacteria in the rumen and decrease the amounts of bacterial proteins flowing to the intestine. Thus, the beneficial effect of tannins on the intestinal flow of ruminal undigested feed proteins can be cancelled by the negative effect on intestinal flow of microbial proteins. The fact that proteolytic bacteria are sensitive to *L. corniculatus* tannins (Min *et al.*, 2002), and that the binding strength of protein complexes formed with this source of tannins is high, explains why *L. corniculatus* can negatively affect the N digestion of ruminants.

Tannins can inactivate rumen microbial enzymes and adversely affect ruminal digestion (Barry and McNabb, 1999). CTs have been shown to cause a shift in the partitioning of nutrients during fermentation in the rumen so that a higher proportion is redirected to microbial mass synthesis at the expense of VFA production (Makkar, 2003), thus increasing the efficiency of microbial protein synthesis. Dietary CTs induce a polysaccharide slime by the rapid release of soluble protein into ruminal fluid that capture of ruminal gases thus effective against ruminal bloat. Severity of bloat legumes is thus lowered in the presence of CTs since gas formation and microbial protein degradation are both decreased (Min et al., 2005). Tannins can also destabilise the structure of plant protein foams by altering surfactant forces in the ruminal liquid phase (Tanner *et al.*, 1995). The structure and content of the tannins in the diet also affect the interactions with dietary and endogenous proteins, and ability of microbes and fungi to attach to substrates within the gastro-intestinal tract (Bento *et al.*, 2005). Such interactions are likely to have a substantial effect on the feed degradation as well as on enzyme activities in rumen.

Feed Safety of Tannins

Tannin toxicity in animals varies with the concentration, chemical structure and type of tannins present. Compared to CTs and HTs usually considered more toxic than CTs (McSweeney *et al.*, 2001; Odenyo *et al.*, 2003; Waghorn and McNabb, 2003). Toxicity of HTs is due to intestinal absorption of small molecular weight substances originating from their microbial degradation, whereas CTs, are not extensively degraded in rumen by rumen microbes (Makkar *et al.*, 1995) and then absorbed (Terrill *et al.*, 1994). Since the activity of tannins depend on their molecular structure, and especially on their molecular weight hence optimum level of tannins are difficult to derive in animal feeding (Meagher *et al.*, 2004). An optimal dose of tannin, even for the same source of tannins, is difficult to determine, as the effect of tannins with source and content of feed protein varies. In general the dietary concentrations of CTs must be less than 50 g/kg dietary DM to have beneficial effects (McMahon *et al.*, 2000; Herva´s *et al.*, 2003; Min *et al.*, 2003). These values obtained with *Lotus* tannins, but are not applicable to all sources of tannins. For example, large amounts of CTs (80 g/kg DM) can promote animal production when supplied from sulla (*Hedysarum coronarium*) or sainfoin (*Onobrychis viciifolia*), whereas doses as low as 25 g/kg DM have a strong negative effect on animal growth rate when the tannins are from carob pulp (Mueller-Harvey, 2006). Chestnut tannins had no significant effect on feed intake, feed efficiency, daily gain or fattening period of lambs from 15 to 25 kg when tested at 21 g/kg soya bean (3.9% of crude protein) (Frutos *et al.*, 2004) but same tannins at 0.4% of crude protein of wilted grass silage increased duodenal non-ammonia N flow and intestinal N digestibility. Further, 80 g/kg of chestnut tannins sprayed onto hay significantly decreased DM digestibility in sheep and goats (Zimmer and Cordesse, 1996). Lactating dairy cows diet containing tannins 15 to 29 g/kg of peanut skins showed positive effects on production or live-weight (West *et al.*, 1993). The optimal concentration of CTs to reduce rumen proteolysis is about 400 mg CTs per ml (Aerts *et al.*, 1999) but this concentration reduced growth of several bacterial strains (Molan *et al.*, 2000). If the condensed tannin level in diet vary between

50 and 100 g/ kg DM, it depress live weight gain, milk yield (Makkar, 2003; Wang *et al.*, 1996a,b) and reproduction (Barry *et al.*, 2001) at 20 to 40 g CT/ kg diet DM, while 10-11 g CT/ kg DM had no deleterious effect (Yildiz *et al.*, 2005). The Oak CT at 11.2 g/kg in lamb diet DM increased microbial nitrogen supply though CP digestibility and N retention reduced but fermentation characteristics did not change, showing that 10-11 g CT in ruminant diet DM can be considered safe for animal feeding.

Adaptation to Tanniferous Feeds

Many tropical plants contain very high levels of tannins and often grazed/ browsed profusely by ruminants to meet their nutrient need. Under these conditions, rumen microbes can develop defence mechanisms to resist the toxic effects of tannins. They protect their cell membrane by external deposition of lipid layers, secrete extracellular polysaccharides (Brooker *et al.*, 2000) and (or) produce glycoproteins with high tannin affinity (Nicholson *et al.*,1986). Rumen micro-organisms of tropical ruminants are able to degrade HTs (Ephraim *et al.*, 2005) and even CTs (Odenyo *et al.*, 1999), and are more tolerant to higher levels of tannins. Differences between ruminants in their ability to degrade tannins have also occurred (McKie *et al.*, 2004). Micro-organisms that have been shown to develop adaptive responses to the presence of tannins includes *Streptococcus caprinus* which is able to grow and increases the activity of gallate decarboxylase and synthesis of extracellular polysaccharides as a response to the presence of tannins (Brooker *et al.*, 1999). Micro-organisms such as *Ruminococcus albus* are able to adapt and proliferate in the presence of phenolics with time (Akin et al., 1988), *Selenomonas ruminantium* degrades TA into gallic acid (Skene and Brooker, 1995).

Tannins Detoxification

Adaptation to high tannin diets and / or microbial detoxifying capabilities of ruminants sometimes cannot overcome the tannin toxicity. The preventive measures to minimize deleterious effect on animal, treatments of forages are necessary to eliminate or degrade tannins. The treatment of tanniferous feeds includes oxidation of phenolics at alkaline pH, storage of dried chopped plants or foliage, chemical treatments and PEG incorporation. Polyethyleneglycol, and to a lesser extent polyvinylpyrrolidone, compete with proteins to complex tannins. The higher affinity of tannins for PEG than for proteins, which protects proteins from an excessive complexation, has been extensively used in tropical and arid countries (Getachew *et al.*, 2001), but the higher the level of dietary proteins, the weaker is the effect of these tannin-complexing agents (Makkar and Becker, 1996). Soaking of tanniferous foliage in water or dilute acid for 24 h removes sizable amount of tannins that improve feed efficiency and animal production (Wina *et al.*, 2005). Wood ash treatment is another effective method for tannins deactivation, high alkalinity of wood ash solution at 180 g/kg exerts optimum tannin deactivation and effective to improve quality of tanniferous foliages (Ben Salem *et al.*, 2005a, b). An aqueous urea (4 g urea in 10 ml water/ 100 g DM) solution treatment for 30 days is quite effective in increasing N content of fodder and reduce significant amount of total phenolics, total tannins and condensed tannins (Vitti *et al.*, 2005), thus improving fodder value for animals.

Tannins and Health Benefits

Recent research has demonstrated that CTs have significant post-ruminal effects, they can be an alternative to chemical anthelmintics to control gastro-intestinal parasitic nematodes (Butter et al., 2000; Athanasiadou and Kyriazakis, 2004; Hoste *et al.*, 2006) in grazing animal production systems. Farmers traditionally used plants for deworming animals, and also that animals select certain plants for self-medication during grazing (Hutching *et al.*, 2003). Tannins have been shown to act through a direct antiparasitic activity, but they may also stimulate host resistance as a result of an increase in intestinal protein supply (Coop and Kyriazakis, 2001). The anthelmintic efficacy of tannins is closely related to their chemical structure (Hoste *et al.*, 2006). Grazing pastures with mixtures of grasses and legumes rich in tannins could be a practical way of preventing nematode parasitism in ruminants (Ramirez-Restrepo and Barry, 2005). Forage CT have ability to inhibit the development of nematodes egg, increase fecal shedding of eggs, breaks the life cycle of sheep nematodes and reduce the contamination of pasture (Min *et al.*, 2003).

Relationship between Tannins and Gas/ Methane Production

Gas production used as a measure of microbial activity, gives a good estimate of the ability of the adhered microbial population to degrade the feed in rumen, a high relationship found between gas production and degradation of DM (Getachew *et al.*, 1998).The inclusion of tannins in ruminant diets decreased the amount of gas produced (m) per unit of adherent microorganisms and related with the effect of tannin on the microbial population pattern attached to feed. Mimosa tannin exerted their effect by reducing microbial numbers and MT may have changed the pattern of the adherent population towards non-gas producing bacterial strains (Bento *et al.*, 2005) and/or tannin may have inhibited the activity of the microbes attached to cellulose. This may reduce the degradation of cellulose.

Table: Total Phenol and Tannin (g/ kg) Content of Different Feeds and Fodders

	Total phenol	Total tannin	Condensed tannin	HT	Reference
Acacia villosa leaves	173.0	162.0	12.7		Wina *et al.*, 2005
Acacia cyanophylla leaves	24.8	16.6	27.3		Ben Salem *et al.*, 2005a
Acacia cyanophylla leaves	39.2	33.1	54.5		Ben Salem *et al.*, 2005b
Acacia angustissima leaves	13.8-15.5	12.4-14.1	0.52-0.70		Mlambo and Makkar, 2005 (PPC, 181-262mg/g)
Acacia karoo leaves	7.5	6.5	1.6		Mlambo and Makkar, 2005 (PPC, 96mg/g)
Quercus coccifera leaves	43.7	38.0	19.9		Ben Salem *et al.*, 2005b
Anadenanthera macrocarpa leaves	119	113	11		Vitti *et al.*, 2005
Astronium urundeuva leaves, wet	167	154	132		Vitti *et al.*, 2005
Astronium urundeuva leaves, dry	201	189	83		Vitti *et al.*, 2005
Capparis flexuosa leaves dry	63	56	2		Vitti *et al.*, 2005
Mimosa hostilis leaves, wet	182	165	107		Vitti *et al.*, 2005
Mimosa hostilis leaves, dry	157	140	49		Vitti *et al.*, 2005
Sida cordifolia leaves, wet	114	8	0.2		Vitti *et al.*, 2005
Sida cordifolia leaves, dry	19	13	0.4		Vitti *et al.*, 2005
Manihot pseudoglaziovii leaves, wet	42	34	18		Vitti *et al.*, 2005
Albizia procera leaves dry	119	61.2	11.0	24.3	Alam *et al.*, 2005 (PPC, 133mg/g)
Albizia procera leaves wet	66.4	26.5	5.1	13.3	Alam *et al.*, 2005 PPC (128 mg/g)
Lespedeza striata leaves			151		Animut *et al.*, 2008
Quercus hartwissiana leaves	70.5	63.7	11.2	0.6	Yildi *et al.*, 2005
Prosopis cineraraia leaves	-	-	91	3	Vaithiyanathan *et al.*, 2007
Carissa spinarum leaves	75.0	54.0	54.0	0.0	Singh *et al.*, 2005 (PPC, 430 mg/g)
Ficus roxburghii leaves	28.0	23.0	23.0	0.0	Singh *et al.*, 2005 (PPC, 194 mg/g)
Hippophae rhamnoides leaves	122.0	103.0	14.0	89.0	Singh *et al.*, 2005 (PPC, 593 mg/g)
Leucaena leucocephala leaves	41.0	28.0	14.0	14.0	Singh *et al.*, 2005 (PPC, 122 mg/g)
Melia azedarach leaves	2.20	0.57	0.26	0.31	Singh *et al.*, 2005 (PPC, 15 mg/g)
Quercus incana leaves	57.5	50.0	34.0	16.0	Singh *et al.*, 2005 (PPC, 424 mg/g)
Robinia pseudoacacia leaves	18.0	9.8	9.8	0.0	Singh *et al.*, 2005
Zizyphus jujube leaves	67.0	54.0	37.0	17.0	Singh *et al.*, 2005 (PPC, 307 mg/g)
Zizyphus nummularia leaves	5.0	48.0	47.0	1.0	Singh *et al.*, 2005 (PPC, 296 mg/g)
Salix spp.	36.53		38.3	2.85	Pitta *et al.*, 2007

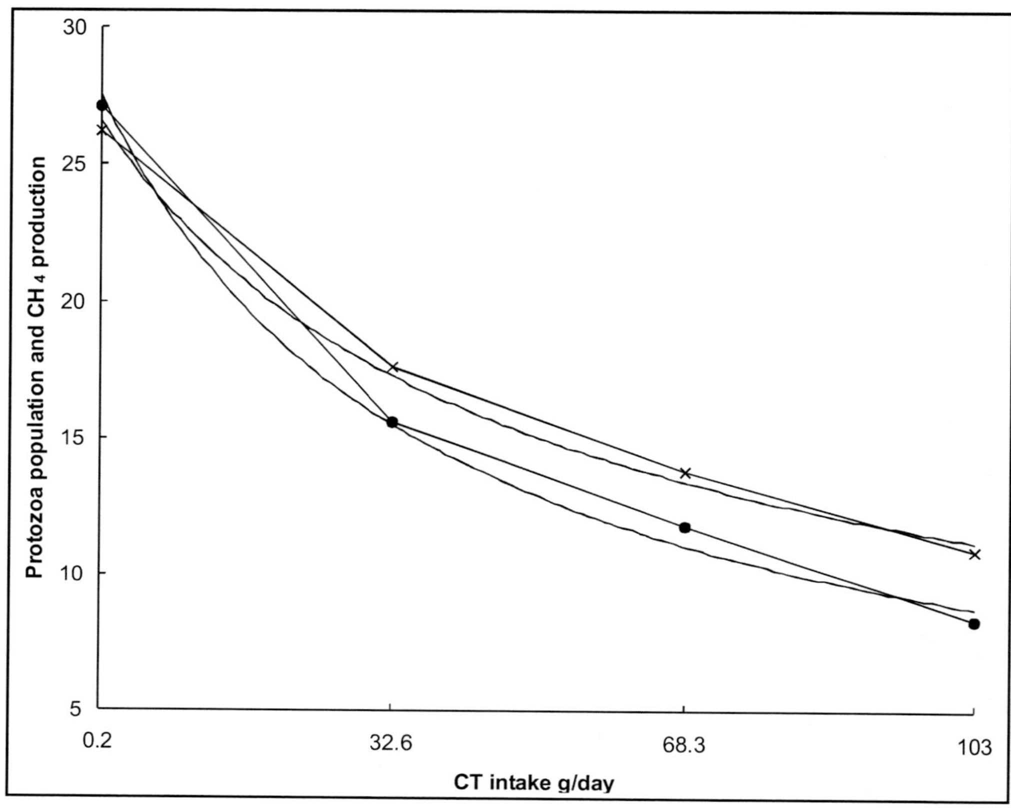

Fig 1. Relationship of condensed tannins (CT) intake with methane and protozoa population. CH_4 (x, l/day) = 26.586 x (g CT intake)$^{-0.6213}$, $R^2 = 0.9948$; Protozoa (?, 10^5/ ml) = 27.528 x (g CT intake)$^{-0.8274}$, $R^2 = 0.9909$

The CT-containing forage decrease CH_4 emission by goats regardless of its feeding level, although the effect per unit of *Lespedeza striata* increased with decreasing *L. striata* levels. The effect of CT on CH_4 emission appeared to have direct impact on activity of methanogenic bacteria, although alterations of protozoal activity could have been involved. Animut *et al.* (2008) suggested that relatively low dietary levels of CT could be employed to lessen CH_4 emission without a marked detrimental effect on other conditions such as total tract protein digestion. Several in vivo experiments have shown that the amount of methane produced per unit of digestible dry matter (DM) intake is reduced (20-30%) when ruminants are fed forages rich in CTs such as *L. pedunculatus* or *L. corniculatus* (Pinares-Patino *et al.*, 2003; Ramirez-Restrepo and Barry, 2005), but no clear indication was given of the mechanisms involved. However, tannins were strongly suspected since addition of

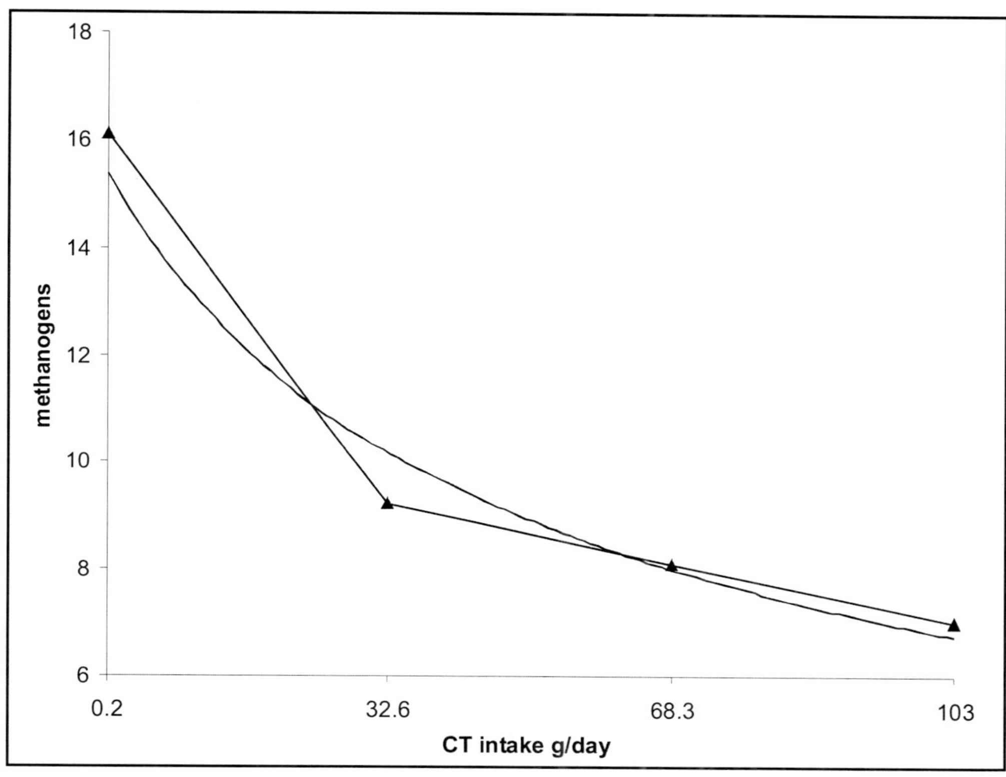

Fig 2. Relationship between condensed tannins (CT) intake and Methanogens population (10^5 ml^{-1}) = 15.358x (g CT intake$)^{-0.595}$, R^2 = 0.9649

polyethylene glycol (PEG), a tannin-binding reagent, caused recovery of methane production. Tavendale *et al.* (2005) stressed that CTs extracted from *L. pedunculatus* have a direct inhibitory effect on archaea methanogens, which impairs ruminal fermentation. Toxicity of HTs and CTs towards methanogens was also demonstrated in methanogenic digesters (Field *et al.*, 1989). CT of *L. corniculatus* at level of 25.9 g (Woodward *et al.*, 2001) and CT of *Lespeza cuneata* at 17 g/ kg DM (Puchala *et al.*, 2005) reduced methane emission but fibre digestion depressed. Supplementation of concentrates with high tannin diets fibre digestion improves and methane production decreases (de Oliveira *et al.*, 2007). Though CTs reduce methanogen numbers but eliminating rumen methanogens may reduce the performance of the rumen and of the animal. The process of methanogenesis serves as the exhaust system of rumen fermentation in which the <1% of rumen organisms which produce methane, eliminate hydrogen gas which would otherwise inhibit the activity of the bulk of other organisms that do not produce methane. Elimination of the

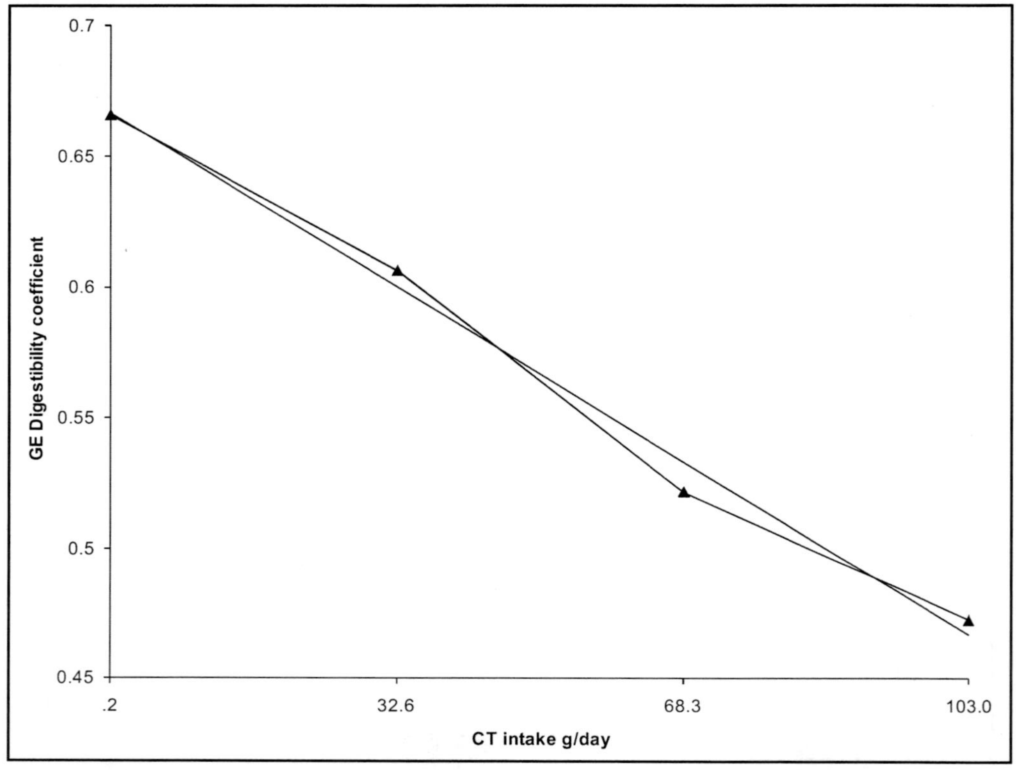

Fig 3. Relationship between condensed tannins (CT) intake and GE Digestibility Coefficient = -0.0666×g CT intake + 0.733, $R^2 = 0.9914$

rumen methanogen populations without establishing alternate hydrogen users will cause the accumulation of hydrogen and have catastrophic effects on inter-related organisms, stopping fermentation and causing livestock to stop eating. Approximately 37% of rumen methanogenesis is due to methanogens living in or on rumen protozoa (Finlay *et al.*, 1994), there has been no comprehensive assessment of the species of protozoa nor methanogens involved. Considering methane production in rumen, reduction of organisms which provide hydrogen to methanogens, or increase of those micro-organisms which compete for hydrogen with methanogens, is likely to improve the stability of methane emissions reduction (Kreuzer 1986). Methane producing organisms normally remove hydrogen gas from the rumen by converting it to methane. Hydrogen is a waste product of fermentation and if hydrogen accumulates, rumen digestion is suppressed (Wolin *et al.*, 1997). The goal in addressing livestock methane emissions then, is not simply stopping methane emissions, but rather, redirecting hydrogen into more beneficial pathways.

Conclusions and Recommendations

Though tannins have shown anti-methanogenic properties still sound evidences of animal responses with regard of reduced methane emissions require. If CT reduce methane emission in ruminant diets, simultaneously depressed fibre degradation was achieved thus it is not evident that reduced methane emission is due to CT or depressed digestion of fibre. Considering the recommendations 5 g CT/kg DM reduce blot cases, 22 g CT improve wool production and 20- 40 g CT/kg DM reduce ruminal degradation of protein, improve live weight gain, milk yield and reproduction. while 10-11 g CT/ kg DM increased microbial nitrogen supply though CP digestibility and N retention reduced but fermentation characteristics did not change, showing that 10-11 g CT in ruminant diet DM can be considered safe limit for animal feeding.

References

Aerts, R.J., Barry, T.N. and McNabb, W.C. 1999. Polyphenols and agriculture: beneficial effects of proanthocyanidins in forages. *Agriculture, Ecosystems and Environment*. 75, 1-12.

Akin, D.E., Rigsby, L.L., Theodorou, M.K., Hartley, R.D. 1988. Population change of fibrolytic rumen bacteria in the presence of phenolic acid and plant extracts. *Animal Feed Science and Technology*. 19, 261-275.

Alam, M.R., Kabir, A.K.M.A., Amin, M.R. and McNeill, D.M. 2005. The effect of calcium hydroxide treatment on the nutritive and feeding value of *Albizia procera* for growing goats. *Animal Feed Science and Technology*. 112, 135-148.

Animut, G., Puchala, R., Goetsch, A.L., Patra, A.K., Sahlu, T., Varel, V.H. and Wells, J. 2008. Methane emission by goats consuming diets with different levels of condensed tannins from lespedeza (*Lespedeza striata*). Animal Feed Science and Technology (in press).

Athanasiadou, S. and Kyriazakis, I. 2004. Plant secondary metabolites: antiparasitic effects and their role in ruminant production systems. The *Proceedings of the Nutrition Society*, 63, 631-639.

Barry, T.N. and McNabb, W.C. 1999. The implications of condensed tannins on the nutritive value of temperate forages fed to ruminants. *The British Journal of Nutrition*, 81, 263-272.

Barry, T.N., McNeill, D.M. and McNabb, W.C. 2001. Plant secondary compounds; their impact on forage nutritive value and upon animal production. Proc. XIX International Grassland Congress, Brazil, 445-452.

Ben Salem, H., Abidi, S., Makkar, H.P.S. and Nefzaoui, A. 2005a. Wood ash treatment, a cost-effective way to deactivate tannins in *Acacia cyanophylla* Lindl. foliage and to improve digestion by Barbarine sheep. *Animal Feed Science and Technology*, 112, 93-108.

Ben Salem, H., Saghrouni, L. and Nefzaoui, A. 2005b. Attempts to deactivate tannins in fodder shrubs with physical and chemical treatments. *Animal Feed Science and Technology*, 112, 109-121.

Bento, M.H.L., Acamovic, T. and Makkar, H.P.S. 2005. The influence of tannins, pectins and polyethyleneglycol on attachment of 15N-labelled rumen microorganism to cellulose. *Animal Feed Science and Technology*, 112, 41-57.

Brooker, J.D., O'Donovan, L., Skene, I. and Sellick, G. 2000. Mechanisms of tannin resistance and detoxification in the rumen. In: Brooker, J.D. (Ed.), Tannins in livestock and human nutrition (In: ACIAR Proc., Adelaide, Australia, pp. 117-122.

Brooker, J.D., O'Donovan, L., Skene, I. and Sellick, G., 1999. Mechanism of tannin resistance and detoxification in the rumen. In: Brooker, J.D. (Ed.), Tannins in Livestock and Human Nutrition, ACIAR Proc., Adelaide, pp. 127-132.

Butter, N.L., Dawson, J.M., Wakelin, D. and Buttery, P.J. 2000. Effect of dietary tannin and protein concentration on nematode infection (*T. colubriformis*) in lambs. *Journal of Agricultural Science, Cambridge* 134, 89-99.

Coop, R.L. and Kyriazakis, I.K. 2001. Influence of host nutrition on the development and consequences of nematode parasitism in ruminants. *Trends in Parasitology*, 17, 325-330.

de Oliveira, S.G., Berchielli, T.T., Pedrira, M. –dos S., Primavesi, O., Frighetto, R. and Lima, M.A. 2007. Effect of tannin levels in sorghum silage and concentrate supplementation on apparent digestion and methane emission in beef cattle. *Animal Feed Science and Technology*, 135, 236-248.

Diaz-Hernandez, A., Nixon, M.J., Ball, F.M., Leng, R.A. and Rowe, J.B. 1997. Protein-tannin complexes are susceptible to proteolytic degradation. *Proceeding of the NewZeland Animal Production*, 57, (116) Abstract.

Ephraim, E., Odenyo, A. and Ashenafi, M .2005. Screening for tannin degradation by rumen and faecal samples of wild and domestic animals in Ethiopia. *World Journal of Microbiology and Biotechnology*, 21, 803-809.

Field, J.A., Kortekaas, S. and Lettinga, G. 1989. The tannin theory of methanogenic toxicity. *Biological Wastes*, 29, 241-262.

Finlay, B.J., Esteban, G., Clarke, K.G., Williams, A.G., Embley, T.M. and Hirt, R.P. 1994. Some rumen ciliates have endo-symbiotic methanogens. *FEMS Microbiology Letters*, 117, 157-162.

Frutos, P., Raso, M., Hervas, G., Mantecon, A.R., Perez, V. and Giraldez, F.J. 2004. Is there any detrimental effect when a chestnut hydrolysable tannin extract is included in the diet of finishing lambs?. *Animal Research*, 53, 127-136.

Getachew, G., Blumell, M., Makkar, H.P.S. and Beker, K. 1998. *In vitro* gas measuring techniques for assessment of nutritional quality of feeds: a review. *Animal Feed Science and Technology*, 72, 261-281.

Getachew, G., Makkar, H.P.S. and Becker, K. 2001. Method of polyethylene glycol application to tannin-containing browes to improve microbial fermentation and efficiency of microbial protein synthesis from tannin-containing browses. *Animal Feed Science and Technology*, 92, 51-57.

Haslam, E. 1988. Plant polyphenols (syn. vegetable tannins) and chemical defense - a reappraisial. *Journal of Chemical Ecology*, 14, 1789-1805.

Hervas, G., Frutos, P., Giraldez, F.J., Mantecon, A.R. and del Pino, M.C.A. 2003. Effect of different doses of Quebracho tannins extract on rumen fermentation in ewes. *Animal Feed Science and Technology*. 109, 65-78.

Horvath, P.J. 1981. The nutritional and ecological significance of Acer- tannins and related polyphenols. MSc thesis, Cornell University, Ithaca, NY.

Hoste, H., Jackson, F., Athanasiadou, S., Thamsborg, S.M. and Hoskin, S.O. 2006. The effects of tannin-rich plants on parasitic nematodes in ruminants. *Trends in Parasitology*, 22, 253-261.

Hutching, M.R., Athanasiadou, S., Kyriazakis, I. and Gordon, I.J. 2003. Can animals use foraging behavior to combat parasites? *Proceedings of the Nutrition Society*, 62, 361-370.

Jones, E.T. and Mangan, J.L. 1977. Complexes of the condensed tannins of sainfoin (Onobrychis viciifolia Scop.) with fraction 1 leaf protein and with submixilary mucoprotein, and their reversal by polyethylene glycol and pH. *Journal of the Science of Food and Agriculture*, 28, 126-136.

Kreuzer, M. 1986. Methodik und Anwendung der Defaunierung bein wachsenden Wiederkauer. *Journal of Veterinary Medicine (A)*, 33, 721-745.

Makkar, H.P.S. 2003. Effects and fate of tannins in ruminant animals, adaptation to tannins, and strategies to overcome detrimental effects of feeding tannin-rich feeds. *Small Ruminant Research*, 49, 241–256.

Makkar, H.P.S. and Becker, K. 1996. Effect of pH, temperature, and time on inactivation of tannins and possible implications in de-tannification studies. *Journal of Agricultural and Food Chemistry*, 44, 1291-1295.

Makkar, H.P.S., Becker, K., Abel, H.J. and Szeletti, C. 1995. Degradation of condensed tannins by rumen microbes exposed to *Quebracho* tannins (QT) in rumen simulation technique (RUSITEC) and effects of QT on fermentation processes in the RUSITEC. *Journal of the Science of Food and Agriculture*, 69, 495-500.

McKie, M.R., Brown, D.L., Melesse, A. and Odenyo, A.A. 2004. Rumen microbes from African ruminants can degrade Acacia angustissima diamino acids. *Animal Feed Science and Technology*, 117, 179-195.

McMahon, L.R., McAllister, T.A., Berg, B.P., Majak, W., Acharya, S.N., Popp, J.D., Coulman, B.E., Wang, Y. and Cheng, K.J. 2000. A review of the effects of forage condensed tannins on ruminal fermentation and bloat in grazing cattle. *Canadian Journal of Plant Science*, 80, 469-485.

McSweeney, C.S., Palmer, B., McNeil, D.M. and Krause, D.O. 2001. Microbial interactions with tannins: nutritional consequences for ruminants. *Animal Feed Science and Technology*, 91, 83-93.

Meagher, L.P., Lane, G., Sivakumaran, S., Tavendale, M.H. and Fraser, K. 2004. Characterization of condensed tannins from Lotus species by thiolytic degradation and electrospray mass spectrometry. *Animal Feed Science and Technology*, 117, 151-163.

Mehansho, H., Butler, L.G. and Carlson, D.M. 1987. Dietary tannins and salivary proline-rich proteins: interactions, induction, and defense mechanisms. *Annual Review of Nutrition*, 7, 423-440.

Min, B.R., Attwood, G.T., Reilly, K., Sun, W., Peters, J.S., Barry, T.N. and McNabb, W.C. 2002. Lotus corniculatus condensed tannins decrease in vivo populations of proteolytic bacteria and affect nitrogen metabolism in the rumen of sheep. *Canadian Journal of Microbiology*, 48, 911-921.

Min, B.R., Barry, T.N., Attwood, G.T. and McNabb, W.C. 2003. The effect of condensed tannins on the nutrition and health of ruminants fed fresh temperate forages: a review. *Animal Feed Science and Technology*, 106, 3-19.

Min, B.R., Pinchak, W.E., Fulford, J.D. and Puchala, R. 2005. Wheat pasture bloat condensed tannins. *Journal of Animal Science*, 83, 1322-1331.

Mlambo, V. and Makkar, H.P.S. 2005. Calibration and validation of the 14C-labelled polyethylene glycol-binding assay of tannins in tropical browse. *Animal Feed Science and Technology*, 122, 29-40.

Molan, A.L., Waghorn, G.C., Min, B.R. and McNabb, W.C. 2000. The effect of condensed tannins from seven herbages on *Trichostrongylus colubriformis* larvae in vitro and on pasture. *Veterinary Parasitology*, 105, 269-283.

Mueller-Harvey, I. 2006. Unravelling the conundrum of tannins in animal nutrition and health. *Journal of The Science of Food and Agriculture*, 86, 2010-2037.

Nicholson, K., Butler, L.G. and Asquith, T.N. 1986. Glycoproteins from Colleptrichum graminicola that bind phenols: implications for survival and virulence of phytopathogenic fungi. *Phytochemistry*, 76, 3824-3830.

Odenyo, A.A,, McSweeneym, C.S., Palmer, B., Negassa, D. and Osuji, P.O. 1999. In vitro screening of rumen fluid samples from indigenous African ruminants provides evidence for rumen fluid with superior capacities to digest tannin-rich fodders. *Australian Journal of Agricultural Research*, 50, 1147-1157.

Odenyo, A.A., Osuji, P.O., Reed, J.D., Smith, A.H., Mackie, R.I., McSweeney, C.S. and Hanson, J. 2003. *Acacia angustissima*: its anti-nutrients constituents; toxicity and possible mechanisms to alleviate the toxicity - a short review. *Agroforestry Systems* 59, 141-147.

Pinares-Patino, C.S., Ulyatt, M.J., Waghorn, G.C., Lassey, K.R., Barry, T.N., Holmes, C.W. and Johnson, D.E. 2003. Methane emission by alpaca and sheep fed on Lucerne hay or grazed on pastures of perennial ryegrass/white clover or birdsfoot trefoil. *Journal of Agricultural Science*, 140, 215-226.

Pitta, D.W., Barry, T.N., lopez-villalopbas, N. and Kemp, P.D. 2007. Willow fodder block-Alternative forage to low quality pasture for mating ewes during drought. *Animal Feed Science and Technology*, 133, 240-258.

Puchala, R., Min, B.R.,Goetsch, A.L. and Sahlu, T. 2005. The effect of condensed tannins containing forage on methane emission by goats. *Journal Animal Science*, 83,182-186.

Ramirez-Restrepo, C.A. and Barry, T.N. 2005. Alternative temperate forages containing secondary compounds for improving sustainable productivity in grazing ruminants. *Animal Feed Science and Technology*, 120, 179-201.

Shimada T 2006. Salivary proteins as a defense against dietary tannins. *Journal of Chemical Ecology*, 32, 1149-1163.

Singh, B., Sahoo, A., Sharma, R. and Bhat, T.K. 2005. Effect of polyethylene glycol on gas production parameter and nitrogen dis appearance of some tree foliages. *Animal Feed Science and Technology*, 123-124, 351-364.

Skene, I.K. and Brooker, J.D. 1995. Characterization of tannin acylhydrolase activity in rumen bacterium *selemonas ruminantium*. *Anarobe Ecology*, 1, 321-327.

Stienezen, M., Waghorn, G.C. and Douglas, G.B. 1996. Digestibility and effects of condensed tannins on digestion of sulla (*Hedysarum coronarium*) when fed to sheep. *New Zealand Journal of Agricultural Research*, 39, 215-221.

Tanner, G.J., Moate, P.J., Davis, L.H., Laby, R.H., Yuguang, L. and Larkin, P.J. 1995. Proanthocyanidins (condensed tannin) destabilise plant protein foams in a dose dependent manner. *Australian Journal of Agricultural Research*, 46, 1101-1109.

Tavendale, M.H., Meagher, L.P., Pacheco, D., Walker, N., Attwood, G.T .and Sivakumaran, S. 2005. Methane production from in vitro rumen incubations with Lotus pedunculatus and Meticago sativa, and effects of extractable condensed tannin fractions on methanogenesis. *Animal Feed Science and Technology*, 123/124, 403-419.

Terrill, T.H., Waghorn, G.C., Wooley, D.J., McNabb, W.C. and Barry, T.N. 1994. Assay and digestion of 14C-labelled condensed tannins in the gastrointestinal tract of sheep. *British Journal of Nutrition*. 72, 467-477.

Vaithiyanathan, S., Bhatta, R., Mishra, A.S., Prasad, R., Verma, D.L. and Singh, N.P. 2007. Effect of feeding graded levels of *Prosopis cineraria* leaves on rumen ciliate protozoa, nitrogen balance and microbialprotein supply in lambs and kids. *Animal Feed Science and Technology*, 133, 177-190.

Vitti, D.M.S.S., Nozella, E.F., Abdalla, A.L., Bueno, I.C.S., Silva Filho, J.C., Costa, C., bueno, M.S., Longo, C., Vieira, M.E.Q., cabral Filho, S.L.S., Godoy, P.B. and Mueller-Harvey, I. 2005. The effect of drying and urea treatment on nutritional and anti-nutritional componenets of browses collected during wet and dry seasons. *Animal Feed Science and Technology*, 112, 123-133.

Waghorn, G.C. and McNabb, W.C. 2003. Consequences of plant phenolic compounds for productivity and health of ruminants. *Proceedings of the Nutrition Society* 62, 383-392.

Wang, Y., Douglas, G.B., Waghorn, G.C., Barry, T.N., and Foote, A.G. 1996a. Effect of condensed in *Lotus conriculatus* upon lactation performance in ewes. and lucern (*Medicago sativa*). *Journal of Agriculture Science* (Camb.) 126, 353-362.

Wang, Y., Douglas, G.B., Waghorn, G.C., Barry, T.N., Foote, A.G. and Purchas, R.W. 1996a. Effect of condensed tannins upon the performance of lambs grazing *Lotus conriculatus* and lucern (*Medicago sativa*). *Journal of Agriculture Science* (Camb.) 126, 87-98.

West, J.W., Hill, G.M. and Utley, P.R. 1993. Peanut skins as a feed ingredient for lactating dairy cows. *Journal of Dairy Science*, 76, 590-599.

Wina, E., Tangendjaja, B. and Susana, I.W.R. 2005. Effects of chopping, and soaking in water, hydrochloric acid and calcium hydroxide solutions on nutritional value of *Acacia villosa* for goats. *Animal Feed Science and Technology*, 112, 79-92.

Wolin, M.J., Miller, T.L. and Stewart, C.S. 1997. Microbe-microbe interactions. In. The rumen microbial ecosystem, 2nd edn (Eds. P.N. Hobson, C.S. Stewart). Blackie Academic and Professional. pp. 475-491.

Woodward, S.L., Waghorn, G.C., Ylyatt, M.J. and Lassey, K.R. early indications that feeding lotus will reduce methane emissions from ruminants. *Proc. New Zealand Society of Animal Production*, ACIAR, Adelaide, pp. 23-26.

Yildi, S., Kaya, I., Unal, Y., Aksu Elmali, D., Kaya, S., Cenesiz, M., Kaya, M. and Oncuer, A., 2005. Digestion and body weight change in Tuj lambs receiving oak (*Quercus hartwissiana*) leaves with and without PGE. *Animal Feed Science and Technology*, 112, 159-172.

Zimmer, N. and Cordesse, R. 1996. Digestibility and ruminal digestion of non-nitrogenous compounds in adult sheep and goats: effects of chestnut tannins. *Animal Feed Science and Technology*, 61, 259-273.

❑❑❑

Chapter **36**

Rangelands and their Improvement Strategies

M. M. ROY
Indian Grassland and Fodder Research Institute, Jhansi-248003 (U.P.), India

Introduction

According to Webster's Dictionary, rangeland is an open region over which livestock may roam and feed. In the Indian context rangeland is a piece of natural or native vegetation where-in grazing by livestock occurs or can occur. This includes natural pastures, grazing lands, savanna and forests.

The rangelands cover about 40 per cent of the earth's continental surface where one-third of man's domesticated flocks and herds survive. They are also habitat of most of the world's wild life. Over the centuries, rangelands were used by wild life and man, the hunter and food gatherer. During this century, these lands have been encroached upon by the developing society and this has become the greatest environmental threat faced by the developing countries. After advent of civilization; the pastoral economy gradually transformed into an agricultural economy where livestock rearing emerged as a major component. As the human population increased, grazing lands were brought under the cultivation of food crops. Thus grazing pressure on forest areas increased, the palatable and nutritious grasses got replaced by less acceptable grasses. Most of the rangelands have thus been reduced to the category of degraded lands, especially in the tropics. The few rangelands we have today have survived mainly because of their topography and thus escaped intensive agricultural encroachment.

In India, the rangelands have been very important since ancient times as is evident from Vedic literature. However, now the rangelands in the country are largely in an advanced stage of deterioration due to many factors such as large bovine population, free range grazing, lack of management practices and deforestation for development activities. The situation is even more alarming in Himalayan region where the rangelands experience severe biotic pressure of both sedentary and migratory grazier.

Despite poor condition of Indian Rangelands, they still meet a major share of present forage supply of domestic animals. Also, with almost no chance of increase in the area under fodder crop cultivation, the rangelands will continue to be a major support for the livestock. Considering these, after independence, efforts have been made by several ICAR institutions to evolve the packages for improving/restoring rangeland productivity in India. These are being discussed in this paper.

Rangeland Restoration

The problems associated with degradation of rangelands and forests have attracted attention of ICAR for the last five decades. Besides the ICAR institutes like CAZRI, IGFRI, CSWCR&TI and VPKAS; several traditional and state agricultural universities (SAUs) have generated a good deal of information on various restoration aspects of Indian Rangelands. Consequently, the technology is now available for the improvement/ conservation of degraded rangelands in the arid/ semiarid/humid tropics and temperate alpine regions.

As the primary functions of rangelands vary, the objectives of restoration of a degraded rangeland will vary as per the intended major use. For example, if the major emphasis is placed on livestock productivity, the output per animal and the productivity per unit of land are to be increased. Even for normal grazing use provision for drinking water, stands for salt licks and shade trees for grazing animals are required. The restoration package will vary if the objective is other than livestock rearing. For example, to provide recreational opportunities, landscaping and beautification aspects are to be looked into. the use of wildlife conservation will require protection and creation of wildlife habitats, and development of water and food resources for particular wildlife communities.

Basically, the range restoration techniques are based on ecological approaches e.g. protection for vegetation recovery, soil and water conservation, reseeding, improved range management techniques and grazing management.

Offering Protection

The rangeland degradation due to overgrazing leads to a situation where nutritionally low grade and unpalatable species dominate the scene. The vegetation becomes sparse and soil working in such situation is likely to further accentuate the erosional process. It has been found that just by mere protection through fencing a lot of benefit in terms of vegetation recovery can be obtained. Rangelands can be protected through barbed/woven/chicken wires or chain links supported by angle iron/cement/stone/wooden poles or through thorn fencing/live hedges/cattle proof ditches/bunds etc. Efficacy and cost effectiveness of various kinds of fencing will require consideration at local levels.

The duration of the exclosure of a habitat is determined by the intended use of the developed vegetation. The time required for natural recovery of vegetation will vary from one habitat to another and from one climatic zone to another. For example, the duration of protection of vegetation recovery in arid regions will vary from one habitat to another; the recovery is faster on alluvial plains compared to rocky gravelly piedments. Once the desired level of vegetation recovery has been achieved, the rangelands should be released for grazing in the deferred rotational grazing management system by compartmentalizing with the help of inner fencing into paddocks.

Soil and Water Conservation

Measures related to soil and water conservation in range and pasture management in dry areas are of crucial importance. The measures are especially important on degraded sloppy lands. Most useful such measures are described below:

Engineering Measures

Bunding

Along the contour lines and periphery bunds are made, either manually or using tractor mounted implements. The details *viz.*, cross section and interval of location of bunds are decided according to topography. In dry degraded areas, bunds are required to be stabilized by sodding with suitable grass species *viz.*, *Dichanthium annulatum*, *Cenchrus ciliaris* and *Panicum antidotale*. The excess runoff may be safely dispersed off through grassed waterways or pipe outlets.

In areas having better soil and rainfall, technique of graded bunds along with grassed waterway and drop structure for safe disposal of water has been found to be better than contour bunds.

Water Storage/Discharge Structures

The construction of water storage/discharge structures *viz.*, checkdams, anne-cuts and embankments, of desired engineering specifications, in areas like ravines and watersheds etc. are required to be constructed for efficient conservation and utilization of already scarce water resources in dry areas. In some ravinous hilly areas gully plugging may also require erecting structure of various engineering specifications.

Biological Measures

Various grass and legume species are known to play an important role in soil and moisture conservation. The runoff and soil loss under such systems is reduced considerably in every agro-climatic region. In excessive windy areas a combination of trees may be planted as wind/shelter belt, in specific designs, to check the wind impact. The spreading of dry grasses or shrubby growth on bare ground along the contours or terraces has the potential to check soil and water loss. Various chemicals may also be used for mulching effect and soil stabilization.

Enrichment Through Reseeding

The grasslands in dry region are usually of poor quality, containing less than seven percent crude protein (CP) for most of the time. Also, because of excessive grazing pressure on most of these grasslands, the productivity of such lands has deteriorated to considerable extent.

In order to increase the pasture productivity in dry areas, it is essential to replace low yielding annual grasses with high yielding perennial grasses which are adaptable to the prevailing condition of that region. Similarly, as the use of fertilizers in these lands is of little practical value on account of high costs, introduction of suitable pasture legumes may be one of the best ways to improve the quality.

Some of the most useful such measure/reseeding techniques are as follows:

- The important species identified for reseeding is provided in Table 1. Selection of such species depends on the site conditions. *Dichanthium annulatum* and *Cenchrus ciliaris* have wide range of climate and soil in areas having rainfall up to 750 mm. In such situation suitable legumes include *Alysicarpus rugosus, Stylosanthes hamata, Styzolobium deeringianum, Indigofera endecaphylla* and *Lablab purpureus.*

Table 1: Important Pasture Species for Reseeding in Dry Areas

Condition	Species
Extreme arid situations	*Lasiurus sindicus*
Sand dunes	*Panicum turgidum*
Arid and Semi-arid sandy loam soils	*Cenchrus ciliaris* *Cenchrus setigerus*
Arid and Semi-arid red gravelly soil/ sloppy lands	*Chrysopogon fulvus*
Semi-arid clay soil with higher moisture status	*Iseilema laxum* *Panicum maximum*
Semi-arid loamy soil with higher moisture status	*Dichanthium annulatum*

- Premonsoon broadcasting of *Stylosanthes* seeds in *Sehima eteropogon* grasslands was found best in terms of legume establishment. In case of *S. hamata* broadcasting of seeds after harrowing gave maximum establishment as compared to broadcasting alone or line sowing at 50 cm apart. Also, use of pelleted seeds gave better establishment.
- Significant increase in nitrogen fixation is possible by appropriate selection of *Rhizobium* strains and by appropriate seed pelleting practice. By using a seed rate of 4-6 kg/ha after required scarification pretreatment and premonsoon sowing/broadcasting have provided the nitrogen equivalence to the tune of 40-60 kg/ha in natural grassland.

Management Practices

The improved range and pasture management practices may be discussed under the following headings:

Bush Cleaning

The undesirable bushes are required to be removed, either manually or by mechanical felling and removal of stumps or application of selective weedicides on the cut stamps to kill them to stop coppicing. The leaf fodder yielding shrubs may, however, be retained in appropriate proportion. It has been found that in rangelands of Indian deserts medium stock density level of

Zizyphus nummularia (14 percent of land area covered by shrub canopy) was optimum for high forage as well as leaf fodder production.

Application of Fertilizers

Application of fertilizers on productivity of grasslands has been extensively studied in arid and semi-arid regions. The response is maximum in case of nitrogen doses followed by phosphorus, almost no response with potassium. The timing of fertilizer application is also important, maximum response has been obtained when applied about 30 days after the start of monsoon. Similarly, application of 60 kg N/ha in three split doses at 25 days interval was found to be superior. The forage yield increases were in the range of 40-70 percent over control.

Controlled Burning

Controlled burning, if used judiciously, is a powerful range/pasture management tool. It is preferable to use this tool in combination with moderate grazing management. Burning is recommended during wetter periods with longer intervals of about three years.

Grazing Management

In pasture grazing systems, a period of rest is essential for the grasses to recoup and rejuvenate. In *Sehima* dominated grasslands at Jhansi, deferred rotational grazing system has been found to be superior in providing more number of animal days (2925) as compared to continuous grazing system (2097 days). In *Lasiurus* dominated arid grassland the same system was superior in terms of average body weight gain and calving of *Tharparker* breed of heifers over continuous grazing system. As at the low altitudes in the Himalayan rangelands, rotational and deferred rotational grazing are not possible at many places, the local practice of enclosing rangelands for 5-6 months (June - October) followed by controlled grazing and later on cutting of grasses for stall feeding has been recommended by various workers. In high altitude regions also the grazing pressure can be reduced by restrictive grazing on selected sites on yearly basis, rotation of grazing sites in different years.

Other Interventions

Rangeland degradation is considered as the greatest challenge to the rangeland users in India. In order to meet the various demands of the growing population, this process needs to be halted and methodologies developed for their sustainable development and utilization.

Sown & Improved Pastures

In a study on plant-soil-animal interface at IGFRI, Jhansi, maximum average biomass production was obtained under sown pasture (Table 2). The introduction of stylo added substantially (40%) to the biomass in the second year but reduced to almost negligible in third year due to grazing. It showed the necessity of introducing stylo every third year. Natural grasslands, however, proved more economical in terms of output per rupee invested. Body weight gain in goats was similar in sown and improved pastures but it decreased in natural grasslands. Sheep showed best performance in sown pastures. Cattle gave better performance in improved pasture.

Table 2 : Biomass Yield and Economics Under three Different Rangeland Management Systems in India

System	**Yield (DM t/ha)**						**Return**
	Year1	**Year2**	**Year3**	**Year4**	**Year5**	**Average**	**(B/C)**
Sown Pasture	2.7	10.9	8.3	4.3	3.6	6.8	1.27
Improved Pasture	2.3	9.7	8.0	3.7	6.1	6.2	1.17
Natural Pasture	1.3	6.1	6.8	3.8	3.7	5.1	2.65

Agroforestry Based Fodder Production Systems

Agroforestry is defined as a "Landuse system that involves deliberate retention, introduction of trees or other woody perennial in crop/animal production to benefit from the resultant ecological and economic intersections". The basic ideas emerging from the definition are (1) agroforestry is a landuse system including combination of agriculture, agrostology, forestry, horticulture, animal husbandry etc. (2) integration of trees/woody perennial is with the main objective of reducing risk and increasing total productivity (3) ideal agroforestry systems are stable and sustainable (4) integration of trees with pastures/crops may result in more efficient use of sunlight, moisture and plant nutrients than is generally possible by mono-cropping of either agriculture or forestry crops - there may be competition between trees and crops for these resources but net effect is positive, if such systems are designed and managed scientifically.

In India, because of various factors such as competition for land due to intense population pressure, rapidly depleting fuel wood resources, time-scale for orchard operations, dispersed distribution of benefits from forestry and seasonal shortage of labour; the growing of trees did not find a suitable place in rural economy and rangeland development programmes in the past.

However, due to the ecological and social cost of rangeland degradation besides highly adverse land-man and land-livestock ratio that has placed enormous pressure on non-renewable land resources, emphasis is now shifting towards agroforestry based landuse systems. Agroforestry based production systems are capable of yielding leaf fodder, wood and food at the same time besides conserving and rehabilitating the ecosystem. The woody perennial/tree is one of the main components in such systems.

Silvopastoral Systems

Such systems integrate pasture and/or animals with trees. Here, woody perennial, preferably of fodder value, are introduced deliberately and systematically and managed scientifically. Under poor soil, water and nutrient situations where cropping is not possible such systems can serve the twin purposes of forage and firewood production and ecosystem conservation. In India such systems have great potential on wastelands, especially in the arid and semi-arid region (Singh and Roy 1993). A list of forage grasses, legumes, shrubs and trees for rangelands in different agro-ecological regions of India is presented in Table 3.

Hortipastoral Systems

Such systems integrate fruit trees with pastures. The objective here is to introduce pasture species under established orchards so as to supply forage for livestock. It is important to recommend only those pasture species that is beneficial to or at worst neutral to the tree crop. There may be competition between trees and pasture but the extent and effect will depend upon spacing and other management.

Boundary Plantings

Farm boundaries, banks, river banks etc. may also be suitably utilized for lean period fodder. Fast growing fodder species like *Leucaena leucocephala* have the potential to yield quality fodder (27.4 and 2.6 t per row km firewood and forage, respectively at 2 m) at 4-5 years.

Livehedges

These are lines of trees or shrubs planted on farm boundaries or on borders of home compounds, pastures, fields or animal enclosures. The purpose is different from other agroforestry technologies based on trees planted in lines such as boundary plantings, contour strips or hedge row intercropping. Besides their main function to control human and animal movement - they provide fodder, fuelwood and food; act as wind breaks; and enrich the soil depending on the species used.

The fodder species selected for this purpose should be adaptable to local agro-climatic conditions, of fast growth rate, of high coppicing ability and should provide higher economic returns. Depending on the site conditions *Gliricidia sepium*, *Erythrina* sp., *Zizyphus mauritiana*, *Prosopis juliflora*, *Dichrostachys cinerea*, *Acacia* sp. and *Albizia* sp. may be used.

Conclusions

The rangelands may be effectively improved through use of available range restoration technologies and also adopting site specific agroforestry options for a variety of situations in India.

References

Dabadghao, P. M. and Shankarnarayan, K. A. 1973. *The Grass Cover of India*. ICAR, New Delhi: 713 p.

Singh, P. and Roy, M. M. 1991. Forage production through agroforestry: constraints and priorities. *Range Management and Agroforestry* 12, 169-178.

Singh, P. and Roy, M. M. 1993. Silvipastoral systems for ameliorating productivity of degraded lands in India. *Annals of Forestry* 1, 61-73.

Yadava, P. S. and Singh, J. S. 1977. *Grassland Vegetation: Its Structure, Function, Utilization and Management*. Today and Tomorrow's Printers and Publishers. New Delhi.

Roy M.M. 2003. Grassland and pasture management systems in arid and semiarid region. *In: Sustainable Animal Production* (eds R. C. Jakhmola and R. K. Jain). Pointer Publications, Jaipur: 54-82.

Roy M.M. 2003. Forage based agroforestry systems for conservation of land resources. *In: Agroforestry: Potentials and Opportunities* (eds P. S. Pathak and Ram Newaj). *Agrobios* (India). Jodhpur, 267-279.

contd...

Table 3 : List of forage grasses, legumes, shrubs and trees for rangelands in different agro-ecological regions of India

Grasses	Legumes	Shrubs/Trees
Region 1: Western Himalaya, cold arid with shallow skeletal soils		
Agrostis spp., *Poa alpina, Trisetum spicatum*	*Medicago sativa*	*Hippophae rhamnoides*
Region 2: Western Plains and Kaccha Peninsula, hot arid with desert and saline soils		
Cenchrus ciliaris, Cenchrus setigerus (Sandy plains); *Lasiurus sindicus* (Sandy interdunal plains); *Panicum turgidium* (Sand dunes); *Chloris gayana, Sporobolus marginatus* (Salt affected lands)	*Casssia rotundifolia*	*Acacia nilotica, A. tortilis, Ailanthus excelsa, Dichrostachys cinerea, Prosopis cineraria, Ziziphus nummularia, Prosopis juliflora, Salvadora oleoides, S. persica*
Region 3: Deccan Plateau, hot arid with red and black soils saline soils		
Andropogon gayanus, Chrysopogon fulvus (Red soil*);* *Dichanthium annulatum, Bothriochloa intermedia* (Black soil)	*Clitoria ternatea, Stylosanthes hamata, S. scabra*	*Acacia nilotica, Albizia amara, A. lebbeck, Desmenthus virgatus, Leucaena leucocephala, Tamarindus indica*
Region 4: Northern Plains and Central Highlands including Aravallis, hot semiarid with alluvium		
Bothriochloa intermedia, Cenchrus ciliaris, Chrysopogon fulvus, Dichanthium annulatum, Sehima nervosum	*Macroptelium atropurpureum, Stylosanthes hamata, S. scabra*	*Acacia nilotica, A. holosericea, Albizia amara, A. lebbeck,A. procera, Azadirachta indica, Dichrostachys cinerea, Hardwickia binata, Leucaena leucocephala, Sesbania grandiflora, S. sesban*

Region 5: Central (Malwa) Highlands,Gujrat Plains & Kathiawar Peninsula, hot semiarid with red loamy soils

Bothriochloa intermedia, Cynodon dactylon, Cenchrus ciliaris, Panicum maximum	*Arachis hagenbackii, Stylosanthes hamata, S. scabra*	*Albizia lebbeck, Artocarpus lakoocha, Dendroclamus strictus, Gliricidia sepium, Faidherbia albida, Holoptelia integrifolia, Pithecellobium dulcee*

Region 6: Deccan Plateau, hot semiarid with shallow black soils

Bothriochloa intermedia, B. decumbense, Cenchrus setigerus, Dichanthium annulatum, Pennisetum pedicellatum, Panicum maximum	*Stylosanthes hamata, S. scabra*	*Acacia nilotica, Albizia procera, Anogeissus pendula, Bauhinia variegata, B. purpurea, Leucaena leucocephala, Moringa oleifera, Pterocarpus marsupium, Sesbania sesban, Terminalia arjuna*

Region 7: Deccan (Telangana) Plateau and Eastern Ghats, hot semiarid with red and black soils

Andropogon gayanus, Bothriochloa intermedia, Chrysopogon fulvus, Pennisetum pedicellatum, Dichanthium annulatum	*Atylosia scrabaeoides, Macrotyloma axillare, Macroptilium atropurpureum, Stylosanthes scabra*	*Albizia lebbeck, Gliricidia sepium, Faidherbia albida, Holoptelia integrifolia, Leucaena leucocephala*

Region 8: Eastern Ghats, TN Uplands and Deccan (Karnataka) Plateau, hot semiarid with red and black soils

Brachiaria decumbens, B. ruiziensis, Cynodon dactylon, Dichanthium annulatum, Bothriochloa intermedia	*Arachis hagenbackii, A. glabrata, Stylosanthes guinensis,*	*Ailanthus malabarica, Albizia falcataria, Erythrina variegata, E. poeppigiana S. hamata*

Region 9: Northern Plains, hot subhumid (dry) with alluvium derived soils

Bothriochloa intermedia, Cynodon dactylon, Chloris gayana, Dichanthium annulatum, Pennisetum pedicellatum	*Clitoria ternatea, Macroptilium atropurpureum, Stylosanthes hamata*	*Albizia stipulata, Desmanthus virgatus, Azadirachta indica, Ficus racemosa, Leucaena leucocephala*

Region 10: Central Highlands (Malwa, Bundelkhand & Satpura), semiarid with black and red soils

Andropogon gayanus, Pennisetium pedicellatum (Red soil); *Bothriochloa intermedia, Chrysopogon fulvus, Sehima nervosum, Dichanthium annulatum*	*Atylosia scrabaeoides, Macroptelium atropurpureum, Stylosanthes hamata, S. scabra*	*Albizia amara, A. lebbeck, Anogeissus latifolia, A. pendula, Dichrostachys cinerea, Hardwickia binata, Leucaena leucocephala*

Region 11: Eastern Plateau (Chhatisgarh), hot subhumid with red soil

Bothriochloa intermedia, Cynodon dactylon, Dichanthium annulatum, Panicum maximum, Pennisetum pedicellatum, Moringa oleifera Setaria sphacelata	*Arachis hagenbackii, Stylosanthes hamata*	*Bauhinia variegata, Dalbergia sissoo, Leucaena leucocephala,*

Region 12: Eastern (Chhotanagapur) Plateau and Eastern Ghats, hot subhumid with red and laterite soils

Andropogon gayanua, Bothriochloa intermedia, Chrysopogon fulvus, Pesnnisetum pedicellatum, Urochloa mosambicensis	*Atylosia scrabaeoides, Macroptilium atropurpureum, Stylosanthes hamata*	*Artocarpus heterophyllus, A. lakoocha, Leucaena leucocephala, Moringa oleifera*

Region 13: Eastern Plain, hot subhumid (moist) with alluvium derived soils

Brachiaria brizantha, B. decumbens, B. mutica, Cynodon dactylon, Papsalum notatum	Arachis glabrata, A. hagenbackii Desmanthus virgatus, Pterocarpus marsupium	Bauhinia variegata, Dalbergia latifolia, D. sissoo,

Region 14: Western Himalayas, warm subhumid with brown forest and podzolic soils

Dactylis glomerata, Festuca rubra, Lolium perenne, Poa spp.	*Trifolium pratense, T. repens, Lotus corniculatus*	*Quercus incana, Robinia pseudocacia, Grewia optiva, Fagus sylvatica, Celtis australis, Morus alba*

Region 15: Bengal and Assam Plains, hot subhumid (moist) to humid with alluvium derived soils

Brachiaria decumbens, B. mutica, Paspalum notatum	*Desmodium uncinatum, D. heterophyllum*	*Artocarpus heterophyllus, A. lakoocha, Ficus hookeri, F. nemoralis, Parkia roxburghii, Morus alba*

Region 16: Eastern Himalayas, warm perhumid with brown and red soils

Coix lacryma-jobi, Pennisetum clandestinum, Tripsacum dactyloides	*Desmodium sp., Pueraria phasceoloides*	*Celtis australis, Ficus hookeri, F. nemoralis, F. semicordata*

Region 17: North-eastern Hills (Purvanchal), warm perhumid with red and laterite soils

Brachiaria decumbens, Pennisetum clandestinum, Tripsacum dactyloides	*Arachis sp., Desmodium uncinatum*	*Dendrocalamus hamiltonii, Parkia roxburghii, Morus alba, Robinia pseudoacacia*

Region 18: Eastern Coastal Plain, hot humid per humid laterite and alluvium derived soils

Chloris gayana, Cynodon dactylon, Dichanthium annulatum, Pennisetum pedicellatum, Stenotaphrum dimidiatum, Urochloa mosambicensis	*Stylosanthes guinensis*	*Ailanthes malabarica, Erythrina variegata, E. poeppigiana Ficus retusa*

Region 19: Western Ghats and Coastal Plain, hot humid per humid laterite and alluvium derived soils

Cynodon dactylon, Dichanthium annulatum, Panicum maximum, Pennisetum clandestinum, P. polystachyon, Setaria sphacelata	*Clitoria ternatea, Desmodium heterophyllum, Pueraria thunbergiana, Stylosanthes hamata, S. guinensis*	*Ailanthes malabarica, Erythrina variegata*

Region 20: Islands of Andman Nicobar and Lakshdweep, hot humid to per humid island with red loamy and sandy soils

Andropogon gayanus, Cynodon dactylon, Cenchrus ciliaris, Pennisetum pedicellatum, P. polystachyon, Brachiaria ruziziensis, Tripsacum laxum	*Centrosema pubescens, Clitoria ternatea, Macroptilium atropurpureum, Stylosanthes guinensis, S. scabra*	*Bauhinia purpurea, Erythrina variegata, Leucaena leucocephala, Trema tomentosa, Pithecellobium dulcee, Gliricidia sepium*

Chapter 37

Grassland / Rangeland Management and Livestock Production

B. K. TRIVEDI AND J. P. SINGH
Indian Grassland and Fodder Research Institute, Jhansi-284003 (U.P.), India

Introduction

Vast areas mostly under natural vegetation dominated by grasses, which are unfenced, unfertilized and non-irrigated, where animals (wild and domestic) grazed are called grassland/rangeland. The grassland vegetation consists of a number of perennial and annual grasses mixed with legumes and forbs.

Animal husbandry plays a vital role in the agriculture as well as in rural economy. It is estimated that roughly 49 per cent of the total geographical area which includes forests, permanent pastures, uncultivable wastes and fallow lands is available for grazing to about 445 million livestock heads over the country. In context of the grassland management it will be important to understand some important term:

Aftermath: Recovery growth of forage plants after harvesting either by animal or by machine.

Agrostology: Study of grasses, their classification, management and utilization.

Annual Pastures : The term usually applied to pastures that are seeded ever year to take place wholly or in part of permanent pasture.

Band : Any number of sheep or goats handled as a single unit under range conditions, usually considered as synonymous with herd as it relates to such livestock.

Brush : Those covered largely with brush and shrubs, where a considerable portion of feed obtained by the livestock comes from browsing woody plants.

Carrying capacity or grazing capacity : In range management, the ability of range unit, exclusive of severe drought years, to adequately support a constant number of grazing animals for a stated period each year, without deteriorating with respect to grazing values or other proper land uses. Normally expressed in number of livestock per acre of a given kind or kinds of vegetative cover or in number of acres of such land per specified livestock unit.

Enclosure : An area within which animals may be confined.

Feed : Harvested forage such as hay or other fodder and grain or other processed food for livestock. When grazed and not harvested it is called forage.

Forage : Forage is un-harvested plant material of any kind available for animal consumption. It may either be used for grazing or be cut for feeding. When cut it becomes feed.

Forbs : A non-grass herbaceous vegetation including all weeds except legumes.

Grass : Botanically any plant of a family graminae is a grass. But generally in grassland agriculture the term grass includes legumes and forbs also.

Herbaceous : Plant growth which is relatively free from woody tissues.

Humus : The plant and animal residues of the soil (litter excluded) which are under going for decomposition.

Indigenous : Plants native to specific environment are indigenous.

Legume : In agriculture, it is considered as the plant, which is a member of Leguminosae family with the character of nitrogen fixing nodule on its roots. But botanically it only means a kind of fruit of taxa of Leguminosae family.

Litter : Accumulation of leaves, stems, fruits, branches and other plant parts on the surface covered by range, pasture or forest vegetation.

Meadow : An area that receives more moisture than adjacent areas. Two classes of meadows are recognized; wet meadows and dry meadows. Wet meadows are characterized by sedges and remain wet and moist throughout the summer. Dry meadows are dominated by grasses and occur as moist meadows-line areas or intermittent meadows, both of which become moderately dry by mid summer.

Paddock: A small fenced field or plot used for grazing purposes.

Palatability: The relative relish with which forage plants are consumed by domestic livestock of other animals. This may vary with consumption of the plant cover or the season of grazing.

Plant community: Any group of plants growing together.

Range condition: The state of health or productivity of both soil and forage of a give range, in terms of what, it could or should be under normal climate and best practicable management.

Range: A very extensive natural pasture.

Savannah: An area of grassland containing scattered trees.

Tiller: A branch originating at the base of the stem of a grasses. Tiller can be induced by grazing where the dormant buds at the base are stimulated into activity to put forth new shoot/roots.

Weed: To the farmer a weed is a plant out of place, specially an aggressive and pestiferous, often coarse, usually herbaceous.

Woodland: Wooded area with grass and other grazing growing in species and among trees.

Now these grasslands are the main resource for the maintenance of our huge livestock population but due to over grazing and mismanagement, these have deteriorated to the extent. At this stage these resources need scientific planning and proper management for sustained and quality production. Thus efforts have been made through this lecture note.

Protection and Bush Control

Grasslands can be provided protection using live fencing of thorny and unpalatable bushes viz. *Pithecolobium dulce, Lawsonia alba, Parkinsonia aculeate, Lantana camera, Euphorbia nirulia, E. niruri, E. tirucalii and Ipomea fistulosa.* As a result of protection for eight years the plant population of desired perennial grasses viz. *Sehima nervosum, Heteropogon controtus* etc. increased many times (Trivedi & Kanodia, 1982). This ultimately affects the production potentials of the native grasslands and 3.5 t/ha dry forage may be harvested against 0.08 t/ha of unprotected area.

Bushes and unpalatable herbaceous species compete forage species for soil moisture, nutrients and space, adversely affecting the herbage production. Garlon 2 (Trichlopyr) at the rate of 800 ml/10 litres of diesel oil could be applied on freshly cut stumps of these bushes.

Effect of Bush Intensity on Grass Production

Bush intensity (No. of bushes/ha)	Dry forage production (t/ha)
3775	0.55
2175	1.25
Zero	4.20

Introduction of Forage Species

Depending on the site, soil types, micro climate and need etc. the forage species have been identified suiting to a number of sites i.e. *Chysopogon fulvus* for red gravelly and sloppy lands, *Cenchrus ciliaris* for sandy soils, *C. setigerus* for flat lands with sandy loam soils and more arid areas, *Dichanthium annulatum* for loamy soils with better moisture status and *Iseilema laxum* and *Panicum maximum* for clayey soils with higher moisture. Legumes are rich in protein and also improve soil fertility by fixing atmospheric nitrogen varying from 100 to 200 kg/ha. Many of these legumes can serve nearly equivalent to 40kg N/ha. The legumes like *Stylosanthes hamata*, phasey bean (*Macroptilium lathyroides)* for arid and semi-arid and white clover (*Trifolium repens)* for temperate conditions were found most suitable for the introduction in grasslands subjected to frequent grazing and cutting for introducing the forage species in grasslands, a brief procedure is as follows:

Field preparation	Light soil working
Seed	Grasses- 4-8 kg/ha depending on the seed size Legumes- 5-12 kg/ha depending on the seed size
Sowing	Broadcasting/line sowing
Period	Ist week of July/ after Ist heavy shower
Fertilizer	30 kg N + 20 kg P_2O_5/ha (basal)

Fertilizer Application

Grasses like other species, require nutrients for their growth. Practically non of the natural grasslands in our country receive fertilizers in any form except dung and urine added by grazing animals. This leads to continuous depletion of plant nutrients resulting in a poor growth of forage species. The average nutrient removal is reported to exceed 300 kg/ha under grassland conditions in UK. The optimum dose of fertilizer for these grasslands depends on the soil nutrients status, pH and plant species. Increase in dry matter yield of over 150 percent has been recorded from grasslands.

Application of 50 kg N/ha in 3 equal splits at 3 weeks interval during monsoon period proved to be superior in term of quality and quantity as compared to those of two splits or single dose. As regards to different sources on nitrogen, the ammonium sulphate was superior to urea, calcium ammonium-nitrate and farm yard manure. The application of 30 kg P_2O_5/ha is recommended in single dose but for legume component the dose increases to its double. The fertilizer application is more effective, when applied in 3rd week of July as compared to 2nd and 1st week of August and September respectively. But for temperate conditions the optimum time, for fertilizer application is from end of March to May. In grasslands which are not easily accessible, aerial broadcast of fertilizers would be desirable.

Utilization Management

Generally uncontrolled continuous and over grazing is prevalent in the grasslands in most of the regions of the country. Through proper utilization in terms of carrying capacity and period of these grasslands influences the productivity and sustenance of these areas. The utilization of forage has to be manipulated so as to allow sufficient time to the plants to rebuild their root reserves.

(a) *Grazing systems*: For *in-situ* utilization of the herbage there are four systems of grazing namely continuous, rotational, deferred rotational and no grazing (cut and fed). In *Sehima* dominated grasslands deferred rotational grazing system was found superior by providing more number of animal grazing days (2925 days) as compared to those of continuous grazing system (2097 days). The average body weight gain was recorded to be 191 g/heifer/day, yet not being detrimental to the range health, although the body weight gain of heifers was slightly higher in continuous grazing system (Upadhyay et al. 1971). The carrying capacity of different types of grasslands at Jhansi was found to be 10.5 sheep/ ha/year on *C. setigerus* 7.0 in *C. ciliaris* as well as in *H. contortus* and 6.2 in *D. annulatum* grasslands (Upadhyay et al. 1971).

(b) *Harvesting method*: The harvesting schedule of grasslands for stall feeding during lean periods is important for their efficient management. Studies conducted at Jhansi on intensity and intervals of cutting on large number of forage species revealed that in almost all cases the highest production was maintained when harvesting of grasslands is carried out at 60 days intervals and at 10 cm height above the ground level with 30 kg N/ha. However, maximum crude protein content was recorded when harvested at 30 days intervals.

Silvopasture

Integrated silvipastoral system of farming has been recognized as low input technology for utilization of wastelands for meeting the shortage of forage leaf fodder besides fuelwood and timber. Long term studies conducted on certain pastures revealed that if properly established and managed the pastures could be reasonably productive even upto 8 to 9 years (Debroy, 1986). During establishment year pasture production is normally low varies from 1.8 to 2.5 t/ha. Generally tree samplings are planted during the monsoon after a good shower followed by sowing/planting of pasture grass and legumes. The details of transplanting of woody species are given below:

Pit size	
Shrub	0.3X0.3X0.3 m
Tree	0.45X0.45X0.45 m
Spacing	
Shrub	5X5 m
Tree	10X8 m
Sapling age	6 months to 1,5 year
Planting season	July/ after break of monsoon

References

Debroy, R. 1986. Studies on some aspects of renewable energy (fuel) and pasture production in an ecosystem of *Albizia procera* and *A. lebbeck* with grass anbd legume Ph. D. Thesis, Giwaji University, Gwalior.

Trivedi, B.K. and Kanodia, K.C. 1982. Succession of grassland vegetation on semi-rocky substratum at Jhansi. Indian Journal of Rangemanagement, 3, 1-10.

Upadhyay, V.S., Davadghao, P.N. and Sankaranarayan, K.A. 1971. Annual Report, Indian Grassland and Fodder Research Institute, Jhansi.

❐❐❐

Chapter 38

Genetic Improvement in Range Grasses and Legumes for Increasing Forage Productivity from Degraded Lands

AJOY KUMAR ROY
Indian Grassland and Fodder Research Institute, Jhansi-284003 (U.P.), India

Introduction

Forages constitute the base of a complete food chain and are the source of protein and fat *i.e.* meat, egg, milk and other dairy products that become available to human beings through intermediaries like cattle, sheep, goats, poultry etc. Since feeding alone accounts for 60-70% of the total cost of milk production, availability of adequate nutritious fodder coming from cheaper sources assumes greater importance.

Out of total Indian geographical area of about 328 million hectares, only 161.3 million hectares (roughly 52.7 percent) is arable. Examining the land resources available for fodder and forage crops, it is estimated that the average cultivated area devoted to fodder production is only about 4.4% of the total area. Out of the net effective cultivated area only 70 million hectares i.e. about 46% is irrigated and the rest 54% is rainfed only. About 158 m ha of lands is estimated to be under wastelands of different kinds *viz.*, sand dunes, ravines, saline-alkali, water-logged, water eroded, degraded forests riverine and land affected by shifting cultivation.

Rangelands are extensive areas which are unfit for arable farming and are mostly under natural vegetation where animals graze. The vast arid and semi arid tracts including the Great Indian Desert, the Himalayan rangelands involving the seasonal pattern of animal migration and other forest grazing areas depict the true nature of Indian rangelands. These vast areas could be put to use through efficient and speedy management through various means and it is in this context that the potential underutilized plants assume special significance and warrant urgent attention.

The availability of green forages from various sources is only 40% of the required quantity. It is a matter of prime concern to bridge this gap. This is also to note that the area under forage production has not increased considerably in the last few decades and our natural grazing lands and pastures are fast degrading. Hence, efforts should be directed to intensify forage production per unit area per unit time, which can be achieved through improved high yielding varieties and better management practices.

Tropical grassland are very important in livestock production, therefore, immediate attention needs to be given to varietal improvement of grasses and legumes through suitable breeding strategies. Forage crop improvement has its own limitations. Many aspects related to forage breeding, plant genetic resource, plant protection measures, quality, seed production and palatability need to be addressed in an integrated approach. The pasture species pose more complex problems in genetic improvement as compared to those encountered in cultivated forages.

India has a vast variety of rangelands depending upon the physical features and climate. However, most of these rangelands are at various stages of degradation. In India, permanent pastures and other grasslands are only 12 m ha . Another 20 m ha of land is reported to be barren and uncultivable.

Major problems associated with the improvement and management of rangelands/ degraded lands: The rangelands/ pasturelands can be effectively utilized for fodder resources. However, the poor degraded nature of these areas offers very little herbage of poor quality. Many problems exist in the proper management or improvement of these areas. Each of these areas in different localities has its own unique problems and remedial steps. No problem or solution is applicable to all types of lands. However, some of the common remedial steps are given below which can be used in many of such cases.

Control of grazing animals : Grazing restricts the growth rate to maturity of germinating seedlings and limits the chances of survival for seedlings over the first summer period.

Lack of protection : These areas are vast in nature and are open to biotic interference in large scale which restricts the growth and development of herbs thereby affecting the optimum productivity levels.

Poor people's participation: Rangelands or community land development needs different stakeholders of which the people/ graziers are most important. Their awareness and effective participation is a must for such task.

High cost of development: Development of uncultivable areas require many operations such as fencing, stone removal, unpalatable shrubs removal, water and soil conservation measures, seeding etc. which involve a huge cost.

Large grazing population / livestock pressure: In India, the grazing pressure is many times more than the carrying capacity of the land which accelerates the deterioration process. Excess grazing beyond the carrying capacity has caused severe degradation of our rangelands. They are now unable to provide nutritious fodder for the grazing livestock.

Poor grazing system/ Anna pratha: Often the farmers are not well informed about the production potential of the grazing lands and excess grazing affects the seedlings and early stages of growth, these affects the seeding process also thereby no next year germination is possible. Furthermore, excessive grazing causes soil erosion also.

Unpalatable and under nutritious shrub infestation: Different phenomenon increases the density of unpalatable shrubs as they are not grazed by the animals. These in turn affect the production of grasslands.

Development and proper management of rangelands: There are various ways that can improve the productivity of grazing lands and can provide high yield of quality nutritious herbage. Some of these steps include limited grazing in rangelands; introduction of rotation grazing; sowing of seeds of nutritious grasses with protection of sown areas; essential soil conservation works like bunds, check dams, fencing; Cultivation treatments to increase the amount of moisture stored in the soil profile providing an improved environment for plant establishment; seed lodgments in favored positions; increasing soil fertility to promote the germination and persistence of seedlings, providing adequate machinery for land development.

Genetic Improvement of Forage Grasses and Legumes

With growing realization of importance of forage crops in agriculture and livestock industry, it is becoming pertinent to pay more attention on the improvement of forages. From breeding point of view, forage crops have not been exploited to the extent as in the case of other crops.

Major problems faced in grass breeding: Some of the general constraints faced with forage production and improvement are as follows

- Non-availability of sufficient quantity of quality fodder seeds as the crop is harvested before seed set – need for dual purpose varieties.
- Non-synchronous flowering / anthesis and spikelet maturity, abscission of spikelet after maturity in grasses and presence of large number of sterile glumes in grasses.
- Apomictic nature of most of tropical forage grasses limit their genetic improvement
- Inbreeding depression occurs in them upon selfing and it is not feasible to develop and maintain inbred lines and therefore hybrid varieties.
- Grasses have small floral parts, making hand emasculation tedious and difficult job and requires highly skilled hand.
- Cytoplasmic male sterile system has not yet been developed.
- Most of the grasses are polyploid which complicate inheritance of traits. Most traits are controlled by numerous genes. Very few genes have been characterized or mapped due to these complex inheritances and the inability to self-pollinate plants.
- Polyploid nature of tropical forage grasses makes them very difficult material for genetic manipulation.
- Individual plants in population are highly heterozygous. Quantitative genetic studies indicate substantial additive genetic variation exists in most grasses for many agronomic traits.
- Usually grain and vegetative biomass have an inverse relationship

Possible Methods to Produce Hybrids of Forage Grasses

- First generation chance hybrids
- Self incompatible hybrids
- Apomictic hybrids

Development of Stress Tolerant Cultivars

It may be through conventional breeding method viz., identification of potential donors, hybridization, and evaluation of progenies and selection of desirable types.

Alternatively, it can be done through biotechnological tools

- Screening of germplasm
- *In vitro* screening
- Embryo rescue
- Somaclonal variation
- Gene transfer

The Major Breeding Objectives in Grasses

- High dry matter yield
- Promise in quality components- crude protein content, in vitro dry matter digestibility, low percentage of neutral detergent fibers
- High response to inputs
- Tolerance to adverse soils- acidic/saline soils, tolerance to extreme weather conditions – high rainfall/ low moisture regime
- Resistance to diseases and insect pests.

Leguminous Forages - Breeding Objectives

- Range legume varieties with high competitiveness and growth
- Efficient nitrogen fixing types
- Cultivars having high phosphate response for high dry matter and protein concentrates
- Transfer of genes from wild taxa for adoption, and disease resistance
- Varieties with less antiquality factors
- Varieties with synchronous flowering and with high seed retention ability.
- Overcoming overlapping of vegetative and reproductive growth phases, uneven pod setting, maturity and shattering in forage legumes.
- Greater persistence- summer persistence in annual multicut forages, e.g., berseem
- Exploitable regeneration ability in perennial multicuts e.g. Lucerne,
- Greater aggressiveness/fast growth and competing ability or complementation with the companion crop.

Breeding Strategies

- Germplasm should be exchanged and evaluated at different locations. The documentation should be done on priority.
- Information should be gathered for estimating the area under different forage species in different states preferably at the district level for setting priorities.
- Hot spot/sick nurseries should be developed for screening of germplasm where the insect-pest, disease infestation is causing severe damage.
- Nurseries should also be maintained for different problem soils and efforts should be made for testing and releasing varieties for problem areas.

Cereal Forages and Grass Pastures

- Direct introduction of high yielding adaptable biotypes
- Isolation of sexual lines in apomictic grasses
- Use of apomixis for rapid fixation of heterosis in cereal forages

Leguminous Forages

- Efficient nitrogen fixation types
- Wide tolerant lines for different soil conditions
- Disease and pest resistant lines.
- High protein and high dry matter content lines

Thrust Areas in Selected Range Grasses and Legumes

Range grasses such as Dichanthium, Cenchrus, Sehima, Chrysopogon etc.

- Collection, evaluation and documentation of all indigenous diversity.
- Identification of sexual lines.
- Good quality seeds

Dinanath Grass

- Tolerant to moisture stress
- High protein content
- Adapted to poor fertility soils

Table 1: Important grass varieties with area of adaptation and yield levels

Crop and Variety	Areas of adaptation	Green forage (q/ha)
Hybrid Napier (Napier- Bajra hybrid) (*Pennisetum.purpureum x P. americanum*)		
Pusa Giant	Whole of India and tropics abroad	1000-1300
NB-21	Whole of tropical humid part	1200-1500
CO-1	Tamil Nadu and Karnataka	1100-1200
Swetika-1 (IGFRI-3)	U.P., M.P., NE hills, Punjab and hills of North India	1100-1200
IGFRI-6	U.P., H.P., NE hills, Punjab an d hills of North India (intercropping)	1200-1300
PBN-83	Punjab	1250-1700
CO-2	Tamilnadu and Southern part	1200-1800
CO-3	Tamilnadu and Southern part	1300-2000
IGFRI-7	Whole of India (acid soils sub-temperate regions)	1300-1500
IGFRI-10	Whole of India (acid soils sub-temperate regions)	1300-1600
Yeshwant (RBN -9)	Whole of India	1300-1400
Guinea grass (*Panicum maximum*) Jacq.		
Macuenii	Kerala	600-700
Hamil	Kerala, Tamil Nadu, A.P., West Bengal, Bihar and North- Eastern States	700-800
PGG-1	North-West	900-1100
PGG-19	Punjab	750-1300
PGG-101	Punjab	800-1450
PGG-3	Northern, North-West and Central India	800-1000
PGG-9	Norther, North-West and Central India	900-1100
Deenanath (*Pennisetum. pedicellatum)*		
P.S. -2	Bihar, West Bengal, Orissa, North region of Maharashtra and A.P.	600-700
Bundel-1 (IGFRI-43-1)	Bihar, West Bengal, Orissa, North region of Maharashtra and A.P.	600-700
IGFRI-3808	Bihar, West Bengal, Orissa, North region of Maharashtra and A.P.	500-600
Bundel-2 (IGFRI-4-2-1)	Bihar, West Bengal, Orissa, North region of Maharashtra and A.P.	550-700
TNDN-1	Tamil Nadu	550-780
Buffel Grass *(Cenchrus ciliaris)*		
Bundel Anjan –1 (IGFRI - 3108)	Arid and semi-arid region	220-400
Bundel Anjan -2 (IGFRI 727)	Arid and semi-arid region	220-400
Neel kolu ka ttai (Co -1) Blou buffel (FS-391)	Tamil Nadu and other semi-arid areas	280-470
Setaria Grass (Nandi Grass)		
PSS-1	Sub –temperate hill region	750-1100
Tall Fescue Grass		
Him –1	Himachal Pradesh	400-450
***Sehima nervosum* (Sen grass)**		
Bundel Sen ghas 1 (IGS 9901)	Entire grass growing area	200-250
***Chrysopogon fulvus* (Dhawalu Grass)**		
Bundel Dhawalu ghas 1 (IGC 9903)	Entire grass growing area	200-250
***Heteropogon contortus* (Lampa Grass)**		
Bundel Lampa ghas 1 (IG HC 03-4)	Entire grass growing area	200-250

Range Legumes Such as Stylos, Siratro

- Collection, evaluation and documentation of available genetic diversity.
- Identifying strains for different ecological situations in country.
- Anthracnose resistant lines in Stylosanthes.

□□□

Chapter 39

Utilization of Degraded Lands through Hortipastoral System

SUNIL KUMAR, S. N. RAM AND A. K. SHUKLA
Indian Grassland and Fodder Research Institute, Jhansi-284 003 (U.P.), India

Introduction

In India, Out of 329 m ha geographical area 142.5 is under agriculture. 57% suffering from land degradation due to poor management of natural resources. 70 % area of cultivated land is rainfed comes under arid and semi-arid zone. In these areas crop husbandry is a big gamble due to uncertain of natural rainfall. Under such situation incorporating of fruit trees along with animal husbandry in common farming system is advisable to improve lively hood security and economic stability of people. In India 16% of human population of world has to sustain and progress only 2.33 % of total geographical area. Due to increasing population pressure, utilization of degraded lands or rainfed land is essential to feed nutritious diet with protective food (fruit) is essential. Similarly, our country is the largest livestock holder in the world and support 55% of its buffaloes, 16 % of the cattle, 20 % of the goats and some 4% of the sheep. These scenario of demographic pressure exert pressure on available land most of which is used for arable farming and food production. The pressure is bound to increase in future. Only 8.33 million ha of the total cropped area of the country is under forage crops and there is little scope for increasing are under forage due to pressure on land for food and commercial crops. The solution therefore, lies in maximizing forage production in space and time, through integration of forage crops in established or grown up plantations as well as establishing orchard.

Hortipastoral system: Integration of fruit trees with pasture (grasses and or legumes) in the same unit of land. This system acts as one of the best and economic alternative system for class V and VI (Singh, 1996). It can supply the protective food (fruit) for human being and fodder for animal and thus help in bridging the wide gap between the supply and demand of fruit and fodder. However, the success of developing hortipasture land use system depends largely on selection of fruit plants and pasture species. The selection depends upon objectives besides soil and climate.

Establishment of Horti-Pastoral System

1. *Protection of area from Animals*: The area should be protected from biotic interferences. This may be done with any fencing material as per its availability such as barbed/woven/chicken wire with angle iron/stone wooden pole, cattle proof trenches (2x1x0.3 m), live hedge fencing etc.
2. *Soil and water conservation through water harvesting in- situ*: Water is always a critical factor in rainfed situation. Survival can be generally enhanced by reducing runoff and collects at the root zone by any conventional practices of soil and water conservation.
3. *Selection of suitable fruit tree:* Selection of fruit tree for hortipastoral system under semi-arid/raifed environment is quite difficult task. Prior to selection of fruit species, it is pre-requisite to select a particular variety for that region. The tree should be chosen which are complementary or less competitive for understorey pastures. In continuous graging system, custard apple could be an ideal option as it is not browsed by livestock and hardy too. Fruit like ber will grow very well in combination with pasture, as the foliage is very notorious and palatable. Suitable fruit species with suitable varieties for raifed situation suggested by Kumar and Kumar 2001 (Table 3).
4. *Selection of suitable perennial grasses*: For specific agro climatic condition it is also important for increasing forage as well as animal productivity. Besides pasture grasses, some legume species such as *Lablab purpureus, Clitoria ternatea* and *Stylosanthes hamata* are also suitable for pasture development. For different agro-climatic conditions, suitable species of pasture for horti-pastoral suggested Korwar *et al*, 1988 (Table 4).

5. *Establishment of fruit tree in hortipasture system*: After protection and Eselection of suitable species, planting and sowing may be done during monsoon season. For fruit trees pit size $1m^3$ should be prepared at 6-10m apart depending upon fruit trees species before onset of monsoon The pits should be filled with well rotten FYM (30-50 %) of the total volume of the pits) and good soil along with termicide. In between row of fruit trees transplanting / sowing of grasses and legume should be done in 1:1ratio at 50 cm spacing. In sodic soil use of 3 kg gypsum + 8 kg FYM or 30 kg sand + 30 kg FYM along with original soil has showed promising results. Apart from FYM or compost, paddy husk, groundnut shell, green leaves of *neem* or any other trees could also be used as filling mixture. Chemical fertilizer generally recommended are DAP, Single super phosphate with urea. Small amount (50 g) of BHC/ Aldrin 10% dust may be applied in light textured soils to prevent termite attack. Water holding capacity in pits improves with the amount of FYM/ compost added and it should occupy minimum 25 % volume. Water holding capacity may be further improved by filling bentomite clay or tank silt (5 cm thick layer) at the bottom which will check the deep percolation losses. In order to get higher forage production from pasture species 30 kg N + 20 kg P_2O_5 /ha /year should be given during monsoon season.
6. *Establishment Techniques of grass*: A moderately good tilth seedbed is required for better establishment of pastures and must be sown at the beginning of the rainy season on a drizzling day. Seeds could be broadcasted, since the seed is small in size, seeding at shallow depth is important. However, raking of the soil after the broadcasting improves the germination considerably; this may be done by running a brush wood (branch with twigs) over the area. For larger areas broad casting of seed pallets (seed and soil) are also recommended. In hortipastoral system pasture establishment are recommended through transplanting of rooted sleep or root tussock along with legume row in 1:1.

Horti-Pastoral Work in India

Ber-based Hotipastoral System

In Ber based hortipastoral system during first phase (1990- 98) involving various combination of *C.ciliaris*, *Stylosanthes hamata* (Sharma 2004), four jujube (*Ziziphus mauritiana* Lamk.) cv. Gola plants per treatment combination were planted at 6x6 m spacing in 1991 in the combinations. The fruit trees and pasture components were planted in the following seven combinations, viz.,(i) Jujube alone (iii) Jujube + Buffel; (iii) Jujube + Stylo; (iv) Jujube + Buffel + Stylo; (v) Buffel sole; (vi) Stylo sole; and (vii) Buffel + Stylo without jujube trees. The ber tree started fruting 3rd year on ward (0.125 to 0.164 t/ha) under various treatment and increased up to 8th year (5.37-6.55 t/ha).The fruit yield was not affected significantly by growing pasture as intercrops in the inter space. Pasture production, in the first year, the fodder production was low in all treatment combinations, e.g., buffel (1.71 t DM / ha), stylo (1.86 t DM /ha) and mixture crop of buffel + stylo (2.22 t DM /ha). Pasture production showed a general trend of increase in dry matter yield upto 4th-5th years. In fourth year, the forage production was 4.59, 3.23 and 4.27 t DM / ha in buffel, stylo and mixed pasture respectively.

In 2nd phase of experiment (1999-2000) ber tree were pruned at bifurcation point of main trunk at three height + bajra (fodder) with four doses of nitrogen. It was found that pruning of main branches of *Ber* trees at 90 cm length from bifurcation point recorded significantly higher fruit yield as compared to pruning at 60 and 30 cm length from bifurcation point during both the years. Nitrogen application @ 80 kg/ha (24.52t/ha green fodder)significantly increased growth parameters and green forage yield of pearl millet as compared to control and 40 kg/Nha (20.57t/ha green fodder)and it was at par with 120 kg/ha(26.05T/ha green fodder) during both the years. However, fruit yield of Ber7.02,10.02,12.58,14.32t/ha increased significantly with increasing levels of nitrogen from 0to 40, 80 and 120 kg N/ha respectively (Kumar and Ram 2007).

In the 3rd phase of experiment (2001-05), the ber tree , secondary branches were pruned at 20,40 and 60 cm height and under storey space of grown up ber plantation was utilized with four pasture combination *i.e.* Guinea grass +*S. hamata*, Dinanath grass +*S. hamata*, Guinea grass +*Dinanath grass* +*S. hamata*. The medium pruned tree produced significantly higher fruit yield 7.3t/ha as compared to severe (5.07 t /ha) and light pruned (6.48 t/ha) tree. The fruit quality except fruit diameter and vitamin C content was influenced significantly with pruning intensities (Kumar *et al*, 2006). The pasture production under severely pruned tree was significantly higher4.02 t DM/ha as compare to light pruned tree 3.44t DM/ha. Among pasture combination under grown up ber orchard Guinea grass +*Stylosanthes hamata* produced significantly higher yield (6.29 t DM/ha) followed by Guinea grass +Dinanath grass + *Stylosanthes hamata* (4.68 t DM/ha), Dinanath grass + *Stylosanthes hamata* (2.82 t DM/ha) and natural biomass(1.26 t DM/ha).

Aonla-based Hortipastoral System

In *Aonla* (cv.NA-7) base hortipastoral system (1996-06) involving seven treatments viz. T_1 (*Aonla* pure as control), T_2 (*Aonla* + *Dichanthium* + 20 kg N/ ha to grass plot), T_3 (*Aonla* + *Dichanthium* + 40 kg N/ha to grass plot), T_4 (*Aonla* + *Dichanthium* 60 kg N/ha to grass plot), T_5 (*Dichanthium* + 20 kg N/ ha) , T_6 (*Dichanthium* + 40 kg N/ha), T_7 (*Dichanthium* + 60 kg N/ha). The fruit production was started from 5th year of planting. The average maximum fruit production 9.50 t/ha was recorded with 60 kg N /ha applied to the grass plot .The understorey pasture production with aonla tree was continuously increased from 1st year to 5th year of experiment and after ward it was declined with growth of tree. This decline was 13.3 % and 34.3 % respectively in the 7th and 8th year and 45 % in 10th year in association with tree. Similar trend of decline in pure pasture treatment was also observed during this period in the 10th year it was 53 % in pure pasture as compared to maximum production in 5th year. Among nitrogen treatments higher dose of N (60 kg/ha) gave significantly higher pasture yield (4.04 t DM/ha) with tree when compared to pure pasture condition (3.73 t DM/ha). The pasture production was always higher in association with tree as compared to pure pasture but spastically at par and N higher dose showed significant effect as compared to lower doses of N and control (Kumar and Chaubey, 2008).

Plate 1. Aonla-based Hortipasture System at IGFRR, Jhansi

Hortipasroral System on Bund

Field bunds are used in most parts of India to check runoff loss of rainwater and top soil. It has been estimated that 6-10 % of cultivatable land area is occupied by bunds.Khan *et al* (2001) reported tha these bund could be utilized through hortipastoral system. Grasses and trees are often planted to reduce water runoff, increase infiltration and reduce soil loss through their barrier effect.

CRIDA, Hyderabad

Guava and custard apple based hortipastoral systems have been developed at Hyderabad and after 5 year highest survival of fruit trees(Table2) was observed in control followed by trees in association with Stylo and *Cenchrus* respectively (Mohammed Osman 2003).

CAZRI, Jodhpur (Rajasthan)

Sharma and Diwakar 1989 reported that among the various hortipastoral combinations has been attempted on growing of different pasture in inter-space of productive and unproductive ber orchard. Hortipastoral studies on sandy rangelands of Rajasthan revealed that *Cenchrus ciliaris-Ziziphus mauritiana* system produced 1.2 t DM/ha as forage and did not affect the fruit production.

In an study in sandy rangelands at Samadri (Western Rajasthan) have shown that plantation of two jujube species, viz, *Ziziphus rotundifolia* and *Z. nummularia* each with densities 280, 140, and 70 plants /ha, and *Cenchrus ciliaris* as pasture grass gave forage yield varying from 0.624 to 0.824 t /ha in different treatments. Different population densities of jujube in the pasture had no significant effect on the yield of pasture indicating thereby that even 280 trees /ha can safely be planted. *Z. numularia* started top feed and brush wood production after 4 years. Sharma and Diwakar (1989) worked out the economics of hortipastoral system and found the return to be remunerative even with the higher discount rate of interest (29%) and increased investment cost (25%) due to hike in prices of the inputs (Sharma and Vashistha, 1985).

In a grazing study, *Z. nummularia* with grass strips in 1:2 ratio led to higher economic returns with higher weight gain of the animal and higher wool production. Growth and top feed production of *Z. nummularia* and *Z. rotundifolia* in association with established pasture of *Cenchrus ciliaris* at Samadri (Western Rajasthan) showed that after 5 years of establishment , growth of jujube species was more than two times and pala production was more than 6 times to control (no pasture) as compared to established pasture. However, the total forage production varied from 0.135 to 1.269 t DM /ha and the difference (with or without jujube) were non significant. Planting of jujube in pasture at 6 m apart was recommended (Sharma and Vashishtha 1985).

CPCRI, Kasargod (Keral)

Subramaian *et al*, (2006) conducted a trial with using husk/coir (byproduct of coconut cultivation) as soil and water conservation in coconut orchard and understorey space was utilizing as intercropping by Napier Bajra grass cv. Co3. he got 100 t green fodder which was sufficient to maintain 8-10 milking animal. Further these soil and water conservation measures resukted in higher crude protein and crude protein yield.

CHES, Godhara (Gujarat)

Raturi and Hiwale (1993) conducted a trial on horti-silvi-pstoral system at Central Horticlture Experiment Station Godhara Gujarat and reported that despie failure of monsoon for three consecutive years, the system gave encouraging result. *In Situ* budded Ber started from very first yea. In second year an average yield of 16.10 q/ha was recordred which increase to 49.00 q/ha in third year along with *Cenchrus ciliaris* gave38.87 q/ha green fodder in second year and 55 q /ha green fodder in third year where as *Stylosanthes hamata* gave 31.9 t /ha green fodder in second year and 38.88 q /ha green fodder in third year. Through ber pruning they reported 2.2q /ha fuel wood in second year 13.00 q /ha fuel wood in third year.

CSWCT&RI, RRS, Chandigarh

Aonla (*Cv Chakaiya*) – based hortipastoral land use system was established in 1991-92 on degraded lands in Shiwalik foot hill in combination with *Chrysopogan fulvus* (Dholu), *Pennisetum purpureum* (Napier), *Cajanus cajan* (Perennial pigeon pea) and sole Aonla (Prasad *et al*,2007). Fruit yield in aonla Cv. Chakaiya started on regular basis from 5th year of plantation. Summarized finding of 14 year old system are as under:

Understorey pasture growth indicated that Dholu grass was most compatible in initial pre bearing stage, and behaves as the fertility restorer and affects the synergetic growth of Aonla in post bearing stages. Hybrid Napier grows very fast and attained height up to 2 meter under annual cut conditions. It produces large biomass thus deprecates soil, water and nutrients, and affects the growth of Aonla.

Fruit yield assessment (per plant) revealed average yield 84.02 kg over a period of 9 year of bearing with minimum at 1st year (25.4 kg) and maximum at 9th year (150.75 Kg) respectively. Also the variation in yield (%) was higher at the 1st year (68.33) and minimum at 3rd year (34.5 %) of bearing respectively.

Table 1. Fodder /Grasses Under Horti-pastoral System on Bund (Dry fodder productin on bund at Jhansi)

S.No.	Treatment combinations (tree+grass+ legume)	Total dry fodder yield of two cust (kg/15 meter length of bund)
1	*Emblica officinalis +Chenchrus ciliaris +Stylo*	35.0
2.	*Emblica officinalis + Cynodon dactylon +Stylo*	20.8
3.	*Emblica officinalis+Napier +Stylo*	44.5
4	*Emblica officinalis+Pennicum tri hybrid +Stylo*	33.0
5	*Ziziphus Mauritian + Chenchrus ciliaris +Stylo*	27.4
6	*Ziziphus Mauritian + Cynodon dactylon +Stylo*	13.1
7	*Ziziphus Mauritian+ Napier +Stylo*	47.1
8	*Ziziphus Mauritian+ Pennicum tri hybrid +Stylo*	22.3
9	*Carissa carandus+ Chenchrus ciliaris +Stylo*	28.4
10	*Carissa carandus+ Cynodon dactylon +Stylo*	25.8
11	*Carissa carandus+ Napier +Stylo*	48.1
12	*Carissa carandus+ Pennicum tri hybrid +Stylo*	32.7
	LSD (0.05)	12.5

(Source: Khan *et al.*,2001)

Table 2. Growth Parameters of Fruit Trees (18 months old) and Yield of dry Grass in Hortipastoral System

Pasture species	Guava		Custard apple		Yield of dry grass (t/ha)
	Plant height (m)	Diameter at knee height (cm)	Plant height (m)	Basal diameter (cm)	
S. hamata	1.67	1.81	1.04	2.45	*
C. ciliaris	1.78	2.04	0.98	2.15	5.0-9.6
Control (No pasture)	1.61	2.24	1.08	2.78	-

*Stylo not harvested in the first year to encourage self-seeding and productivity
(Source: Annual Report, CRIDA, 1987)

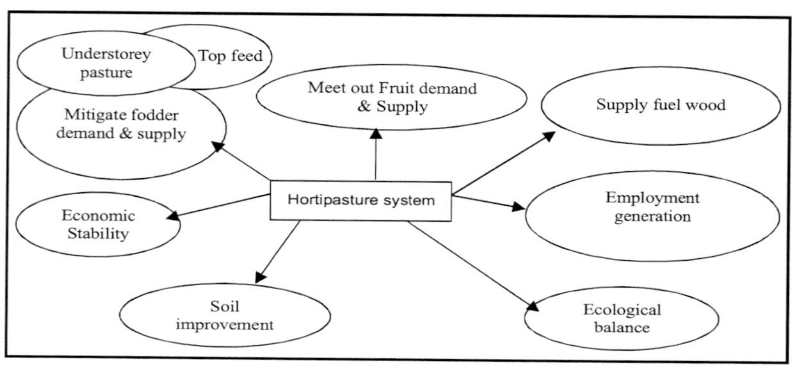

Flow-chart of Advantages of Hortipastoral System

Table 3. Suitable Fruit Crops and Cultivars for Different Rainfall Zones.

Rainfall (mm) zone	Fruit crop	Cultivars
Less than 350	Ber *(Zizyphus mauritiana)*	Seb, Gola, Mundia
	Gonda *(Cordia myxa)*	Local strain
	Ker *(Capparia decidua)*	Local strain
	Pilu *(Salvadora oleoides)*	Local strain
350-500	Ber *(Zizyphus mauritiana)*	Katha, Maharwali, Bagwadi, Seb, Gola, Mundia
	Aonla *(Emblica officinalis)*	Banarasi, Hathijhool, Chakaiya,
	Khirni *(Manilkara hexandra)*	Local strains
	Jamun *(Syzygium cuminii)*	Local strains
	Mulberry *(Morus alba)*	Local strains
500-700	Bael *(Aegle marmelos)*	Faizabad selection
	Mango *(Mangifera indica)*	Seedlings
	Custard apple *(Annona spp.)*	Balanagar, Mammoth, Washington
	Sour lime *(Citrus aurantifolia)*	Baramasi
	Aonla *(Emblica officinalis)*	NA 7, NA 6, Krishna, Kanchan
More than 700	Mango *(Mangifera indica)*	Bombay green, Chausa, Safeda, and local cultivar
	Pomegranate *(Punica granatum)*	Jalor seedless, Jodhpur red
	Guava *(Psidium guajava)*	Allahabad Safeda, L-49
	Jackfruit *(Artocarpus heterophyllus)*	Local strain
	Wood apple *(Limonia acidissima)*	Local strain
	Tamarind *(Tamarindus indica)*	Local strain
	Aonla *(Emblica officinalis)*	NA 6, NA 7, Kanchan, Chakaiya.

Source: Kumar and Kumar, 2001

Table 4. Suitable Pasture Species for Drylands

Rainfall (mm) zone	Grass/ Legume Name	Soil type	Dry forage yield (t/ha)
Less than 350	Buffel grass (*Cenchrus ciliaris*)	Sandy to sandy loam	4.0
	Bird wood grass (*C. setigerus*)	Sandy to sandy loam	3.0
	Sain (*Sehima nervosum*)	Red, gravelly to sandy loam	2.5
	Stylo *(Stylosanthes scabra)*	Sandy to sandy loam	2.5
	Butterfly pea *(Clitoria ternatea)*	Sandy loam to Silty clay loam	3.0
	Bankulthi *(A. scarabaeoides)*	Gravelly, sandy loam well drained	2.0
350 - 500	Blue panic (*Panicum antidolate*)	Loam to sandy loam	3.0
	Marvel (*Dichanthium annulatum*)	Loamy to clay loam	2.5
	Dinanath grass (*P. pedicellatum*)	Loamy to clayey loam	3.0
	Bahia grass (*P. notatum*)	light	3.5
	Sabi grass (*U. mosmabicensis*)	Sandy to sandy loam	3.0
	Stylo *(Stylosanthes hamata)*	Sandy loam to loam	3.5
500-700	Rhodes grass *(Chloris gayana)*	*Sodic, loam to sandy loam*	2.5
	Stylo *(Stylosanthes hamata)*	Sandy loam to loam	3.5
	Siratro (M. atropurpureum)	Sandy to sandy loam	1.8
More than 700	Napier grass (*P.* purpureum)	sandy loam to clay loam	4.0
	Marvel (*Dichanthium annulatum*)	Loamy to clay loam	2.5

Source: Korwar *et al*,1988

Conclusions

Based on above discussions, through the development of hortipastoral system in degraded land of India particularly under rain fed situation we can make possibility of imparting stability to farmer's income by promoting integration of fruit trees in their farming system as well as animal husbandry occupation. Hortipastoral systems provide nutritional security through fruit as well as animal product, maximize biomass productivity and sustained productivity.

References

Korwar, G. R., Usman Mohd., Tomar,D. S. and Singh, R. P. 1988. *Dryland Horticulture,* CRIDA Hederabad 34 pp.

Kumar,S Singh , J. B. and Ram ,S. N. 2006. Influence of pruning and weather on productivity of ber based hortipasoral system in rainfed conditions. *Indian Journal of Horticulture* 63, 260-263.

Kumar, S. and Chaubey, B.K. 2008. Performance of aonla (*Emblica officinalis*)- based hortipasture system in semi-arid region under rainfed situation.*Indian Journal of Agricultural Sciences* 78, 748-51.

Kumar, S. and Kumar, S. 2001. Fruit trees for sustainable agriculture through hortipastoral system in semi-arid environment. *Range Management & Agroforestry* 22, 33-42.

Kumar,S. and Ram S.N. 2007. Effect of pruning and nitrogen levels on fruit and fruit and forage yield of ber + pealmillet inter cropping system under rainfed condition. *Farm Science Journal* (In press).

Osman, Mohammad 2003. Alternate land use system for sustainable production in rainfed areas In : P.S.Pathak and Ram Newaj (Edts.) Agroforestry Potential and Opportunities pp 170-181. *Agrobios* (India) Jodhpur.

Prasad, Ram; Yadav,R.P., Agrawal,R.K., Agnihotri, Y., Mittal , S.P., Tiwari A.K. and Samara, J.S. 2007. *Chandigarh,Prospects of Aonla cultivation on degraded Shivalik land.* Technical Bulletin No. T-55/ C- 13. Publisher P.P. Advertisers & Printer Pvt. Ltd. 140. Industrial Area Phase-1 .

Rao,J.V. and Rao, V.S. 2004. *Raifed horticulture in different AESRs of A.P. in rainfed Agriculture Technologies for different agro-eco- regions of Andhra Pradesh* (eds.P. K. Mishra *et. al*), Central Research Institte for Dryland Agriculture , Hyderabad ,pp 86-93.

Raturi, G.B. and Hiwale, S.S. 1993. Horti-silvi-pastoral system for increased productivity of marginal and degraded lands under rainfed condition. *Advances in Horticulture and Forestry* 3, 179 186.

Sharma, S. K. 2004. Hortipastoral based land use systems for enhancing productivity of degraded lands under rain fed and partially irrigated conditions. *Uganda Journal of Agricultural Sciences* 9, 320-325.

Sharma, S. K. and Diwakar, G.D. 1989. Economic evaluation of hortipastoral system on arid region of western Rajasthan. *Indian Journal Range Management* 10,119-122

Sharma, S.K. and Vashishtha, B.B. 1985. Evaluation of jujube-buffel grass hortipastoral system under arid environment. *Annals of Arid zone* 24 :303-309

Singh, R.P. (1996). Alternate land use system for sustainable development. *Range Management & Agroforestry* 17, 155-177.

Subramanian, P.,Dhanapal,R.,Palaniswami,C and Sebastian Joseph . 2006. Napier bajara grass as intercrop in coconut garden .*SAIC Newsletter* 16,10-11.

❑❑❑

Chapter **40**

Role of Weather Based Agro-Advisory Services in Decision Making for Forage and Livestock Production

J. B. SINGH AND PRADEEP BEHARI
Indian Grassland and Fodder Research Institute, Jhansi-284003 (U.P.), India

Introduction

Inspite of technological developments for the last few decades, Indian agriculture is still subject to vagaries of weather. Weather is crucial to food and fodder production. A study of WMO reports that the weather produced variability of food production is more than 10 per cent. The weather induced variability can be as high as 59 per cent of the normal production in respect of smaller area. In India, the losses in crop production are even of greater magnitude as the weather impacts are magnified by poor management practices. Agriculture is thus largely weather and climate sensitive. Recently increased activity of the extreme weather events such as drought, cold wave and heat wave has been observed. Deficit rainfall in many parts of the country during 2004 and excess rains in 2005 (100 cm in Mumbai, floods in MP, Gujarat, Orissa, AP, Tamilnadu and Karnataka) affected the food grain production. Thus timely and accurate weather based agro advisories is the need of the hour for sustainable agricultural production.

The country has about 4.9 % of the total cropped area under cultivated fodder. The fodder production in the country is not sufficient to meet the requirement of the growing livestock population. Even for meeting the current projected (Hazra, 1995) green fodder requirement (1136 mt) about 10% of the cultivated area (14 m ha) needs to be diverted for the production of the cultivated fodder crops, which is a difficult task to achieve. Meeting fodder demands has become major challenge now than ever before. Therefore, a significant effort is necessary to increase fodder and livestock productivity by proper use of agro-advisories especially in dryland agro-ecosystem of the country. The effective dissemination of advisories to farmers will create awareness across larger farming community on the importance of weather-based agriculture management and making farmers to adapt weather-based advisories. This will ultimately leads to increase the quality and quantity of produce, minimizing losses due to adverse weather and reducing cost of cultivation to enhance profits

Weather forecasting is a specialized field to serve a segment of national economy. Because of its economic importance, there is definite need to expand the sphere of weather forecasting knowledge. It plays a vital role for making much operational decision in agriculture both strategic and long range and tactical or short range decisions (Singh and Tomer, 2002). Weather decision which influences fodder and livestock production can be forecasted for different time span. Weather forecast can be classified into three categories viz., i) short range forecast-covering a period of few hours to three days ii) medium range forecast-covering a period of 3-10 days iii) long range forecast-covering a period of a month or a season.

Short Range Forecast

Short range forecast (SRF) has a period of validity from a few hours to three days. The short range forecasting is usually approached by subjective or objective methods. Short range forecast utilizes synoptic methods, statistical methods and numerical weather prediction (NWP) techniques. The SRF is based on the synoptic situation prevailing at the time of forecast. Forecast are issued for various for various meteorological elements e. g. i) distribution of rainfall & warning for heavy rainfall ii) day, night temperature, western disturbances, storms, depression, heat and cold wave and special weather hazards like thunderstorm, hailstorm, snow, frost etc. Besides above elements expected cloudiness, probability of rain and expected rainfall, wind speed and direction and dew duration are also included in the forecast. These forecasts are used by agricultural scientist for agricultural operations. This helps farmers to make tactical decisions such as cultural practices and decisions involving crop protection system to minimize input losses resulting from adverse weather condition and can markedly improve the fodder yield.

Medium Range Forecast

The medium range forecasts (MRF's) are considerably important for agricultural operations. Various techniques of MRF based on statistical methods, a combination of synoptic and statistical methods and NWP are available. MRF aims at weather elements e.g. rainfall, temperature from long term averages rather than the instantaneous weather. Keeping in view the importance of MRF, Govt. of India had set up (Fig.1) national centre for medium range forecast (NCMRWF) in New Delhi. Once the primary objective (Rathore *et al.* 2004) of NCMRWF is to prepare medium range (3-10 days in advance) weather forecast on real time basis. This centre is equipped with a super computer CRAY X/MP 14 with VAX gateway and VAX based front-end processors. Online dedicated and fast telecommunication linkages (Fig. 2) with the meteorological centre of IMD and Agro-met field units are envisaged to disseminate MRF's for issue of agro-advisories to farmers in different climatic zones of the country. Presently 107 of the 127 agroclimatic zones of the country are covered under this programme.

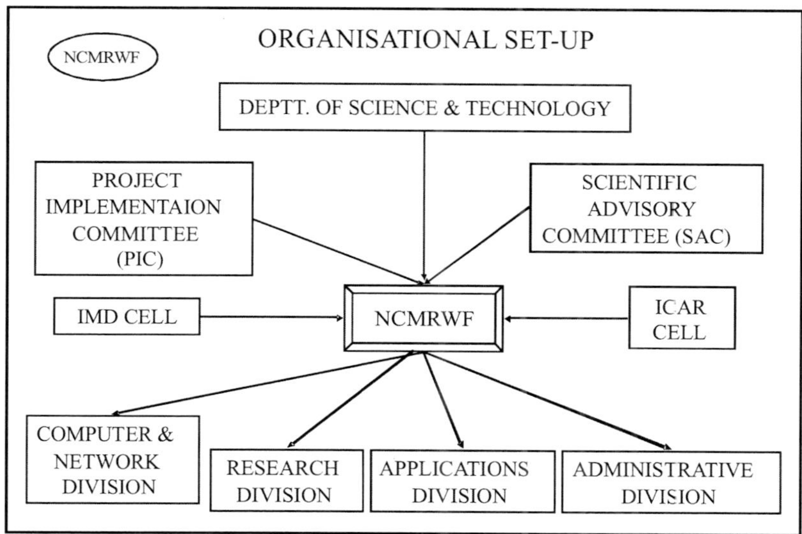

Fig. 1. Organizational Set up of NCMRWF

Developing weather forecasting models in the Indian context has always been a formidable task due to the uniqueness of the Indian sub continent and he associated monsoon phenomenon. Scientists from IMD and NCMRWF are making sincere efforts to develop suitable medium range forecast model. It is hoped that operationally viable mathematical model for specific use will be developed very soon and weather based agro-advisories will become an integrated part of farming system approach.

Short and medium range weather forecasts are extremely useful in a number of agricultural operations viz; sowing, irrigation, chemical application, and control of plant disease, frost protection, harvesting and transportation. These decisions which would be realistic, can help in use of resources to optimal level i.e. there need to be over supply of labours, power, water, chemicals, machinery etc.

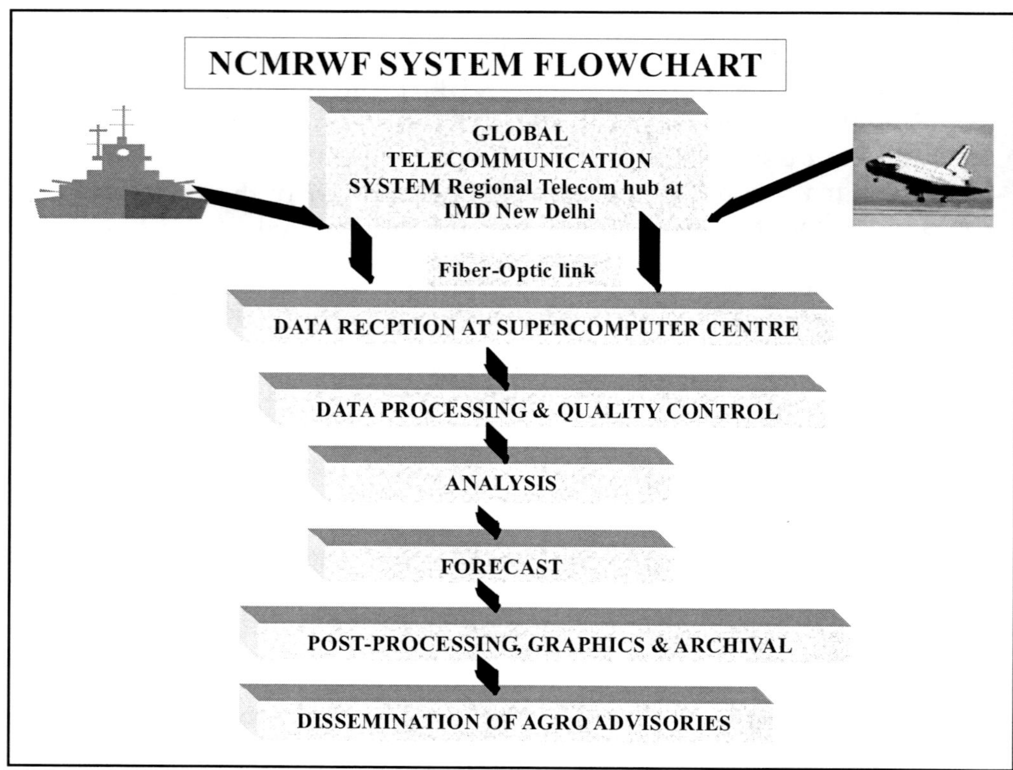

Fig. 2 Flowchart for NCMRWF System

Long Range Forecast

Weather forecast valid for more 10 days is known as long range forecast (LRF). In the field of LRF, India enjoys a pioneering position. Recently, the India Meteorological Department (IMD) has achieved a major breakthrough in making all India long range forecast of southwest monsoon rainfall accurately on the basis of parametric and power regression model (Gowariker,

1991), which take into account the impact of 16 regional and global weather parameters. The accuracy of prediction of the model has been witnessed for a number of years in succession since its introduction in 1988. Long range forecast on all India summer monsoon rainfall can be highly useful in predicting the likely trends in food grain production in the country before the beginning of the *kharif* season.

Agromet Advisory Services

Agro-met advisory bulletins are issued bi-weekly and it is fully operational in India. The IMD regulates and coordinates the AAS unit. These units prepare detailed weather charts for meteorological elements of specific relevance to agriculture to agriculture like rainfall, number of rainy days, max. and min. temperature, cloud cover and relative humidity for the week. To cater the agricultural need up to district level, the IMD has been issuing regular forecast known as Farmers Weather bulletin (FWB). The objective of FWB is to provide better planning, improved management and reduced inputs. These FWB contains district wise forecast for next 36 hours with an outlook for the subsequent two days. Agromet advisory has application at each stage in the planning and management for farming. Agromet advisory may recommend certain practices to help effectively or minimize possible weather related crop damage or loss, for example:-

i) Spraying advice based upon past and forecast weather conditions to combat crop disease and insects.

ii) Harvesting advices to obtain optimum crop and fodder quality

iii) To initiate cultural practices which are weather sensitive

District Level Agro-meteorological Advisory Service (DAAS) of IMD

The weather based Agromet services of National Centre for Medium Range Weather Forecasting (NCMRWF) and India Meteorological Department (IMD) were integrated into IMD to provide DAAS for the benefit of farming community in July 2006. This is multi-institutional project involving stakeholders such as State Agricultural Universities (SAUs), Indian Council of Agriculture Research (ICAR), Krishi Vigyan Kendras (KVK), Department of Agriculture and Cooperation, State Department of Agriculture/ Horticulture/ Animal Husbandry/ Forestry (up to district level officers), NGOs, Media Agencies etc. The DAAS system is implemented under the five-tier structure to set up (Fig. 3) different components of the service spectrum and is in operation since 1st June 2008. IMD is generating and issuing district level Medium Range Weather Forecast (MRWF) from current monsoon season i.e.

June 2008 for issuance of district level agromet advisory. District level weather forecast (rainfall, temperature, wind direction and speed and cloudiness) for next 5 days with cumulative weekly rainfall are being prepared at Regional Meteorological Centres (RMCs)/ Meteorological Centres (MCs) based on the NWP products (Table 1) received from the NHAC, IMD, New Delhi. These weather forecasts are communicated to the 130 Agromet Field Units (AMFUs) in the country on every Tuesday and Thursday for preparation of the crop specific advisory at district scale. A network of 130 AMFUs (Agro-meteorological Field Units) has been set up covering the agro-climatic zones of the country. These units are responsible for recording agro meteorological observations, preparing medium range weather forecast based Agromet advisories for the districts. The Agromet bulletins include advisories on field crops, horticultural crops and livestock.

Table 1 : NCMRWF Forecast Products Disseminated to AAS Units (source: Rathore, 2008).

Forecasting elements	• 24 hr precipitation (mm) • Average cloudiness • Average wind speed (Km/hr) • Predominant wind direction • Maximum temperature trend (^{0}C) • Minimum temperature trend (^{0}C)
Frequency of forecast	Twice a week
Dissemination	On Tuesday and Friday
Period covered	4 days/Outlook for week

Agro-Advisory and Information Dissemination System

These weather based advisories would be disseminated to the farmers at district/block/village level primarily through District level intermediaries, Mass media dissemination, Internet. The advisories would also be communicated through other multi-channel dissemination system like~ All India Radio (AIR) and Doordarshan, Private TV and radio channels, Newspaper, Internet , Krishi Vigyan Kendras (KVKs)/Kisan Call Centres / ICAR and other related Institutes / Agricultural Universities / Extension network of State / Central Agriculture Departments. Regular feedback would be made based on the quality of forecast, quality & relevance of advisories, problem solving through interactive mode, answering questions of common interest through bulletins, accessibility to information via ICT, accessibility to experts & video conferencing. The flow chart for implementation of District

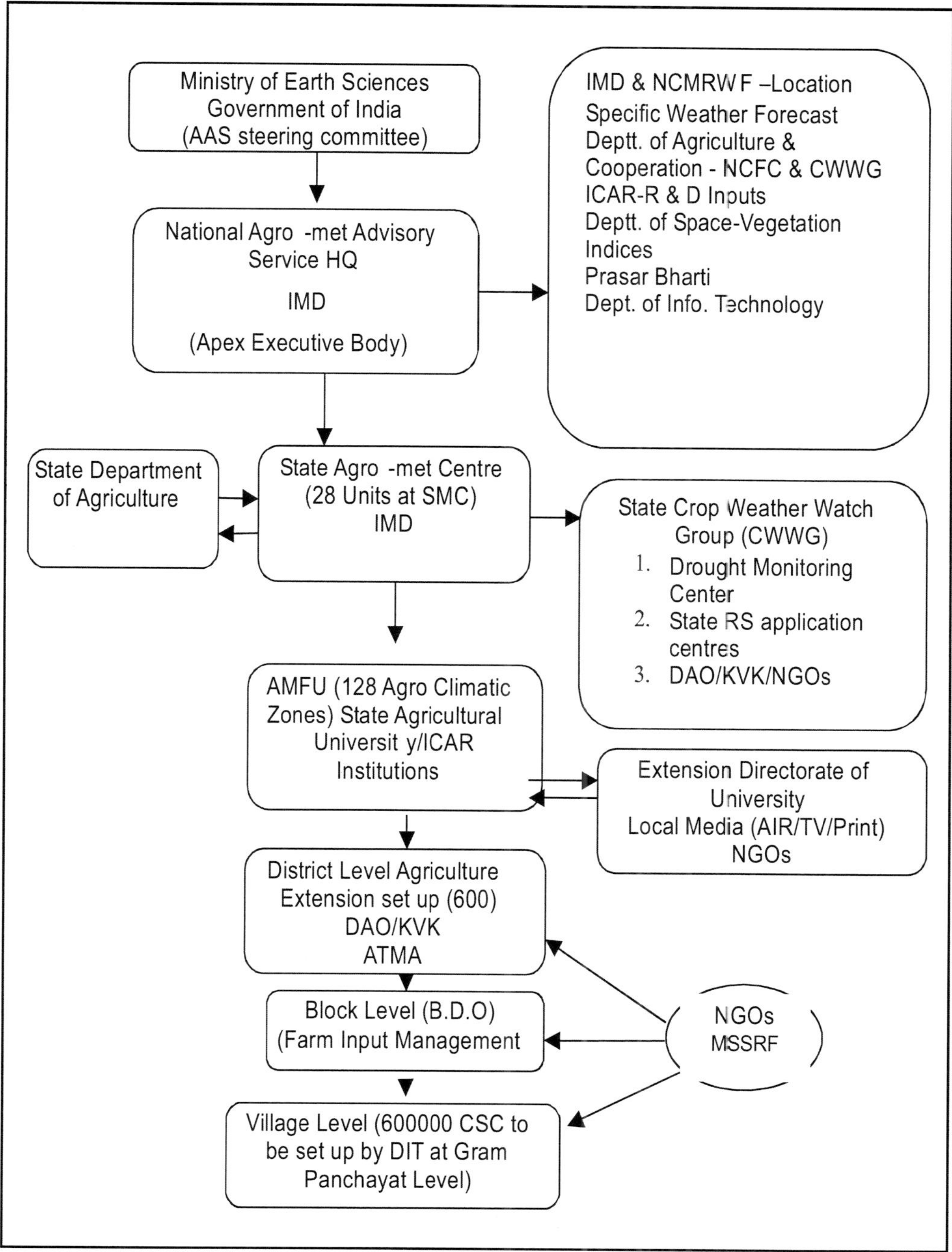

Fig. 3: Collaborating Organization and their Linkages under Integrated Agromet Services (Source: IMD, 2007).

Level Agro-Met Advisory Service (DAAS) System in the country is shown (fig. 4) as below:

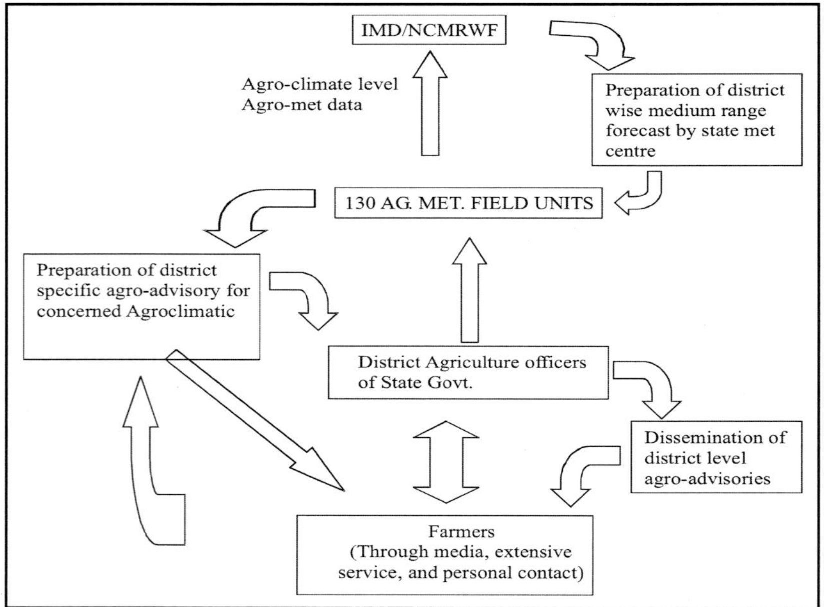

Fig.4 District Level Agro-Met Advisory Service System (Source: IMD 2007)

Weather Based Agro Advisory Programs in ICAR

At present 25 centers of AICRPAM (Rao, 2008) and 107 field units of NCMRWF along with regional agro advisories of IMD are involved in the advisory services (fig 5). Agro advisories are available at Crop weather watch group at the national level. Dissemination of agro advisories are also done through Website (www.cropweatheroutlook.ernet.in) and IT enabled 'e-sagu(IIIT-CRIDA).

Role in Farm Management

The weather informations through agro advisories has crucial role in decision making on the following aspects

- Cultivars selection
- Choosing windows for sowing/harvesting operations
- Irrigation scheduling - optimal water use
- Mitigation from adverse weather events such as frost, low temperature, heavy rainfall - at critical crop stages

- Fertilizer application
- Pesticide/fungicide spraying schedules
- Feed, health and shelter management for livestock [Optimal temperature for dairy/hatchery etc]

Advisory Bulletin Format

The agro advisory bulletin contains following information

Weather Information

- Preceding week weather summary
- Climatic normals for the week
- Weather forecast

Crop Information

- Type and Phenological stage of the crops
- Information on the pest and disease and crop stresses

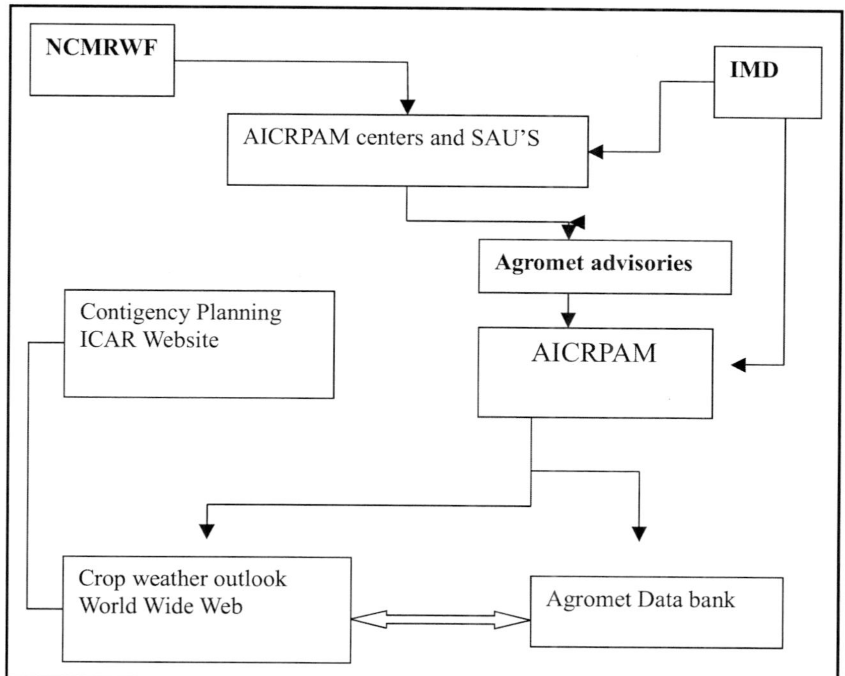

Fig. 5 Agro-Advisory Services in ICAR (Source: Rao, 2008)

Advisory Bulletin

- Crop wise farm management information like sowing, irrigation, fertilizer use etc.
- Spraying conditions for insect, weed and their products
- Livestock management information for housing, health and nutrition etc.

The efficiency of Agro advisories depends on (i) reliability of forecasts issued by NCMRWF (ii) knowledge on crop-weather relationships (iii) adaptation of technologies for the anticipated changes in weather and farmers capabilities to carryout timely operations as mentioned in the AAB. The advisory have recorded many success stories of achieving economic benefits through appropriate weather forecasts and related farm operation strategies in various parts of the country. The information gained from weather based agro advisory should be used in strengthening national food-fodder and livestock security. Forecasting ability particularly medium range forecasting and ability to communicate the client-oriented information to the users needs to be perfected. In the backdrop of rapid changes in technology, one can hope that India can meet the fodder need of its growing livestock population for better human prosperity in twenty-first millennium.

References

Gowariker, V., Thapliyal, V., Kulshrestha, S.M., Mandal, G. S., Sen Roy, N and Sikka, D.R. 1991. Power regression model for long range forecast of southwest monsoon rainfall over India. *Mausam*, 42, 125-130.

Hazra, C.R. 1995. Advances in forage production Technology. Technical Bulletin, IGFRI, Jhansi. pp.4

Rao, G.G.S.N. 2008. Weather based Agro advisory Research Programs in ICAR. *In*: (oral presentation) International Symposium on Agrometeorology and food security, Feb. 18-21., CRIDA, Hyderabad.

Rathore, L. S. 2008. Application of weather forecasting for managing Agriculture. *In*: (oral presentation) International Symposium on Agrometeorology and food security, Feb. 18-21., CRIDA, Hyderabad.

Rathore, L. S., Singh, K. K. and Gupta, A. 2004. National center for medium range weather forecasting: Activities, current status and future plans. *Journal of Agrometeorology* 6, 258-264.

Singh, J. B. and Tomer, P.S. 2002. Weather forecasting and its implication in Agriculture. *Indian Farming*, 7-8.

❑❑❑

Chapter **41**

Underutilized Feed Resources for Enhancing Animal Production

K. K. SINGHAL
National Dairy Research Institute, Karnal-132 001 (Haryana), India

Introduction

Currently, total DM requirement for Indian livestock is 623 million tones (MT) (62% for cattle and 38% for buffaloes), which will be about 670 MT (53% for cattle and 47% for buffaloes) in the year 2020. DM availability through all the feed resources in the country was 509.5 MT in the year 2003 which is likely to be 633.1 MT in the year 2020. CP requirement in year 2003 was 59 MT, which is likely to grow to66 MT by the year 2020. The TDN requirement in corresponding years is estimated as 354 and 387 MT against the availability of 238.5 and 371.0 MT, respectively. The critical assessment indicated that straw and stovers supplied 69% of the available DM while it is envisaged that these will contribute 76% in the year 2020. The CP availability during 2003 was about 32 MT and likely to be 45 MT by the year 2020 (D. K. Jain, Personal communication). This analysis showed that presently there is a shortage of nutrients and it will continue to persist in coming years. Therefore, concerted efforts have to be made to bridge the gap between supply and demand of nutrients for improving the animal productivity as animals maintained on the poor plane of nutrition can not exhibit their production potential.

To bridge the gap between supply and demand of nutrients we have to explore new feed resources in the country, which are not being utilized due to one reason or other. For example abundant availability of monsoon grasses, and some industrial by products such as cottonseed hulls, sugarcane bagasse. Monsoon grasses are utilized to some extent by allowing the animals for grazing but large chunk remained unutilized because of lack of management and proper method of their preservation at right stage of maturity. Similarly, industrial byproducts are not being utilized because of their poor palatability and low nutritional value these are not being utilized for animal feeding. Similarly, by products of food processing industry such as fruit and vegetable wastes are available in abundance in many pockets in the country but due to their high moisture content there is a problem for their handling and transportation therefore, these are not being utilized optimally to make use of their nutrients. Traditional crop residues such as straws of wheat and paddy and stovers of sorghum and maize are also not being properly utilized as these are not harvested completely due to the mechanized harvesting and the residual crop residues are being burnt for the early preparation of land for next sowing, which not only reduce the availability of straws but also create environmental pollution.

Monsoon Grasses

After the scorching heat of may and June, monsoon seasonn starts and the empty field fill with natural grasses such as *dub* grass (Cynodon dactylon). During the summer season, animals loose their productivity and also the body weight due to the non availability of adequate nutrients as they usually maintained on straws and stovers. Fresh early vegetative dub grass, also known as Bermuda grass, contains 21.9 % CP, which decreased up to 10% on maturity ((Ranjhan, 1998). Other grass such as Baru (*Sorghum helepens*), wild *cholai, kamakki, Gadbad* (*Trianthema protulascastrum*) etc are also available in plenty. Grazing on grasses provide the nutrients and resulted in recovering the loss of body weight and productivity. In addition to this grass various other vegetation also provides the nutrients and wide choice for the animals to select the most nutritious part of the available vegetation. In dry areas including the desserts the monsoon grasses are the sole source of nutrients for animals to maintain them. In addition to the grasses, tree leaves also form the source of nutrients during monsoon season, especially for the goats and sheep. In certain areas, where stall feeding is practiced, 'cut and carry' grasses in are the source of nutrients in monsoon season. However, keeping in view the abundance, the grasses, harvested at the right stage of maturity may be preserved in the form hay, to feed the animals in scarcity period. Through this approach appears simple but it involves labour and hence this is the big impediment. In certain pockets tree leaves are preserved, known as *pala*, for the supplementation of green biomass to the animals.

Industrial Byproducts in Animal Ration

(i) Sugarcane Byproducts

India is one of the major sugarcane producing countries and many a times it is difficult to manage the bumper crop for crushing. Sugarcane tops are being utilized for animal feeding in many areas in India, in spite of its poor nutritive value and palatability. Chopped sugarcane tops are mixed with wheat straw and water soaked concentrate mixture/ oilcake and this mixed ration is offered to animals to overcome the palatability problem. Sugarcane bagasse (SB) is another crop residue, which is abundantly available (about 35.55 MT) from sugar mills. Though the nutritive value of SB is poor than the straws due to its higher lignin content, however, efforts have been made to utilize it as a source of roughage with varying success. Recently, experiments at National Dairy Research Institute, Karnal have shown that 40% of WS in complete feed (roughage: concentrate 60:40) can be replaced with SB without any adverse affect on the animal performance (Table 1).

Table 1. Average DM Intake, Growth, Feed Efficiency and Cost of Feeding of Crossbred Calves on Two Types of Feed Blocks.

Attributes	Wheat straw based	Wheat straw & sugarcane bagasse based
Initial b. wt. (kg)	203.24 ± 8.07	200.54 ± 7.69
Final b. wt. (kg)	261.50 ± 9.40	262.07 ± 10.99
Total gain (kg),120d	58.26 ± 3.07	61.53± 5.80
Average b. wt. gain (g/d)	485.46 ± 43.51	512.76 ± 35.81
DMI (feed block, kg/d)	5.82 ± 0.30	5.90 ± 0.31
Total DMI (kg/d)[1]	6.21 ± 0.30	6.29 ± 0.31
Feed efficiency (kg DMI/kg gain)	12.79 ± 0.83	12.27 ± 0.80
Cost of feed block (Rs./ kg)	3.85	3.54
Total cost in 120 days (Rs.)	2785.94±120.88	2602.73±109.79
Feed cost/day/animal (Rs.)	23.22±1.01	21.69±0.91
Feed cost/ kg gain (Rs.)*	47.82^{a}±1.28	42.29^{ab}±2.09

[1]Additionally 0.39 kg DM was offered to each calf of all the groups daily through green maize fodder.

a, b values in a row bearing different superscripts differ significantly ($P<0.05$)

SB have higher bulk density than WS and its incorporation in complete feed followed by its densification using hydraulic pressure (about 3000 psi for 30 seconds) turn the feed in block form, which increased the bulk density by 4.5 times thereby the transportation, storage and handling cost of feed can be reduced besides reducing the feed cost (Hozhabri, 2005).

Addition of banana tops to basal ration of either chopped whole sugarcane or of chopped cane stalk the intake of dry matter increased by approximately 20% without altering the apparent digestibility of the ration. Higher intake was due to increased efficiency of microbial fermentation of dietary dry matter in rumen which in turn was brought about through improved availability of substrates (mainly protein and amino acids) to these microorganisms.

Cottonseed Hulls

Cottonseed hulls can be utilized for the feeding of ruminants as a source of roughage as it is having low protein and high fibre contents. Though the availability of CSH in the country is limited due to the lack of cotton decortication facilities, however, it can be utilized for animal feeding. Complete feed system for utilizing the CSH opens a new vista to balance the nutrients in the diet of animals. CSH does not need any processing to reduce its particle size and it can be mixed with other feed ingredients with the held pf normal mixers. Moreover, molasses, a cheap and easily available energy source, can be absorbed in higher quantity over CSH as it is having higher molasses absorption capacity than other feed ingredients. In an experiment at National Dairy Research Institute, Karnal, complete feed containing CSH at 60 % level was evaluated. Growth rate of the calves was found better than those fed on wheat straw based complete feed (Table 2). Voluntary feed intake and growth rate of CSH based complete feed were significantly higher than straw based feed, however physical form of CSH based feed did not make any significant difference on the feed intake as well as growth rate of calves (Ramachandran and Singhal, personal communication).

Other hulls: Availability of groundnut shell, the outer covering of groundnuts, is abundant as India is a leading producer of groundnut. Groundnut shells are fibrous and poor in protein and mineral contents. Unlike cottonseed hulls, these are very brittle, hence can be turned in to powder form and can be utilized for animal feeding at lower level in the ration, when fiber is not a limiting factor.

Table 2. Performance of Calves Fed on Cottonseed Hulls based Complete Feed.

Particular	**Wheat straw based Control**	**Cottonseed hulls based Mash**	**Cottonseed hulls based Flakes**
B. wt. gain (g/day)	245.4	585.8	533.3
DMI (kg/day)	3.99	5.20	5.19
DMI (% b.wt.)	2.78	3.55	3.55
Feed efficiency %	15.61	8.49	9.29
DM digestibility%	58.7	54.8	59.1
CP digestibility%	69.3	54.2	56.3
NDF digestibility%	46.4	43.9	51.3
DCP %	8.8	6.9	7.4
TDN %	54.3	53.6	58.9

Horticultural Residues in Animal Feeding

India is the second largest producer of vegetables in the world and its share is over 13 percent of the produce globally. It is estimated that the annual yield of fruit and vegetables in India is about 64.6 and 86.0 million tones, respectively. Horticultural byproducts as feed can be classified in the following categories

1. Byproducts of fruits processing industry
2. Byproducts of vegetable processing
3. Byproducts of other horticultural crops.

All the categories of horticultural products are being processed to some extent and during the processing varying type of residues are obtained. These residues contain sizable nutritional value (Table 3) besides good quality of fiber, however, these contain high moisture content, which make them perishable. Therefore they must be utilized quickly after processing to make best use in animal rations. Keeping in view their low protein and high carbohydrate contents, they must be enriched with protein supplements. Since the availability of these residues is seasonal, therefore, these may be preserved by adopting suitable technologies.

Table 3. Chemical Composition of Various Industrial B yproducts (% DM basis)

	CP	CF	Ash	EE	NFE	Ca	P
Brewer's yeast	49.9	1.5	8.5	1.3	38.8	0.13	1.56
Spent barley	27.8	12.6	4.9	8.0	46.7	0.16	0.65
Cottonseed hulls	3.9	30.0	3.6	8.8	43.7	0.13	0.06
Groundnut shell	4.9	68.4	7.4	0.6	18.7	0.25	0.06
Groundnut haulm	14.7	30.9	11.8	3.6	38.9	2.27	0.39
Rice hulls	3.8	43.9	21.6	1.7	29.0	0.08	0.08
Cassava leaves	29.0	16.7	8.6	6.2	39.5	1.17	0.62
Potato haulm	10.9	27.0	13.5	4.3	44.3	0.05	0.15
Orange pomace	6.9	13.1	7.1	2.8	70.1	-	-
Orange peels	6.8	6.2	3.7	1.9	81.4	1.30	0.12
Cabbage waste	20.0	10.3	27.3	3.5	38.9	5.95	0.46
Tomato waste	24.8	27.6	6.6	22.0	19.0	-	-
Apple pomace	5.5	16.4				0.03	0.12
Grape pomace	13.4	33.3	6.6	7.0		-	-
Sunflower head (seed free)	7.2	15.8	11.4	2.9	62.6	-	-
Coconut fiber	1.8	19.3	5.4	2.9	70.6	0.3	0.1
Pea pods	15.4	23.4	4.6	2.2	54.4	-	-
Carrots waste	16.7	8.7	8.1	1.9	64.6	-	-
Okra waste	20.0	34.5	7.6	9.8	28.1	-	-

To give the information about the scope of horticultural residues in the animal ration, following examples are cited.

1. *Apple pomace*: Apple is a main fruit crop in the country and sizable part of this fruit is processed for manufacturing the juice, jam and other products. Apple pomace, the residual material from pressing the apples for juice extraction, contains pulp, peels and cores. Its moisture content is about 85%. Apple yields about 25 to 35 % of fresh pomace, which is highly palatable for animals and can be fed as fresh or after preservation or processing. Its nutrients content in terms of DCP and TDN is 1.62, and 64.84% respectively. Apple pomace can be preserved by sun drying and it takes 52 h for complete drying with occasional stirring (Sharma et al., 1989).

Urea treatment of apple pomace under air tight condition for 2 days following its drying increased its nitrogen content, which was found similar to the maize grain. When the urea treated apple pomace replaced the maize grain in the concentrate mixture of growing and lactating cows, the growth and milk production performance remained uninfluenced and cost of feeding reduced significantly (Table 4).

Table 4. Performance of Growing and Lactating Cows on Feeding Urea Treated Apple Pomace in Place of Maize Grain in Concentrate Mixture

Parameter	Control	Urea treated AP
Growth		
DM intake (kg)	483.6	4834
Live wt gain (kg)	53.0	51.9
Daily wt gain (g)	541.6	529.8
Cost/kg gain (Rs)	13.8	10.75
Lactation		
DM Intake (kg)	6154	6268
Milk production (kg)	5137	5248
Total fat production (kg)	233.7	232.0
DM intake/kg milk yield (Kg)	1.2	1.2
Feed cost/ kg milk (Rs.)	1.25	1.04

Sharma et al. (1989)

Apple pomace can be preserved following ensiling process. Apple pomace and straw can be mixed in equal quantity and mixed with 4 kg urea. The mixture can be kept under air tight conditions after covering it with polythene sheet for 3 to 6 months (Singhal et al., 1991). This technology not only preserves the apple pomace but also improves the nutritive value of straw. Thus, the utilization of apple pomace not only improves the financial viability of apple processing industry, but enhances the feed availability in the country. Similar technology may be applied for the residues obtained from the pineapple, orange and other similar residues having high moisture content and low protein value for the feeding of animals.

Other residues: Maize bran, a residue obtained from the wet milling of maize grain for starch production, is also a moist residue. This can also be fed as such and in case of abundant availability, can be preserved by ammoniation along with straws (Singhal and Sharma., 1991, Singhal and Grant, 2000).

Spent grain, a residue from brewery industry, having sizable nutritional value can also be preserved by the ammoniation process (Singhal and Sharma, 1991).

Sea Weeds for Animal Feeding

India is having a 2000 km long coastal line and there is an abundant availability of sea weeds, which have sizable nutritional value, however, instead of boon these are considered problem as they block the water ways. Sea weed is a novel phyto-nutrient resource with great potentialities. The kind of sea weed used as feed includes *Ascophyllum nodosum, Alaria esculanta, Fucus vansecan Laminaria* and Sargassum. These algae contain abundant and balanced nutrition which can enhance animal performance (Table 5), immune status and productivity of animals.

Sea weed meal of good quality has constituted up to 10% of cattle feeds with good results, and 35 g a day have been fed experimentally to sheep with increased gains as compared to control animals. Sea weeds imbibe many rare minerals from the sea. These rare minerals are transformed to chelated organic matter which can be absorbed by the animal.

Table 5. Chemical Composition of Seaweeds (DM basis)

	Ascophyllu m nodosum	***Laminaria digitata***	***Alaria esculenta***	***Palmaria palmata***	***Porphyra sp.***	***Porphyra yezoensis***	***Ulva sp.***
Type	Brown	Brown	Brown	Red	Red	Red	Green
Water (%)	70-85	73-90	73-86	79-88	86	nd	78
Ash (%)	15-25	73-90	73-86	15-30	8-16	7.8	13-22
Protein (%)	5-10	8-15	9-18	8-25	33-47	43.6	15-25
Fat (%)	2-7	1-2	1-2	0.3-0.8	0.7	2.1	0.6-0.7
Tannins	2-10	0.1	0.5-6.0	ND	ND	ND	ND
Potassium	2-3	1.3-3.8	ND	7-9	3.3	2.4	0.7
Sodium	3-4	0.9-2.2	ND	2.0-2.5	ND	0.6	3.3
Magnesium	0.5-0.9	0.5-0.8	ND	0.4-0.5	2.0	ND	ND
Iodine	0.01-0.1	0.3-1.1	0.05	0.01-0.1	0.0005	ND	ND

ND - Not detected

Supplementation of livestock ration with sea weeds improve feed utilization and productivity, provide iodine and other trace elements, builds immunity and reduce incidences of mastitis (due to its active biological ingredients) besides improving the growth rate in heifers by boosting cobalt and selenium level and conception rate. In a recent study at National Dairy Research Institute, Karnal it was demonstrated that supplementation of sea weed (*Sargassum wettai*) @10% of total ration on DM basis can supplement adequate minerals and improved the milk production in Sahiwal cows. Sea weeds have great potential for animals feeding, however, their harvesting and processing are the major constraints in their utilization.

Non Conventional Oilcakes for Animal Feeding

Oil cakes of mustard, groundnut, cottonseed, sesmum etc are the conventional protein supplements in animal ration; however, some non-conventional oil cakes are also available in sizable quantity in the country as a result of oil extraction from oilseeds of mainly tree origin such as neem, mahuwa, karanj etc. The availability of these oil cakes and areas where these are available in plenty are presented in Table 6. These oilcakes are non-conventional as these are containing one or other incriminating factor, which restricts their utility as a feed resource, however, research work carried out to develop the techniques to make them edible in animal ration by removing their incriminating factor completely or partially.

Table 6. Availability of Non-conventional Oilseeds in the Country*.

Oilseed	Availability 000 ton	Oil %	States	Incriminating factor
Mahwa	2176	35	AP, UP, MP, Bihar, Guj., Orrisa, TN, Karnataka,	Sapoglucoside (mowrin), Sapogenin
Karanj	111	27	AP, Karnatake, TN, Bihar,	Karanjin
Kosum	90	33	Bihar, UP, MP, Orissa	Prussic acid
Neem	418	20	UP, AP, MP,TN, Karnataka, Guj., Raj.	Nimbin and its derivatives
Sal	5504	13	MP, UP, WB, Bihar, Orissa	Tannins
Mesta (Thumba)	153	30	WB, Guj., Bihar	Fiber
Piludi	46	33	Guj., Raj., UP	Not identified
Nahar	6	40	Assam, WB, Orissa, Kar., Kerala, Maha,	Not identified
Kokam	2	40	Maha., Karnataka	Not identified
Undi	4	30	Kerala, Kar., Maha., AP, WB, TN	Not identified
Dhupa	10	25	Karnataka, Kerala, TN	Not identified
Mango kernel	15	7	Whole India	
Rubber seed	3360	20	Kerala, TN	Prussic acid
Castor	407	37	Guj., AP, Bihar, Orissa, Karnataka	Ricin, Ricinin

*Mudgal and Singhal (1993)

The elaborate techniques for the removal of incriminating factors in non-conventional oilcakes have been described by Mudgal and Singhal (1993) besides presenting the data on the performance of animals fed on treated oilseed cakes based diets. In addition of these oilcakes, information has been provided about other protein rich non-conventional feeds such as oak kernel, tobacco seed cake, Palash seeds, Coffee seeds residue, babool seed, *cassia tora* seeds, taramira seeds etc.

The non-conventional feed resources are important for livestock feeding and these must be preserved and processed to make their best use, wherever it is possible and feasible.

Future Projections

CP availability in cattle and buffalo rations would continue to be deficient in future, however, magnitude of deficiency is like to narrow down in future. The projections revealed that the overall deficiency of CP was 46.5% in the year 2003 and likely to be narrow down to the magnitude of 31.3% by the year 2020. Similarly, the TDN deficiency was about 32.7% of requirement in the year of 2003 and likely to be about 4.2% by the year 2020. Keeping in view these deficiencies, we have to mobilize protein resources for the livestock feeding to bridge the gap between requirement and availability for enhancing the livestock productivity. Since importing of the animal feed ingredients have remote possibility, therefore, we have to augment our feed resources and utilize them judiciously to improve the deficiency of nutrients. Sea weeds are having great potential as these are not being utilized currently. Besides this, we have to develop the systems for the preservation and storage of our feed resources and educate the farmers for the benefits of balanced feeding to avoid the feeding excess or deficient nutrients.

References

Hozhabri, F. (2005) Effect of roughage sources and protected protein in complete feed on fiber digestion kinetics, nutrient utilization and growth performance of crossbred calves. Ph.D. Thesis submitted to NDRI, Karnal, India.

Mudgal, V.D. and Singhal, K.K. (1993) Agro-industrial byproducts in Animal Nutrition (Hindi) Rajasthan Hindi Granth Academy, Jaipur

Ranjhan, S.K. (1998) Nutrient Requirements of Livestock and Poultry. Indian Council of Agricultural Research, New Delhi.

Sharma, D.D., Bakalkar, R.K. and Singhal, K.K. (1989) Apple pomace: a new feed resource for animals, Publ No. 149, NDRI, Karnal

Singhal, K.K., Thakur, S.S. and Sharma, D.D (1991) Nutritive value of dried and stored apple pomace and various methods of its utilization for ruminant feeding. *Indian J. Anim. Nutri.* 8(3): 213-216

Singhal, K.K. and Sharma, D.D. (1991) Effect of ammoniation of brewer's spent grain along with crop residues for preservation and utilization of the product for ruminant feeding. *Indian J. Anim. Sci.* 61(6):620-623

Singhal, K.K. and Sharma, D.D (1991) Effect of urea added or ammoniated wet maize bran and paddy straw on nutrient utilization in buffaloes. *Indian J. Anim. Nutri.* 8(3): 209-214

Singhal, K.K. and Grant, R. (2000) Effect of ammoniation of wheat straw alone or in combination of wet corn gluten feed on fibre digestion kinetics. *Indian J. Anim. Sci.* 70(5): 513-517.

❑❑❑

Chapter 42

Scope of Utilization of Unconventional Feeds in Livestock Feeding

S. K.TOMAR
National Dairy Research Institute, Karnal-132001 (Haryana), India

Introduction

The approach paper for India's Eleventh Five Year Plan (2007-2012) focuses on faster and more inclusive growth of the economy at an average rate of 9 percent leading to doubling of per capita GDP over the next 10 years so that the aspiration of the entire population can be met. The goal for the growth of agricultural sector as a whole is about 4 percent and for the dairy sector, a growth rate of about 5 percent is envisaged so that the expected contribution of the livestock sector to the nation in terms of milk, meat, wool and other animal products availability, income and employment is ensured The livestock sector has helped national economy by emerging as the highest milk producing country in the world. According to FAO's Economic and Social Development Report, India's milk production has increased from a mere 17 million tones in 1951 to 100 million tones in 2007; with a share of close to 15% of world milk production (Mathur, 2008). India ranks first in bovine population holding of about 51 percent of Asia and about 19 percent of world bovine population but the productivity of these animals is very low. This situation is mainly due to their state of nutrition and low genetic potential.

In India, inadequate supply of nutrients is always a problem primarily due to shortage of grains, fodders and overdependence on crop residues for the feeding of dairy animals. The gap between availability and demands of feeds is very wide. There is a deficit of about 48% for concentrate, 31% for green fodder and 17% for dry fodder (Aneja *et al.* 2002). Use of unconventional feeds is an important alternate to mitigate the shortage of grains. Since various unconventional feeds are produced in huge quantities every year and available locally at cheaper rate, these must be preferred in order to have maximum benefit from locally available unconventional feeds. But some of these unconventional feeds adversely affected the performance of animals when fed as such, particularly at higher levels of inclusion for longer periods. These adverse effects were due to one or the other indiscriminating or toxic factor (s) present in them. The present paper deals in brief the availability, toxic factor(s) present and scope of utilization of the important unconventional feeds available in India in livestock ration.

Agro-Industrial Byproducts as Unconventional Feeds

Before considering the different groups of byproducts it is necessary to examine the contrasting types from the point of view of a feed value. This is particularly important in relation to least cost ration formulation when different ingredients are evaluated as source of energy, protein, minerals and vitamins in relation to their cost and then included in livestock ration subject to constraints in relation to specific nutrients or anti-nutrient properties.

Ørskov (1980) proposed that byproducts should be considered in four main categories in relation to their contents of metabolizable energy and nitrogen. This is particularly relevant in relation to ruminants, where there is a degree of interdependence between these two major nutrients. The microbial population of the rumen thus has a requirement for nitrogen which varies according to the proportion of the organic matter digested in the rumen.

A. Byproducts of Low Digestibility and Low Content of Nitrogen

By products of this category include straws, husks, pods and haulms. Very often the nitrogen content in these by-products is so low that supplementary degradable nitrogen (e.g urea) is required to speed up both rate and extent of digestion in the rumen. If given as a high proportion of diet, these byproducts have slow rates of digestion and long retention times in the rumen consequently levels of voluntary intake are low.

Rice hulls : Rice hulls or husk is the largest byproducts of rice milling industry and most of it is either burnt or used as bedding for poultry and animals. It contains 3.2% CP, 1.4% EE, 32.7% CF, 3.8.7% NFE, 23.6% ash, 16.0% silica, 69.4% NDF and 38.0% ADF (Ranjhan, 1990). Ground rice husk is often mixed to adulterate the rice polishing / bran (Singh and Kundu, 2003). Husk when substituted at 0, 33, 66 and 100 percent of rice straw in growing crossbred calves ration, the growth rate was depressed with the increase of rice hulls, however at 33% level the depression was not significant as compared to that in the control. The DCP and TDN of rice hulls were 0.16 and 11.7 per cent, respectively.

Maize husk: It is superior to even gram husk in its nutritive value. it has about 8% CP. Fifty per cent of the dry roughage portion in complete feed of adult non-producing cattle and buffaloes can be replaced by this byproduct.

Sugarcane bagasse: Bagasse is one of the most important byproduct of sugar industry. It is a source of cellulose but poor in protein and high in lignin. Cattle and buffaloes can consume about 80% of their expected dry matter intake through bagasse impregnated with 1 and 2 per cent urea. About 60 kg compelete feed consisting of 20 kg bagasse, 6 kg molasses, 0.34 kg comman salt, o.34 kg mineral mixture and 65 litres of water forms a maintenance ration along with 800 g of groundnut cake for adult animals weighing 400 kg. Sengar *et al.* (1993) reported that sodium chloraite treatment of bagasse had decreased the lignin content and increased *in situ* dry matter digestibility. Hozhabri and Singhal (2006) concluded that 40% replacement of wheat straw with dried sugarcane bagasse in feed blocks improved their physical attributes such as bulk density, post compression expansion and durability besides decreasing feed cost. They also reported that 40% wheat straw in a complete feed (roughage : concentrate, 60:40) can be replaced without any adverse affect on its *in vitro* nutrients digestibility.

Babul pods : Babul (*Acacia* spp.) pods contain aout 13% CP and 50% TDN (Punj,1988). The main degraded product of *Acacia nilotica* tannin in ruminants are phloroglucinol and resorcinol. The use of babul pods as animal feed can be used as a substitute of energy source like cereal grains in conventional rations for livestock, but contains 18.71% tannins (Barman and Rai, 2006). There are possibilities of achieving beneficial effects of tannins using within threshold level of tenniniferous feeds on animal performance by means of enhancing protection of dietary protein as well as value addition of milk (Rai, 2007). Tannins may affect the digestibility of nutrients either by binding down digestive enzymes or by binding feed nutrients and also reduce the bioavailability of several minerals specially that with divalent metal cations (Barman, 2004). Therefore, the incorporation of babul pods beyond safe level requires inactivation / detoxification of tannins for better utilization in livestock.

Efforts have been made to improve the utilization of tanniniferous feeds for ruminants by reducing the tannins levels through physical treatments (Vitti *et al.*, 2005), chemical treatments (Barman and Rai, 2006; Aemiro and Rai, 2006; Alam *et al.*, 2005; Bayssa, 2005) and biological treatments (Aemiro, 2002; Makkar, 2003)

Cocoa pod husk : The dried cocoa (*Theobroma cacao*) pods are brittle and the husk contain 6% CP, 31% CF, 3 % EE, 12% ash. It can be used upto 10-15% in the complete feed of ruminants.

Cotton straw: Cotton straw in ground form could be used as roughage source upto 30% in complete diets for lactating cattle and buffaloes. The palatability and nutrients utilization of cotton straw was improved by expandor method without enhancing the cost of feed on production basis (Nagalakshmi *et al.*, 2004).

Cassava (*Tapoica*) *waste*: Cassava (*Manihot esculenta*) feeds and roots are poor in CP (2%) but rich in NFE (85%) where as tapoica starch waste has 2% DCP and 64% TDN. It has been recommended that tapoica waste can be safely included in the complete feed with a considerable economy for maintaining body weight, milk yield and butter fat production in cattle and buffaloes.

Coconut meal: India's coconut meal production meal production is 0.28 million tones which is about 15.4% of world production (FAO, 2004). The feed consists of nearly 20-35% of oil cakes / meals as a protein source. Since the coconut meal is available locally in south India, it is recommended to include the coconut meal upto 10% in egg type ration for better egg production in poultry (Moorthy and Viswnathan, 2008).

B. Byproducts of Low Digestibility but Relatively High in Nitrogen

By products of the food industry, such as coffee residues, fruit pulp and cocoa meal are of low energy value, but are relatively high in nitrogen. In addition, animal excreta tend to have an excess of nitrogen in relation to available energy.

Coffee waste : Coffee (*Coffea arabica* and *C. robusta*) is cultivated commercially in coastal regions of India and is availiable about 70,000 tonnes which yield about 45,000 tonnes of coffee husk. It has about 9% CP, 40% CF, 10% EE and 5% ash. It has 3.4% DCP and 42.2% TDN and can be incorporated in the compelete diet of large ruminants (Ranjhan,1990).

Leguminous pulse straws: The residues of pulses are composed of husk of pod with leaves and tender stems which are more nutritious than the cereal straws and stovers. They contain 8.0-12.5% CP and 40-50% NFE. The voluntary intake varies from1.8-2.5 per cent of body weight in cattle and buffaloes. Groundnut straw along with wheat straw (35%) and wheat bran (1.0 kg) can meet the nutritional requirement of a buffalo weighing 450 kg and producing 5 kg milk.

Pineapple waste: Pineapple (*Ananas camosus*) caunery waste consists of outer peels (shells), the crown and bud ends of fruit, fruit trimmings, the inner core and the pomace of the fruit from which the juice has been extracted. The pineapple waste contains 80-90% moisture and 5% CP. Two per cent of dry matter through this waste can be consumed in complete feeds by cattle and buffaloes.

Jackfruit waste : It contains 13.3% CP, 5.09 % EE, 12.4% CF, 0.21 % Ca and 0.35% P. It can be incorporated in ruminant's compelete diet upto 10-15%. Banana peels, orange peels, mango peels, lemon peels have about 6-13% CP and 55-74% NFE and can be included in the compelete feed of cattle and buffaloes. Cauliflower leaves, papaya leaves, cabbage leaves and pea pods have 12-27% CP, 38-55% NFE, 1.0 - 4.5% Ca, 0.05-0.60% P and can be included in the complete diets of ruminants.

Citrus waste : In complete feeds citrus waste can be fed to large ruminants upto 40 kg/head/day without any harmful effects (Umoh,1982). After processing oranges, lemons or grapes for juice, the left over which comprises the pod seeds and rag etc. amounts to 40-60% by weight of the fruit could be very palatable to cattle and buffaloes. Citrus pulp contains a relatively high amount of pectin and soluble carbohydrates and for this reason dried citrus pulp has been used to replace cereals in ruminants diet (Bhattacharya and Harb, 1973).

Seaweed: Historical records show that the use of sea weeds in agriculture is very old and some species are used as feed and fodder for pig, horse, chicken, cattle and buffaloes. Vitamin content is higher as compared o wheat and milk. There is possibility of using these seaweeds either after drying or preserved with chemicals before using as ruminant feed. Agar agar is being extracted from Sargussam seaweed and its byproducts can be used in complete feeds for livestock feeding. This byproduct contain about 10% CP and 67% organic matter (Tomar, 2004).

C. Byproducts of High Energy Value but Low in Nitrogen

The byproducts of the sugar industry (molasses, sugar beet pulp), the horticultural industry and products such as stock feed potatoes and bakery waste products are in this category. Because of the rapid rate of their digestion, there is usually a substantial requirement for supplementary sources of readily available nitrogen to be given to balance the excess of energy. With monogastric animals this may take the form of skim milk, blood meal, fish meal, soy bean meal or dried yeast. In ruminants where the need in for a highly degradable source of nitrogen in the rumen, heat treated proteins are unsuitable.

Fats and oils are the best examples of byproducts in this category. Animal fat can be included in the diet of pigs and poultry but at levels, which are usually less than 5% of complete diet (Cooke and Pugh, 1980). At higher levels of use, these may be problems of physical structure and adverse effects on carcass composition. In the case of pigs, the use of vegetable oils containing linoleic acid may lead to soft fat, particularly if the linoleic acid content of the diet exceeds 16 g per kg (Wood *et al.*, 1984). With ruminants, there may be problems of maintaining cell wall digestion in the rumen if the proportion of fat in complete diet exceeds 5 percent.

Sugarcane molasses: Molasses is about 3% of the sugarcane or about $1/3^{rd}$ of the quantity of sugar produced. It contains about 61% sugar and 3% protein and has been used in complete feeds of ruminants at 5-10% levels. Molasses is mainly used as (a) carrier for urea impregnation of poor quality roughages (b) as binder for commercial pelleted feeds and (c) as sweetener for increasing the voluntary intake of unpalatable feeds.

Mango seed kernels : The annual availability of mango seed kernels is about 5,00,000 tonnes (SEA, 2000). It has a high content of starch i.e.92% of its total carbohydrates content. It contain about 68-70% natural lipids of the total lipids (Dhingra and Kapoor, 1985). Stearic and oleic acids are the main components and saturated and unsaturated fatty acids were 40 and 60%, respectively. Histidine and cystine were in greater amount as compared to other amino acids in the mango seed kernel (Barman, 2004). Growth rate was found to be 440 g/d/ animal when used 20% mango seed kernel in the concentrate mixture of calves.

Potato waste: Potato waste is the byproducts remaining after extraction of starch with cold water. 16-20% of starch is produced and 3-3.5% dried potato waste is obtained for feeding in complete diets to cattle and buffaloes ((Boucque and Fiems, 1988).

Tomato pomace : Dried and ground pomace contained 22.36% CP and 28.11 % CF with a nutritive value of 13.7-15.8% DCP and 60.8-73.3% TDN. Tomato waste can be incorporated upto 35-40% of the concentrate mixture in the compelet diet of cattle and buffaloes.

Bakery wastes: The use of bakery waste as animal feed has been reported in India (Ganai *et al.*, 2000) and abroad (Waldroup *et al.*, 1982) as a substitute for comparatively high priced cereal brans. Ganai and Sharma (2008) reported that replacement of high priced wheat bran upto 35% in concentrate mixture by bakery waste is safe, effective and economical for raising of lambs for meat and wool production.

Byproducts of category (B) are therefore well suited to being used in mixture with those of category (C). They are also suitable for use with other feeds of high energy content but which are low in N, such as cereal grains.

D. Byproducts High in Energy and Nitrogen

Byproducts in this category include products such as blood meal and poultry offal meal, fish meal, extracted oil seed meals/cakes and waste vegetables such as carrots. Care is needed to ensure that potentially toxic compounds are recognised and that the inclusion of the byproduct in the complete diet will not adversely affect flavour and acceptability. Thus it is not recommended to use blood meal in calf complete diets (Cooke and Pugh,1980). There is also a need to ensure that the byproducts do not deteriorate during storage. Those based on leaves,(e.g cauliflowers, cabbages, pea pods) may have a short shelf life and in consequence are better suited for use near the place of their production.

Blood meal: Blood meal is a major slaughterhouse byproduct but much of it is wasted. Coagulating the blood by steaming or by boiling, then drying and milling usually makes it. Blood can also be soaked on wheat bran/rice bran and then dried. This can be an ingredient in the poultry complete feeds. Blood meal contains over 80% protein but poor is in Ca and P. As such it is unpalatable to animals. However, supplementation of basal cereal free complete diet with 3,6, and 9 per cent of blood meal gave performance similar as the control ration containing 10% fish meal.

Poultry offal meal: Offal meal, a byproduct of poultry dressing plant is available for use as a meal. It consists of head, feet, lungs, wind pipe, oil glands and intestines. After processing it in a valuable protein source and contains about 58% protein and 2910 kcal ME/Kg. Offal meal is poor in methionine and lysine. Its inclusion in poultry complete feed beyond 2.5% lowers the overall performance of birds.

Fish meal: Fish meal is the most valuable by product and has been accepted as balanced source of protein for poultry, swine, newborn growing calves, lambs and kids.

Frog meal : Frog meal is a left over of the frog leg industry. About 1000 tonnes of this byproduct are available annually alone in Kerala. It contains about 60% protein and can replace fish meal twice by weight in the poultry complete diets both for growth and egg production.

Oil seed meals/cakes: Groundnut, soybean, mustard, lineseed, til, cotton seed and coconut cakes /meals are conventional protein supplements and costly items for animal production. All these cakes had 30-45% CP and high energy value. Feeding of higher levels (60%) of cotton seed in the diet of Surti buffaloes improved the milk yield and milk fat percentage (Chaudhary and Jat, 2008). In addition to annually cultivated crops, there are many oil bearing perenial trees, for example sal, rubber, neem., mahua, castor, kosun, niger, karanj etc. All these fall into the category of unconventional oil seed cakes / meals. In India, the potential of unconventional oilseeds available from these trees is estimated to be 6.7 million tonnes to yield about 1.0 million tonnes of oil and remaining available as cakes (Ranjhan,1990). However, due to lack of facilities and awareness only about 3.5% of total potential is collected and utilized.

Waste vegetables : Vegetable wastes like carrots,cabbage, cauliflower, lady's finger, empty pea pods can be included in the compelete diets of cattle and buffaloes (Rai and Barman, 2004).

Matching Byproducts to System of Animal Production

It is important to assess the feed value of byproducts in relation to the type of animal and the system of production in which the material is to be used. Thus low quality byproducts in categories A and B are best suited to use with animals which have relatively low nutrients requirements such as dry dairy cows, and goats. Such feeds may aso be used tactically at times of feed shortage to reduce the cost of the diet. By products in categories C and D are most appropriately used in more intensive systems such as pig and poultry production. They also have wide application in ruminant livestock operations. Thus oilseed cakes and animal byproducts are regularly part of compounded feeds for high yielding dairy cows. The fact that many are heat treated means that the protein is less readily degraded in the rumen and more likely to enhance the total supply of amino acids to the animal.

Conclusions

These are clear limitations to the extent to which certain byproducts may be recycled directly into the complete feeds of livestock. Apart from a low feed value, there may also be logistical problems of matching their availability to the supply of animals for their consumption. This problem may be exacerbated by a short life of the byproduct and restrictions on use imposed by compositional factors such as toxins or physical factors which might be associated with rapid deterioration. Strategic supplementation is justified because of regular feed shortages that occur and the fact that ruminants subsist for most of their life on agro-industrial byproducts on small farms. Added costs of detoxification and supplements often discourage byproducts use. There is a need for future systematic research on the optimization of the use of agro-industrial byproducts in complete feeds and develop strategies for their optimum supplementation in different animal species.

Table : Important Unconventional Feeds, their Recommended Levels of Inclusion in Complete Feeds and Antinutritional Factors.

S.No.	Particulars	Recommended level of inclusion in complete feed			Antinutritional factor (s)
		Calves	Milch cattle/buffalo	Poultry	
1.	Rice bran		30 %	20 %	
2.	Wheat bran	15 %	30 %	15 %	
3.	Deoiled Groundnut cake	10 %	15 %		Aflatoxin
4.	Deoiled Mustard cake		20 %		Sinalbin
5.	Linseed cake	10 %	20 %		Linamarin & NCN
6.	Cotton seed cake		20 %		Gossypol
7.	Mahua seed cake	15 %	25 %		Mowrin
8.	Rubber seed cake	20 %			HCN
9.	Kosum cake	35 %		17 %	Cyanogens
10.	Babool seeds	30 %	15 %		Tannins
11.	Water hyacinth		15 %		
12.	Acacia seed chuni	30 %	45 % (bullock)		Tannins
13.	Mango seed kernel		10 %		
14.	Tea spent waste	20 %		10 % (broilers)	Tannins
15.	Panewar seed (*Cassia tora)*		15 %		Crysophonic acid
16.	Safflower cake		10-15 %		
17.	Niger seed cake		20-25 %		
18.	Karanj cake	10 %	15 %		
19.	Guar meal	5 %	10 %		Gum, Trypsin
20.	Bone meal	2.5 %	5 %	2.5 %	
21.	Blood meal		2 %	2 %	
22.	Poultry offal meal	1.0 %	2.5 %	2.5 %	

References

Aemiro, K. 2002. Studies on chemical and biological inactivation of tannins in animal feedstuffs. M.Sc. thesis, National Dairy Research Institute, Karnal (Haryana) India.

Aemiro, K. and Rai, S.N. 2006. Optimization of treatment with different levels of Ca(OH)2, moisture content and reaction period of Mango seed kernel, Proc. : 12th Animal Nutrition Conference held at Anand Agricultural University, Anand, (Guj) from Jan 7-9, 2006, pp 142.

Alam, M.R., Kabir, A.K., Amin, M.R. and McNill, D.M. 2005. Effect of calcium hydroxide treatment on nutritive and feeding value of Albezia procera for growing goats. *Animal Feed Science and Technology* 122, 135-148.

Aneja, R.P., Mathur, B.N., Chandan, R.C. and Banerjee, A.K. 2002. Technology of Indian Milk Products. *A Dairy India Publ.*, Delhi, India, pp- 34.

Barman, K. 2004. Biodegradation of tenniferous feeds and their influence on nutrients utilization and productivity in dairy animals. Ph.D. thesis submitted to NDRI, Karnal.

Barman, K. and Rai, S.N. 2006. Utilization of tenniferous feed 2. Effect of tenniferous feed on nutrient digestibility, gas production and tannin degradation products in cattle in vitro. *Indian Journal of Animal Scence.* 76, 829-837.

Boyssa, M. 2005. Detoxification of tannins in *Acacia nilotica* pods on *in vitro* nutrient digestibility and milk production in lactating goats. M.Sc. thesis, National Dairy Research Institute, Karnal (Haryana) India.

Bhattacharya, A.N. and Harb, M. 1973. Dried citrus pulp as a grain replacement for Awassi lambs. *Journal of Animal Scence* 36, 1175-1180.

Boucque, C.H.V. and Fiems, L.O. 1988. Vegetable byproducts of agro-industrial origin. *Livestock productiton Science* 19, 97-135.

Chaudhary, J.L. and Jat, H.R. 2008. Effect of feeding different levels of cotton seed on the reproductive performance of Surti buffaloes. *Animal Nutrition and Feed Technology* 8, 81-88.

Cooke, B.C. and Pugh, M.L. 1980. Slaughter waste in animal feeding. *In* : E.R. Ørskov (Editor), *By-products and Wastes in Animal Feeding.* Occasional Publication No. 3, BSAP, pp. 79-83.

Dhingra, S. and Kapoor, A.C. 1985. Nutritive value of mango seed kernel. *Journal Science of Food and Agriculture* 36, 752-756.

FAO. 2004. *Production Yearbook* . Food and Agricultural Organization of United Nations, Rome, Italy.

Ganai, A.M., Purohit, G.R. and Mathur, B.K. 2000. Economics of protein protection and feed replacement in sheep. *Annals of Arid Zone* 39, 97-99.

Ganai, A.M. and Sharma, T. 2008. Performance of lambs fed formaldehyde treated mustard oil cake and bakery waste based concentrate mixture. *Animal Nutrition Feed Technology* 8, 203-212.

Hozhabri, F. and Singhal, K.K. 2006. *In vitro* evaluation of sugarcane bagasse in complete feed. *Indian Journal Animal. Nutrition* 23, 88-93.

Makkar, H.P.S. 2003. Effect and fate of tannins in ruminant animal, adaptation to tannins and strategies to overcome detrimental effects of feeding tannin rich feeds. *Small Ruminant Research* 49, 241-256.

Mathur, B.N.2008. Prespective planning and R&D support toward future challenges confronting the dairy industry in India. A lecture delivered in 'Golden Jubilee Celebrations of Dairy Science College & National Seminar on Dairy Education', from March 12-13, 2008. pp 15-21.

Moorthy, M. and Viswanathan, K. 2008. Extracted coconut meal in white leghorn layer diet. *Animal Nutrition and Feed Science* 8, 227-236.

Nagalakshmi, D., Reddy, D.N. and Kumar, M.K. 2004. Performance of Murrah buffaloes fed expander pelleted cotton straw based diets. *Animal Nutrition and Feed Science* 4, 23-32.

Ørskov, E.R. 1980. Nutritional principles in utilization of wastes and by-products. *In* : E.R. Ørskov (Editor), *By-products and Wastes in Animal Feeding. Occasional Publication* No. 3, BSAP, pp. 13-17.

Punj, M.L. 1988. Availability and utilization of non-conventional feed resources and their utilization by in South Asia. *In* : Non-conventional Feed Resources and Fibrous Agricultural Residues, Strategies for Expanded utilization, Hisar, India, March 21-29, 1988, pp. 50-81.

Rai, S.N. 2007. Use of tenniferous feed (*Acacia nilotica* pods) to enhance milk production and value addition of milk and milk products. Lecture delivered in Model Training Course on "Feeding Strategies for Upkeep of High Yielding Dairy Animals" 26-11-2007 to 3-12-2007, at NDRI, Karnal, pp 89-100.

Rai, S.N. and Barman, K. 2004. New animal feed resources : problems and potential. Proc. XI Animal Nutrition Conference on Nutritional Technologies for Commercialization of Animal Production System, held at J.N.K.V.V. Jabalpur (M.P.) from January, 5 – 7, 2004.

Ranjhan, S.K. 1990. *Agro-industrial Byproducts and Nonconventional Feeds for Livestock Feeding*. ICAR, New Delhi, pp. 7-91.

SEA (Solvent Extraction Association of India) 2000. *Millennium Hand Book on Indian Vegetable Oil Industry and Trade* (7th ed.) SEA Publication.

Sengar, S.S., Tomar, S.K., Lall, D. and Mehra, U.R. 1993. Effect of sodium chlorite treatment of straws on their cell wall constituents and *in situ* dry matter disappearance. *Indian Journal Dairy Scence* 46, 130-132.

Singh, R. and Kundu, S.S. 2003. Quality evaluation of some animal feedstuffs available in Haryana. *Animal Nutrition and Feed Technology* 3, 143-150.

Tomar, S.K. 2004. *Unconventional feed resources for buffaloes in various agro-climatic regions. In* : Kundu, S.S., Mishra, A.K. and Pathak, P.S. (Editors), Buffalo Production Under Different Climatic Regions. International Book Distributing Co., Charbagh, Lucknow, India, pp. 299-320.

Umoh, J.E. 1982. The use of concentrate and byproducts in beef cattle production : Emphasis on use of concentrate on pasture. *Proc. National Conf. on Beef Production*, July 27-30, Kaduna.

Vitti, D.M., Nozella, E.F., Abedlla, A.L., Bueno, I.C.S., Sila Filho, J.C., Costa, C., Bueno longo, C., Viera, M.E., Carbral Filho, S.C., Godoy, P.B. and Mueller Harvey, I. 2005. The effects of drying and urea treatment on nutritional and anti nutritional components of browses collected during wet and dry season *Animal Feed Science and Technology* 122, 123-133.

Waldroup, P.W., Whelchel, D.L. and Johnson, Z.B. 1982. Variation in nutrient content of samples of dried bakery product. *Animal Feed Science and Technology* 7,419-421.

Wood, J.D., Brown, A.J., Enser, M. and Jones, R.C.D. 1984. Tissue quality in lean pigs. *In* : A. Cuthbertson and R. E. Gunn (Editors), *Matching Production to the Markets for Meat.* Occasional Symposium, BSAP, pp. 97.

❑❑❑

Chapter **43**

Livestock Biodiversity in India

B. P. KUSHWAHA
Central Institute for Research on Buffaloes, Hisar-125 001 (Haryana), India

Introduction

India is considered a mega bio-diversity center where domestication of species for agricultural production was practiced much earlier than the many European countries and almost all the major species of livestock and poultry exists in India. Animal genetic resources are the most valuable and strategically important assets of the country, which are intricately associated with the social, cultural and traditional values of the region of their origin. They also serve as vital source for food, fibre, draught power, manure and provide much needed self employment to small and marginal farmers as well as weaker sections of the society. Over 140 breeds of livestock species including cattle, buffaloes, sheep, goats, pigs, camels, horses, donkeys, yak and Mithun have been distributed over a large area spread in different ecological zones of the country. There is a long list of domesticated avian species apart from poultry, such as ducks, geese, quails, turkey, pheasants and partridges exit in India.

India is still fortunate in possessing some of the wild ancestral species of sheep like *Ovis musimon, O. vignis, O. orientalis, O. amon* , wild goats like

Himalayan Tahr, Nilgiri Tahr and Markhors, Wild yals (*Bos mutus*), Gaur (*Bos gaurus*), Red Jungle fowl and Snow partridge (Ahlawat & Gupta, 2006) in natural habitat, that make the subcontinent a treasure of farm animal bio-diversity.

Animal genetic resources take a million of years to evolve through the process of natural selection. Various species of livestock have been domesticated throughout the world to meet the local or regional needs. Animals have been selected by the farmers for their particular characteristics and at the same time they were also adapting genetically to local conditions, diseases, available feeds, climate, predators and many other variables imposed by the surrounding environment where they have been evolved. Hence these breeds commonly possess valuable traits such as adaptation to harsh conditions, including tolerance of parasitic and infectious diseases, drought and poor quality feed. Various breeds of different species of livestock have been described on the basis of their external phenotypic characteristics such as color, type of horn, face, tail, eyes and body measurement. They have also been described on the basis of their phenotypic biometry as well as production and reproduction performance. The livestock census is conducted every five years. Though the information on flock structure, sex, age, production etc. are taken into consideration, they do not reflect the factual situation about the status of breed in terms of their numbers and performance in their breeding tract. Hence, the actual number of each breed and their distribution is not very clearly known.

India possesses about 7.75 percent of world cattle breeds, 26.39 per cent of buffalo breeds and 8.26 per cent of goat breeds (Joshi 2008). The total livestock which includes cattle, buffaloes, sheep, goats, pigs, yaks, mithuns, camels, horses, ponies, mules and donkeys has increased over the years. As per the 2003 livestock census data, there were 185.18 million cattle, 97.92 million buffaloes, 61.47 million sheep, 124.36 million goat, 13.52 million pig, 0.75 million horse, 0.63 million camel, 0.28 million Mithun, 0.07 million yak and 440.69 million poultry. Population of all the major livestock species is increasing except the cattle, which have declined by (-) 6.9% over 1997 census figure. India ranked first in the world in cattle and buffalo population, second in goat, third is sheep and seventh in poultry population. Livestock population and population trend over the last five years has been presented in Table 1.

Table 1. Livestock Population in India.

Species	No. of Breed	Livestock population (million)		Trend (%) (1997- 2003)
		1997	2003	
Cattle	30	198.88	185.18	-6.9
Buffaloes	10	89.91	97.92	8.9
Sheep	42	57.49	61.47	6.9
Goat	22	122.72	124.36	1.4
Pig	3	13.29	13.52	1.7
Donkey	-	0.88	0.65	-26.1
Horse	6	0.83	0.75	-9.6
Mule	-	0.22	0.18	-18.2
Camel	8	0.91	0.63	-30.8
Yak	-	0.06	0.07	16.7
Mithun	-	0.18	0.28	55.6
Poultry				
Chicken	18	315.43	408.02	29.3
Duck	5	32.18	32.67	1.5

Cattle

India has 30 recognized breeds of cattle. However, there is different opinion about the total number of cattle breeds *(as per conservative estimate, it is 26)*. These breeds have been evolved over centuries to suit to the local climatic conditions of the region where they are bred and reared. These breeds can be broadly classified, based on their utility, as dairy breeds, draught breeds and dual-purpose breeds. Dairy type cattle breeds are Sahiwal, Red Sindhi, Gir and Rathi. Cows of these breeds are high milk producers but bullocks are poor draft quality. Draft breeds, majority of the breeds belongs to this category, the milk production of these breeds is very low and these breeds are kept mainly for agricultural operations. Draft type cattle breeds are Nagori, Bachaur, Kenkatha, Malvi, Kherigarh, Hallikar, Amritmahal, Khillari, Bargur, Kangayam, Ponwar, siri, etc. The dual purpose cattle breeds viz. Haryana, Ongole, Tharparkar, Kankrej, Krishna valley, Deoni, Gaolao, Vechur etc. are fairly good milkers and bullocks provide a good draft power.

Well-defined breeds constitute less than 20% of total cattle population. Besides the well-defined breeds there are about 19 (lesser known) strains of cattle. The remaining cattle are non-descript, named after their location and they do not possess uniform characteristics. The non-descript cows have poor growth rate (150g/day), late maturity (first calving at age of about five years) and low milk production (1 to 2 kg per day).

Indian cattle breeds are known for their hardiness, their ability to digest coarse fibres, endurance for heat tolerance, and resistance to tropical diseases and adaptability to the tropical harsh climatic conditions. For example, breeds like Tharparkar, Kankrej, Sahiwal, Red Sindhi and Gir have been evolved in arid regions of the country where agriculture is not well developed and the animal husbandry is the main source of livelihood for the farmers. Gir the main cattle breed of Kathiawar region of Gujrat is known for stress tolerance capacities. Rathi is other important breed of desert widely found among the nomads who maintain their animals on ranges. Nellore cattle are highly fertile and resistant to parasite and to tropical disorders, precocious and have extraordinary maternal ability. Sindhi animals are small in size, good looking and useful in areas where there is a lack of forage (Mariante et al, 1999). Tharparkar is another breed which is well adapted to withstand the scorching heat of Thar Desert. The Tharparkar cattle have developed the ability to change their body coat colour, in summer they change their coat colour to white, which reflects most of the sun rays and protect them from scorching heat and in winter their coat colour becomes bluish grey which absorbs the sun light and keep the animal worm during extreme winter. Sahiwal calves were found to be more resistant to tick borne tropical theilerioses infection as compared with Holstein calves. (Glass et al, 2005). Skin of Dangi breed cattle exudes an oily secretion which protects the animals form heavy rains (Sadana, 2007). Khillari bullocks are famous for travel very fast without any fatigue. Vechur is smallest cattle breed of the world and very famous for production of maximum milk proportionate to its body weight. Similarly every other breed has developed their own specific characteristics to face the climatic stress of the region where they are being reared. Breeds along with their breeding tract, few important characteristics & production performance have been presented in Table 2.

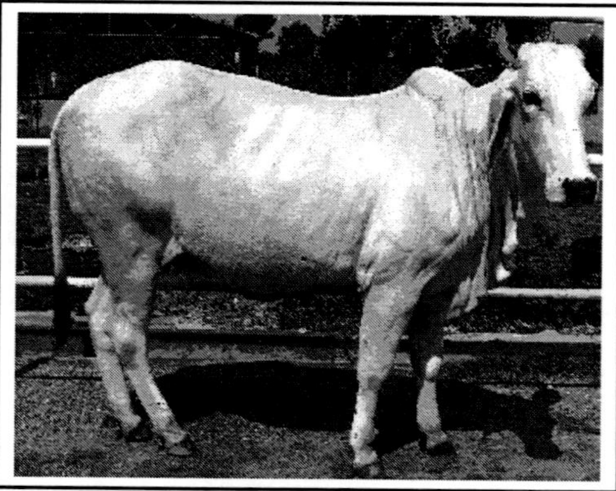

Fig. 1. Tharparkar Cow Photo credit- IGFRI Jhansi

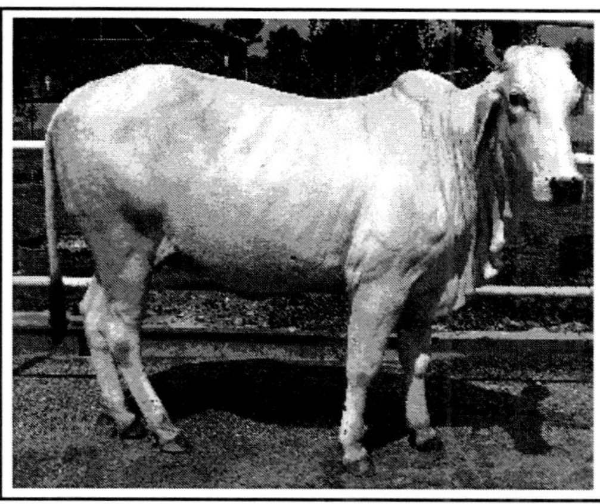

Fig. 2. Tharparkar Cow Photo credit- IGFRI Jhansi

Table 2. Cattle Breeds of India.

Name of Breed	Home Tract	Physical Characteristcs
Dairy Breed		
1. Sahiwal	Montogomery district (Pakistan) but some animals are found in India Ferozpur and Amritsar (Punjab)	Color: usually reddish but pale red or brown mixed with white spot is not uncommon. Cows give 7 to 8 milk per day.
2. Red Sindhi	Pakistan. But few animals are found in Orissa, Tamil Nadu, Bihar, Kerala.	Color is red with patches of white. Cows produce 6 kg milk per day.
3. Gir	Junagarh, Bhavnagar and Amreli (Gujrat)	Red color. Milk yield : 7 kg per day.
4. Rathi	Bikaner, Ganganagar and Jaisalmer(Rajasthan).	Brown color with some white patches. Milk yield: 5 kg per day.
Draught Breed		
5. Amritmahal	Hassan, Chikmagalur, Chitradurga (Karnataka)	Grey color with shades varies form white to black. Horns are very long.
6. Bachaur	Sitamari, Madhubani, Darbhangha, Samistipur and Muzaffarnagar (Bihar)	Grey color. Medium sized horn.
7. Bargur	Erode (Tamil Nadu)	Brown color with white marking.
8. Dangi	Nasik and Ahmednagar (Maharashtra)	White coat color with red or black spots .
9. Hallikar	Mysore, Mandya, Bangalore, Tumkur, Hassan, Chitradurga (Karnataka)	Color is grey to dark grey with deep shade in the fore- and hind –quarters. Long horn.
10. Kangayam	Kangayam, Dharampur, Erode, Karur (Tamil Nadu)	Coat color is red at birth but changes to grey after 6 months of age.
11. Kenkatha (Kenwaria)	Lalitpur, Hamirpur, Banda (UP), Tikamgarh (MP).	Grey on the barrel to dark grey on the rest of body.

Contd...

Table 2. Contd...

12. Kherigarh	Lakhimpur , Kheri (UP)	White coat color.
13. Khillari	Kolhapur, Solapur, Sangli, satara(Maharashtra), Belgaum, Bijapur and Dharwad (Karnataka)	Greyish white to white color. Horns are long and pointed.
14. Malvi	Indore, Dewas, ujjain, Shajapur, Mandsaur, Ratlam, Rajgarh (MP), Jhalawar (Rajasthan)	There are three stains: light, medium and heavy. Color of adult animal is white.
15. Nagauri	Nagaur, Joshpur, Bikaner (Rajasthan)	White to light grey color.
16. Nimari	Khandwa, Khargon and Barwani (MP)	Red color with large splashes of white on the body.
17. Ponwar	Pilbhit , Lakhimpur (UP)	Brown or black color with white patches.
18. Punganur	Chittor (Andhra Pradesh)	Colour: White, grey or light brown to dark brown. This breed is almost on the verge of extinction.
19.Red-Kandhari	Nanded, Ahmadpur, Parli, Latur (Maharashtra)	Dark red color
20. Siri	Darjeeling (WB), Sikkim	White or light grey color.
21. Umblachery	Thanjavur, Thiruvarur, Nagappattinam (Tamil Nadu)	Red or brown at birth; grey after 6 monhts. White marking on face and legs.
22. Alambadi	Tamil Nadu, Karnataka	Grey or white color
Dual-purpose Breeds		
23. Deoni	Deoni in Latur, Parbhani, Nanded, Osmanabad (Maharashtra), Bidar (Karnataka)	Three strains: complete white without any spot (balankya), white with black face(wannera) and black and white spot (Shevera)
24. Gaolao	Wardha (Maharashtra), Balaghat, Chhindwara, Durg, Seoni (MP)	White or light grey color.
25. Hariana	Rohtak, Hisar, Jind and Gurgaon (Haryana)	White or light grey color
26. Kankrej	Mehsana, Kutch, Ahmedabad, Kaira, Sabarkntha, Banaskantha (Gujrat)	Color varies from silver-grey to iron-grey to steel-black,
27.Krishna-valley	Solapur, Sangli, Satara (Maharashtra), Belgaum, Bijapur and Raichur (Karnataka).	Common color is grey white with a darker shade on fore-quarters and hind-quarters.
28. Mewati	Alwar, Bharatpur (Rajasthan), Mathura, Kosi (UP)	White color
29.Ongole (Nellore)	Nellore, Vizianagaram, Chittor, Kurnool, Cuddapah, Anatapur, Nalgonda, Mahabubnagar, Khamman (Andhra Pradesh)	This animal was brought into Indiaby Aryans more than 4000 year ago. Glossy white color (padakateeru)
30.Tharparkar	Pakistan. Some animals in Rajathan border and Rann of Kutch (Gujrat)	White or light grey.
31. Vechur	Kottayam, Kasargod (Kerala)	Colour: Light red, black or fawn and white. Milk yield 2 kg per day
Lesser Known Strains		
1. Bengali	West Bengal	Grey or white color
2. Binjharpuri	Cuttack (Orissa)	Crosses of Hariana and local cattle
3. Frieswal	Military Dairy Farms	Crosses between exotic breeds with Sahiwal
4. Ghumsuri	Cuttack (Orissa)	Silver grey or white color

Contd...

Table 2. Contd...

5. Jellicut	Chennai	Very small population
6. Jersind	Allahabad (UP)	Crosses of Jersey and Red Sindhi
7. Karan Fries	NDRI (Karnal)	Crosses of Friesian and Tharparkar
8.Karan-Swiss	NDRI (Karnal)	Crosses of Brown Swiss with Sahiwal / Red Sindhi
9. Kumauni	Uttar Pradesh	Jet black or pale to dark red color
10. Ladakhi	Ladakh (Jammu & Kashmir)	Local hill cattle
11.Malnad-Gidda	Karnataka	Good draught animal
12. Mampati	Madhya Pradesh	Local cattle
13. Manapari	Chennai	Local cattle
14.1Motu	Orissa, Madhya Pradesh	Small cattle with red color.
15.Purnea	Bihar	Local cattle
16.Shahabadi	Bihar and Uttar Pradesh	Similar to Hariana cattle
17. Sunandini	Kerala	Developed with crosses of imported Brown Swiss and Jersey bulls.
18.Tarai	Uttar Pradesh	Local cattle.
19. Taylor	Patna (Bihar)	Crossbred cattle.

Buffalo

Buffalo is believed to be originated in the Indian sub-continent from the wild arni buffaloes (*Bubalis arnee*), which are still found in forests of Assam. Buffaloes have been a blessing to India. Their huge number, superior genetic potential, docile temperament, less demanding in terms of management and nutrition, disease resistance, thriving under stress, their contribution towards milk production, draught and manure and above all their being part of our mythology speaks of their integration with human beings. The buffaloes are normally classified into river and swamp type, depending upon variation in their habitat and genomes. River buffaloes are massive in size mostly with curved horns, prefer to enter clear water, have 50 chromosomes, are primarily used for milk production and also used for meat production and draught purpose. Swamp buffaloes are stocky animals with marshy land habitats and have 48 chromosomes. They are primarily used for draught power in paddy fields and haulage and are also used for meat and milk production. Most of the Indian buffaloes are river type and they are found where clean water of rivers, irrigation canals and ponds are available to wallow. Swamp types of buffaloes are found in the northeastern region. Buffaloes produces more than 50% of total milk production of the country. India has some of the best milch breeds of buffaloes in the world. e.g. Murrah, Nili-Ravi, Jaffrabadi and Surti. These breeds have high potential for milk and fat production besides, draughtability and meat production. There are several other breeds which have regional importance and economic value, e.g. Bhadawari, Nagpuri,

Pandharpuri, Tarari, Parlakhemundi, Manda, Jerangi, Kalahandi, Sambalpur, Toda and South Kanara etc. Mehsana and Godavari are the breeds developed from grading up of Surti buffaloes in Measana Distt. of Gujrat and local and non-descript buffaloes in Krishna and Godavari district of Andhra Pradesh, respectively. Sethi (2003) has estimated that the number of pure bred animals of specified breeds is about 20-25 % of the total buffalo population in the country while rest of the buffaloes are non descript. The name of the breed/ strain along with home tract and some important characteristics are given in Table 3.

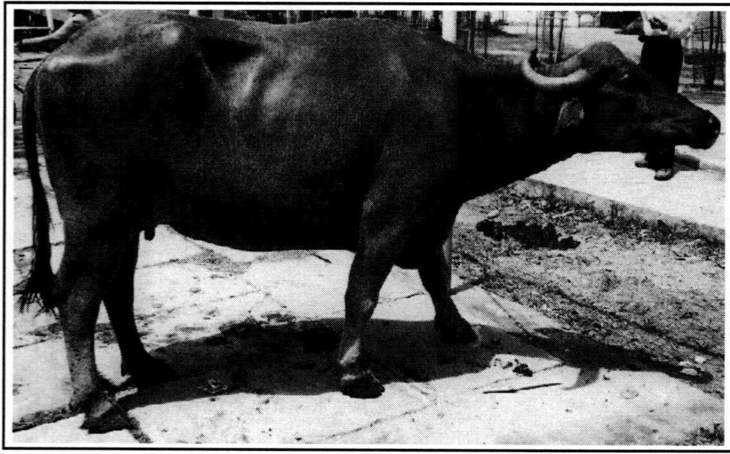

Fig. 3. Bhadawari Buffalo, Photo Credit, Network Project on Bhadwari Buffaloes, IGFRI

Fig. 4. Toda Buffalo

Table 3. Buffalo Breeds of India

Name of Breed	Home Tract	Physical Characteristics
1. Bhadawari	Agra, Etwah (UP), Morena, Bhind (MP)	Medium size (425 to 475 kg), Body color blackish-copper to light copper color with white color in leg. Famous for high fat (6 to 12%) milk.
2. Jaffarabadi (Gir, Bhavanagari)	Junagrh, Bhavnagr, Amreili (Gujrat)	Large sized (700 to 1000 kg). Body colour Black or grey, few animals have white spots on forehead, feet and tail switches. High milk yield (7 to 10 kg per day).
3. Marathwada	Parbhani, Nanded, Bid and Latur (Maharashtra)	Small-sized (300 to 400 kg). Color varies from greyish black to jet-black with occasional white marking on forehead and lower parts of the limb. Low milk yield (3 to 4 kg per day)
4. Mehsana	Mehasana, Banaskantha, Sabarkantha, Gandhinagar Ahmedbad (Gujrat)	Medium-sized (500 to 600 kg), Color is mostly black. Milk yield 7 to 10 kg per day with 7% fat.
5. Murrah	Rohtak, Jind, Hissar (Haryana), Delhi. However, this breed has now spread to almost all of the country.	Medium-sized body (450 to 800 kg), Colour jet black. Horns are short and tightly curved in a spiral form (hence called murrah). Milk yield 5 to 8 kg per day. Murrah males are extensively used for meat purpose.
6. Nagpuri	Nagpur, Wardha (Maharashtra)	Small to medium -sized animal (350 to 500 kg), black colured, horns are long (50-65 cm) and reach nearly to shoulder. Milk yield 4 to 5 kg per day.
7. Nilli-Ravi	Pakistan. In India some animals are found in Ferozepur and Amritsar (Punjab)	Same as Murrah except some white marking on extremities of foot and forehead.
8. Pandharpuri	Kolhapur, Solapur, Sangli and Satara (Maharastra), Belgaum (Karnataka)	Medium-sized animal (400 to 450 kg), Colour varies from light black to dark black. Horns are very long and extend beyond the shoulders.
9. Surti	Kheda, Vadodara, Bharuch and Surat (Gujrat)	Small-sized animal (300 to 400 kg). Colour varies from rustry brown to silver-grey. Horns are flat and sickle shaped. Milk yield 4 to 7 kg per day.
10. Toda	Nilgiri (Tamil Nadu)	Medium-sized. Colour varies from fawn to ash-grey. Horns are long (55 to 70 cm). Milk yield 2 t0 3 kg per day.
Lesser Known Breeds		
1. Assamese	Assam	Used for draught purpose
2. Godavari	Andhra Pradesh	Hardy animal (450 to 550 kg)
3. Jerangi	Orissa & Andhra Pradesh	Small-sized animal. The buffalo is useful for ploughing in water-logged paddy fields.

Contd...

Table 3. Contd...

4.Kalahandi (Peddakimedi)	Orissa, Andhra Pradesh	Hardy in work
5. Kujang	Cuttack (Orissa)	Milk yield: 3 kg per day
6. Manda	Orissa	Good draught animal
7.Paralakhemundi	Orissa, Andhra Pradesh	Excellent draught animal
8. Sambalpuri	Orissa, Andhra Pradesh	Used for draught
9. Sikamese	Sikkim	Poor milker
10.South Kanara	Karnataka	Used for draught
11. Swamp	Assam	Animals are hardy and quite adapted to hot and humid conditions
12. Tarai	Tanakpur, Ramnagar(UP)	Milk yield: 2-3 kg per day.

Goat

There are 23 well-described goat breeds in India. Most of these breeds have been evolved over ages through geographical isolation and natural selection to suit the specific agro-ecological conditions. Most of these breeds are adapted to harsh climates, long migration, tropical diseases, poor nutrition and shortages of water. However, majority (about 75%) of the total goat population are considered as mixed and non-descript. Goats are important livestock species of the country, mainly on account of short generation interval, higher rates of prolificacy, and ease which the goats as also their products can be marketed. Goats are known to be especially useful to difficult zones, where they can sustain themselves on sparse forage and extreme climatic conditions where other species of animal may perish. Goats are the main meat animals in India, their meat is most preferred and hence the costliest of all meats. The goat breeds can be classified on the basis of size (Acharya, 1992) in to large, medium and small sized breeds. Large sized group include Jamunapri, Beetal and Jhakrana. Medium sized group include the goat breeds like Sirohi, Marwari, Zalawadi, Kutchi, Surti, Sangamneri, Osmanabadi, Gaddi, Ganjam, Mehsana etc. Black Bengal and Assam local breeds are very small in size and put under small size group. Barbari and Changthangi breeds, although classed as medium sezed breeds, tend to be smaller than other breeds of their group. The goats are also classified according to the major produce associated with them. Although, goats are the main meat animals, but some breeds are known for milk production, like Jhakrana and Jamunapari and fibre production, like Chegu and Changthangi that produce elite pashmina fibres and Angora (imported to India) that produces mohair fibre. Goat hair also fetches good price and there has been demand for goat hair of such breeds as Marwari of Rajasthan. The name of breeds along with their home tract and some important characteristics are given in Table 4.

Table 4: Indian Goat Breeds Grouped Under Prime Production Trait(s).

Breeds	Home tract	Characteristics
A. Meat and skin		
1. Black Bengal (Syn. Bengal)	Whole of South Bengal, typical specimen at the Nadia districts of W.B.	Dwarf in size (12-20 kg), legs short, straight back, beard in both sexes. Coat color: predominantly black, brown/gray (30%) and white (10%) ex st; soft, glossy and short hairs. Ears are short and neatly upright Most prolific among Indian breeds, Multiple births common: 2, 3 or 4 kids born at a time. Kidding twice a year.
2. Ganjam	Ganjam, Koraput and Phoolbani district of Orissa	Compact, tall and leggy (30 to 45 kg) Coat color: Brown, black and gray, white rare, small hair. Beard often lacking in does. Ears are medium sized, relatively straight and horizontal; horns are pointed backward upward and straight. Kidding: once a year, two kidding seasons and twins mostly.
3. Kannaiadu	Ramnathpuram and Tirunelveli districts of Tamil Nadu	Tall and stout (25-35 kg). Coat color: Predominantly black or black with white spots. Males are horned and females are polled. Kidding: Twice in 18 months, single or twins.
B. Meat, milk and skin		
4. Barbari (syn. Guinea goat)	Agra, Aligarh, Mathura, Mainpuri, Etah districts of U.P.	Small sized (20-40 kg) with short legs, Coat color: Varies greatly but white with red or tan spots considered typical. Ears: small tubular, erect, pricked, up wards and outwards. Horns: Medium and slightly twisted, directed upwards and backwards. Kidding: twice in 16 months, twin or triplet is common
5. Malabari (syn. Tellicherry)	North Malabar hills, Kottayam, Calicut districts of Kerala	Medium sized animal (30-40 kg). Coat color: varies from white to complete black, black and white, black and brown and white with colored patches occur; both long and short haired varieties available.Ears: Medium in size, directed outward and downward. Horns: mostly horned; small, slightly twisted and pointed. Kidding: once a year, twins
6. Osmanabadi	Around Osmanabad, Maharashtra	Tall and large size (40-50 kg). Coat color: Predominantly black, white, brown and spotted occur; long and short haired type, based on presence or absence of long hair on the thighs and hind quarters. Ears: Long, pendulous about 20 cm in length. Kidding: Once a year, twins are common.
7. Sirohi	Sirohi district Ajmer & surrounding Aravalli ranges of Rajasthan	Compact and medium sized(25-50 kg): tail twisted and carries coarse pointed hair: nostrils short and pointed; wattles often present. Coat color: Brown, white and admixture (common): off colors in typical patches: hair coarse and short. Ears: Long and flabby. Horns: small and pointed, curved upward and backward. Kidding: Once a year, twins common.

Contd...

Table 4. Contd...

8. Surti	Mahi river valley of Saurashtra Baroda, Gujarat	Medium sized (25-35 kg). Coat color: Predominantly white but often black and brown hair in different parts of the body. Ears: Medium sized and ha nging. Horns: Small, running backwards and slightly curved. Kidding: once a year, single or twins.
9. Mehsana	Banaskantha, Mehsana, Baroda and Gandhi Nagar districts of Gujarat	Medium sized (30 to 40 kg), Roman nose and broad muzzle: tall short and abrupt . Coat color: Black with white spots at the base of the ear, long hair; small beard and gray black skin. Ears: The ears are white, leaf like and drooping. Horns: Slightly twisted and curved upwards and backwards. Kidding: once a year, mostly single.
10. Zalawadi	Surendranagar, Rajkot, Zalwad districts of Gujarat	Large sized (30-40 kg), broad nostrils; neck long and throat carries lobular appendages; tail short and curved. Coat color: Black with white spots, skin pinkish blue, hair, coarse black and lustrous the staple length 10-12 cm (used for preparing rope). Ears: Long, wide, leaf like and drooping. Horns: Long and straight, twisted and crew like. Kidding: Once a year; the age at first kidding 23-24 months
11. Jhakhrana	Nagaur districts, Rajasthan	Large sized (40 to 60 kg). Coat color: Predominantly black, spotty white. Ears: Resembles that of Beetal though smaller in size. Kidding: Once a year, mostly singles.
C. Meat, hair and skin		
12. Bakharwal	Hilly tracts of Kashmir	Very hardy large sized goats (65-70 kg). Coat color: White brown or black with white markings; fine silky hair (about 10-12 cm long); average yield of hair per annum 1.5 kg. Hair used for ropes, Namada etc.
13. Gaddi (syn. Chamba)	Kangra, Shimla, Chamba, Kinnaur, Kulu districts of H.P.	Medium sized (25 to 30 kg) with Roman nose Coat color: Predominantly white, but gray and red occur; long (12 cm) drooping ears. Kidding: once a year single (rarely twins). Meat is and devoid of fat. Milk yield : 1.5 kg per day. Hairs are used for r opes, snow boots and coarse rug making
14. Gohilwadi	Bhavnagar, Amreli and Junagarh districts of Gujarat	Medium sized (35 to 45 kg). Coat color: Black coarse and long hairs. Kidding: Once a year, single.
15. Kangan	Hills of H.P.	Food and cottage (ropes, bags)
16. Kutchi (syn. Cutchi, Kathiwadi)	Kutch, Kathiawad district of Gujarat	Medium sized (35 to 45 kg), Coat color: predominantly black, white or speckled at ears, neck and face. Kidding: once a year, single or twins.
17. Khasi (syn. Assam Hill)	Khasi, Naga, Lushai Hills, Meghalaya, Nagaland	Medium sized (20 to 30 kg). Coat color: white with silver or gray shades; beard and long haired. Kidding: commonly once a year, twins common.

Contd...

Table 4. Contd...

18. Marwari	Arid western plains of Rajasthan (Jodhpur, Marwar, Pali, Bikaner, Jalore, Jaisalmer, Barmer)	Medium sized (25 to 35 kg). Coat color: predominantly black, long shaggy hair coat: hair jet black, lustrous and growth 10-12 cm/year. Kidding: once a year single, twins with advancing age.
19. Sangamneri	Poona and Ahmed Nagar districts of Maharashtra	Medium sized (30 to 40 kg). Coat color: Highly variable, white, brown and black with spots of 3 colors. Kidding: Once a year; single or twins.
D. Meat, pashmina and skin		
20. Changthangi	Changthang region of Leh and Ladakh, J&K	Small sized (20-25 kg), strong body and powerful legs.Coat color: Predominantly white (50%) and rest brown, gray and black. Undercoat white/gray: yields warm delicate fibre, pashmina (cashmere, pashm).Horns: Large turning outward, upward and inward forming a semicircular ring. Kidding: Once a year normally single.
21. Chegu	Lahaul and Spiti valley, H.P. Uttarkashi, Chameli and Pithoragarh districts of Uttranchal	Medium sized (25 to 40 kg). Face and muzzle tapering Coat color: Predominantly white but grayish red and mixed colors seen. Long hair below which under coat of delicate fibre (cashmere or pashmina). Horns: Bent upward backward and outward, one or more twists. Kidding: once a year and mostly single. Animals are also used for draught (pack) to carry salt and small loads.
E. Milk, meat and skin		
22. Beetal	Gurdaspur, Amritsar and Ferozpur districts of Punjab	Resemble Jamunapari but smaller (40 to 65 kg); face convex, lips black, eyes blue black with white or brownish corneal surroundings Coat color: Predominantly black: red tan or black, heavily spotted on white also occur: males usually possess beard. Ears: Long pendulous drooping beetal leaf shaped pinna. Kidding once a year, single or twins
23. Jamunapari (syn. Etawah)	Chambal ravines, Chakarnagar, Batpura, Etawah district of U.P.	Largest(45 to 65 kg)and most-elegant of the long- legged goats of India; Coat color: White with tan or black markings at neck and ears; beard in both sexes, tuff of long hair in the buttocks. Roman nose having a tuff of hair which results in parrot mouth appearance. Ears: Long and pendulous (26-28 cm): tubular and tube opening is towards the front. Kidding: once a year single or twins. Large udder and big teats: average yield I kg per day, highest recorded 4 kg a day.

Sheep

According to the second World Watch List (1995) there are 57 breeds of indigenous sheep in India. However, most of our text books refer their number to be 42 (Acharya, 1982, and Bhat & Acharya, 1984), they constitute about

25% of the total sheep population. About two-third of the animals are non-descript. Inter-mixture amongst the breeds is very common in areas where more than one breed exists. Indian sheep are well adapted to harsh climate, long migration, tropical diseases, poor nutrition and shortage of drinking water and its poor quality. These breeds are classified on the basis of major product i.e. a) Apparel wool b) Carpet wool, c) meat and d) Meat and carpet type. The sheep breeds can also be classified on the basis of Agro-ecological regions (Bhat et al, 1980; Acharya, 1982; Acharya and Bhat 1984) viz. a) North temperate region, b) North Western arid and semi arid region c) Southern penninsular region and d) Eastern region. Sheep breeds classified region wise and product wise are presented in Table 5.

Table 5. Breeds of Sheep in Different Agro-ecological Regions in India and their Major Products.

North Temperate	North-Western Arid and Semi Arid	Southern Peninsular	Eastern
Bhakarwal (CW)	Chokla (CW)	Bellary (MCW)	Balangir (MCW)
Changthangi (CW)	Hissardale (AW)	Coimbatore (MCW)	Bonpala (MCW)
Gaddi (CW)	Jaisalmeri (MCW)	Daccani (M)	Chottanagpuri (MCW)
Gurez (CW)	Jalauni (MCW)	Hassan (M)	Ganjam (MCW)
Karnah (AW)	Kheri (MCW)	Kanguri (M)	Garole (M)
KashmirMerino(AW)	Magra (CW)	Kilakarsal (M)	Tibetan (CW)
Poonchi (CW)	Malpura (MCW)	Madras Red (M)	
Rampur Bushair(CW)	Marwari (MCW)	Mandya (M)	
	Muzaffarinagri(MCW)	Mecheri (M)	
	Nali (CW)	Nellore (M)	
	Pattanwadi (CW)	Nilgiri (AW)	
	Pugal (MCW)	Rammand White (M)	
	Sonadi (MCW)	Tiruchy Black (M)	
	Munjal(M)	Vembur (M)	

Within parenthesis is the major product from the breed. Apparel wool (AW); Carpet wool (CW); Meat and Carpet wool (MCW) & Meat (M)

India is endowed with large and biologically diverse sheep population comprising of sheep producing fleeces having fibre diameter below 30 μ to totally hairy sheep. Some of the breeds from the adjoining areas have been identified as separate breeds although they may be similar to each other genetically. The phenotypic differences may be due to selection for adaptation to specific agro-climatic conditions. The Madras Red in Tamil Nadu, Red Nellore in Andhra Pradesh and Ganjam sheep in Orissa although identified as separate breeds appear to be similar but their actual genetic differences are not known. Similarly in Rajasthan Marwari, Jaisalmeri and

Kheri are black face sheep. The so called Bikaneri sheep include brown face breeds, viz. Chokla, Magra and Nali and the phenotypic differences might have been due to assortative mating. Real genetic differences, if any, among similar groups need to be established.

There are more than one strains in some of the breeds viz. Deccani, Patanwadi, Nellore etc. The sheep of Deccani breed are not uniform and their coat colour varies from black to white and the mixture of the two. It has been estimated that about 57 percent of Deccani are black, 28 per cent white and 15 per cent have a mixed coat colour. The varieties of Deccani are Lonad, Sangamneri, Solhapuri (Sangola) and Kolhapuri. Sangamneri type sheep has a stronger body constitution and higher wool yield. The wool is coarser and of longer staple. Sangola type sheep have higher proportion of black colour sheep and produce finer fleeces (Jangnure, 1992).

Similarly, Patanwadi sheep has three types (a) non-migratory; small bodied and fine fleeced Patanwadi, (b) Migratory; leggy with long tubular ears and relatively coarse fleeced-Kathiawadi and (c) low stuttered, meat type, very coarse fleeced-Charotari also called Sonadi in Rajasthan (Mason, 1981). However, looking to the phenotypic appearance of Sonadi and Patanwadi their genetic similarity needs to be verified.

Kheri sheep evolved in the farmers flock under the field conditions especially in the areas that are important for sheep migration in Rajasthan appears to be a need of migratory sheep breeders (Bohra *et al*, 1993). The animals of this breed have established owing to their ability to walk long distances and ability to survive on limited amount of coarse feed. The animals of this breed can sustain stress and on return of favourable condition regain faster resulting in better reproduction rate and growth of lambs as compared to prevalent breeds in the area.

Garole sheep are reputed for multiple birth and are found in the Sunderban area of West Bengal. These sheep are reported to have contributed prolificacy to the Boorola Merino sheep. Sundarben area has a sheep population of about 0.16 m. The breed tract is tropical humid. The climate of the area is hot and humid. Litter size at first lambing is two and at subsequent lambings is 2 to 3. Prolificacy reported is singles 25-30%, twins 55-60%, triplets 15-20% and quadruplets 1-2% (Singh and Bohra, 1996).

Munjal sheep is the least discussed among Indian sheep breeds. Information on population and production performance of this breed is scanty. Munjal is one of the important mutton type sheep. Animals of this breed are quite big and tall. Average adult body weight is around 45 kg. Population of this breed is around 90 thousand (Kushwaha et al 1999). Animals of this breed are found around Muktsar, Bhatida, Firozpur and Faridkot district of Punjab, Hisar and Karnal in Haryana and Ganganagar in Rajasthan..

The breeds like Nilgiri, Kashmir Merino, Hissardale, Avivastra, Bharat Merino and Avikalin are developed by crossing exotic wool breeds with indigenous breeds and their number is very limited. The home tract and some important characteristics of each breed are given in Table 6.

Fig. 5. Chokla sheep, Photo credit- CSWRI, Avikanagar

Fig. 6. Garole sheep with triplet, Photo credit-Dr. R.C. Sharma equine Production Centre, Bikaner

Table 6. Sheep Breeds of India

Breed	Home tract	Characteristics
Chokla	Churu, Bikaner, Jhunjhunu, Jaipur and Nagaur {Rajasthan)	Light to medium sized animals (25 – 35kg). The face devoid of wool, reddish brown and the color may extend up to the middle of the neck: skin pink. Ears small to medium in length and tubular. Both the sexes polled. Tail thin and of medium length. Coat dense and relatively fine, covering belly and greater part of legs.
Nali	Ganganagar & Jhunjhunu (Rajasthan), Hisar and Rohtak (Haryana)	Medium sized animals (25-35 kg). Face light brown, skin pink. Both sexes are polled. Ears large leafy, Tail short to medium and thin. Fleece white, coarse, dense and long stapled. Forehead, belly and legs covered with wool.
Marwari	Jodhpur, Jalore, Nagaur, Udaipur Pali (Rajasthan) & Jeoria region of Gujarat	Medium sized animals (25-30 kg). Black face; color extending to lower part of neck. Ears extremely small and tubular. Both sexes are polled. Tail short to medium and thin. Fleece white and not very dense.
Magra	Bikaner, Nagaur, Jaisalmer and Churu (Rajasthan)	Medium to large animals (25-35 kg). White face with light brown patches around the eyes characteristic of this breed. Skin color pink. Ears small to medium and tubular. Both sexes polled. Tail medium in length and thin.
Jaisalmeri	Jaisalmer (Rajasthan)	Tall, well built animals (25-35 kg). Black or dark brown face; color extending up to neck. Typical Roman noses. Long drooping ears, generally with a cartilaginous appendage. Both sexes are polled. Tail medium to long..
Pugal	Bikaner and Jaisalmer (Rajasthan)	Fairly well built animals (25-35 kg). Black face with small light brown stripes on either side above eyes; lower jaw typically light brown. Black color may extend up to neck. Ears short and tubular. Both sexes polled. Tail short to medium and thin.
Malpura	Jaipur, Tonk, Sawaimadhopur, Ajmer, Bhilwara & Bundi (Rajasthan)	Fairly well built animals (25-40 kg), long legs. Face light brown. Ears short and tubular with a small cartilaginous appendage on the upper size. Both sexes polled. Tail medium to long and thin. Belly and legs devoid of wool.
Sonadi	Udipur, Dungarpur and Chittorgarh (Rajasthan)	Fairly well built (25-40 kg), somewhat smaller than Malpura with long legs. Light brown face, with color extending to the middle of neck. Ears generally have a cartilagenous appendage. Tail long and thin. Both sexes polled. Udder fairly well developed. Belly and legs devoid of wool.
Patanwadi	Saurashtra, Kadi, Kutch, Patan, Kalol, Sidhapur and Mehsana (Gujarat)	Medium to large animals (25-40 kg), with relatively long legs. Typical Roman rose. Face brown and spotted tan. Ears medium to large tubular with a hiry tuft. Tail thin and short. Both sexes polled.

Contd...

Table 6. Contd...

Muzaffarnag-ari	Muzaffarnagar, Bulandshahar, Saharanpur, Meerut, Bijnor, Dehradun (Uttar Pradesh) and parts of Delhi and Haryana	Medium to large animals (35-50 kg). Face line slightly convex. Face and body white with occasional patches of brown or black. Ears occasionally black. Both sexes polled. Males occasionally show rudimentary horns. Ears long and drooping. Tail extremely long and reaches fetlock. Belly and legs devoid of wool.
Jalauni	Jalaun, Jhansi and Lalitpur (Uttar Pradesh)	Medium sized straight nose line (30-40). Both sexes polled. Ears large flat and drooping. Tail think and medium in length. Belly and legs devoid of wool.
Munjal	Bhatinda, Patiala (Punjab) Hisar, Karnal, Ambala (Hryana), Ganganagar (Rajasthan)	It is mutton type sheep, big in size, Tall, body color predominantly white with dark brown face. Wool is very coarse and hairy. Adult male may weigh 50-80 kg. ,
Hissardale	Hissar (Haryana), Kullu, Kangra (Himachar Pradesh)	Small animals with short legs (35-55), giving a low set appearance, leaf like medium sized ears. Most animals polled. Color predominantly white, although some brown patches can be observed.
Deccani	Maharashtra, Andhra Pradesh and Karnataka	Medium sized (30-40 kg), color black with white marking. Ears medium in length, flat and drooping. Tail short and thin.
Nellore	Nellore Prakasham and Ongole (Andhra Pradesh)	Relatively tall animals (30-35 kg) with little hair except at brisket, withers and breech. Arms horned, ears long and drooping. Tail short and thin, sex of the animals carry wattles.
Bellary	Bellary (Karnataka)	Medium sized animals (25-35 kg), body color ranging from white through various combination of white and black, 1/3 of the males are horned, females generally polled. Ears medium long, flat and drooping. Tail short and thin. Belly and legs devoid of wool.
Hassan	Hassan (Karnataka)	Small animals (20-25 kg) white body with light brown or black spots. Ears medium long and drooping, 39% of the males horned. Females white, extremely coarse and open legs and belly generally devoid of wool.
Mandya	Mandya (Karnataka)	Relatively small animals (20-30 kg). Color white, but in some cases face light brown and this color may extend to the neck. Compact body with a typical U shaped conformation from the rear. Ears long, leafy and drooping. Tail short and thin. Large animals carry wattles. Slightly Roman rose. Both sexes polled. Coat extremely coarse and hairy.
Mechari	Salem and Coimbatore (Tamil Nadu)	Medium sized animals. Light brown ears medium sized. Both sexes polled. Tail short and thin. Body covered with very short hairs.
Kilakarsal	Ramnathpuram, Madurai, Thanjavir and Ramnad (Tamil Nadu)	Coat dark tan, with black spots on head, belly and legs. Ears medium sized. Tail sm all and thin. Males with thick twisted horns. Most animals have wattle. CHECK

Contd...

Table 6. Contd...

Vembur	Vembur Melakharandhai, Nagalpuram, Achangulam (Tamil Nadu)	Coat dark tan, with black spots on head, belly and legs. Ears medium sized. Tall small and thin. Males with thick twisted horns. Most animals have wattle. CHECK
Coimbatore	Coimbatore, Madurai (Tamil Nadu)	Medium sized animals (20-25 kg), white with black or brown spots, Ears medium sized and directed outward and backward. Tail small and thin, 30% of males poll ed.
Ramnad	Ramnad (Tamil Nadu)	Medium sized animals (20-30 kg), predominantly white, some animals have fawn or black markings over the body. Ears medium sized and directed outward and down ward. Males with twisted horns. Females polled. Tail short and thin.
Madras Red	Chingalpet and Madras (Tamil Nadu)	Body weight 25-35 kg, color predominantly brown, the intensity varying from light tan to dark brown; some animals with white marking on forehead, inside the thigh and on the lower abdomen. Ears medium, long and drooping. Tail short and thin. Rams have strong corrugated and twisted horns, the ewes are polled. Body covered with short hair.
Tiruchi	Tiruchy, Arcot, Salem and Dharampuri (Tamil Nadu)	Small animals (20-25 kg). Body completely black. Males horned, ewes polled. Ears short and directed downward and forward. Tail short and thin.
Kenguri	Raichur (Karnataka)	Medium sized animals 25-35 kg). Body color mostly dark brown but colors ranging from white to lack with spots or different shades also observed. Males horned. Females generally polled.
Tibetan	North Sikkim and Kameng in Arunachal Pradesh	Medium size, mostly white body color with black or brown face. Belly, legs and faces are devoid of wool.
Bonapala	Sikkim	Animals are tall, leggy. Colour ranges from white ot black. Ears are small and tubular. Both sexes are horned. Tail is thin and short. Belly and legs are devoid of wool.
Rampur Bushair	Shimla, Kinnaur, Nahan, Bilaspur, Solan and Lahaul and Spiti (Himachal Pradesh), Dehradun, Rishikesh, Chakrota and Naini Tal (Uttaranchal)	Medium sized. Coat color is white or brown or and black. Ears are long and drooping. Face line is conves, giving a typical Roman nose. Males are horned; most females are polled. Legs, belly and face are devoid of wool.
Karnah	Karnah (Kashmir)	Large animals. Males have large curved horns and prominent nose line. Coat color is White.
Gurej	Gurej (Kashmir)	Large sized. Coat color varies: mostly white, some are brown or black. Ears are long, thin and pointed. Tail is thin and short.
Changthangi	Changthang (Ladakh)	Large sized with strong built.

Contd...

Table 6. Contd...

Balangiri	Balangir, Sambalpur and Sundargarh (Orissa)	Medium sized, white or light brown or black. Ears are small and stumpy. Males are horned and females are polled. Tail is of medium length and thin. Legs and belly are devoid of wool.
Ganjam	Koraput, Phulbani, Puri (Orissa)	Medium sized with coat color ranges from brown to dark tan. Some animals have white spot on face and body.
Bhakarwal	It has no home tract as the animal are entirely migratory.	Animals are reared by the nomads. Medium sized and have typical Roman nose. All animals are spotted, fawn or grey. Males are horned and females are polled. Ears are long and drooping ; tail is small and thin
Sahabadi	Shahabad, Patna and Gaya (Bihar)	Medium sized and leggy. Coat color is grey with black spots. Ears of medium size and drooping. Tail is extremely long and thin. Both sexes are polled. Legs and belly are devoid of wool.
Gaddi	Kistwar, Bhadarwah (Jammu);Ramnagar, Udampur and Kullu (Himachal Pradesh);Dehradun, Nainital, Tehri-garhwal and Chamoli (Uttaranchal)	Medium sized. Coat color is usually white, although tan, brown and black and their mixtures are also seen. Males are horned. Some females are also horned. Tail is small and thin.
Poonchi	Poonch and rajori (Jammu)	Animals are similar to Gaddi but are smaller is size. They are mostly white. Tail is short and this. Legs are also short.

Pig

Domestic pigs found in India are believed to be have a common origin with the wild pig *Sus scrofa cristatus* (Macdonald, 2001). The presence of long dark brown longitudinal strips on the newborn piglets, which gradually vanish as animal grows, is a proof of its wild ancestry. Phenotypic variations among sub groups are very low and they are generally called "local pigs or deshi pigs". Traditionally Indian pigs produce bristles and meat was by-product. But with the increase use of nylon in place of bristle, pig is now primarily reared for production of pork (pig meat) in our country. Indian pigs are able to survive in most adverse condition of malnutrition and health management. In the urban areas they are raised on garbage and kitchen wastes. In rural areas they are let loose to fend for themselves. Pig farming is livelihood of many rural poor belonging to the lowest socio-economic strata and they have no ideas about the scientific pig breeding cum rearing for meat production. There are two types of pigs in the country.

Wild Pig: Wild Pig (Indian Wild Boar) mainly found in the Himalayan jungles, Eastern India and Southern Peninsula. Its colour is rusty at young age and becomes dark chestnut brown in adult animal. Its main diet is root, tuber, insects, snakes etc.

Domesticated indigenous pig (Desi Pig): There are three main Indian breeds of pig.

Deshi Pig: Mainly found in U.P, Bihar, M.P and Punjab. This is the largest sized pig among Indian breeds and weighs about 150 kg. The colour of the animal is gray or brown or black. The sow (female pig) gives birth of 8-12 piglets at a time under good management conditions.

Ankamali Pigs: It is found in Kerala, Karnataka, Tamil Nadu and Maharashtra. It is black/gray in colour with white patches and weighs about 40 to 70 kg. Although female produces more number of piglets (about 10), but mortality is very high (about 50%).

Ghori (Dome Pig, Pigmy Pig): These small sized pigs are found in all the Northeastern states and Himalayan Tarai region. It weighs about 50-60 kg and its colour is mainly black with an admixture of brown. Like the wild pigs, these animals also have a row of coarse and straight bristle starting from neck to back.

Another breed of pig with distinctively superior productive and reproductive characteristics has recently been identified in the eastern sub-Himalayan region of West Bengal and adjoining Nepal. The breed is commonly known as *Ghoongroo* are black in colour with a compact body, long thick coarse hair, a long tail and upwardly curved snout. The face is broad and flattened with large heart shaped ear resembling that of an elephant. Average litter size is 11.92±0.08 and a litter size up to eighteen is common (Pan *et. al.*, 2005).

Camel

According to National Commission on Agriculture (1976) the important indigenous camel breeds available in the country are Bikaneri, Jaisalmeri, Kachchhi, Marwari, Mewari, Sachori, Sindhi and Revarine. But some of the workers are of the opinion that there are only 4 major breeds Bikaneri, Kachchhi, Jaisalmeri and Mewari while others are perhaps the crosses of these major breeds, which have derived their names from their habitat. The Bikaneri breed well known for draught potential and long distance travel. The camels Bikaner and Jaisalmer breed are adapted to the climatic condition of the Thar desert, where temperature gets very high during the summer and very low in winter. The Mewari breed is known for the production of milk and is adapted to the hilly terrains of Mewar area in Rajasthan. The Kachchhi breed has a medium level of milk production and draught capabilities, and is adapted to the marshy land of the Ran of Kuchch in Gujrat state (Rathore, 1986; Mehta et al, 2006).

The population of camel in India numbers 6.32 lakh. Rajasthan state has highest population of 4.58 lakh followed by Gujrat with 53,000 and Madhya Pradesh with 8,000. There has been 29.65% decline in the overall population in the last five years (Livestock census 2003).

The Indian double humped camels are found in cold desert of Ladakh and Jammu & Kashmir where temperature ranges from 10 to -40 ^{0}C in winter and 40 to -10 ^{0}C in summer (Sahni & Mehta, 2005).

Fig. 7. Bikaneri Camel, Photo credit-NRC for cannels, Bikaner

Fig. 8. Double hump camel, Photo credit-NRC for camels, Bikaner

Equine

The equines (Horse/pony, Mule & donkey) are the important race and draught species having an edge over oxen, buffales and camel with multipurpose use of draught, transport, pack, pleasure, and patrolling and traditional ceremonies. However, due to mechanization and road network the horse has become obsolete in its primitive activities. Horse has unique place in history of civilization. Equines are still integral part of Indian army on mountain terrain. Pony, mule and donkeys are primary source of livelihood of sizable section of society. Donkeys and ponies are well known to perform under extreme cold and hot conditions and utilize low quality feed resources, which other species hardly consumed. India has 6 breeds of horse viz.

Fig. 9. Mewari Horse

Fig. 10. Donkey at work

Kathiawari, Marwari, Manipuri, Spiti, Bhutia and Janskari, evolved in different agro-climatic zones and representing both Arabic and Mangolian types (Singh and Yadav 2005). FAO has also recognized 3 other breeds of Indian horse, namely Chumarti, Deceeni and Sikang. Besides horse, assess (*Equine a sinus*), Asiatic Indian wild assess namely Kulan in Ranof Kutchh in Gujrat and Kiang in upper Himalayan region are also important unique germ plasm.

Yak

Yaks are believed to be domesticated by Qiang people in the Qiangtang and other areas of Northern Tibet. In India yaks are mainly found in Himalayan states of Himanchal Prdesh, Jammu and Kasmir, Sikkim and Arunanchal Pradesh. The wild yak (brown and black in color) lives in the coldest and most remote parts of Himalaya. Domesticated yaks are relatively smaller in size and variable in colour, being often pure white to piebald (mixture of two colors). The yak is massively-built animal with a drooping head, high humped shoulders, a straight back and short sturdy limbs. Yak supplies milk, meat, wool and hide. An adult yak weights around 550 kg. In summer yaks often eat and drink melting snow, apart from shrubs.

Pal et al (1994) categorized Indian Yaks into three types (i) Common yaks (ii) Bisonian and (iii) Bareback yaks and their respective population are 56.20%, 29.8 % and 14.0%. The common yaks are comparatively smaller in size, the appearance of the head and body conformation are like cattle. The Bisonian type are heafty animals with compact head connected by strong and short neck to the main trunk. The Bareback yaks are bodily rectangular with short hairs on rump and wither. The rosette hairs hang down the belly line and almost touching the ground. Das (2003) reported that there are five different types of Yaks in India, and they are Common, Bisonian, Bareback, White and Hairy fore head types. For centuries Yak has been mated with local cows to produce hybrids. Two hybrids (Horned hybrid (Zo), Hornless hybrid (Zum)) are now available. Hybrids have good beef conformation particularly when castrated, the quality of meat and hide is superior to that of yak.

Mithun

It is believed that mithuns have originated along the Assam-Mynamar border by crossing the wild cattle (Gaur) with local cows. Mithun is slightly smaller than than Gaur. It has a muscular ridge upon its shoulder which slopes down to the middle of the back where it ends in an abrupt dip. It has strong big size legs, narrow face, ears are erect and small to medium in size. An adult Mithun weighs about 400- 500 kg. The animals are reared by tribal community for meat purpose and are considered as a sign of prosperity and superiority of an individual in the society. The animals are reared under free-range condition at various altitude between 300 and 3000 M msi. In India Mithuns are found in Arunalchal Pradesh, Nagaland, Manipur and Mizoram. There are two distinct types (Nagami and Arunachali) of mithun in India. However, C Rajkhowa (2005) have reported four strains of Mithuns namely Arunanchal, Nagaland, Mizoram and Manipur.

Fig. 11. Wild Yak

Fig. 12. Arunanchal Mithun

Nagami : It is mostly found in Nagaland and Manipur. They are heavier in size; body weight ranges from 400 to 500 kg. Forehead is broad, horns emerge from the top of the poll and extend to the sides. The animals are primarily used for beef production

Arunachali: The mithun of this type is found in Arunachal Pradesh. They have comparatively short and compact body; body weight ranges from 350 to 450 kg. Forehead is broad and flat. Horns emerge from top of the poll and extend on the side. These animals are used both for meat and milk.

Conclusions

Most of indigenous breeds are unique in their adaptation to agro-climatic conditions of their habitat and management practices. Indigenous stock is not only able to thrive under harsh climatic conditions and very low input system but also produce a subsistence level. So there is a need of hour to exploit the full potential of our indigenous livestock by identifying their unique characteristics. Over the years, with adopting increasingly intensive agriculture production system, mechanization of agricultural operations and deforestation has forced the Indian farmer to change his priorities. As a result there has been genetic erosion in the population of indigenous livestock breeds as well as shrinkage in his herd size. Some important breeds have declined at faster rate where as many others are in the process of loosing their existence. There is a need for the conservation of indigenous livestock breeds which far the present, some of them, may not be economical but their unique characteristics adaptability and survival under specific environment can be exploited in future. Many of these breeds also have their own social and cultural utility to the community raising them and their extinction will be a permanent loss of human heritage bestowed on.

References

Acharya, R.M. 1982. Sheep and goat breeds of India. Rome : *FAO Animal Production and Health paper No. 30.*

Acharya, R.M and Bhat, P.N. 1984. Livestock and Poultry Genetic Resources in India. *IVRI, Publication.*

Ahlawat, S. P. S. and Gupta, S. C. 2006. Domestic animal biodiversity and breeding policies – issues and perspectives. In proceedings National Symposium on Conservation and Improvement of Animal Genetic Resources under low input system: Challenges and Strategies, February 9-10, 2006, NBAGR Karnal.

Bhat, P.N., Bhat, P.P., Khan, B.U., Goswami, O.B. and Singh, B. 1980. Animal Genetic Resources in India. Publication No. 192, NDRI, Karnal.

Bohra, S.D.J., Jain, A. and Sharma, S.C. 1993. Kheri- A new type of sheep in Rajasthan. *Wool & Woollens of India*.

Das, D. N. 2003. Genetic studies on certain physical, biochemical and performance traits of Yak on Arunanchal Pradesh. Ph.D. thesis submitted to Assam Agricultural University, Jorhat, Assam.

Glass, E. J., Preston, P. M., Sprinhett, A., Craigmile, S., Kirvar, E., Wilkie, G. and Brown, C. G. 2005. Bos Taurus and markedly different levels of acute phase proteins. *International Journal of Parasitology*,35(3): 337-347.

Jangnure, N.M. 1992. Sheep Development in the state of Maharashtra. A paper presented at one day Brain storming session on strategies for sheep development in 1990's CSWRI, Avikanagar, 29th January, 1992.

Joshi B K (2008). Livestock bio-diversity lossess: Economic and Ecological Issues. In proceedings National Symposium on Redefining role of Indigenous animal genetic resources in rural development, February 15-16, 2008, Bangalore.

Kushwaha, B.P., Abhay Kumar and Singh, R .N. 1999. Characteristics of Munjal Sheep. *Animal Genetic Resources Information* (25): 27-31.

Macdonald D. (Ed.) 2001. The Encyclopedia of mammals. Oxford University press.

Mariante, A. da S.M do, S. M. Albuquweque, A. A. do Egito & C MaManus 1999. Advances in Brazilian animal genetic resources conservation programe. *Animal Genetic Resources* (25): 109-123.

Mason, L.L. (1981). A world dictionary of Livestock breeds types and varieties pp. 348 CAB International Walling Ford, Oxford, U.K.

Mehta, S. C., Mishra, B. P. & Sahni, M. S. 2006. Genetic differentiation of Indian camel (*Camelus dromedaries*) breeds using random digonucliotide primers. *Animal Genetic Resources Information* (39): 77-88.

Nivsarkar, A. E., Viz, P. K. and Tantia, M. S. 2000. Animal Genetic Resources of India: Cattle and Buffalo. *ICAR publication*, New Delhi, India.

Pal, R. N., Barari, S. K. and Basu, A. 1994. Yak and its type- a field study. *Indian Journal of Animal Sciences*, 64: 853-856.

Pan, S., Misra, S. K. and Kundu, M. S. 2005. Ghoogroo pig: A new found animal genetic resource of sub-Himalayan West Bengal India. *Animal Genetic Resources Information* (37): 91-96.

Rajkhowa, C. 2005. Genetic Improvement of Mithun (*Bos frontalis*) in India- An evaluation and future breeding policy. In proceedings of VIII National Conference on Animal Genetics and breeding: National Livestock Breeding Policy, March 9-10, 2005, CIRG Makhdoom, Farah (Mathura)

Rathore, G. S. 1986. Camels and their management. ICAR publication, New Delhi, India.

Sadana, D. K. 2007. Indigenous cow breeds of India. *The Indian cow*, 13, 6-25.

Sahani, M. S. and Mehta, S. C. 2005. Genetic improvement programs for camel in India: Performance evaluation and breeding policy. In proceedings of VIII National Conference on Animal Genetics and breeding: National Livestock Breeding Policy, March 9-10, 2005, CIRG Makhdoom, Farah (Mathura).

Sethi, R. K. 2003. Buffalo breeds of India. Paper presented in 4th Asian buffalo congress held at New Delhi, February 25-28, 2003.

Singh, M. K. and Yadav, M. P. 2005. Equine genetic resources in India and strategy for their conservation. In proceedings of VIII National Conference on Animal Genetics and breeding: National Livestock Breeding Policy, March 9-10, 2005, CIRG Makhdoom, Farah (Mathura).

Singh, R.N. and Bohra, S.D.J. 1996. Garole sheep: A profile (Bengal breed of sheep locally known as Garole). *Indian Journal of Small Ruminants,* 2(2)38-42.

World Watch List for Domestic Animal Diversity, second Edition (Eds Beate D.Sherf), F.A.O. Rome, September, 1995.

❏❏❏

ABOUT THE EDITORS

Dr. N. Das is currently working as Principle Scientist and Head, Plant Animal Relationship Division at Indian Grassland and Fodder Research Institute, Jhansi (UP). He has 25 years of research experience in the field of Livestock Production and management. As deputy advisor Planning Commission, Government of India, he was associated with the identification of policies and strategies for development of animal husbandry sector in India. He was also associated with the Mid-term appraisal of the Ninth (1997-2002) and formulation of the Tenth Five Year Plan (2002-2007) programme in the livestock sector. He has guided many postgraduate and PhD student at IVRI, Izatnagar. Dr. Das has 50-research publication of national and international repute. He is one of the pioneer worker in the field of animal behaviour.

Dr. A.K. Misra, Principle Scientist in the Plant Animal Relationship Division at Indian Grassland and Fodder Research Institute, Jhansi (UP) has 20 years of research experience in the field of Animal Nutrition. He also served at Central Sheep and Wool Research Institute, Avikanagar (Rajsthan) and actively contributed in the development of complete feed block utilizing unconventional feed resources for ruminants and bioprocessing of crop residues for animal feeding. He has successfully operated many research projects funded by ICAR, NATP etc. Dr. Misra has 80-research publication of national and international repute. He is also member of many scientific societies.

Dr. S.B Maity is working as Senior Scientist in the Plant Animal Relationship Division at Indian Grassland and Fodder Research Institute, Jhansi (UP). He has 18 years of research experience in the field of Livestock Production and management. His main areas of research are ethology, livestock housing, organic forage and Livestock Production. He has been engaged in many research projects funded by ICAR, NATP etc. He has 30-research publication of national and international repute. He is also member of many scientific societies.

Dr. K.K. Singh is working as Senior Scientist in the Plant Animal Relationship Division at Indian Grassland and Fodder Research Institute, Jhansi (UP). He has 16 years of research experience in the field of Animal Nutrition. His main areas of research are forage evaluation and mineral nutrition. Dr. Singh has undergone advance training in the field of forage evaluation in UK. He has successfully completed many research projects funded by ICAR, NATP etc. He has 50-research publication of national and international repute. He is also member of many scientific societies.

Dr. M.M. Das is working as Senior Scientist in the Plant Animal Relationship Division at Indian Grassland and Fodder Research Institute, Jhansi (UP). He has 15 years of research experience in the field of Animal Nutrition. His main areas of research are improvement of low-grade roughages and development of complete feed block for ruminants. He has successfully undertaken many research projects funded by ICAR, NATP etc. He has 40-research publication of national and international repute. He is also member of many scientific societies.